Active Volcanoes of the World

Series editors

Corrado Cimarelli, München, Germany
Sebastian Müller, Mainz, Germany

More information about this series at http://www.springer.com/series/10081

Ulrich Kueppers · Christoph Beier
Editors

Volcanoes of the Azores

Revealing the Geological Secrets of the Central Northern Atlantic Islands

Springer

Editors
Ulrich Kueppers
Department für Geo- und Umweltwissenschaften
Ludwig-Maximilians-Universität (LMU) München
Munich
Germany

Christoph Beier
GeoZentrum Nordbayern
Friedrich-Alexander Universität Erlangen-Nürnberg
Erlangen
Germany

ISSN 2195-3589 ISSN 2195-7029 (electronic)
Active Volcanoes of the World
ISBN 978-3-642-32225-9 ISBN 978-3-642-32226-6 (eBook)
https://doi.org/10.1007/978-3-642-32226-6

Library of Congress Control Number: 2017963851

Printed on acid-free paper

This Springer imprint is published by the registered company Springer-Verlag GmbH, DE part of Springer Nature
The registered company address is: Heidelberger Platz 3, 14197 Berlin, Germany

Foreword

The Northern Mid-Atlantic Ridge (MAR) between 33°N and 41°N exhibits a gradient in seafloor depth, from 3500 m just north of the Hayes Transform (near 33°N) to less than 1000 m near 39°N. Together with this depth gradient, the MAR is characterized by a broadening with a funnel shape, best represented by the 2000 m bathymetric contour. This large bathymetric structure representing a prominent submarine topographic high is called the Azores platform. The maximum bathymetric expression of this platform is given by the emergence of a group of nine volcanic islands, forming the Azores archipelago.

The Azores platform represents a tectonic peculiarity as it is located at the triple junction between the American, Eurasian and African plates. In this context, the tectonic evolution of this region and the nature of the boundary between the Eurasian and African plates have always been contentious, with several models proposed to describe the overall plate kinematics, but this issue is still being debated.

Another highly disputed subject is related to the origin of the Azores platform and its islands. The geochemistry of the Azores lavas (either from land or from submarine samples) shows incompatible trace element enrichment coupled with high Sr–Nd–Pb radiogenic isotope ratios (a typical signature of oceanic island basalts) and has been used to suggest the existence of a compositional mantle anomaly beneath the Azores. Additionally, this geochemical enrichment was found to correlate with other geophysical characteristics, such as negative gravity anomaly, anomalously low S-wave velocities in the 100–200 km depth range or crustal thickness 60% higher than normal, all of them suggesting high magma productivity. These observations and inferences led to the generation of alternative models by suggesting the existence of a thermal anomaly in the mantle ("hotspot"), a volatile enrichment anomaly ("wetspot") or the presence of a mantle plume. To date, origin, size, depth and accurate location of this mantle anomaly are still a matter of debate.

The chapters included in this book provide a comprehensive and multidisciplinary overview of the geoscientific knowledge of the Azores platform. The scientific core themes of this volume encompass the geophysical, geochemical and petrological subjects, addressing not only the two aforementioned main controversial issues in the Azores, but also related topics such as hydrogeology and palaeontology. Additionally, Chapter "A Portrait of the

Azores: From Natural Forces to Cultural Identity" by Beier and Kramer deals with the Azorean cultural identity that has been strongly conditioned by the forces of nature, expressed in this region mainly by earthquakes and volcanic eruptions. Since the first permanent settlements in the Azores in the fifteenth century, about thirty volcanic eruptions, both on land and offshore, have been documented. Besides, the Azores are located in an area of elevated seismic activity. The majority is usually expressed through microearthquakes; however, periodically, the Azores are shaken by moderate to strong earthquakes that have caused destruction and negative economic impact. Cumulatively, more than 5000 people have perished and this, undoubtedly, impacted on the socio-economic and cultural development of these Portuguese people living isolated and enclosed by volcanoes in the middle of the Atlantic Ocean.

The six following chapters constitute the geophysics section in this book. The second Chapter "The "Azores Geosyndrome" and Plate Tectonics: Research History, Synthesis, and Unsolved Puzzles" by Vogt and Jung addresses the anomalous morphological, geological and geophysical framework present in the so-called Azores Triple Junction. It integrates and compares these characteristics with those present in a broader and more regional context such as the encompassing Mid-Atlantic Ridge, in a way to clearly understand the peculiarities existent in the Azores. These anomalies are identified through water depth and gravity/geoid variation, crustal thickness, upper mantle seismic structure, plate boundary morphology and rock geochemistry. The authors also review the evolution of the present-day understanding of the Azores and discuss the implications of their findings for other geologically recent plate boundaries. Chapter "The Contribution of Space-Geodetic Techniques to the Understanding of the Present-Day Geodynamics of the Azores Triple Junction" by Fernandes et al. focuses on the major contributions obtained with Space Geodesy (Global Navigation Satellite Systems (GNSS) and Interferometry Synthetic Aperture Radar (InSAR) techniques) in the Azores through a review of the published results since the late 1980s. Global Navigation Satellite Systems (GNSS) have been applied to modelling and understanding large-scale processes such as the relative movements and angular velocities of the three tectonic plates but, more recently and as a result of denser distribution of GNSS stations throughout most Azorean islands, more detailed studies concerning intra-island deformations have been carried out. Both techniques addressed have been contributing significantly to better understand the tectonic dynamics and the volcanic processes occurring in the Azores. In Chapter "Crust and Mantle Structure Beneath the Azores Hotspot—Evidence from Geophysics", O'Neill and Sigloch assess the geophysical constraints on crustal and mantle structure beneath the Azores hotspot discussing the possible existence of a traditional mantle plume. The data obtained through surface wave models, suggesting a shallow origin (250–300 km of the mantle) to the Azores hotspot or, alternatively, that the plume is waning, contrast with recent finite-frequency body-wave tomography which indicates that the Azores plume may extend to the core–mantle boundary. The authors suggest a common origin under West Africa for the Azores, Canary and Cape Verde hotspots. Chapter "The Tectonic Evolution of the Azores Based on

Magnetic Data" by Miranda et al. reviews the progress made in the geophysical research of the Azores, based on geophysical observations complemented by numerical modelling, and presents an updated interpretation scheme for the genesis and evolution of the Azores Triple Junction. Whereas on land investigations with detailed topographic maps, aerial photographs and satellite imagery as well as rock samples are straightforward and allow the study of geological structures at all scales, at sea things are quite different and have only advanced recently with technological developments. Mitchell et al. (Chapter "Volcanism in the Azores: A Marine Geophysical Perspective") present and discuss the Azores sea bottom morphology obtained through different types of sonar data (GLORIA, TOBI and multibeam bathymetric data). The studied sonar datasets allowed them to view and understand the topographic structures of the ridges (that constitute important extrusive structures in the Azores region) and interpret the morphologies in terms of the volcanic and tectonic features present on them. Fontiela et al. (Chapter "Characterisation of Seismicity of the Azores Archipelago: An Overview of Historical Events and a Detailed Analysis for the Period 2000–2012") give an overview of the existing historical and instrumental seismic catalogues, describe the seismicity of the region since early 1915 and analyse the features of the observed seismicity in the period 2000–2012, which is typically characterized by high number of events with relatively low magnitude.

Chapters "The Marine Fossil Record at Santa Maria Island (Azores)" and "Surface and Groundwater in Volcanic Islands: Water from Azores Islands" deal with direct observations or measurements. Ávila et al. (Chapter "The Marine Fossil Record at Santa Maria Island (Azores)") describe the unique exposure of sedimentary rocks containing marine fossils in the Azores. These outcrops on Santa Maria Island are interpreted as the result of tectonic processes. The authors also discuss the need and importance of preserving this palaeontological heritage. The magmatic origin of the Azores is first approached by Larrea et al. (Chapter "Petrology of the Azores Islands") with a detailed compilation of mineralogy and major element chemistry of xenoliths and volcanic rocks. These geochemical data are used to give an overview of the petrology of these igneous rocks, namely insights into magmatic evolution paths/processes and depth of melting, in a geochronological context (where appropriate temporal data are available), to better understand the origin of the specific geochemical characteristics present in each island. In Chapter "Melting and Mantle Sources in the Azores", Beier et al. detail the geochemistry of igneous rocks (mainly from land but also including submarine samples) by reviewing the trace element and radiogenic isotope (Sr–Nd–Pb–Hf systems) data. The alkaline character of the lavas and the large compositional range from basalts to trachytes identified in the volcanoes is emphasized, and elemental and isotopic comparisons inter-islands are made. These findings are used to address the mantle source characteristics (composition, degree of heterogeneity and temporal evolution) of the magmas from the Azores islands as well as their chemical evolution and the processes involved during magma ascent and differentiation. The geochemistry study is complemented by Moreira et al. (Chapter "Noble Gas Constraints on the Origin of the Azores Hotspot") which discuss the origin of the Azores archipelago by making comparisons between the different islands and integrating the findings in a wider regional scheme along the MAR between 28°N and 53°N. Antunes and Carvalho (Chapter "Surface and Groundwater in Volcanic Islands: Water from Azores Islands") characterized surface waters (lakes, springs and rivers) and revealed the degree of direct input by volcanic activity. The authors constrained by the volcano-tectonic activity that influenced today's geomorphological characteristics and emphasize the importance of volcanic lakes as a freshwater resource.

The final Chapter "Where to Go? A selection and Short Description of Geological Highlights in the Azores" by Kueppers et al. describes a selection of geological outcrops from each of the Azores islands that the authors considered worthwhile visiting, to get an overview of prominent geological features. Some of the selected outcrops have been subject to detailed geological descriptions and are regularly visited by researchers and students, and data from the peer-reviewed literature, wherever available, are used to complement the descriptions and integrate them in a larger regional geological context.

I am happy to see such a compilation of national and international chapters that cover a wide range of geoscientific questions about the Azores.

Alfragide, Portugal

Pedro José Lopes Tavares Ferreira
LNEG—Laboratório Nacional de
Energia e Geologia

Acknowledgements

The editors Ulrich Kueppers and Christoph Beier acknowledge Springer and in particular Johanna Schwarz and Annett Buettner as well as the series editors Corrado Cimarelli and Sebastian Müller for their invaluable and patient help during the difficult and even rocky way through science and politics. The topical structure has in part been developed by Ulrich Kueppers and José Pacheco (Instituto de Investigação em Vulcanologia e Avaliação de Riscos, IVAR). Ulrich Kueppers and Christoph Beier finalized the structure and worked hard—sometimes effortlessly—towards this book. Christoph Beier acknowledges help and support by Jim McEwan, Simon Coughlin, Mark Reynier and Adam Hannett. Both editors have learned that the compilation and handling of such a complex and large topic is subject to both scientific and political discussions. Both editors take this opportunity to thank all authors of this volume as well as the reviewers for their valuable contributions and also thank the Deutsche Forschungsgemeinschaft (DFG) and the Bundesministerium für Bildung und Forschung (BMBF) for support of the RV Meteor cruise M128 to the Azores Plateau during which we could expand our scientific knowledge of this unique island group.

Contents

Contributors

Paulo Antunes Department of Geosciences, University of Massachusetts Amherst, Amherst, MA, USA; Centro de Investigação e Tecnologia Agrária dos Açores (CITA-a), Azores University, Ponta Delgada, Portugal

Sérgio P. Ávila InBIO Laboratório Associado, CIBIO, Centro de Investigação em Biodiversidade e Recursos Genéticos, Azores, Portugal; Departamento de Biologia, Faculdade de Ciências e Tecnologias, Universidade dos Açores, Ponta Delgada, Azores, Portugal; MPB - Marine PalaeoBiogeography Working Group of the University of the Azores, Ponta Delgada, Azores, Portugal

Christoph Beier GeoZentrum Nordbayern, Friedrich-Alexander Universität Erlangen-Nürnberg, Erlangen, Germany

Rudolf Beier Department of English, American and Romance Studies, RWTH Aachen University (retired), Aachen, Germany

Philipp A. Brandl GeoZentrum Nordbayern, Friedrich-Alexander Universität Erlangen-Nürnberg, Erlangen, Germany; Research School of Earth Sciences, The Australian National University, Acton, ACT, Australia; GEOMAR Helmholtz-Zentrum für Ozeanforschung Kiel, Kiel, Germany

Diogo Caetano LabGeo—Engenharia E Geotecnologia, Azores, Portugal

João Catalão Instituto Dom Luiz (IDL), Faculdade de Ciências, Universidade de Lisboa, FCUL, Lisbon, Portugal

Rui M. S. Fernandes Instituto Dom Luiz (IDL), University of Beira Interior (UBI), Covilhã, Portugal

João Fontiela Instituto Ciências da Terra (ICT), Escola de Ciências e Tecnologia (ECT), Universidade de Évora, Colégio Luís Verney, Évora, Portugal

Zilda França Geosciences Department, Azores University, São Miguel, Portugal

Felix S. Genske Institut für Mineralogie, Westfälische Wilhelms-Universität Münster, Münster, Germany

Karsten M. Haase GeoZentrum Nordbayern, Friedrich-Alexander Universität Erlangen-Nürnberg, Erlangen, Germany

Jörg M. Habermann Faculdade das Ciências Humanas e Sociais, Interdisciplinary Center for Archaeology and Evolution of Human Behaviour (ICArEHB), Universidade do Algarve, Faro, Portugal; Institute of Cognitive and Evolutionary Anthropology (ICEA), University of Oxford, Oxford, UK; GeoZentrum Nordbayern, Friedrich-Alexander-Universität Erlangen-Nürnberg, Erlangen, Germany

Woo-Yeol Jung Marine Geosciences Division, Naval Research Laboratory, Washington, DC, USA

Johannes Kramer Department of Romance Studies, Trier University, Trier, Germany

Ulrich Kueppers Department für Geo- und Umweltwissenschaften, Ludwig-Maximilians-Universität (LMU) München, Munich, Germany

Marceliano Lago Department of Earth Sciences, Faculty of Sciences, Universidad de Zaragoza, Zaragoza, Spain

Patricia Larrea School of Geological Sciences and Engineering, Yachay Tech, Urcuquí, Ecuador; Department of Geology and Environmental Earth Science, Miami University, Oxford, OH, USA

Marco Ligi Istituto Scienze Marine, Consiglio Nazionale delle Ricerche, Bologna, Italy

Nuno Lourenço Instituto Português do Mar e da Atmosfera, IDL, Lisbon, Portugal; University of Algarve, IDL, Faro, Portugal

J. Freire Luis University of Algarve, IDL, Faro, Portugal

Pedro Madureira Estrutura de Missão para a Extensão da Plataforma Continental, Paço d'Arcos, Portugal; Instituto de Ciências da Terra/Dep. Geociências da Universidade de Évora, Évora, Portugal

João Mata Instituto Dom Luiz, Faculdade de Ciências, Universidade de Lisboa, Lisbon, Portugal

J. Miguel Miranda Instituto Português do Mar e da Atmosfera, IDL, Lisbon, Portugal

Neil C. Mitchell School of Earth and Environmental Sciences, University of Manchester, Manchester, UK

Manuel A. Moreira Institut de Physique du Globe de Paris, Sorbonne Paris Cité, Université Paris Diderot, Paris, France

Craig O'Neill Department of Earth and Planetary Science, CCFS ARC Centre of Excellence, Macquarie University, Sydney, Australia

Ricardo S. Ramalho Faculdade de Ciências, Instituto Dom Luiz, Universidade de Lisboa, Lisbon, Portugal; School of Earth Sciences, University of

Bristol, Bristol, UK; Lamont-Doherty Earth Observatory, Columbia University, Palisades, NY, USA

Philippe Rosset ICES, International Centre for Earth Simulation Foundation, Geneva, Switzerland

M. Rosário Carvalho Departamento de Geologia, Faculdade de Ciências da Universidade de Lisboa and Instituto D. Luiz, Lisbon, Portugal

Karin Sigloch Department of Earth Sciences, University of Oxford, Oxford, UK

Carlos Sousa Oliveira Instituto Superior Técnico, CEris, Universidade de Lisboa, Lisbon, Portugal

Rachelle Stretch Department of Geography, University of Cambridge, Cambridge, UK

Jürgen Titschack MARUM – Center for Marine Environmental Sciences, University of Bremen, Bremen, Germany; Marine Research Department, Senckenberg am Meer, Wilhelmshaven, Germany

Fernando Tempera Departamento de Oceanografia e Pescas, Universidade dos Açores, Horta, Açores, Portugal

António N. Trota Instituto de Investigação em Vulcanologia e Avaliação de Riscos, University of the Azores, Azores, Portugal

Peter R. Vogt Marine Science Institute, University of California at Santa Barbara, Santa Barbara, CA, USA

Elisabeth Widom Department of Geology and Environmental Earth Science, Miami University, Oxford, OH, USA

Introduction

Ulrich Kueppers and Christoph Beier

The archipelago of the Azores comprises nine inhabited islands and is an autonomous region of Portugal. It stretches more than 600 km in an E–W direction and is separated by the Mid-Atlantic Ridge (MAR). Two of the islands (Flores and Corvo, the Western Group) are situated on the American plate while the seven eastern islands of the Central (Faial, Graciosa, Pico, Terceira and São Jorge) and Eastern Groups (Santa Maria and São Miguel) are located on (sub-)parallel transform faults extending eastwards from the MAR that define the complex plate boundary between Eurasia and Africa (see Vogt and Jung, Chapter "The "Azores Geosyndrome" and Plate Tectonics: Research History, Synthesis, and Unsolved Puzzles").

Volcanism started at different times for these islands, manifested by significantly different ages constrained for the oldest rocks of individual islands (Santa Maria 8 my, Pico 0.25 my). Since the earliest permanent settlements in the 15th century, approx. 30 volcanic eruptions have taken place in the archipelago, half of which from subaerial vents. Important examples took place on Faial (1672 and 1957/8), Pico (1562, 1718, 1720), São Jorge (1580 and 1808), São Miguel (1563/4 and 1652) and Terceira (1761). In submarine settings, eruptions repeatedly produced new islands, most of which were eroded within few weeks or months, as e.g. at the Banco D. João do Castro (1720) or the Ilha Sabrina (1811). As of today, Corvo, Flores, Graciosa and Santa Maria are considered extinct while some volcanic center on the other islands are considered dormant. The last confirmed eruption took place between 1998 and 2001 at a water depth of 200–300 m, few kilometres off the Western tip of Terceira. Today, all islands are monitored in detail for indications of tectonic but more importantly magmatic activity by the *Centro de Informação e Vigilância Sismovulcânica dos Açores* (http://www.cvarg.azores.gov.pt/civisa/Paginas/homeCIVISA.aspx).

The book aims at presenting a comprehensive view of the Azores geology, reviewing the scientific literature of past decades. We do not aim at covering the entire geological history and knowledge of the archipelago for which the reader is referred to the individual peer-reviewed publications in detail in each respective chapter. It rather covers a wide range of geological subjects that have been addressed on these islands in the past and for which these islands are famous. This book addresses the fields of Geochemistry, Geology, Geophysics, Palaeontology, Petrology and Volcanology, Chapters "The "Azores

U. Kueppers (✉)
Department für Geo- und Umweltwissenschaften, Ludwig-Maximilians-Universität (LMU) München, Munich, Germany
e-mail: u.kueppers@lmu.de

C. Beier
GeoZentrum Nordbayern, Friedrich-Alexander Universität Erlangen-Nürnberg, Erlangen, Germany
e-mail: christoph.beier@fau.de

U. Kueppers and C. Beier (eds.), *Volcanoes of the Azores*, Active Volcanoes of the World, https://doi.org/10.1007/978-3-642-32226-6_1

Geosyndrome" and Plate Tectonics: Research History, Synthesis, and Unsolved Puzzles "–"Surface and Groundwater in Volcanic Islands: Water from the Azores Islands". We have also added a chapter on interesting outcrops (see Kueppers et al., Chapter "Where to Go? A Selection and Short Description of Geological Highlights in the Azores") which will be of interest to those readers travelling to the Azores, accompanied by a chapter addressing the social and linguistic evolution within the Azores (see Chapter by Beier and Kramer, Chapter "A Portrait of the Azores: From Natural Forces to Cultural Identity").

The impact of volcanism on the formation of the Azores, on their geological and biological evolution and on the socio-economic development of the islands is unique amongst the world's volcanoes due to the isolated positioning in the Atlantic albeit close connection to mainland Portugal and the United States and certainly will remain a topic of intense scientific debates in the future as much as is has been in the past. All this has made all of the Azores islands a worthwhile destination for both holiday and work trips which both of the editors and many of the authors of this volume have experienced or are experiencing during finalizing their manuscripts.

A Portrait of the Azores: From Natural Forces to Cultural Identity

Rudolf Beier and Johannes Kramer

Abstract

A comprehensive description of Azorean cultural identity does not exist, and would certainly deserve thorough investigation on the islands themselves and a monograph of its own. What we can do here is to take a look at some single landmarks of this unmapped territory, selected mainly from the perspective of how people live with the forces of nature. How do people deal with them, and how are these forces mirrored in their behaviour, beliefs, rituals, and symbolic and imaginative expressions? Is there an Azorean culture? Before we address this question, presenting some evidence from emigration, religion, and literature, we will provide some historical and economic background.

R. Beier (✉)
Department of English, American and Romance Studies, RWTH Aachen University (retired), Aachen, Germany
e-mail: rudolfbeier@me.com

J. Kramer
Department of Romance Studies, Trier University, Trier, Germany

1 Prologue: Once upon a Time…

Once upon a time in the lost kingdom of Atlantis there ruled a king whose name was Graywhite. He had married the beautiful Queen Rosewhite. They lived in a magnificent palace, but it was a sad place because there were no little children in it…

This is the beginning of the folk tale about the origin of the Azores, entitled *Princess Bluegreen of the seven cities* (Eells 1922/2011, 17ff.). In the course of the story, the king and the queen are visited by a fairy who promises them that a princess would be born, but only if they agreed that the princess would be withheld from them for twenty years, to give the King the chance to make up for his cruel behaviour towards his subjects. They agree, but in the end the King becomes impatient and breaks the vow after eighteen years, trying to free Princess Bluegreen by force. And here is the 'geological ending' of the tale:

... King Graywhite marched on and on. It was a long and perilous journey and the army suffered many hardships on the way. It seemed as if they would never arrive, but at last they drew near to what everybody knew to be the most beautiful part of the whole kingdom, where the fairy had taken the Princess Bluegreen to conceal her. Storms raged; lightning flashed; ominous roarings and rumblings sounded from the depths of the earth. ... "On! On!" cried the king."Do you think I would abandon this expedition now?" The words were hardly out of his mouth when a huge rock fell from its place near where he stood and rushed away down the mountainside. The earth trembled

U. Kueppers and C. Beier (eds.), *Volcanoes of the Azores*, Active Volcanoes of the World, https://doi.org/10.1007/978-3-642-32226-6_2

violently beneath their feet. Fearful rumblings and roarings sounded all about them. ... King Gray-white struck his royal sword against the great wall. At that moment the walls fell. The earth beneath their feet rose. Great flames swept up towards the sky and rushed over the land, sweeping everything before them. Then the sea raged over the earth in violence until it had covered the whole kingdom of Atlantis. The fairy's curse had been fulfilled. The king was dead. His kingdom was consumed by fire.

When at last the waters grew calm again all that remained of the great rich kingdom of Atlantis was the group of nine rocky islands which to-day is called the Azores. ...

We have a storm, a landslide, an earthquake, an eruption, and we have a tsunami. This, as it were, is punishment by the forces of nature …

2 Historical Outline

Our look into the history of the Azores [based mainly on Goulart (2008), Marques (2001), and Santos (1995)] will present the islands and their inhabitants as torn, in a unique way, for about three centuries, between the powers of Portugal and Spain, and, more recently and more fundamentally, as a bridge connecting Europe and the Americas. Section 2 will mainly deal with the Azores from a political perspective and less with nature and economy (for these, see Sect. 3), while Azorean culture and the influence of natural forces will be the focus of Sect. 4.

Mysterious coins and mythical lands

All regions of Portugal take pride in their Roman origins, with the exception of the Azores: They, in ancient times, were still *terra incognita*. However, in the early days of research, there were attempts to imagine, invent or fabricate connections to antiquity. For example, it was claimed that coins from Carthago (fourth century BC) were found in 1749 on the western island of Corvo. However, they were declared to be fakes as early as 1836 by Alexander von Humboldt, maybe corresponding to the wishes of local people, but not to historical reality.

Antique and medieval stories and early maps contain many mythical lands and legendary islands, such as Atlantis, Brasil, the Fortunate Islands, or the Isles of Saint Brendan. For these phantom lands, virtually all islands we know in the Atlantic Ocean, including the Azores, have served as possible matches. From a historical perspective, what the old sources tell us about islands west of the *Columns of Hercules* (Gibraltar) can, basically, be considered as literary topoi about unknown worlds and strange creatures, leading us to the realm of dreams and myths. The most common assumption seems to be that there has, up to now, been no convincing evidence that human beings set foot and lived on the islands before they were explored by the Portuguese (e.g. Santos 1995, 3; Wakonigg 2008, 122). There are researchers, however, who try to refute this: Rodrigues et al. (2015) claim to have found evidence of pre-Portuguese presence in the Azores, a man-made rock-basin with engravings produced on Terceira during or before the eleventh century. Apart from this controversy, what we can safely say is: the islands were uninhabited when the Portuguese occupied them, and there must have been some knowledge of them before the Portuguese exploration (see below).

Medieval manuscripts and maps

The actual history of the discovery of the Azores starts at the beginning of the fourteenth century, possibly with the Genoese, but there are vague messages of speculative sightings during ocean exploration dating back to the thirteenth century. We find the oldest available notices about the Azores in the manuscripts of he famous *El libro de conoscimiento de todos los reinos* (*The book of knowledge of all kingdoms*, Marino 1999), a kind of geographic encyclopedia, or more precisely, a medieval fictional travel book written by an anonymous author (some say a Spanish

Franciscan, others doubt this). It was composed in the last quarter of the fourteenth century and was based on contemporary portolans (maritime charts) and mappaemundi (special medieval maps of the world, for details Marino 1999, XXVIIIff.). The author mentions the Canary Islands, Madeira, Porto Santo, and eight further islands, which must be the Azores, although most of the names used differ from the ones we know today (Marino 1999, 50): *la Isla del Lobo* (Santa Maria), *la Isla de las Cabras* (São Miguel), *la Isla del Brasil* (Terceira), *la Isla Colunbaria* (Pico), *la Isla de la Ventura* (Faial), *la Isla de Sant Jorge* (São Jorge), *la Isla de los Conejos* (Flores), and *la Isla de los Cuervos Marines* (Corvo). Examples of famous contemporary maps, which, well before the Portuguese exploration, showed seven or eight islands and, though in different languages, used strikingly similar names, were the *Medici Atlas* (the *Portulano Medicea Laurenziano*, or *Medici-Laurentian Atlas*, of 1351, for details: Marino 1999; Santos 1995), the *Catalan Atlas* (most probably from 1375, Marino 1999), and *L'atlante Corbitis* (1384–1410). The *Corbitis Atlas* arranges and names eight islands from North to South (see Fig. 1 and *l'atlante nautico 'Corbitis'*): *y de corui marini* (Corvo), *liconiGi* (Flores), *s(an)c(t)o zorzi* (São Jorge), *y la uentura* (Faial), *li colonbi* (Pico), *y de brazil* (Terceira), *caprara* (São Miguel), and *louo* (Santa Maria). It was atlases like these, or possibly their predecessors, that provided the information which *El libro de conoscimiento de todos los reinos* was based on (Marino 1999).

Henry and the hawks
Portuguese exploration

The first comprehensive attempts to describe the islands and to document local Azorean history were made by two Jesuits: Gaspar Frutuoso's (ca. 1522–1591) work *Saudades da terra* (*Longings for the native country,* 1586–1590) was not printed during his lifetime, but in 1873. His friar, António Cordeiro (1641–1722) used Frutuoso's work in his *História insulana das Ilhas a Portugal sugeytas no Oceano Occidental* (*History of Portuguese islands in the Western Ocean*), which was published in Lisbon in 1717. This book is a detailed description of the nine islands, their discovery, places, important people, their nature, including geological observations, e.g. "furnas, fogos, & tremores" (*caves, fires, tremors/quakes*) on São Miguel (Cordeiro 1717, 151). According to Frutuoso and Cordeiro, the discovery of the islands is attributed to Gonçalo Velho Cabral (1400–1460), who in 1431 was sent on a reconnaissance voyage to the West by Prince Enfante Dom Henrique, or Henry the Navigator, the leader of the *Holy Order of Christ* and founding father of modern navigation and, as it were, also of the Azores (Santos 1995). On this journey, Cabral only found a series of volcanic rocks (about 25 miles off the nearest 'Azorean' island), which he named *Formigas* (*ants*). Today, the *Ilhéus das Formigas* are almost completely submerged (see also Larrea et al., Chapter "Petrology of the Azores Islands"). During his second journey, on 15 August 1432, the day of the feast of the *Assumption of our Blessed Mother*, or *Santa Maria*, he discovered an island, which he named accordingly, and found suitable for settlement. The second island, on which the Portuguese landed in 1444, was São Miguel. Later on, the seafarers discovered Terceira, the 'third island' chosen for settlement, with Gonçalo Velho Cabral as the first administrator, or 'captain-donatory'. After the Portuguese explorers landed on the central islands of Graciosa, São Jorge, Pico, and Faial, the sighting of the western islands, Flores and Corvo, in 1452, concluded the discovery of the Azores.

Other sources attribute the discovery of the Azores to Portuguese navigators "around the period of 1427 to 1432, probably by Diogo de Silves", who is said to have found Santa Maria and São Miguel in 1427 (Farjaz 2008, 25; Wakonigg 2008, 122). His biography is obscure, his name is assumed to appear in a chart drawn by a Catalan cartographer in 1439.

The name 'Azores', Portuguese *Açores*, is derived from Port. *açor*, Span. *azor* (<Lat. *acceptor*), English *hawk*. The hawk, however, is not a native bird on the islands. It has been assumed that in the discovery phase some other

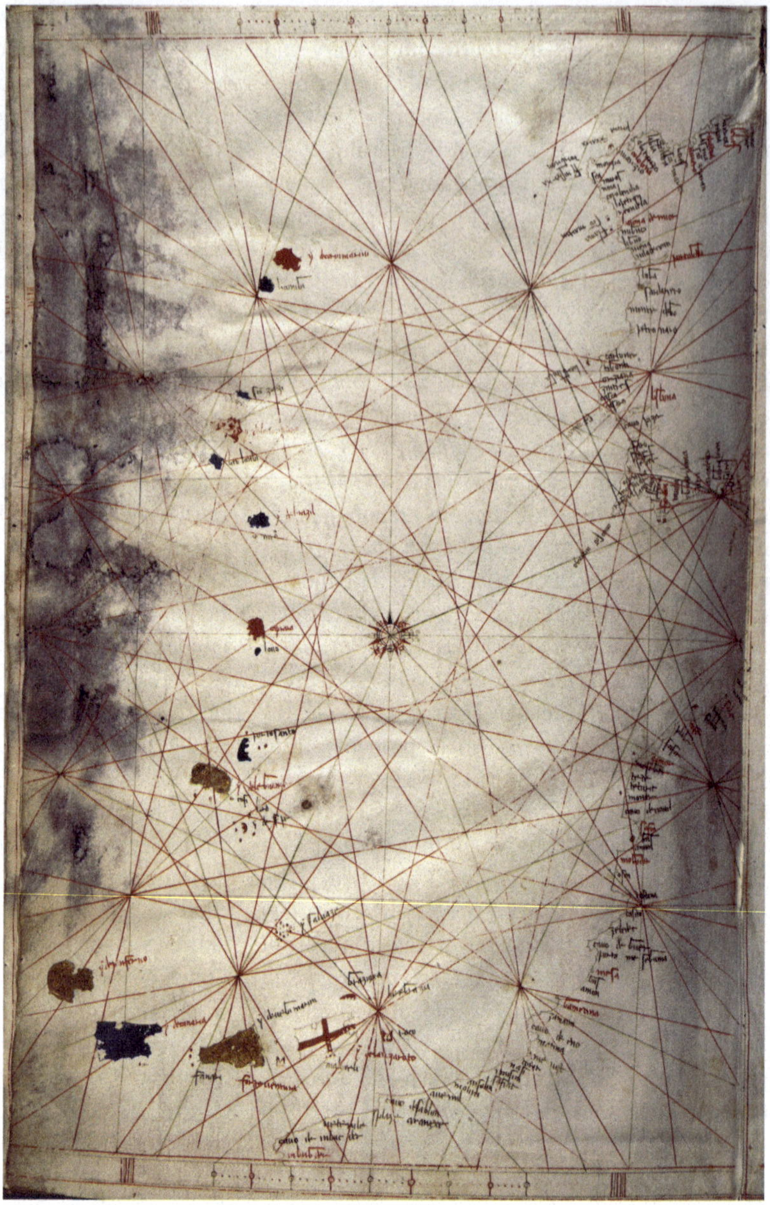

Fig. 1 *L'atlante nautico 'Corbitis'*, page 4 (courtesy of Biblioteca Nazionale Marciana di Venezia). *Su concessione del Ministero dei Beni e delle Attività Culturali e del Turismo—Bibliotheca Nazionale Marciana. Divieto di riproduzione.* (By kind permission of Ministero dei Beni e delle Attività Culturali e del Turismo—Bibliotheca Nazionale Marciana. Do not reproduce)

bird (such as a large shearwater, or the Azorean buzzard) was confused with the hawk. Other explanations are quoted by Santos (1995).

Sheep and settlers

The first 'settlers' of the Azores were sheep (and other domestic animals) released in order to provide meat for people still to come. On 4 July 1439, King Adonso V granted permission to settle seven islands: São Miguel and Santa Maria in the east, and the central islands of Terceira, Pico, Faial, São Jorge, and Graciosa. On 8 January 1453, the Portuguese king donated the (still unsettled) island of Corvo to his uncle Afonso de Bragança. In 1456, the first settlers came to Terceira, and up to 1470 the other islands were populated. The first settlers lived in natural accommodations, in rock caves and improvised dwellings (*cafuas*) protecting them from bad weather and storms.

In addition to the Portuguese immigrants, many Flemings came to settle on the islands. The name of *Horta* on Faial is a corruption of the Flemish name *Joss van Hurtere*, and the islands were also called *Insulae Flandricae* or *Ilhas Flamenga*, which survives in the place name *Flamengos* on the island of Faial.

Apart from Portuguese and Flemish settlers, there were smaller groups of Moorish prisoners, African slaves, and people from Madeira, France, Italy, Scotland, England, Morocco (esp. jews), and Spain (esp. clergymen, Santos 1995).

Portugal and Philipp

Although early Spanish attempts to seize the Azores had failed, the relations between the Portuguese and Castilian kingdoms remained tense. After all, they were competitors in the exploration and acquisition of new territories to be colonized. For Portugal, discoveries helped to assert "its independence and emancipation from Spain" and helped to stabilise Portugal as a nation (Ramos Villar 2006, 16; for sociocultural similarities and differences between Portugal and Spain today, see Hofstede and Hofstede 2005). When in February 1493 Columbus returned from his discovery voyage to America and anchored in the *Baia dos Anjos* on Santa Maria to resupply his ship (the *Niña*) with provisions and water, he was given a hostile reception. However, nowadays the city honours him with a monument, and Christóvão de Aquirar devoted a poem to him entitled *Navigator of the Island Seas* (Almeida and Monteiro 1983, 56ff.).

Nominally, the territorial conflicts between Portugal and Spain were solved by the *Treaty of Tordesillas* (7 June 1494), which was mediated by Pope Alexander VI: It drew a line through the Atlantic from the North to the South Pole; this line ran 370 Spanish miles (1770 km) to the west of the Cape Verde Islands and split the world into a western Spanish half and an eastern Portuguese half, by awarding everything west of this imaginary line to the Spaniards and everything east of it to the Portuguese. In the European area this arrangement was not valid, but the affiliation of the Canary Islands to Spain and of Madeira and the Azores to Portugal created a kind of temporary balance between the two powers.

As for the whole of Portugal, the year 1580 brought major change for the Azores: On 15 January 1580, the so-called Cardinal-King Henrique I was killed in the Battle of Almeirim in North Africa. He was childless and in his will left his kingdom to the Spanish King Philipp II. The attempt to establish a rival king in Portugal in the person of António of Crato (from an illegitimate line of the royal family of Avis) was quickly thwarted by Spanish troops. In the Azores, however, António's claims to power were still supported, and it was not until the summer of 1583 that the last resistance was ended by Spanish troops in a bloodbath at Angra.

For the time being, belonging to the Spanish Empire seemed positive for the Azores: Terceira became an important trading centre for the Spanish galleons full of gold, silver, and other treasures from America on their way from the Caribbean to Europe. Admittedly, the power of Spain weakened after the defeat of the armada in August 1588, and especially after the Armada was ultimately defeated and destroyed by the Dutch in the Battle of Gibraltar on 25 April 1607. Afterwards, English privateers took the chance to plunder the single islands of the Azores, which individually were hard to defend.

During the sixty years of Spanish government, the foreign rule became more and more unpopular in Portugal, as it became increasingly obvious that all profits were diverted to Madrid. Thus, it is not surprising that when in Portugal the revolution of the nobility led to the installation of João IV from the house of Bragança as King of Portugal, this met with strong sympathies in the Azores. Still, it lasted until 4 March 1642, before the Spanish occupation of the Castelo de São João Baptista on Monte Brasil in the outskirts of Angra ended.

Maria and Miguel

The French and the British, the liberals and the absolutists

The riots of the French Revolution affected Portugal, but had no direct and immediate impact on the Azores: Portugal remained loyal to its traditional ally Great Britain, and did not join the anti-British alliance. In 1806 King João VI refused to follow the Continental System ordered by Napoleon to stop all trade between Europe and Great Britain. In the *Peace of Fontainebleau* (27 October 1807) Spain offered the French troops the right to march through to Portugal. The French general Jean-Andoche Junot occupied the whole of Portugal, but on 29 November 1807, just before French troops occupied Lisbon, the royal family, under the protection of the British fleet, escaped to Brazil, where they remained until their return to Portugal in 1821. In Portugal, there was a liberation struggle against the French occupants, supported by the later Duke of Wellington, and finally the troops had to leave (*Peace of Sintra*, 30 August 1810). From 1810 to 1820, Portugal was *de facto* occupied by British troops.

During this turbulent period, the connections between the Azores and Portugal were interrupted, with the British fleet securing the intactness of the islands from external enemies. Brazil became an independent empire in 1822 under emperor João I, who at the same time was King João VI of Portugal. When he died in 1826, his son Pedro I, the rightful successor, and since 1822 Emperor of Brazil, abdicated the Portuguese throne to his eldest daughter, Maria (Maria II, *Maria da Glória*), who at that time was seven years old. Nevertheless, his exiled and absolutist-minded brother Miguel laid claim to the Portuguese throne. In a coup d'état he deposed Maria and proclaimed himself King. Miguel abolished the first Portuguese constitution, which had been passed by the Cortes of 1822 and to which João IV had already sworn allegiance.

During this struggle of liberal versus absolutist forces in Portugal, the Azores entered the stage as an important political agent: The liberals of Terceira removed the *Capitão geral* loyal to Miguel, and replaced him by a follower of the young queen. On Terceira, a *Junta provisória* was proclaimed, and Angra was proclaimed the capital of Portugal. The so-called *Regencia de Angra* (1828–1832) beat off all attacks of Miguel's followers and won the Battle of Praia on 11 August 1829, which put the Portuguese fleet out of action. In 1831, all islands were in the possession of the liberals, and Pedro I arrived on 22 February 1832 in Ponta Delgada on São Miguel, driving Miguel's troops out of the country. After the important role the Azores had played in the power struggle between Pedro and Miguel, queen Maria da Glória honoured Angra by giving it the name *Angra do Heroísmo* in 1837.

Captains and captaincies

Azorean administration

The internal administration of the islands was based on the division into *Capitanias* (*captaincies*), as part of the *Captain-Donatory* system introduced by Henry the Navigator, allowing absentee landowners to control their property and receive payments from the peasant tenants. Apart from other privileges, the *capitão*, a hereditary post, and the *donatórios* (*donatories*) had the monopoly on all mill products, salt and bread, and granted land for tenant farming. This feudal heritage of tenant farming must be considered a major reason for Azorean emigration (Almeida 1980; Santos 1995, see below).

In 1766, Azorean administration was centralised: The *Capitanias* of the single islands were subordinated under the *Capitania Geral*, placed in Angra do Heroísmo on Terceira, which was declared the capital of the Azores. This general captainship had significant functions, it concentrated all administrative, military, and judicial power, managed local economic life, and controlled the regional governors of the single islands.

The *Capitania Geral dos Açores* was abolished in 1832 in favour of a *Provincia dos*

Açores with Angra as capital, but little later (in 1833), motivated by the displeasure of the inhabitants of São Miguel to be subordinate to Terceira, the new province was split into two, the *Provincia Oriental dos Açores* with Ponta Delgada on São Miguel as capital and the *Provincia Occidental dos Açores* with the capital Angra de Heroísmo on Terceira.

Resupplying, refuelling, relaying

Ships, planes, and cables

Throughout their history, the Azores repeatedly served, for certain periods of time, as an essential strategic place linking Europe and America via ship, plane, or cable. These connected the islands more closely to the world, until technological innovation reduced their importance as a base of support or refuelling or telecommunications. "Between these transitory intervals of keeping pace with the world, the islands endured, in silence and quietude …" (Almeida 1983, 20).

The Azores were an ideal trading station and a place to resupply ships. During Spanish occupation, they served as a port of call for their galleons. Since the thirties of the nineteenth century, American ships were seen more frequently in the rich whaling grounds around the islands, and they often landed to stock up on provisions or to be repaired. The British warship *HMS Styx*, under captain Alexander Thomas Emeric Vidal, carried out surveys of the Azores (1841–1845), on which the drawing of the first reliable map of the Azores was based.

For steamships, the Azores became an important stopover in transatlantic trade. Angra de Heroísmo, Horta and Ponta Delgada developed into important ports, although the importance of the Azores for ship traffic diminished again from the last third of the nineteenth century onwards, with the shift of traffic to the northern part of the Atlantic. For some time, planes between the continents needed the Azores for refuelling. Technological progress made the stopover obsolete in the mid sixties, and put an end to still another opportunity Azoreans, temporarily, had profited from.

With the development of telegraphy, the Azores became a major transfer station. In 1893, a first cable was installed between Horta on Faial and Carcavelos near Lisbon, soon followed by the connection with US stations. In 1900, Horta became the seat of the American *Cable Company* (CCA), of the British *Europe and Azores Telegraph Company* (EAT) and of the German *Deutsch-Atlantische Telegraphengesellschaft* (DAT), which linked Horta to Borkum. The latter had to cease its operations in 1916, but could re-establish its activities in 1926 until the final end of the Company in 1943. The buildings of the so-called *Colónia Alemã,* where the twenty members of the DAT lived with their families, can still be visited. In the course of time, several cables were destroyed or burned due to volcanic and seismic incidents, esp. submarine eruptions, and had to be repaired, or parts of them had to be replaced entirely (Farjaz 2008). The end of the boom of telegraphy in Horta came in the fifties, when relay stations were made obsolete by the higher capacity of cables, stronger amplifiers, the increasing use of radio, and later of satellites. In 1969, the British closed down the last cable station in Horta.

We would like to point out, however, that it was not only by virtue of technological advances that the Azores linked the Old and the New Worlds. Above all, the connection has been established and fostered by people, i.e. by thousands of migrants (see Sect. 4.2 below). One outstanding builder of bridges and cultural negotiator in the field of literature is Almeida (e.g., 2006).

Capelinhos and Carnations

The Azores in the twentieth century

Politically, the beginning of the twentieth century brought the last years of the Portuguese monarchy, which were characterized by riots and a lack of necessary reforms. The proclamation of the Republic, on 5 October 1910, did not improve this situation: the urgently needed agricultural reform remained unfinished, and social unrest continued to rock the country.

In 1916, Portugal entered into the First World War because of the German submarine menace. Some Portuguese merchant ships sunk off the Azores, but the more serious war events happened rather around Madeira.

A rebellion in 1926 installed a military regime in Portugal, which paved the way to Salazar's *Estado Novo* (also called *salazarismo*). António de Oliveira Salazar, an economist, became a strong and influential finance minister in 1928, and in 1932 was appointed Prime Minister, he ruled until 1968. Portugal was transformed into a corporatist-authoritarian state dominated by Salazar's fascist party *União Nacional.* During the Spanish Civil War (1936–1939), Portugal remained neutral, but indirectly favoured Franco. Though Portugal preserved its neutrality in the Second World War, too, the Azores were involved in the hostilities: Again, they played a decisive role as a link, this time between the American and European allies in their fight against Hitler Germany, esp. in the Battle of the Atlantic.

In August 1943, the Lisbon government allowed Great Britain to use the port of Horta, Faial, the port of Ponta Delgada, São Miguel, and the military installations on São Miguel (*Santana Field*) and Terceira (*Lajes Base*, *Lajes Field*, in the *Plain of Lajes*, in the northeast of the island). Little later, from December 1943 onwards, the *Royal Air Force* on Terceira was joined by the *United States Army Air Force* and the *US Navy.* In a joint effort, they made Lajes Airport a large military facility, which played a key role in World War II, particularly in the Battle of the Atlantic. In November 1944, the Americans were, after difficult negotiations, allowed to build an airfield in Ponta do Monteiro, in the west of the island of Santa Maria, which was closer to the European and North African mainland than Terceira. The British soldiers departed in 1946, while the Americans remained on the base of Santa Maria. American rights to *Lajes Field* were re-negotiated after the war, so that *Lajes Field* remained a strategically important facility for the United States and, later, for the NATO as a whole: In 1949, Portugal joined the NATO, and in 1951 a base treaty permitted the stationing of NATO troops on the Azores (see Consulate of the United States).

The presence of foreign soldiers who stayed longer than the ship crews of earlier days led to a noteworthy change in the traditional Azorean life style. The contact between the strangers and the local population was intense, and at festivals and parties close relations were established. Typical Anglo-Saxon sports such as hockey and football became established, beer consumption rose, and the first brothels opened.

During the 1957/1958 Capelinhos volcanic eruption and the associated earthquakes on Faial, Salazar was still in power. There is disagreement as to how successful the Salazar government actually was in the management of the disaster. Ramos Villar (2006, 14) points out that, in economic terms, the Portuguese government at that time was unable to help, due to the recession and the outbreak of the colonial war. According to de Oliveira (2008, 101) the "Portuguese government did not feel obligated to provide any type of help… ", while according to the analysis by Coutinho et al. (2010, 265) "response, recovery and rehabilitation were generally highly successful", with much of the credit, however, going to Dr. António de Freitas Pimentel, the local Civil Governor, who has been praised for his effective leadership, his highly successful management of the emergency situation, and for the active role he played influencing the US legislation to allow immigration from Faial.

The so-called Carnation Revolution of 1974 overthrew Salazar's successor regime and released the colonies into independence. What followed on the Azores was a period of some political instability and of a new orientation towards independence, which, on 8 September 1976, led to the establishment of the first regional self-government, the *Governo Regional dos Açores,* with its own president and its own parliament. Their new policy is clearly orientated towards regional interests. For instance, it promotes local production sectors such as viticulture, and created a special department to promote Azorean tourism, which is still less developed than tourism on Madeira. In 1986, Portugal joined the European Communities. Today, some areas of Portugal, including the *Região Autónoma dos Açores,* are defined by the European Union as less developed regions and, accordingly, are supported by a number of regional funds (European Commission 2014). Examples of Azorean projects relevant to our context, which have benefited from investment through EU regional policy programs are entitled *The soft energy of the volcanoes* (geothermal energy), *Make the sea our servant, not our master* (wave energy), and the *Capelinhos Lighthouse Environmental Information Centre* (website of the European Commission: ec.europa.eu).

The Carnation Revolution of 1974 and the full autonomy of 1976 put an end to centuries of 'slow communications' between Portugal and the Azores isolated within the ocean, characterized by "neglect, absenteeism and under-development of the resources found within the islands" (Ramos Villar 2006, 9). Since it became a member of the European Communities, the Azorean economy has grown, and there have been major increases in living standards in Portugal as a whole (Coutinho et al. 2010, 278). These three key events of the century (revolution, autonomy, EU membership) have also brought changes in the economic and demographic 'pre-disaster vulnerability' of the islands, because they are no longer the poor and peripheral region of Portugal they still were during the Capelinhos emergency (Coutinho et al. 2010, 268f.). Quantity and quality of support after natural disasters have improved. When an earthquake on 1 January 1980 (see Fontiela et al., Chapter "Characterisation of Seismicity of the Azores Archipelago: An Overview of Historical Events and a Detailed Analysis for the Period 2000–2012"), killed more than sixty people and injured about 600, made thousands homeless and destroyed the historical centre of Angra de Heroísmo, this was the first time that "the Lisbon government shouldered its responsibility and obligations" (de Oliveira 2008, 101). The reconstruction was also helped by generous grants from the USA, which still hold their military base in Lajes. Another earthquake struck Faial in 1998, killing eight people, injuring 150, and making ten times as many homeless. According to Coutinho et al. (2010, 278), it showed that the new system of civil defence infrastructure, which is organized like a cascade, works well. There was considerable financial support and public investment. What still remains vulnerable, however, is the traditional building stock of the Azores.

3 What Azoreans Have Gained from Nature

Throughout Azorean history, the primary sector of the economy (agriculture, fishing, whaling) has played a key role. For hundreds of years, Azoreans have tilled the volcanic soil rich in alkalies from the source rock (see Larrea et al., Chapter "Petrology of the Azores Islands" and Beier et al., Chapter "Melting and Mantle Sources in the Azores"), to produce agricultural products for subsistence and for export. Nature has allowed them to grow, mostly without irrigation, a wide variety of plants we know from tropical, subtropical and moderately climatised regions. The kinds of plants grown, more or less successfully, varied over the centuries.

Colours, crops, and cattle

The first settlers and cultivators in the fifteenth and sixteenth centuries mainly grew sugar cane and wheat (which was needed for the market places in Ceuta, North Africa). Neither of the two crops proved to be successful in the long run. The most important products the Azores exported at that time were plants used as colouring agents (mainly for garments): *Roccella tinctoria* (*orchil*) and *Isatis tinctoria* plants (*woad, glastum*). Both lost their importance in the middle of the seventeenth century, when they were replaced by indigo. From the beginnings, the rich fishing grounds were exploited, and the animals that had come to the islands with the arrival of the Europeans provided food for the settlers. Trees were felled and forests cleared rapidly in the course of the fifteenth and sixteenth centuries for a variety of purposes, mainly ship building.

The sixteenth century marked the end of sugar cane and wheat, and the beginning of the cultivation of sweet potatoes (*batatas*), citrus and other fruits, and wine. In the seventeenth century, the most important crops were oranges, batatas, and wine, and people also cultivated corn and flax. The seventeenth and early eighteenth centuries saw a certain economic boom, because improved boat connections increased exports to mainland Portugal. New crops entering the Azores in the nineteenth century were pineapples (an important crop for export, grown in greenhouses), tobacco, and, since 1883, the famous tea grown at Gorreana on the northern coast of São Miguel. Today, the tea factory, plantation and shop are a tourist attraction. The nineteenth century witnessed disasters in Azorean agriculture: Apart from potato rot, vineyards were destroyed by grape blight in the middle of the century, and at the end of the century, the citrus blight caused a breakdown in the cultivation of citrus fruit, esp. oranges.

In the twentieth century, the sugar beet was introduced. A much more fundamental change, however, as a consequence of Salazar's policy of economic autarchy, was the dramatic transformation of the Azorean primary sector from traditional crop production to cattle breeding, now the most important agricultural branch. The Azores were supposed to be Portugal's main supplier of beef and dairy products. Around 1980, there were about 150,000 heads of cattle, in 2001 it was 231,000, plus 62,000 pigs. About 50,000 heads of cattle are exported to Portugal every year. The breeding of cattle amounts to about 70% of Azorean exports. Cattle and dairy farming are not competitive in the EU and need subsidies. They suffer from the general structural weaknesses of the Azorean agriculture, which according to Wakonigg (2008) are: The traditionally strong contrast between a few large and privileged landholders on the one hand and the large number of businesses too small to be really productive on the other, the lack of processing capacity and effective marketing strategies for high-quality products, and the long distance to the Portuguese and other European markets, which causes high transportation costs and reduces the quality (especially of livestock).

Today, it is estimated that about 52% of the total area of the Azores is used agriculturally, of which almost three quarters are pasture. The remaining area of about 15% of the islands is used for growing crops, esp. corn (to feed cattle), potatoes, sugar beets, oranges, wheat, bananas, and wine. However, Azorean agriculture cannot provide the population with sufficient basic foodstuffs, especially bread grain.

Wines, whales, and whale watching

Traditionally, Azoreans grew three white wine varieties: *Verdelho* and *Terrantez,* both imported from Madeira, and *Arinto de Bucelas,* originating from the area of Lisbon. The main wine cultivation areas producing dessert wines high in alcohol are on Pico, Terceira and Graciosa. The volcanic grounds offer ideal cultivation conditions, because the black basaltic soils store the warmth of the day. In the nineteenth century, wine became the main export article of the Azores. It was exported from Pico to New York, India, Brazil and the Baltic States, and even the Tsarist Court of St. Petersburg ordered the *Verdelho of Pico*. However, in the last third of the century the wine blight (caused by *Phylloxera*), brought in from America, put an end to this climax. The crossing with American wild wines produced new hybrids (*vinho de cheiro, perfumed wine*) low in alcohol, which, according to wine connoisseurs, do not reach the quality of the old wines. Today, there is virtually no wine export any longer. Pico is famous for its network of rectangular vineyards (*currais*, Fig. 2) surrounded by walls made of basalt rock to protect the vines from wind and salty seawater spray. In 2004, the Pico island vineyard culture was classified as a World Heritage site by the UNESCO. It is no surprise that names of wines are motivated geologically, like *Terras de Lava* and *Basalto*. In the Azores, basalt is not only used for the *currais*. As material for buildings and pavements it is ubiquitous. Black basalt often contrasts with white paint or plaster in buildings,

Fig. 2 *Currais* on Pico (*photo* Rudolf Beier)

or with white limestone (on the Azores only found on Santa Maria), such as in the renowned Portuguese pavement (*calçada Portuguesa,* Fig. 3).

As compared to the surrounding Atlantic Ocean, the Azores are a shallow-water region (see Vogt and Jung, Chapter "The "Azores Geosyndrome" and Plate Tectonics: Research History, Synthesis, and Unsolved Puzzles"), the bathymetric swell providing a biologically active area attractive for whales. Whaling was one of the most important activities in the Azores, from its beginnings in 1765, following the American example. The whalers approached the animals in small rowing boats (*canoas*) and caught up to 200 sperm whales a year. The most important product was the oil for oil-lamps, which was won from the blubber, of which a whale delivered between three to four tons. Whale meat is not tasty, so it was usually processed into animal feed or even fertilizers. The so-called *spermaceti*, a waxy mass in the head of a whale, was used for making ointments, candles, and lubricants. The digestive system of sperm whales provided the so-called *ambergris*, a fragrant substance in which the animals envelop indigestible objects. It was used for the production of valuable perfumes, scored fantastic prices on the world market and was later replaced by synthetic substances. The exhibits of the *Scrimshaw Museum* at *Peter Café Sport* (Horta) show how whalebones served as material for artful carving. Once a year, the products of whaling were transported by ship to Lisbon. In the long run, the Azorean whalers' success was limited, because they were not able to compete with American whalers, who had a larger market and better facilities and tools (Ramos Villar 2006, 8). Whaling ended officially in the mid-80s, with the ban on whaling and with the last factory closing in São Roque do Pico due to unprofitability in 1984. The death of whaling was the birth of whale watching, today one of the branches of sustainable tourism.

Fig. 3 *Calçada Portuguesa*, Horta, Faial (*photo* Rudolf Beier)

Construction, cans, and cozido

The secondary sector

The secondary sector of the Azorean economy (industry, construction) has been restricted to the building and construction industries, the supply of energy and water, and the processing of goods from the primary sector. It consists mainly of small businesses, which produce sugar, canned fish and vegetables, beer, wine, spirits, mineral water, tea, tobacco, milk, cheese, and meat (Wakonigg 2008, 206).

The production of cheese on the Azores goes back to the Flemish immigration in the fifteenth century. The most common type of cheese is still called *Queijo Flamengo, Flemish cheese*; it is produced on São Miguel, Terceira and Faial. However, the 'real' local cheese is the *Queijo São Jorge,* which is originally made from a mixture of cows', goats' and sheep's milk, a combination optimally adapted to the livestock of the Azores. This cheese owes its special aroma to the mint that grows abundantly on the pastures of the island. The big whole cheeses are transported to the mainland by means of specially equipped ships. There is also the spicy *Queijo São João do Pico,* which is processed into smaller whole cheeses.

Generally, the secondary sector of the Azorean economy has suffered from emigration and a shortage of qualified labour. It has always been weaker, either than the primary sector, or, more recently, than the tertiary sector (Wakonigg 2008, 235). Elavai (2008) gives the weights of the sectors in the gross regional product for 1980 and for 2004 (primary sector: a decrease from 30 to 13%, secondary sector: a decrease from 26 to 17%, tertiary sector: a rise from 44 to 77%). Apart from the relative weakness and the relative decrease of the secondary sector, this shows us a decreasing weight of agriculture (due to the unfavourable distribution of property, esp. land, and the increase in the less work-intensive production of cattle), and a considerable increase of the tertiary sector (services, esp. tourism).

Like the people in other volcanically active regions of the world (for example Iceland, New Zealand), Azoreans make use of geothermal energy, which according to Wakonigg (2008) and the Sociedade Geotérmica dos Açores (Sogeo 2011) provided about 22% of their energy consumption in the first decade of the 21st century. Plans are to increase the proportion of geothermal energy considerably (Sogeo 2011). Geothermal energy also helps, as it were, to prepare one special dish, a traditional stew called the *Cozido das Furnas*. Beef, pork, chicken, sausages (like chorizo and morcella), potatoes, cabbage and a number of vegetables are all given into a pot and cooked in an adequate hole in the ground for about six hours. Bom apetite!

Travel and tourism

The tertiary sector

Apart from the dairy and cattle industries and fishing (mainly tuna), the modern economic outlooks of the Azores are mainly based on the expansion of tourism. Admittedly, the Azores have not yet arrived in the catalogues of all-inclusive tourism, a fact mainly due to trip difficulties: The traffic is almost exclusively in the hand of the Portuguese air line *TAP* and the Azores regional company *SATA*; there are few charter companies, and only the islands of Terceira, São Miguel and Faial have a daily connection with Lisbon, the usual place of transfer. A regular shipping line between Portugal and the Azores does not exist, and even cruise ships are touching the Azores comparatively rarely. Today, all this does not necessarily have to be a disadvantage. Rather, it may be a good starting point for the development of tourism in a controlled and sustainable way, protecting the Azores and their nature from the pitfalls of mass-tourism.

4 How Azoreans Live with the Forces of Nature

4.1 A Word About Azorean Culture

Is there an Azorean culture? And if there is, is it influenced by the forces of nature, and if yes, how? These questions are delicate, because they require generalizations and invite overgeneralizations. Quite obviously, there are relations between nature and culture, but they are neither direct nor simple. To answer the initial question, it will be helpful to take a closer look at literature. If there is an Azorean literature, this is a very strong piece of evidence that there is an Azorean culture.

The core of culture

Portuguese values

On the one hand, the majority of Azoreans have Portuguese ancestors, their culture has its roots in mainland Portugal. The language they speak is a variety of Portuguese (see Rogers 1948). For most of their history (see Chapter "Introduction"), the Azores, politically and economically, have belonged to Portugal. With this in mind, it is not surprising that most of the characterizations of Azoreans and the 'Azorean character' we can find in the literature fit quite well into the sociocultural pattern that Hofstede and Hofstede (2005) found out about continental Portugal, in their empirically based comparison of 74 countries. In terms of the value of *power distance* (the acceptance by cultures of the unequal distribution of power), Portugal occupies ranks 37 and 38 (a much higher score than the US, ranking 57 to 59, recently the main destination of Azorean emigrants). Concerning the value of *individualism* (vs. *collectivism*), Portugal's ranks are 49 to 51 (with the US scoring highest of the 74 countries). With respect to values of *masculinity versus femininity* (ambition versus quality of life), Portugal has the rank of 65 (US: rank 19) in the masculinity index (which, interestingly, means Portugal's orientation towards life is much more feminine than that of the United States). Even more striking seems Portugal's position in the *uncertainty avoidance* index. It occupies rank 2 (after Greece on rank 1, and with the United States on rank 62). Uncertainty as defined by the Hofstedes refers to "the fact that we do not know what will happen tomorrow: the future is uncertain, but we have to live with it anyway (Hofstede and Hofstede 2005, 165). The three basic 'tools' humans have developed to handle uncertainty are technology, rules, and, particularly relevant in our present context, religion. As the Hofstedian values are rather stable over time, we can assume that, basically, what applies to Portugal is true of the Azores, too. In our context, the fourth dimension mentioned above, uncertainty, clearly invites geology, as it relates to unpredictable events (such as eruptions or earthquakes). As uncertainty avoidance is high in the Azores, the level of anxiety must be high, too. Obviously, ways to deal with anxiety (religion…) must be of particular relevance to the Azores. Indeed, the prominent role of the mystical aspects of the *Holy Order of Christ* and the Franciscans (such as the cult of the *Holy Spirit*) has been a typical feature of Azorean Catholicism (see Sect. 4.3 below). This leads us to the more specifically Azorean side of the cultural coin.

Environment, experience, expression

Azoreans live in an environment that is quite unique in terms of geography, meteorology, and geology. These include the geographical and meteorological phenomena, i.e., the insularity, the sea with its winds, fog, rain, floods on the one hand, and on the other hand, the reason and inspiration for this book: the geological situation, which brings with it volcanic eruptions and earthquakes, which in turn can cause tsunamies. Outlines of eruptions and earthquakes in recorded history are given by Santos (1995) and Farjaz (2008). Different types of landslides (Wakonigg 2008, 241) are very frequent in the Azores, they can be caused both by meteorological phenomena (heavy rain) and volcanic and seismic activities (see Fontiela et al., Chapter "Characterisation of Seismicity of the Azores Archipelago: An Overview of Historical Events and a Detailed Analysis for the Period 2000–2012"). The forces of nature certainly do not have a simple and direct effect on culture, but they do leave their traces, both in the literal and the figurative meanings of the word. And more than that, there appears to be an interrelationship between natural disasters, religion (4.3), and emigration (4.2). As is born out by Ramos Villar's (2006, 84) analysis of the works of Azorean novelists, these three phenomena must be considered as "intrinsically linked in the creation of a perceived Azorean identity".

The forces of nature have more or less predictable (the weather) or unpredictable effects (earthquakes, eruptions) on the lives of individuals and communities. What people experience in their environment, and share with others, and hand on as knowledge to the next generation, forms part of their cultural identity. We may call this their 'shared subjective knowledge'—a concept we borrow from Brutt-Griffler (2002), who coined the term to describe the development of new (postcolonial) identities in English-speaking countries all over the world. Though their situation differs from that of the Azores (with respect to their political and linguistic situation), the mechanism of identity formation seems rather universal and, therefore, comparable. Among other things, it can be observed in language: words may acquire specific meanings in the environment, to express shared knowledge and experience.

A prominent example, which is geologically relevant here, is the word *mistérios*. In the Azores, it is used to refer to peculiar formations of basalt and large-scale fields of lava (e.g., the *Mistério da Prainha*, the *Mistério de Sta. Luzia*). To former generations of Azoreans, the eruptions that created these phenomena were supernatural in nature, mysterious, and frightening. The only feasible explanation for an event as horrible as a volcanic eruption or an earthquake, was a *castigo madado por Deus*, a punishment sent by God (da Luz 2000, 120, for the role of religion see below). According to Farjaz (2008, 32), the first occasion of the use of *mistérios* by priests and people was one of the longest eruptions in the Azores in 1562, in the *Prainha Plateau* area, Pico. On São Jorge, the land burned during the volcanic eruptions in 1580 is still called the *Mistérios of Santo Amaro and Queimada*. Other examples are (brackets give islands and historical eruptions according to Farjaz 2008): the *Mistério do Capello* (Faial, 1672), the *Santa Luzia and Sao João Mistérios* (Pico, 1718), the *Mistério da Silveira* (Pico, 1720), the *Mistérios dos Biscoitos e dos Negros* (Terceira, 1761), the *Urzelina Mistério* (São Jorge, 1808), the *Mistérios dos Capelinhos* (Faial, 1957). Examples of Azorean (place) names with geological motivations are: *Cascalho Negro* (São Miguel, *negro* means *black, cascalho* can stand for English *gravel, pebbles, shingle*), *Furnas* (São Miguel, English *caves*), *Lajes* (Pico, the Portuguese word *lajes* has a meaning similar to English *flagstones*), *Caminho da Pedra* (São Jorge, English *rocky road or path*). Examples of words, the (nongeological) meanings of which differ between their use in mainland Portugal and the Azores, can be found in the introduction to Kinsella (2007). According to Ramos Villar (24), Azorean writers have, from the 1930s onwards, asserted the existence of an 'Azorean language'.

To summarize the discussion so far, we can borrow Ramos Villar's (2006) expression of Azorean culture being part of, but distinct from the mainland Portuguese culture. This means that the above-mentioned cultural core is rather stable over time. What changes more quickly are the other layers of culture (Hofstede and Hofstede 2005), the symbols (such as language), the heroes (appearing, e.g., in legends, folk, and fairy tales, see Eells 1922/2011), and the rituals (like in religion). In the course of almost six centuries, Azoreans have gained and shared subjective knowledge and experience with respect to nature, religion, and emigration. These, and the relations between them, have added to and modified their Portuguese (or other…) cultural roots and heritage. We believe it is this specific Azorean cultural identity that has repeatedly and, it seems, quite adequately, been referred to as *açorianidade* (Almeida 1980; Ribeiro 1964), *azoreanness*, or "the Weltanschauung of the islands" (Almeida 1983, 22). According to Nemésio, who introduced the term in 1932, the concept of *açorianidade* is closely connected with the geographical situation of the Azores (isolation, islandness) and their geology (see Sect. 4.4 below). It expresses a way of life, an identifiable island culture, as opposed to *Portugueseness* (Ramos Villar 2006, 28, 191). In Nemésio's poems, Ramos Villar discovers the same idea according to which prolonged contact with a natural surrounding influences behaviour, and shapes identity. Nemésio's *Stormy Isles* (1994) can be considered an outstanding artistic elaboration of *açorianidade*.

What we must not forget, however, is that we have already been generalising. As small as the islands may appear from a distance, the Azorean situation is still complex and differentiated. There are different social groups in the islands. Some Azoreans came from countries other than Portugal. Observers like Almeida (1983) have claimed a number of inter-island differences, between Micaelense, Terceirense, Picaroto, Mariense… (Almeida 1980). There are even distinguishable island dialects.

Before we present some more concrete evidence and examples from religion and literature, we will consider Azorean emigration. It plays an important role in both religion and literature, and, more than that, it is closely connected with them in a way that is uniquely Azorean.

4.2 Emigration

In terms of personal attitudes, involvement, and decisions, emigration (like religion, see Sect. 4.3) is "an intrinsic part of the Azorean character that the individual must either embrace or reject" (Ramos Villar 2006, 130). It is also one of the topics dominating Azorean literature.

Basically, emigration is "a displacement of people to where they may achieve a better quality of life" (Ramos Villar 2006, 1). Apart from the emigrants who leave the Azores and Portugal, there are people who have been forced to leave their homes and move to other Azorean islands (for example during the *Capelinhos* disaster) or, for different reasons, have decided to go to mainland Portugal, some to work for governmental, others to attend educational institutions. For many Azoreans without relatives or connections outside the islands who could call for them and provide support, the Church provided a means to leave, by joining seminaries and convents in pursuit of a (religious) education or vocation." (Ramos Villar 2006, 87).

Emigration in general and the numbers of people leaving the Azores have depended on a variety of factors, such as the urgency of the immediate causes (e.g. natural disasters, economic crises), immigration laws in the countries of destination (e.g. quota, the necessity of 'invitation letters' and sponsors for the US, strong restrictions by the US between the 1920s to the 1950s), and on whether the Portuguese rulers supported emigration, because they needed them, first as explorers, later as settlers to protect Portugal's overseas possessions, as during the colonization of Brazil in the eighteenth century (Ramos Villar 2006, 10).

Emigration has been a conspicuous social phenomenon in Portugal in general and in the Azores in particular, but with some quantitative and qualitative differences between them: Historically, the Azores as we know them today would not exist without migration. After all, it was migrants who took possession of the uninhabited archipelago almost six centuries ago. As compared to mainland Portugal, the proportion of people who emigrate in relation to the population as a whole has generally been higher in the Azores. As an example, it is estimated that, historically, about 50% of the total Portuguese emigration to the US have come from the Azores (Goulart 2008, XVIII). Coutinho et al. (2010, 270) quote data according to which by the 1950s about 250,000 Azoreans were estimated to live in the United States, with the population of the Azores at that time being about 337,000. From 1900 to 1990 about 333,000 people are estimated to have left the islands, and in the second half of the twentieth century there was a real decrease of the population by about 100,000 people. In 2006, the total population of the Azores was about 243,000 (Wakonigg 2008, 133ff.).

Among the circumstances causing people to emigrate from the Azorean archipelago, the economic situation and natural forces and disasters have played prominent, and usually cooperating roles. Frequent periods of overpopulation made it difficult or impossible for the local population to survive on the basis of local products. Other factors that contributed to emigration of Azoreans have been the unequal distribution of property, esp. land, cyclical unemployment, and a number of plant diseases and pests (esp. the crises in the production of oranges and wine). In general terms, emigration worked as a 'relief valve' for the combined effects of economic problems and population pressure (Goulart 2008, 319). A special motivation for young men to leave was to avoid military conscription (e.g. during the Portuguese colonial war at the beginning of the 1970s). More often than not, the humble living conditions of a majority of Azoreans were further aggravated by the economic and human effects of volcanic eruptions and seismic crises which, again and again, brought unpredictable destruction, impoverishment and misery upon them.

From the seventeenth to the nineteenth centuries, the preferred destinations promising better living conditions to Azoreans were Brazil (for the whole period) and Hawaii (for about two decades from 1878, for work in the sugar-cane plantations). From 1617 onwards, many families were registered settling in the northeast of Brazil (in Maranhão, Ceará and Pará). Along with families who set out legally to the new American homes and figured in the settlers' lists, there were many people who went on their own. Starting in about 1670, Azoreans emigrated to the south of Brazil. Despite the immense distance, contacts between the islands of the Azores and Brazil have never broken off. For obvious (linguistic, historical, religious) reasons, there has always been a certain affinity in the relations between Brazil and Portugal in general, and Brazil and the Azores in particular (Ávila and Mendonça 2008, 19; Rocha and Ferreira 2009, 9ff.). This helps to explain why people from Brazil make up a sizeable proportion of the immigrants who, in a relatively new development since the 1980s, have come from a number of countries (US, Cape Verde, Eastern Europe, EU) to live in the Azores (Coutinho et al. 2010, 277; Rocha 2008, 124f.).

It was not only economic reasons that made people leave for Brazil, and later on, to North America. As early as 1675 and 1719 sizeable groups of Azoreans left Faial and Pico due to volcanic crises (on Faial from 1671 onwards, on Pico starting in 1718, Ávila and Mendonça 2008, 18).

As early as the end of the eighteenth century, the first Azoreans may have gone to the United States as ship crews, especially on whaling ships (Marcos 2008, 106). Many more followed in the decades beginning with 1800, and many of them jumped ship at the American East Coast and in California. Since the turn of the twentieth century, for the majority of Azorean emigrants the destination has been North America (the United States and Canada). Azorean emigration reached its highest numbers during the periods from 1880 to 1920 (with an annual number of about 8000 in the first two decades), and again from 1965 to 1980.

The most recent peak in emigration to North America, quite well documented and analysed, started in December 1958 and lasted for about two and a half decades. It was an excellent example of how the demographic and economic circumstances cooperate with natural disasters: The main problem of Faial before the disaster was demographic and economic: overpopulation. This was aggravated by the disaster, "because the loss of so much productive land meant that the long-term problem of excess population in relation to employment opportunities, particularly in the agricultural sector, was exacerbated" (Coutinho et al. 2010, 276). The Capelinhos emergency, i.e. the volcanic eruption of Capelinhos and a large number of associated earthquakes, began on 16 September 1957 and lasted until 24 October 1958, destroying villages and crops and traumatising the population considerably. During that period, "the distance from the Azores to America grew shorter" (de Oliveira 2008, 100). Post-Capelinhos emigration was considerably facilitated by new immigration laws in the United States, the *Azorean Refugee Acts* of 1958 and 1960, and the *Hart-Celler Act* of 1965, which replaced the quota system and the harsh regulations of the period between 1920 and the 1950s. About 100,000 Azoreans moved to the US, reducing the population of the Azores by about one-third, a "mass exodus…the largest wave in a short period of time, of Portuguese emigration to the United States" (Goulart 2008, XVII). Emigration may have improved the lives of many individuals, and for some, it was a success story, at least in material terms. The islands, however, especially Faial, lost a considerable part of their work force. For the Azores, post-Capelinhos emigration meant a relief of pre-disaster overpopulation and, at the same time, an aging population, because the majority of those who left were younger than 40 years of age (Goulart 2008, 1; de Oliveira 2008, 101). It also meant a tremendous amount of solidarity, generosity, and financial help from the relatives and friends in North America. This is an important aspect of Azorean emigration in general: emigrants have always helped those who stayed in the Azores to survive (Ramos Villar 2006).

There is a relation between the development of the economy and a new stage of demographics, which the Azores entered in the 1980s. The figures we find in the literature for the period between 1980 and 2006 allow the conclusion that the characteristic features of this new development are: a decrease of emigration in general, particularly to the US, a decreasing birth rate, an increasing number of immigrants from Brazil, former African Portuguese colonies, the EU and Eastern Europe.

4.3 Religion and the Church

With respect to their ecclesiastical history, the Azores, at first, fell within the jurisdiction of the *Order of Christ*, like many other overseas territories, under the direction of the vicar of Tomar (*vicarius nullius*). From 1514 onwards, when the diocese of Funchal was founded, the Azores were subordinate to the jurisdiction of the Bishop of Funchal. In 1534, a suffragan diocese was founded at Angra on Terceira, and the church of São Salvador was elevated to become the seat of the bishopric (*Sé Catedral*). It remained under the suffragan of the archbishop of Funchal until 1550, since then it has belonged to the ecclesiastical province of Lisbon. The *Diocese de Angra* (sometimes, unofficially, referred to as the *Diocese de Angra e Ilhas dos Açores*) includes all islands of the Azores.

Since the Azores were colonized under the *Holy Order of Christ*, Catholicism has been a strong, to some authors the strongest pillar of Azorean mental and spiritual tradition and folk culture. Da Silva considers Catholicism as one of the fundamental components of the Portuguese-Azorean cultural identity, of *açorianidade* (da Silva 2008, 72ff.). Catholic faith on the Azores is closely linked to emigration (see Sect. 4.2 above) and natural disasters (below). Both in the archipelago and in the diaspora, it is practised through numerous festivals and processions (Fig. 4), the most typical of them related to the *Império* (the *Holy Spirit*). It was Franciscan missionaries and spiritualists, who made the Azores a stronghold of the veneration and cult of the *Holy Spirit*, and made *Pentecost*, the feast of the *Holy Spirit*, the feast of the Azores (for details: Montez 2007). A look at Azorean maps shows that religion is a very frequent motivation of place names. Apart from the islands of *Santa Maria*, *São Miguel* and *São Jorge*; we find places like *São Roque do Pico* (referring to *Saint Roch*), *São Caetano*, *São João* (all on Pico), we find roads like *Rua Nossa Senhora da Conceiçã*, *Rua do Espirito Santo*, *Rua de Santa Lucia* (all in Ribeira Grande), and geological phenomena, e.g. *Cabeço* (English *hill, hillock*) *do Padre Roque* (Pico), and the many Azorean *mistérios* (English *mysteries*, examples in Sect. 4.1 above).

The forces of nature and the uncertainties of the environment on the Azores provided a fertile ground for the mystical orientation of the Order of Christ and the Franciscan spiritualists, so that, from the beginnings until today, they have played an important role in dealing with disasters. On 22 October 1522, a violent earthquake, followed by a landslide, destroyed Vila Franca do Campo, the first capital of the Azores, and killed several thousand people. The report on the destruction of the city (*Romance que se fez de algumas magoas e perdas que causou o tremor de Villa Franca do Campo em 1522, Lição de Gaspar Frutuoso*, Braga 1869, 335ff., also known as *Romance de Vila Franca*) is the first extensive text of the oral literature. This disaster was followed by a plague, which also killed many thousands of people and is also described in folk literature. Natural disasters like these strengthened the feeling of general threats from the forces of nature, leaving their traces in Azorean religious behaviour and folk beliefs, and reinforced the mystic component of the religiousness stamped by the Franciscans.

Throughout their history, Azoreans have pleaded for divine attendance and protection from earthquakes and volcanic eruptions, in numerous processions (on all islands), fireworks, and church services. The construction of *Holy Ghost chapels* (*impérios*) was common in the Azores during volcanic and seismic crises (Farjaz 2008, 37). As an example, the *Holy Ghost Chapel* in Horta (*Império dos Nobres*) was built in remembrance of the eruptions in the *Mistério do Capelo*, Faial, in 1672. The eruption of Urzelina (São Jorge), in 1808, caused deaths inside the church, shattering the traditional religious belief that churches protected people from the fire rising from the interior of the earth, "a phenomenon never observed in the mainland but found in the Azores beginning with the settlement of São Miguel" (Farjaz 2008, 37).

Religious and folk beliefs and practices merge into one another. A striking example of this comes from the island of Pico. In the communities of Bandeiras and Santa Luzia, people have handed from generation to generation the belief that in the night from the first to the second of February, ghostlike creatures arise from the lava and move upwards towards the top of Pico. To prevent them from getting too close to their houses, people burn rosemary on charcoal. On the following day, they express their gratitude in a procession, for having been saved from the lava during the last eruption of Pico (Sabias que … ? 2012, 41).

Fig. 4 Ready for the *festa*, São Miguel (*photo* Rudolf Beier)

It was priests who often used the religious fervour and folk beliefs of Azorean congregations in order to allow, even encourage "them to believe that the natural disasters that affect the Azores were punishment from God for not staying on the true path. The priest thus becomes His spokesman, gaining and exerting considerable power and control in a society where volcanic activity and earth tremors are very frequent" (Ramos Villar 2006, 86).

Coutinho et al. (2010, 274) report that throughout the Capelinhos emergency many people turned to God for help and protection. "… there were numerous examples of: prayers being offered; processions of sacred objects and votive images being held; intercessions for safe deliverance being made and special Eucharistic services with an accent on safe deliverance being performed … " On the other hand, the Capelinhos disaster showed that people nowadays widely resist notions of divine responsibility by the clergy: "… a sermon preached by the Bishop of Angra, arguing that the earthquake and volcanic eruption represented divine punishment for individual human sinfulness, caused considerable dissent with some of the congregation spontaneously walking out of church… " (Coutinho et al. 2010, 274).

If we consider Azoreans' current attitudes towards religion and the social functions of religious festivals in emigrant communities in the US and the native islands, we can observe interesting differences. In her analysis of novels, Ramos Villar (2006, vii) finds evidence "of the old paradox whereby Catholic conservatism is most evident among American diaspora communities seeking to preserve and show identity (the main function of the *festas* in the diaspora), while the home-based authors of the archipelago tend to be more critical of the Church as an impediment to progress". As an instrument of oppression and social control, the church has helped to stabilise and support economic and political conditions and the rigid social hierarchy that have forced people to emigrate. The other side of the coin is that many people, who were not able to meet the requirements for 'regular' emigration (money, connections, relatives, 'invitation letters'), were allowed to leave the islands through the church (Ramos Villar 2006, 84; see Sect. 4.2 above).

4.4 Literature

If we consider emigration as an essential component of Azorean culture, and if we look into the lives of those who consider themselves as, and are considered to be, Azorean authors, it becomes obvious that it is impossible to define an 'Azorean writer' in narrow geographical and biographical terms, as a person who was born and lives on the islands. Some were born in the archipelago and stayed, many were born in the Azores and went to mainland Portugal (to be educated and/or to live there, or to go somewhere else), or migrated to America, others are descendants of emigrants to America. Some were born outside the Azores, but (have) lived in the archipelago. So being an Azorean writer is rather a question of identification and of themes (Almeida 1983, 24).

In her very detailed and differentiated analysis of different genres, Ramos Villar (2006, iii) proves that the works of Azorean writers portray "a regional, cultural and literary uniqueness that is distinct from, but part of, Portuguese literature". Following her, there are two prominent distinctive features of Azorean literature, its negotiating character, and its main themes: Azorean literature, as a result of the long history of migration, negotiates between cultures in two directions, between the Azores and mainland Portugal, and between the Azores and America. Its main themes are island life and emigration. Emigration means travelling, both mentally and physically. The destination of the journey is 'a utopian end', a myth (born by the negotiation between Azorean and American cultures), the metaphorical *tenth island* (a term originally coined by Almeida, see Almeida 2006).

Against this general background, we can say that the natural environment, forces, and disasters, as parts of island life and as driving forces in emigration, play a key role in Azorean folk stories (see the prologue to this article), novels, and poetry.

We will here take a final look at poetry, which has always been an important genre for the Azores (Kinsella 2007; da Silveira 1977). Azorean poems are clearly dominated by two topics: insularity and the sea (as two sides of the same coin), and very closely related to it, voyage and emigration. To readers interested in a detailed and professional literary study of several anthologies, we recommend Ramos Villar (2006).

We would here like to quote five excerpts from poems, which, in one way or another, refer to the geological reality of the islands. It can simply be part of the poetic scenery described (as in example a).

(a) From Carlos Faria's poem ***Adiados, adiados sempre/Postponements, delays always…*** (Kinsella 2007, 96f.)

Montanha do Pico, with black clouds
covering its tower and coming down the whole
slope which finally looks out over the heights of
Friar Matias' Cave...
Joao waves to me from afar...

The geological forces can appear as 'agents', as elemental, disruptive, unpredictable powers, which can seriously affect the lives of people. This, typically, is achieved by the use of figurative language, as in example (b).

(b) From Vasco P. da Costa's poem ***São Jorge*** (Almeida 1983, 29):

...São Jorge is a dragon stretched out in the channel...
Men are suspicious of the sleeping beast.
They climb its back of cascading rivulets beach tree cliffs
and yams misteries and paths.
On the wings of the beast they erect houses pastures milk
cows make cheese dry fish talk softly
- do not awaken the sleep of the beast.
When the dragon awakens it stretches sticks out its
Serpent's tail – it's the tremor. The earthquake.

Quite obviously, the dragon metaphor in example (b) is invited by São Jorge's shape: it is lengthy and slender, and has steep cliffs. Unlike all the other islands, it does not have a main volcano.

What strikes us about examples (c) and (d) below is the closeness of the natural forces and the human beings and their bodies. The geological phenomena virtually amalgamate with the people (and vice versa), the analogies used here present them as organically, maybe symbiotically, related. This does not seem to be a new idea in Azorean thinking. We can already find it in Némesio's often quoted explication of *açorianidade* from the 1930s: "We are welded historically to this land from which we have come; our roots are by nature within these mountains of lava. Some substance, oozing out from volcanic guts, has penetrated our souls. Geography inviolably means as much to us as history, and it is no idle coincidence that at least fifty percent of our written records are accounts of floods and earthquakes. Like mermaids' ours is a double nature: We are built of flesh and stone. Our eyes submerge themselves in the sea" (Némesio 1932, 79; translation by Almeida 1980, 133).

(c) From João de Melo's poem ***Este povo da ilha/The island people*** (Kinsella 2007, 156f.)

This is the people born in the sea. Their blood came from
salt. Their veins floated at one time
amidst seeweed hair and volcanic spores.
Its mouth opened in the remote oblivion
of shells. Their memory are the deserted conches
the pebble rolling in sandy silence over rocks.
...
So this people
(re)cognizes itself between sand and sea
at the exact moment when stone
and body touch each other and love
the water.

(d) From Victor Rui Dores, ***Canção de Margarida/Margarida's song*** (Kinsella 2007, 220f.)

my island, my people
salty lava, my scent
ah love made serpent
in this ocean of betrayal

in the small flame of my face
burns the force of a volcano
the wall of my trouble
is made of stone and loneliness ...

The final examples below (e, f) are similar to (c) and (d). There are two poems by Manuel

Alegre, in which what merges together are the geological nature of the islands on the one hand and the act of writing and language on the other. One of them is *Ilha do Pico* (Alegre 2007, 30, example e). In (f) we quote the second and third stanzas of his poem *Código*. As for these examples we did not find an expert English translation, we give the original wording and try a (nonprofessional, non-authorised) interlingual translation, only to give the reader an idea of the poetic message.

(e) Manuel Alegre, Ilha do Pico (engl. Pico Island)

Pode escrever-se um poema com basalto
You can write a poem with basalt
com pedra negra e vinha sobre a lava
with black rock and vine on lava
com incenso mistérios criptomérias
with incense mistérios full of cryptomeria
e um grande Pico dentro da palavra.
and a big Pico in the middle of the word.
Ou talvez com gaviotas e cagarras
Or maybe with seagulls and cagarras
cigarras do silêncio que se trilha
cicadas strangling the silence
sílaba a sílaba até ao poema que está escrito
syllable to syllable till the poem is written
lá em cima no Pico sobre a ilha.
up there on Pico on top of the island.

(f) From Manuel Alegre's poem ***Código*** (engl. *Code*, Alegre 2007, 38)

Mas eu oiço o vulcão a pulsação
But I hear the volcano the pulse
a linguagem secreta
the secret language
transmitida
transmitted
de poeta a poeta.
from poet to poet.

Oiço o mistério o cedro a música da lava
I hear the mistério the cedar the music of the lava

o rumor mineral e o canto
the mineral rumbling and the subterranean
subterrâneo. O código e o sinal.
song. The code and the sign.
Tudo é ritmo e palavra
All is rhythm and word
celebração imagem lugar santo.
feast, image, holy place.

5 Epilogue

We would like to conclude this paper with still another poem, by giving the floor to John Updike, voyager and great American writer, and his poetic observations of the Azores, their geology, their landscape, and their insularity (Almeida and Monteiro 1983).

John Updike
Azores

Great green ships
Themselves, they ride
At anchor forever;
beneath the tide

Huge roots of lava
Hold them fast
In mid-Atlantic
to the past

The tourists, thrilling
from the deck,
hail shrilly pretty
hillsides flecked

with cottages
(confetti) and
sweet lozenges
of chocolate (land)

They marvel at
the dainty fields
and terraces
hand-tilled to yield

the modest fruits
of vines and trees
Imported by
the Portuguese:

a rural landscape
set adrift
from centuries ago
the rift

enlarges.
The ship proceeds.
Again the constant
music feeds

An emptiness astern,
Azores gone.
The void behind, the void
ahead are one.

6 Summary

Their eventful history of almost six centuries has shown the Azoreans as builders of bridges between the Old and the New Worlds. Azoreans have tilled the soil of their islands for various crops, but diseases, disasters and the land-holding system have often limited their success. While their culture is basically Portuguese, Azoreans have developed an identity of their own. In this process, the natural environment and the forces of nature have played prominent roles, usually cooperating with other (mainly economic) factors. As amply evidenced by the writers of Azorean literature, major features of Azorean cultural identity are (the role and practising of) religion and the issue of emigration.

Acknowledgements We are grateful to the editors: for inviting us as strangers in the field, and for being patient, and for your confidence. To John M. Kinsella: for your valuable help and advice. To Carmen Ramos Villar: for much of the same, and for this excellent, eye-opening book of yours. To a very good friend, Gerardo Herschung Iglesias: for being so 'loco' (as usual…) as to share it all with us: our enthusiasm for the islands, our search for relevant information, your expertise in Romance literary studies, and in Portuguese, and in translation. For those geologists who read the acknowledgements first: be warned, social scientists (especially those dealing with language, literature, culture) are a different, maybe strange species. And last but not least, should Azoreans read this: do forgive us all remaining errors, and the painful incompleteness of this portrait painted from far away …

References

Alegre, M. (2007). *Escrito no mar. Livro dos Açores.* Lisboa: Sextante Editora.

Almeida, O. T. (1980). A profile of the Azorean. In *Issues in bilingual education.* Cambridge, MA: National Assessment Determination Center. Downloadable from: The Canadian Center for Azorean research and Studies at www.yorku.ca.

Almeida, O. T. (ed.), & Monteiro, G. (Trans.). (1983). The sea within. A selection of Azorean poems. Providence, RI: Gávea-Brown.

Almeida, O. T. (2006). Tales from the Tenth Island. Selection, Translation and Introduction by D. Brookshaw. (University of) Bristol: Seagull/Faoileán.

Ávila, J. I., & Mendonça, L. (2008). Azorean emigration: Causes, waves, and destinations. A brief historical summary (Goulart, Ed., pp. 17–24).

Braga, T. (ed.). (1869). Cantos populares do Archipelago Açoriano. Porto: Typ. da Livraria Nacional. Digitized by Google.

Brutt-Griffler, J. (2002). *World English. A study of its development.* Clevedon, UK: Multilingual Matters Press.

Consulate of the United States Ponta Delgada, Azores: History of Airbase 4. www.azores.usconsulate.gov. Consulted on October 10, 2015.

Cordeiro (original spelling: Cordeyro), A. (1717). Historia insulana das ilhas a Portugal sugeytes no Oceano Occidental. Lisboa Occidental: Antonio Pedrozo Galram (digital version viewed at scribd.com, digitized by Google).

Coutinho, R., Chester, D. K., Wallenstein, N., & Duncan, A. M. (2010). Response to, and the short and long-term impacts of, the 1957/1958 Capelinhos volcanic eruption and associated earthquake activity on Faial, Azores. *Journal of Volcanology and Geothermal Research, 8,* 265–280.

da Luz, A. A. (2000). Uma viagem aos Açores. Do que vi, li e ouvi… In Geosul (Florianópolis) 15, n. 29.

da Silva, H. G. (2008). Portuguese-Azorean cultural identity. Common traits of Açorianidade/Azoreanness (Goulart, Ed., pp. 73–84).

de Silveira, P. (1977). *Antologia de poesia açoriana.* Lisboa: Sá da Costa editora.

de Oliveira, Á. (2008). Emotional ties and humanitarian support by the Azorean communities. 1500 U.S. Visas via a volcano (Goulart, Ed., pp. 99–103).

de Oliveira Marques, A. H. (2001). *Geschichte Portugals und des portugiesischen Weltreichs.* Stuttgart: Kröner.

Eells, E. S. (1922/ eBook 2011). The Islands of magic. Legends, folk and fairy tales from the Azores. Copyright by Harcourt, Brace, and Company 1922, printed by The Quinn and Boden Company, Rahway/NJ, made accessible as eBook in 2011 and downloaded from The Project Gutenberg.

Elavai, A. (2008). The Azores: From Capelinhos to the present day (Goulart, Ed., pp. 327–334).

European Commission (2014). The European Union explained: Regional policy. Luxembourg: Publications Office of the European Union. Downloadable from: europa.eu/pol/pdf/flipbook/en/regional_policy_en.pdf.

European Commission: ec.europa.eu/regional_policy/en/projects/portugal. Consulted on October 10, 2015.

Farjaz, V. H. (2008). Historical eruptions in the Azores (Goulart, Ed., pp. 25–44).

Goulart, T. (Project Coordinator, ed.). (2008). *Capelinhos. A volcano of synergies.* San Jose, CA: Protuguese Heritage Publications of California.

Hofstede, G., & Hofstede, G. J. (2005). *Cultures and Organizations. Software of the mind*. New York: McGraw-Hill.

Kinsella, J. M. (2007). *Voices from the islands. An anthology of Azorean poetry*. Providence, RI: Gavéa-Brown Publications.

L'atlante nautico 'Corbitis', già detto 'Combitis'. Il website cartografico e grafico della Biblioteca Nazionale Marciana di Venezia: http://geoweb.venezia.sbn.it. Consulted on October 8, 2015.

Marcos, D. (2008). The eruption of Capelinhos: A window of opportunity for Azorean emigration (Goulart, Ed., pp. 105–114).

Marino, N. F. (ed.). (1999). El libro de conoscimiento de todos los reinos [The book of knowledge of all kingdoms]. Tempe, AZ: Arizona Center for Medieval and Renaissance Studies (available on archive.org).

Mendes, J. (1974–1979). *Literatura portuguesa* (pp. 1–4). Lisboa: Verbo.

Montez, M. S. (2007). The imperio in the Azores. The five senses in rituals to the Holy Spirit. *Traditiones, 36*, 169–176.

Nemésio, V. (1932). Açorianidade. In Insula, Número Comemorativo do V Centenário do Descubrimento dos Açores, nº 7–8 (julho-agosto). Ponta Delgada.

Nemésio, V. (1994). *Stormy Isles. An Azorean tale*. Providence, RI: Gavéa-Brown Publications (edited, translated, annotated and with an introduction by F. C. Fagundes).

Ramos Villar, C. M. (2006). *The metaphorical 'tenth island' in Azorean literature. The theme of emigration in the Azorean imagination*. Leviston, Queenston, Lampeter: The Edwin Mellen Press.

Ribeiro, L. (1964). Subsidios para un ensaio sobre a Açorianidade. Angra de Heroísmo, Açores: Instituto Açoriano de Cultura (informação preambular e bibliografia por João Agonso; our quotations in English taken from Almeida 1980).

Rocha, G. P. N. (2008). The Azores and emigration: Demographic evolution in the second half of the twentieth century (Goulart, Ed., pp. 321–326).

Rocha, G. P. N., & Ferreira, E. (2009). Azorean emigration in the context of social change: Some notes from the press in São Miguel (1920–1950). *e-journal of Portuguese History 7.2*, 1–17.

Rodrigues, A. F., Martins, N. O., Ribeiro, N., & Joaquinito, A. (2015). Early Atlantic navigation. Pre-Portuguese presence in the Azores Islands. *Archeological Discovery, 3*, 104–113. Published online July 2015 in Scientific Research Publishing www.scirp.org/journal/ad.

Rogers, F. M. (1948). Insular Portuguese Pronunciation: Porto Santo and Eastern Azores. *Hispanic Review, 16*, 1–32. Insular Portuguese Pronunciation: Central and Western Azores (1949). *Hispanic Review 17* (pp. 47–70).

Sabias que…? Em tomo da açorianidade. Pequenas singularidades das gentes e da terra (2012). Currículo regional de educação básica. www.edu.azores.gov.pt. versão.pdf.

Santos, R. L. (copyright 1995). *Azoreans to California. A history of migration and settlement*. Denair/CA: Alley-Cass Publications. Accessible via the Bob Santos Homepage. www.library.csustan.edu.

Sogeo (Sociedade Geotérmica dos Açores, 2011). Use of geothermal energy resources in the Azores islands: A contribution to the energy self-sufficiency of a remote and isolated region. Séminario redes energéticas do futuro: desagios para a Macaronésia. Ribeira Grande, 13 Dezembro de 2011. Via: www.azores.gov.pt.

UNESCO: Landscape of the Pico Island vineyard culture. whc.unesco.org.

Wakonigg, H. (2008). Die ostatlantischen Vulkaninseln. Azoren. Madeira Archipel. Kanaren. Kapverden. Ihr Natur-, Wirtschafts- und Kulturraum. Wien: Lit Verlag.

The "Azores Geosyndrome" and Plate Tectonics: Research History, Synthesis, and Unsolved Puzzles

Peter R. Vogt and Woo-Yeol Jung

Abstract

The Azores volcanic archipelago, the Azores Plateau (AP) and the Azores triple junction (ATJ) between the Eurasia, North America and Nubia plates occupy the summit of a regional feature we refer to as the '*Azores Geosyndrome*'. Included are anomalies in crustal thickness, rock composition, basement depth, plate boundary morphology, seismicity, gravity and geoid, and upper mantle seismic velocity structure, and there are many similarities between the Azores and Iceland geosyndromes. The location of the Azores in the central North Atlantic, technological advances in marine geophysics as well as logistic, geomilitary and geopolitical motivations and advanced research of island geology/volcanology have contributed to make the ATJ the most studied oceanic triple plate junction. However, a unified understanding of the Azores Geosyndrome awaits future deep crustal boreholes (particularly on the AP) and regional sea-floor seismometer arrays to resolve the seismic velocity structure below the AP down to the middle and perhaps lower mantle. Whereas a deep mantle plume appears unlikely to exist below the Azores, it cannot yet be excluded (see O'Neill and Sigloch, Chapter "Crust and Mantle Structure Beneath the Azores Hotspot—Evidence from Geophysics", and Moreira et al., Chapter "Noble Gas Constraints on the Origin of the Azores Hotspot"). What is already clear is that the development and evolution of the Azores Geosyndrome has involved dynamic interactions among the North America-Nubia-Eurasia plates and at least the uppermost mantle below those plates—*even far from the ATJ area*. The plate boundary reorganization that resulted in the triple plate junction jumping from the end of King's Trough south to create the ATJ was largely complete by Chron 6C (23 Ma) and coincided within dating uncertainties with the jump of the spreading plate boundary from the Norway Basin to the new Kolbeinsey Ridge just north of Iceland. Major geological changes in the Pyrenees and Alpine Tethys region at that time have long been known. In fact, the Palaeogene-Neogene boundary, a time of global change in planktonic biogeography, is placed at 23.0 Ma, in the upper part of C6C. Why the ATJ developed where it did and not elsewhere along the MAR suggests the lithosphere and subjacent mantle had already created a region of plate weakness. The subsequent development of the AP, largely

P. R. Vogt (✉)
Marine Science Institute, University of California at Santa Barbara, Santa Barbara, CA, USA
e-mail: ptr_vogt@yahoo.com

W.-Y. Jung
Marine Geosciences Division, Naval Research Laboratory, Washington, DC 20375, USA
e-mail: wooyeol.jung@gmail.com

U. Kueppers and C. Beier (eds.), *Volcanoes of the Azores*, Active Volcanoes of the World, https://doi.org/10.1007/978-3-642-32226-6_3

via Mid-Atlantic Ridge spreading, produced a thick crust and more fertile mantle lithosphere-particularly from ca. 12 to 8 Ma. This mantle lithosphere was and continues to be relatively weak and fertile, favouring transtensional fissuring, formation of central volcanoes, as well as oblique hyperslow spreading along the Terceira Rift—particularly in the last 1.5 Ma.

1 Introduction

The Azores Archipelago of volcanic islands (Forjaz et al. 2010; Figs. 1 and 2) is the focus of this volume—but these islands collectively form the summit of the much broader regional "Azores geosyndrome", a term we use for an ensemble of spatially and probably causally related anomalies in crustal thickness, depth and gravity/geoid anomaly, upper mantle seismic structure, plate boundary morphology, and rock geochemistry. Whereas this paper focuses on the plate boundaries intersecting at the Azores Triple Junction (ATJ), the anomalous characteristics of the plate boundaries near the ATJ can only be fully understood within the context of the Azores Geosyndrome, which in many ways resembles similar regional 'geosyndromes' like those centred at Iceland and the Afar Triangle.

The Azores geological literature has exploded, especially in recent years, as obvious from the reference lists in this volume. This poses a challenge for reviews—key publications may be overlooked, while newer data or models are still unavailable to the reviewer. However, our review offers the easily overlooked research history and —we hope- some fresh perspectives on recently published data and interpretations. We cast a wide net and thus inevitably overlap several other chapters in this volume (Fernandes et al., Chapter "The Contribution of Space-Geodetic Techniques to the Understanding of the Present-Day Geodynamics of the Azores Triple Junction ", O'Neill and Sigloch, Chapter "Crust and Mantle Structure Beneath the Azores Hotspot—Evidence from Geophysics; Miranda et al., Chapter "The Tectonic Evolution of the Azores Based on Magnetic Data"; and Mitchell et al.,

Fig. 1 **a** Computer-illuminated seafloor topography from the Azores archipelago and MAR (Mid-Atlantic Ridge) to Gibraltar, derived from satellite radar altimetry and surface ship echo-sounding (from Smith and Sandwell 1997); earthquake epicentres (teleseisms) from 1973 to 2012. See also Simkin et al. (2006) for a generalised global synthesis of plate boundaries, volcanism and seismicity. Known hydrothermal vent fields are numbered: (1) Moytirra; (2) Menez Gwen; (3) Lucky Strike; (4) Lost City; (5) TAG; and (6) Joao do Castro Bank. **b** Seafloor topography and earthquake epicentres of the Azores platform, including the Terceira Rift, the triple junction area, hypothesised Azores microplate, and adjacent parts of the Mid-Atlantic Ridge. Solid line shows axis of Terceira Rift plate boundary after Vogt and Jung (2004), with letters US, G, F, E, D, C, B and A points along the plate boundary topographic profile (Fig. 3a). (The extrapolation of the TR to the MAR (Vogt and Jung 2004) has been revised, reflecting evidence for a diffuse plate boundary (Marques et al. 2013; Miranda et al. 2014). The age '39 Ma' marks the Monaco Bank rock of Beier et al. (2015). Key magnetic lineations (heavy dashed lines) from Luis and Miranda (2008). Red triangles are volcanic centres active since 10 ka (as in Simkin et al. 2006). AB, Azor Bank; PAB, Princess Alice Bank. Inset map (modified from Vogt and Jung 2004), shows other ultraslow spreading axes, with small circles (on East Pacific Rise) enclosing two postulated microplates at triple plate junctions (See also DeMets et al. 2010). Seismic reflection profile locations (*Vema* and *Gibbs* cruises) correspond to Fig. 2. **c** Seafloor topography, plate boundaries, and earthquake epicentres of the immediate Azores triple junction area. F, Faial; P, Pico; SJ, São Jorge; G, Graciosa; AB, Azor Bank; PAB, Princess Alice Bank. PAR denotes postulated fossil (pre-TR) Prince Alice Rift plate boundary (dashed line). **d** High resolution hillshade seafloor topography of the Azores TJ region (Miranda et al. 2014); image courtesy of J. Luis, Univ. of Algarve. **e** Structural/Tectonic (morpho-structural) interpretation (modified from Miranda et al. 2014) of the Azores TJ region, based on multibeam bathymetry (Fig. **d** and other data), earthquake epicentres, and magnetic anomaly isochrons (numbers denote chrons). Solid lines denote tectonic, and dashed lines, volcanic topographic structures. N1 lineations are abyssal hill fabric exported from the MAR axis; N2 trends are WNW trending normal faults and volcanic fissure eruption trends extending westwards from the islands; N3 denotes NW and conjugate NE trending normal faults attributed to diffuse Eurasia-Nubia motion. Brackets show crust generated ca. 8.2–6.2 Ma, during pole location change and slowdown in Eurasia-North America (Merkouriev and DeMets 2008, 2014b) and Nubia-North America motion (Merkouriev and DeMets 2014a) ▶

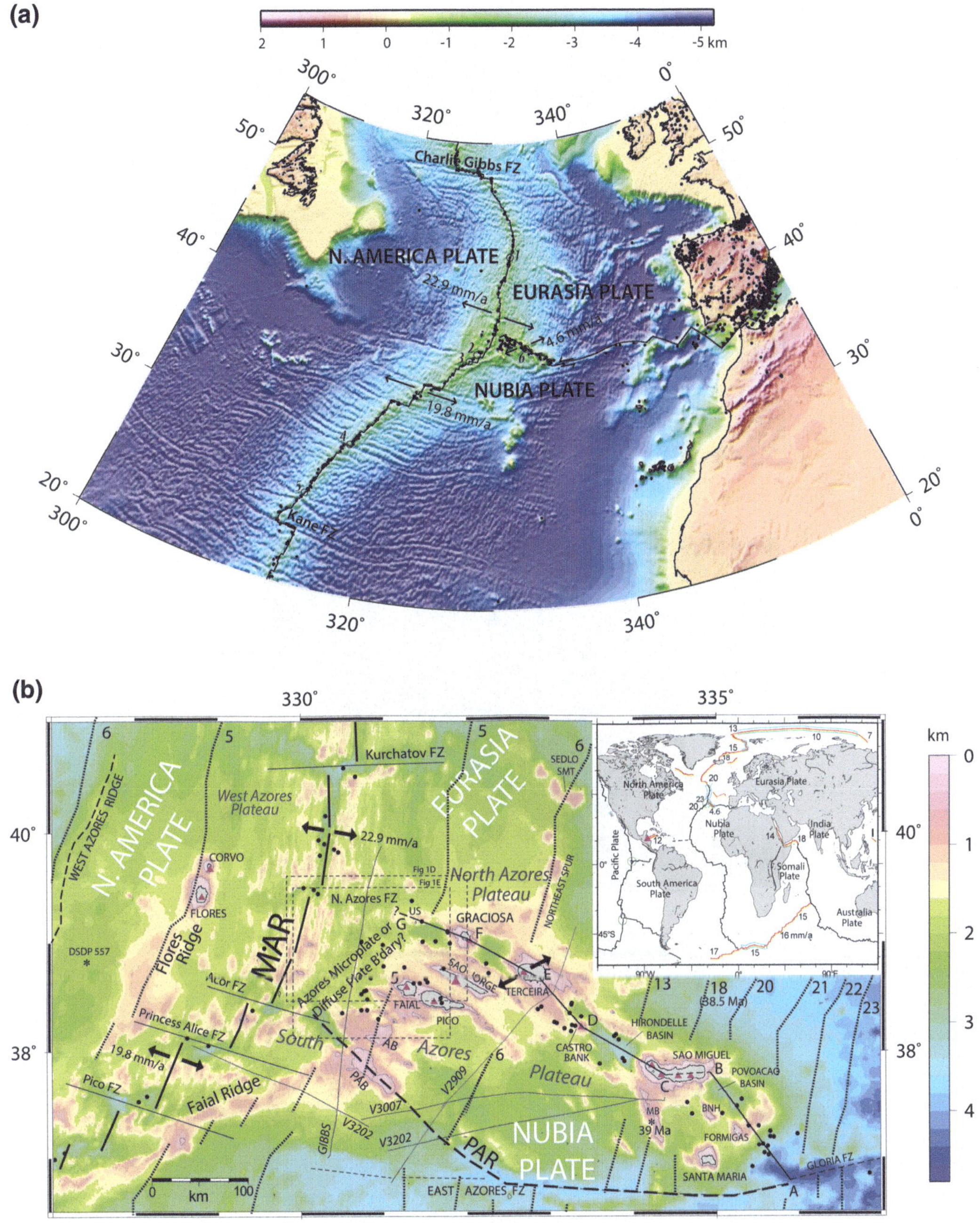

Chapter "Volcanism in the Azores: A Marine Geophysical Perspective"). We trust that any differences in interpretations of the same data will stimulate further discussion and research.

Since the plate tectonic revolution (discussed further below) and the subsequent but still controversial mantle plume/hotspot concept (Morgan 1972), the volcanic Azores islands and plateau have been attributed (e.g., Gente et al. 2003) to the interaction between some type of hotspot/mantle plume/melting anomaly and the triple junction between the North America,

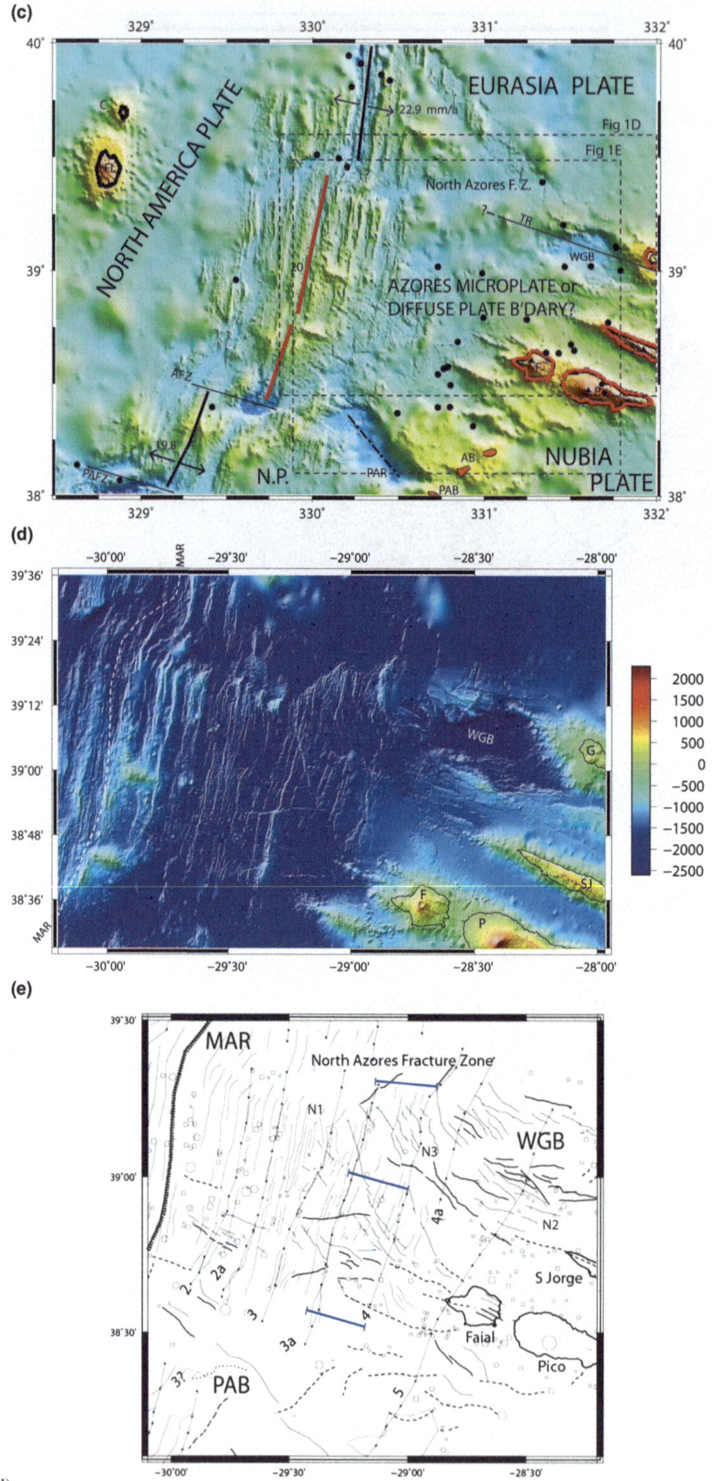

Fig. 1 (continued)

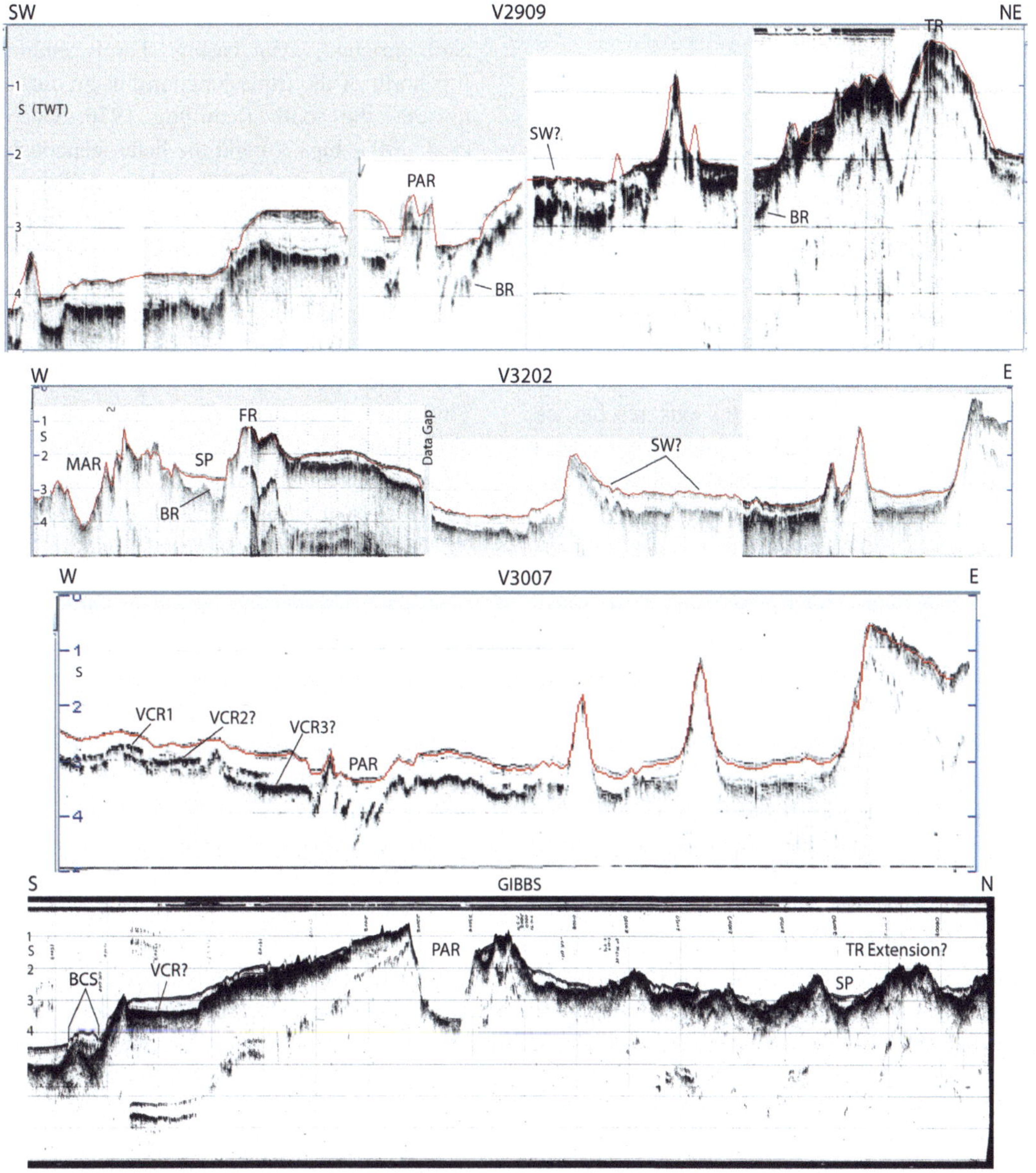

Fig. 2 Single channel seismic reflection profiles collected on transits of the Azores Plateau, with vertical scale in two way reflection time. In the acoustically transparent hemipelagic sediments, 1 s is approximately 1 km. Profile locations are shown in Fig. 1b. *R/V Vema* profiles are from GeoMapApp (Ryan et al. 2009); *R/V Gibbs* profile courtesy of National Geophysical Data Center. MAR, Mid-Atlantic Ridge rift valley; FR, Faial Ridge; PAR, Princess Alice rift; TR, Terceira Rift, BCS, bottom current sedimentation and erosion; BR, basement reflector; SP, sediment pond; SW sediment waves; VCR, volcanoclastic reflector. Multichannel seismic profiles were collected (2009) in areas around São Miguel (Weiss et al. 2015b, 2016)

Eurasia, and Africa/Nubia plates (see also O'Neill and Sigloch, Chapter "Crust and Mantle Structure Beneath the Azores Hotspot—Evidence from Geophysics"). Although plume-based numerical models account for some observations (Adam et al. 2013), other numerical models reproduce some features of the Azores Plateau (particularly the trends of the Terceira Rift (TR) and linear volcanic seamounts and islands SW of the TR (Neves et al. 2013) by boundary forces fracturing the brittle upper plate and providing transtensional conduits for magma ascent.

The anomalous geological and geophysical attributes comprising the "Azores Geosyndrome" include:

- a regional geoid (over +50 m) and corresponding free-air gravity high, which fuses with the highs over Iceland (e.g., Grevemeyer 1999; Rabinowitz and Jung 1986);
- a regional positive oceanic depth anomaly (Louden et al. 2004), which along the Mid-Atlantic Ridge (MAR) axis reaches ca +1.5 km near the ATJ and may extend 3200 km south from the Charlie Gibbs Fracture Zone (FZ) (52°N) to the Kane FZ (Fig. 1a);
- the shallow Azores Plateau (AP), alternatively called platform or rise, with presently thickened (up to 14 km; Escartin et al. 2001; Dias et al. 2007) crust, formed largely from 20 to 7 Ma at the crest of the MAR (Gente et al. 2003; Miranda et al., Chapter "The Tectonic Evolution of the Azores Based on Magnetic Data");
- central volcanoes (Mitchell et al., Chapter "Volcanism in the Azores: A Marine Geophysical Perspective"), rising up to 1105 m above sea level as geologically young (<1 Ma) islands along the Terceira Rift (TR; Graciosa, Terceira, Joao de Castro bank, and São Miguel), and somewhat older ones emplaced in AP crust both west (Corvo, Flores) and east of the MAR axis (Santa Maria, Formigas, Faial, São Jorge, Pico—with its 2351 m summit one of the tallest volcanoes only ca. 150 km from the MAR axis–plus Princess Alice and Azor banks and several guyots);
- a segment of the MAR (ca. 38.7°–39.3°N) near the triple junction lacking a conspicuous rift valley, with comparatively low historical teleseismic activity, and exhibiting, despite the slow (ca. 22 mm/a; Fig. 1c) spreading, well-developed magnetic lineations (Searle 1980; Luis et al. 1994; Miranda et al., Chapter "The Tectonic Evolution of the Azores Based on Magnetic Data") typical of faster-spreading ridges;
- geochemically anomalous (e.g., light-rare-earth-enriched) axial basalts sharply ending just north of the triple junction but gradually towards the south (Schilling 1976; Gente et al. 2003; Fig. 3c) and the heterogeneously sourced (Beier et al. 2008, 2015 and Chapter "Melting and Mantle Sources in the Azores") island volcanoes, and newer data suggesting the Azores overlie a ca. east-west compositional boundary in the underlying mantle (Madureira et al. 2014);
- southward pointing V-shaped Faial and Flores ridges (Vogt 1979, 1986c), formed along the MAR 10–4 Ma (Cannat et al. 1999; Escartin et al. 2001);
- an older set of basement ridges formed along the MAR ca. 20 Ma, plausibly marking the beginning of Azores Plateau formation; and
- a complex low-velocity upper mantle region under AP, and extending from 250 km to at least ca. 400 km depth under or just northeast of the TR according to Yang et al. (2006) and to 250–300 km based on the negative S-wave anomaly (Silveira et al. 2006), with no compelling seismic or other evidence (e.g., Anderson 2005) for a mantle convection plume (Morgan 1972) rising from near the ca. 2900 km deep core-mantle boundary.

Not all features of the ATJ plate boundaries are anomalous, however: The MAR near the ATJ (Fig. 2 of Luis et al. 1994) is segmented at 50–60 km scales, similar to the MAR far from the Azores (Fig. 2 of Vogt 1986a), and includes discontinuities slowly (3 mm/a; Luis et al. 1994) propagating along the axis, of the type also found on the MAR far from the ATJ (Fig. 3 of Vogt and Jung 2005; Schouten et al. 1985), and an axial volcano (minimum depth, 750 m; 38.3°N; Luis et al. 1994) similar to others scattered widely along slow-spreading ridges (Vogt and Jung 2005; Vogt et al. 2008).

The longer (ca. 100 km) segmentation wavelengths and pervasive oblique spreading along the TR are also typical of other ultra-slow spreading ridges far from 'hotspots' (Vogt and Jung 2004). While most papers accept the TR as an accreting plate boundary (but see Luis et al.

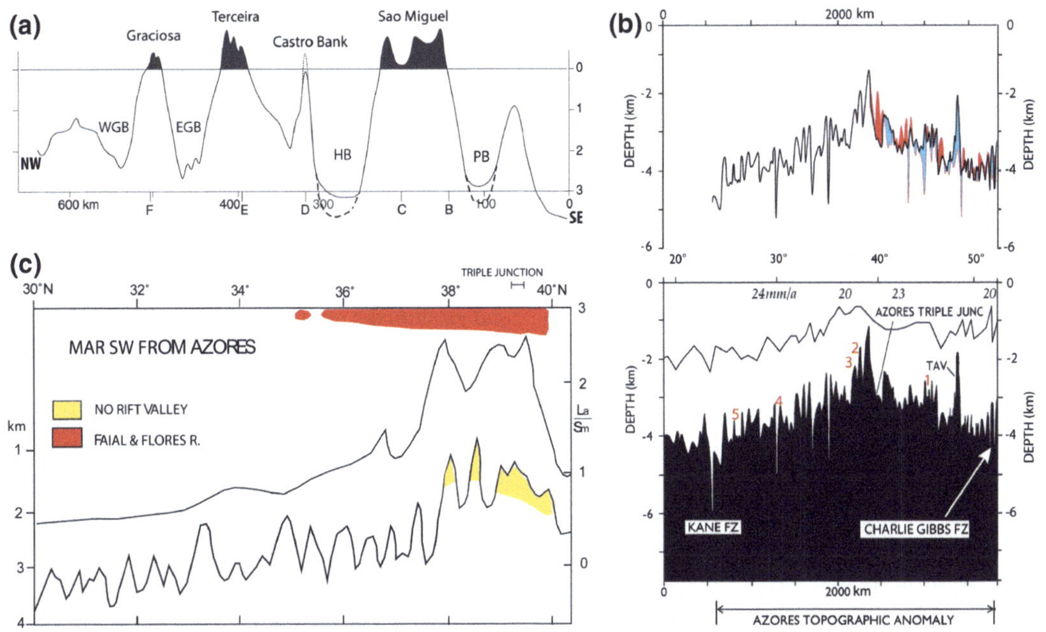

Fig. 3 **a** Along-strike topographic profile along the axis of the Terceira Rift (Fig. 1b), from Vogt and Jung 2004). Letters B, C, D, E and F denote locations where TR plate boundary changes direction. WGB, West Graciosa Basin; EGB, East Graciosa Basin; HB, Hirondelle Basin, and PB, Povoacao Basin. The latter two basins contain at least ca. 500 m sediment (dashed, based on Weiss et al. 2015b). **b** Along-strike topographic profile (lower) along the axis of the Mid-Atlantic Ridge along the Azores and Iceland regional highs (modified from Fig. 9 of Vogt and Jung 2005), suggesting that as much as 3200 km of the MAR (2200 km according to Goslin and Party 1999; Gente et al. 2003) is influenced by the 'Azores Geosyndrome'. Top profiles show southern flank overlapped on northern flank as a test for north-south along-axis symmetry. TAV is Telegraph Axial Volcano (Vogt and Jung 2005). Known hydrothermal vent fields are numbered: (1) Moytirra; (2) Menez Gwen; (3) Lucky Strike; (4) Lost City; (5) TAG; (6) João do Castro Bank. **c** Along-axis topographic (lower) and La/Sm (upper) profiles from 30°N to just north of the Azores triple junction (reproduced from Fig. 10 of Vogt and Jung 2005; based on Gente et al. 2003, originally from Dosso et al. 1999). Red shows the extent of the paired off-axis V-shaped Faial and Flores ridges, formed at the MAR axis progressively from 10 Ma to 4 Ma (Vogt 1979; Cannat et al. 1999)

1998), the detailed depiction of this boundary varies from author to author. Here we follow Vogt and Jung (2004) who approximated the alternation of volcanic islands and intervening deeps as a serious of strait line segments—in detail many segment transitions are curved (sigmoidal), as is true for the Mid-Atlantic plate boundary where mapped in detail by multibeam bathymetry. GPS data from various Azores islands support a diffuse plate boundary, with the TR itself opening even slower than assumed by Vogt and Jung (2004), perhaps accommodating only ca. 2–3 mm/a of the total ca. 4.5 mm/a Nubia-Eurasia motion (Marques et al. 2013), as discussed further below.

In this chapter, we review the evolution of understanding, but beginning with the profound advances that set the stage for seafloor spreading and then plate tectonics, and the role of costly technological innovations, driven by geopolitical (mainly military) needs, which made most marine geophysical advances possible. We conclude by summarising some outstanding research problems which remain.

2 Preludes to Plate Tectonics (1855 to 1968)

Although the Azores islands have been historically known to mankind since at least the early 15th century, what lay below and beyond the shallow banks and island shelves remained a mystery until the abyss was sounded with long wires (cables), the first in the South Atlantic in 1840 (see also Beier and Kramer, Chapter "A Portrait of the Azores: From Natural Forces to Cultural Identity"). The first crude representation of the north-central Mid-Atlantic Ridge (MAR) was that of Maury (1855), who called this shallower region "Middle Ground". Maury's map also showed, but did not name, the even shallower region around the Azores islands, today known as the "Azores Plateau" (also referred to as a rise or platform, divided in this paper into North, West and South (or East) plateaus; Fig. 1b). Wire soundings proved invaluable for laying of transatlantic telegraph cables, which included a shore station in the Azores.

Higher resolution MAR bathymetry had to await the advent of echo-sounding technology—particularly the southern and equatorial MAR profiles collected by the 1925–27 Meteor expedition (Stocks and Wuest 1935). About the same time, earthquake seismologists discovered that the MAR—including a belt from the Azores region to Gibraltar— was seismically active (Tams 1927; Gutenberg and Richter 1941). The narrowness of these belts would not be discovered until the 1960s.

Alfred Wegener, arguably the father of continental drift (beginning ca. 1913), died on the Greenland ice cap in late 1930, and so never had full access to the new bathymetric and seismologic results, dating from the mid-late 1920s, that bolstered his theory. In the final edition of his book (Wegener 1929), he still believed that the Azores, Iceland, and the MAR were sialic debris, continental material left behind as a kind of 'fossil breakup zone'. Yet, he was obviously troubled by continental reconstruction problems such as posed by Iceland, and by the great width of the Mid-Atlantic Ridge. Wegener did at least consider the possibility that the MAR was of basaltic composition. More important, he attributed the shallower averaged ocean depth of younger ocean basins to thermal expansion of what today would be called the 'lithosphere', thus predating by decades the now well-known 'age-depth' relationships (Sclater et al. 1971). Had Wegener applied his thermal expansion argument to the MAR—plus the new knowledge, acquired after his death, of a central rift valley and earthquake seismicity—Wegener might well have also fathered plate tectonics himself. Simple observation of an intermittently widening, freezing lead between two ice floes—the youngest ice the thinnest—might have inspired this polar scientist to envision continents as frozen into the tops of giant plates, rafting apart along narrow crack-like zones at the axis of the Mid-Oceanic Ridge.

Improvements in paleomagnetism provided the first post-Wegnerian evidence for continental drift, but the new polar wander curves (e.g., Runcorn 1956) did not convince many skeptics. Instead, the next big advance was the formulation of 'sea-floor spreading', a term coined by Dietz (1961), which replaced Wegener's 'sialic' MAR debris with elongate upwelling mantle convection cells below the Mid-Oceanic (and thus Mid-Atlantic) Ridge. Drawing on earlier mantle convection conjectures by Arthur Holmes (1929), Hess first presented the spreading concept two years prior to the Dietz publication (Hess 1959). Ewing and Heezen (1956) had already called attention to the earthquake belt along the MAR crest, with a spur extending towards the Mediterranean from the Azores area. Moreover, Heezen et al. (1959; references therein) had identified the MAR rift valley and, with an early marine magnetometer, found a high positive magnetic anomaly over the rift valley, flanked by negative anomalies.

It took less than a decade for 'sea-floor spreading' to evolve into plate tectonics. During this last pre-plate tectonic interval (1959–1968), linear magnetic anomalies, also called lineations or magnetic striping, were attributed to a combination of sea-floor spreading and magnetic reversals (Vine and Matthews 1963; independently by Morley and LaRochelle 1964). The reversals of geomagnetic polarity are recorded, largely in the highly magnetised upper 500 m of oceanic crust, as bands of alternately normally and reversely magnetised basalt. The anomalies, small departures of total field strength, depend in their amplitude on distance from the top of the oceanic crust, on the strength of crustal thermoremanent magnetization, and on the relation of a band of crust relative to the present geomagnetic field. In the North Atlantic including the Azores region, positive anomalies approximately overlie bands of normal magnetization. The reversal time scale (e.g., Walker et al. 2012) is divided into 'chrons', with chron numbers approximately equivalent to magnetic anomaly numbers, but include a conspicuous band of largely positive magnetic polarity at the end of a chron, plus the preceding interval of largely reversed polarity. Given the slow spreading rates and frequent reversals for the time of Azores Plateau formation, we use anomaly and chron numbers interchangeably. Magnetic lineations over the Mid-Atlantic Ridge are mostly rather irregular and difficult to identify, a consequence of slow spreading and lava extrusion rates. Thus, Vogt and Ostenso (1966) concluded from a local magnetic-bathymetric survey of the MAR crest north of the Azores (42°–46°N) that convincing evidence for the Vine and Matthews (1963) hypothesis was lacking there.

Wilson (1965) identified and named transform faults—strike-slip plate boundaries along which crust is neither made nor destroyed. However, few obvious transform faults were evident in the complex MAR bathymetry north of the Azores. Menard (1965) identified the active seismic belt and associated rough topography between the Azores and Gibraltar as an 'active' fracture zone, which to first order it turned out to be.

3 The Azores Region Interpreted During the Plate Tectonic Era

Interpretations of the Azores region evolved rapidly following the publications of Morgan (1968), Isacks et al. (1968), LePichon (1968), McKenzie and Morgan (1969) and others. Morgan (1968) proposed the East Azores FZ as the boundary between the African and Eurasian 'blocks' (now called plates). It was early recognised that the Azores platform was too complex to have evolved from a simple ridge-ridge-ridge (r-r-r) triple plate junction. Starting with Krause and Watkins (1970) and continuing with later studies (e.g., McKenzie 1972; Laughton and Whitmarsh 1974) evolutionary models variously incorporated changes from transform faults to rifts, migration of the triple junction along the MAR, and/or jumps in the spreading axis (Searle 1980). Krause and Watkins (1970) applied the idea of a "leaky transform fault" (Menard and Atwater 1968) to what is now called the Terceira Rift. In more recent publications, the "leaky transform" concept has been applied to the central and eastern part of São Miguel (e.g., Sibrant et al. 2016; Weiss et al. 2015b).

Magnetic data available at the time lacked the density and inherent resolution to test aspects of the three basic evolutionary models (Krause and Watkins 1970; McKenzie 1972; Searle 1980). Work in succeeding decades refined these three triple junction evolution models (Neves et al. 2013; Miranda et al. 2014), but resolution of the magnetic stripes is limited by the slow Nubia-Eurasia opening rates within the "Azores Domain" (Luis and Miranda 2008). However, some key NNE-trending magnetic lineations generated at the MAR axis have been identified over parts of the Azores Plateau (Luis and Miranda 2008; Miranda et al. 2014 and Chapter "The Tectonic Evolution of the Azores Based on Magnetic Data"; Fig. 1b, c, e; Miranda et al., Chapter "Volcanism in the Azores: A Marine Geophysical Perspective").

A detailed aeromagnetic survey of the Azores region provided the first evidence for the intermittent (10–3.85 Ma) existence of a separate "Azores microplate" (Luis et al. 1994). DeMets et al. (2010) concluded, from the gradually northward increasing opening rates (since 3.16 Ma) along the MAR from ca. 38.35 ± 0.15°N to 37.7 ± 0.3°N, that such a microplate must on average have existed since then, perhaps with the historically active volcanic islands Faial, Pico, São Jorge, plus Azor and

Princess Alice banks (Fig. 1c), thus forming its present wide, complex eastern boundary. However, in support of earlier studies (Bastos et al. 1998), GPS data from Terceira, São Jorge, Faial, and Pico support distributed (diffuse) deformation, with no microplate required (Marques et al. 2013, 2014), a result corroborated (Miranda et al. 2014) with detailed multibeam topography and earthquake seismicity (Fig. 1e).

When did the Azores triple plate junction currently located near 38°50′N, 30°25′W (Gaspar et al. 2015), first form? North Atlantic plate kinematic reconstructions have proved challenging, because slow spreading rates generally produce “noisier” magnetic lineations. The MOR on either side of Iceland is an exception (Vogt 1986b). The first major reconstruction (Pitman and Talwani 1972) was successively refined (e.g., Klitgord and Schouten 1986; Srivastava and Tapscott 1986). The latter two papers conclude that the Azores triple junction originated about anomaly 9 time (ca. 27.5 Ma on the Walker et al. 2012 timescale), when the plate boundary from an earlier MAR triple junction last extended via Kings Trough into the Bay of Biscay and the Pyrenees. About that time, the triple junction jumped south, thereby transferring Iberia to the Eurasia plate, as first suggested by Smith (1971). However, magnetic lineations between #13 and #6 are poorly developed, and King’s Trough actually extends west nearly to anomaly #6 (e.g., Fig. 1 of Klitgord and Schouten 1986), so the plate boundary ‘jump’ might have occurred gradually during the period 27–20 Ma, rather than ‘instantaneously’. Geological evidence suggests rifting along King’s Trough lasted until roughly 20 Ma (Kidd and Ramsay 1987). Roest and Srivastava (1991) concluded that the triple junction ‘jumped’ south to the Azores site around anomaly 6C time (23 Ma). The most recent plate kinematic models, based as in earlier models on successively rotating back together crustal isochrons defined by prominent magnetic lineations on both MAR flanks, are by Luis and Miranda (2008) and Merkouriev and DeMets (2008; 2014a, b). We interpret the Northeast Spur and West Azores Ridge, formed around anomaly 6 time (Fig. 1b), as marking the initiation of Azores Plateau formation, i.e., soon after the triple junction jump (see later discussion). Merkouriev and DeMets (2014b) found small misfits (overlaps) when they rotated magnetic lineations from the one MAR flank to the other. These misfits (20 km for anomaly 6 and 5 km for anomaly 4A) suggest some diffuse ‘post-jump’ intraplate deformation continued within the Eurasia plate from the triple junction north to 42°N.

The r-r-r type triple junction model (Krause and Watkins 1970) already included a slow-spreading ‘Terceira Rift’ (TR) crossing the Azores platform. The attempt to measure the TR opening rate using magnetic profiles proved (e.g., Miranda et al. 1991) unconvincing, probably due to the very slow opening rates and wide rift zone, relative to the geomagnetic reversal frequency. If, contrary to Vogt and Jung (2004), the TR is not an established, albeit ‘hyperslow’ spreading axis, no magnetic lineations would be expected in any case. The opening rates across the slow-spreading, variably oblique TR plate boundary had to be estimated by vector subtraction of the well-constrained North America-Africa motion from the North America Eurasia motion, and basically the same method is still used today. The rifting/spreading half-rate Krause and Watkins (1970) estimated for the TR plate boundary (2.0–2.5 mm/a) is closely comparable to the total opening rates computed for about the middle of the TR from more recent global inversions (e.g., 4.9 mm/a, NUVEL; DeMets et al. 1990; 4.6 ± 0.2 mm/a, S64 W ± 1.3° (MORVEL; DeMets et al. 2010). When only Eurasia-North America and Nubia-North America motion is considered, the “geologic” (since ca. 3.16 Ma) MORVEL inversion yields 4.5 ± 0.4 mm/a in the direction S68.1 W ± 2.8°, and the GPS “present” inversion is closely similar (4.6 ± 0.3 mm/a, S87.9 W ± 3.3°) but with a Nubia-Eurasia pole ca. 3000 km south of the MORVEL pole. However, inter-island motions measured by GPS indicate that only about half this Nubia-Eurasia motion currently takes place along the TR (Marques et al. 2013; Fernandes et al., Chapter “The Contribution of Space-Geodetic Techniques to the Understanding of the Present-Day Geodynamics of the Azores Triple Junction”).

All three types of plate boundaries are represented between the ATJ and Gibraltar (Fig. 1a; e.g., Bezzeghoud et al. 2014). In the west, a rifting/spreading boundary (the TR) extends from the ATJ to about 24°W, SE of São Miguel Island. At that point, the rift intersects the East Azores Fracture Zone, part of which comprises the 400 km long right-lateral strike-slip GLORIA fault (Fig. 1b); east of 18°W the plate boundary becomes an oceanic-oceanic plate convergence zone, with rough, complex sea-floor topography (Fig. 1a).

4 Technology and Plate Tectonics —Geopolitical, Economic, Military and Logistic Factors

Whereas the plate tectonics paradigm shift was largely and initially motivated by scientific curiosity, subsequent and very costly technological developments (accurately located earthquake epicentres, focal depths and rupture mechanisms; magnetometers for use on the high seas; multibeam bathymetry; satellite radar altimetry; GPS navigation) that led to our current understanding exemplify 'geopolitical/geomilitary trickledown'. This circumstance contrasts with other areas of geosciences, e.g. magma evolution and geochemistry of Azores volcanic systems (Larrea et al., Chapter "Petrology of the Azores Islands", Beier et al., Chapter "Melting and Mantle Sources in the Azores"), driven mostly by scientific curiosity alone. Marine geophysical understanding of the Azores region was an early and ongoing beneficiary of geopolitically motivated technological advances, most of which had their origins prior to emergence of plate tectonics concepts in the late 1960s.

(1) ***WWSSN*: The *World Wide Standardised Seismograph Network*** was developed during the early 1960s to improve detection capability for nuclear test ban violations. Particularly in oceanic regions, the WWSSN reduced epicentre location and focal depth errors and threshold magnitudes, while improving the accuracy of first-motion (fault plane) solutions for larger shocks. The more accurately located epicentres helped refine active plate boundary locations, including in the ATJ-AP region (Fig. 1a), which is seismologically active, creating seismic hazards (e.g., Gaspar et al. 2011, 2015; Hipolito et al. 2013). The epicentre distributions, focal depths, and first-motion/fault plane solutions helped refine and test plate kinematic models, particularly along the tectonically complex Azores-Gibraltar Nubia-Eurasia boundary (Udias et al. 1976; Hirn et al. 1980; Grimison and Chen 1986, 1988; Buforn et al. 1988; Borges et al. 2007; Navarro et al. 2009; Hipolito et al. 2013; Bezzeghoud et al. 2014; Fontiela et al., Chapter "Characterization of Seismicity of the Azores Archipelago: An Overview of Historical Events and a Detailed Analysis for the Period 2000–2012"). Madeira et al. (2015) reviewed the neotectonics (driven by dextral transgressive stress) of the central and eastern islands, estimating slip rates of a few tenths mm to a few mm/a on faults, and maximum expected earthquake magnitudes ca. $M_w = 6$ to 7. Seismographic stations on the Azores islands, OBS (Ocean Bottom Seismometer) deployments on the Azores Plateau (AP; Miranda et al. 1998) and OBH (Ocean Bottom Hydrophone) deployments on the MAR flanks north and south of the ATJ (http://www.ipgp.fr/rech/lgm/MOMAR/) helped make the plate boundaries in the Azores region among the better studied in the world ocean.

(2) ***Multibeam bathymetry***: Detailed mapping of ocean-floor topography, especially the active spreading and transform plate boundaries, was not practical until the mid-1960s development of "SASS" (Sonar Array Survey Systems), the world's first hull-mounted multi-narrow-beam bathymetric mapping system, developed for use by the US Naval Oceanographic Office (Glenn 1970). This type of bathymetric mapping is commonly referred to simply as 'multibeam' or 'swath-mapping' to distinguish it from conventional single-beam echo sounding. One of the first scientific applications of the SASS system was mapping the FAMOUS area of the MAR axis (36.4°–37.1°N), not far south of the ATJ (Phillips and Fleming 1978). The narrow axial "Neo-Volcanic Zone" rising from the floor of the rift valley was one of the notable discoveries, and

played a large role in planning *ALVIN* submersible dives. A somewhat smoothed version of otherwise classified SASS bathymetry shows the Kurchatov Fracture Zone area (40.3°–40.7°N), immediately north of the ATJ (Fig. 8 of Vogt and Tucholke 1986).

Starting in the later 1970s, multibeam technology spread into the commercial and thereby academic communities, leading by the 1990s to detailed Portuguese multibeam mapping of the ATJ-AP area (e.g., Lourenco et al. 1998). While the gridding-based 'hillshade' images (computer-illuminated perspective-view topography) convey much of the topographic information at grid spacings of ca. 1 km (e.g., Fig. 1b, c), their full spatial resolution (e.g., Miranda et al. 2014) requires grid spacings of ca. 50–100 m on the relatively shallow Azores Plateau. Figure 1b–d illustrate the gaps—some filled in 2009 (Weiss et al. 2015a, b) in detailed multibeam coverage. In the Azores Plateau area and elsewhere near actively spreading plate boundaries, multibeam bathymetry has yielded a treasure of previously unresolved volcano-tectonic features, notably distributions and morphometry of small seafloor cones, fissure eruptions, and faults, and the bathymetric modifications of these features by mass-wasting and sedimentation processes (e.g., Weiss et al. 2016; Mitchell et al., Chapter "Volcanism in the Azores: A Marine Geophysical Perspective").

A high-resolution hillshade map of the ATJ area, derived from gridded multibeam data, is shown in Fig. 1d (courtesy of J. Luis, Univ. of Algarve). Compared to most oceanic regions, the relatively shallow ATJ-AP region requires closer track spacing (typically ca. 4 km) for complete swath coverage, but the spatial resolution is proportionately higher. By 2014, ca. 150,000 km^2 of the Azores Plateau had been swath-mapped by the Portuguese Navy and academic institutions (J. Luis, personal communication, 2014). See also Weiss et al. (2015a, b, 2016), Casalbore et al. (2015), and Mitchell et al. (Chapter "Volcanism in the Azores: A Marine Geophysical Perspective ") for other multibeam mapping.

(3) ***Seismic reflection*** profiles were collected on an opportunity basis on transits across the Azores Plateau in the 1960s–1970s, primarily by Lamont Doherty Earth Observatory (Fig. 2). These single-channel (or two-channel) profiles, using mostly airguns as sources, readily resolve an acoustically transparent, ca. 100–400 m thick hemipelagic sediment cover, which has been ponded in intermontane valleys due to slumping from nearby topographic highs (Fig. 2) and locally sculpted by bottom currents (Weiss et al. 2016). The frequent high earthquake-generated accelerations in the Azores area (e.g., Matias et al. 2007) have likely helped prevent sediments from accumulating on seamounts and any other steep topographic slopes. In 2009, the University of Hamburg collected ca. 1000 km of 2D multichannel seismic reflection profile and multibeam bathymetry data NW, SE, SW, and S of São Miguel Island. The processed data resolved seismic reflectors to ca. 1 km below the ocean floor, revealing a complex history of explosive submarine and subaerial volcanism (Weiss et al. 2015a) and mass wasting (Weiss et al. 2016), and providing a record of TR volcano-tectonic evolution (Weiss et al. 2015b).

(4) ***Sidescan sonar*** mapping, images based on the strength of backscattered or reflected acoustic returns, has higher spatial resolution but cannot be as objectively interpreted as water depth. Sidescan sonar images can be made from surface-towed, deep-towed or deep autonomous systems, but can also be derived from hull-mounted multibeam data if the entire echo, not just the first arrival, of each ping is recorded for each beam. The GLORIA towed sidescan system, developed from World War II antisubmarine sonars, provided an early view of the entire GLORIA fault, a key element of the plate boundary east of the TR (Laughton and Whitmarsh 1974), with the tectonic fabric of the MAR in the ATJ area mapped by Searle (1980). A comparison between GLORIA sidescan and SASS multibeam bathymetric contours over the Kurchatov FZ (Fig. 8 of Vogt and Tucholke 1986) illustrates the complementary nature of the data, with sidescan imagery most diagnostic for MAR axial tectonics (fault-controlled escarpments typically 5 km apart). To date, few multibeam-derived or other sidescan sonar maps

have been published for the ATJ-AP region. Backscatter mapping of Condor Seamount (Tempera et al. 2013) and near São Miguel (Weiss et al. 2015a) are exceptions.

(5) ***GPS (Global Positioning Satellites)***: The application of earth-orbiting satellite technology for accurate surface ship and aircraft navigation, independently by the US and the USSR, was another costly military technology that has come to play large roles in marine geophysics. No marine survey would be conducted today without GPS navigation. Moreover, a network of fixed GPS stations made it possible, beginning in the 1990s, to measure present motions between and among the tectonic plates, and to compare these motions with the 'geologically recent' rates from ca. post-3 Ma magnetic lineations recorded by the oceanic crust. These advances are evident in comparing the latest global model for current plate motions (MORVEL; DeMets et al. 2010) with earlier models (Minster and Jordan 1978; NUVEL model of DeMets et al. 1990). Temporary and permanent GPS stations on various Azores islands have enabled measurement of inter-island (Fernandes et al. 2006; Marques et al. 2013) and intra-island (Terceira: Miranda et al. 2012; São Jorge: Mendes et al. 2013) motions. These new data (Fernandes et al., Chapter "The Contribution of Space-Geodetic Techniques to the Understanding of the present-day geodynamics of the Azores Triple Junction") show that at least in recent decades, the Nubia-Eurasia plate boundary in the Azores area is diffuse—a band ca. 100–140 km wide (Fig. 1b; Marques et al. 2013; Miranda et al. 2014)—and contrary to morphologic predictions (Vogt and Jung 2004) Graciosa is moving with the Eurasia plate.

(6) ***Radar altimetry***: First proposed in the late 1960s, microwave altimetry from earth-orbiting satellites began (late 1970s) to map geoid undulations recorded (upon correction for other signals) by the shape of the sea surface. This originally military technology, used as a navigation aid, began to constrain submarine crustal density structures, especially seafloor topography (seamounts, rift valleys, transform fracture valleys, etc.) and its compensation. Satellite altimetry, combined with surface ship soundings, is integral to modern ocean floor maps over much of the world ocean (e.g., Smith and Sandwell 1997). The spatial resolution, however, is only ca. 10–20 km, ca. 100–1000 times coarser (depending on water depth) than multibeam bathymetry. In the Azores region, areas mapped in part from satellite altimetry are thus easily distinguished from areas fully mapped by multibeam bathymetry (Fig. 1a, b).

(7) ***Exclusive Economic Zones (EEZ)***: The 1958 United Nations Law of the Sea (UNCLOS) Convention and the succeeding LOS treaty (e.g., Knauss 1986), still a 'work in progress', have led to coastal states claiming legal (not necessarily physiographical or geological) extensions of their "continental shelf", in which these nations exercise certain sovereign rights over resources in the water column and on/under the seabed. The large extension of the claimed 'Portuguese Continental Shelf' surrounding the Azores (1.3 million km^2) includes the Azores Plateau and parts of the MAR both north and south of the triple junction. The need to support the claim with data has motivated extensive detailed Portuguese multibeam and magnetic mapping (e.g., Fig. 2 of Luis and Miranda 2008; Miranda et al. 2014; Fig. 3d), making the ATJ region one of the best-charted parts of the Mid-Oceanic Ridge and oceanic-crust-floored parts of the world ocean.

(8) ***Logistics:*** Few parts of the oceanic plate boundaries are as easily accessible to shipborne and airborne geophysical mapping as is the Terceira Rift and the MAR near the ATJ. As a result, the ATJ is the best-studied fully oceanic triple junction. Many research vessels have taken on fuel and supplies, and allowed exchange of scientific parties at Ponta Delgada (São Miguel), and the MAR aeromagnetic surveys reported in Phillips et al. (1975) and Luis et al. (1994) would not have been possible without the Lajes (Terceira) airfield, which was constructed in 1945 and served as a base for anti-submarine-warfare patrols by NATO aircraft during the Cold War. Of course, the fact that central volcanoes have grown above sea level along the TR (Terceira, São Miguel, Graciosa) and elsewhere on the AP has greatly facilitated research (e.g., Catalao et al. 2006; Sibrant et al. 2016; other chapters in

this volume) compared to what would practically be possible had these seamount summits today been submerged below the sea (e.g., Condor Seamount (Guyot), Tempera et al. 2013).

(9) ***Radiometric dating***: Due to excellent subaerial exposures of intrusive and extrusive rock, the Azores Plateau, and particularly the Terceira Rift, have benefited much more from radiometric dating (Holmes 1929; Dalrymple 1991) compared to the average submerged Mid-Oceanic Ridge. Abdel-Monem et al. (1975) published the first radiometric ages of Azores islands (Santa Maria, the Formigas, and São Miguel). Some of their ages, determined by K/Ar, have proven much too old, for example their 4.01 Ma, versus less than 900 ka (Johnson et al. 1998) for the oldest volcanics on São Miguel. Prior to 1975, the only age constraint for pre-historic Azores volcanism was a biostratigraphic 'Mid-Miocene' age (shown by Hildenbrand et al. 2014, to be latest Miocene to earliest Pliocene) for a coquina on Santa Maria (see Avila et al., Chapter "Petrology of the Azores Islands"). A good geochronology of Azores island volcanism has been developed in recent years, as summarised below. In recent decades, K/Ar whole-rock dating, which often yielded erroneous ages, has been complemented by $^{39}Ar/^{40}Ar$ ($^{39}Ar/^{40}Ar$) dating (e.g., Larrea et al. 2014; Beier et al. 2015). However, improved K/Ar dating methods are still used (e.g., Hildenbrand et al. 2014), especially when combined with geochemical analysis.

(10) ***Marine magnetometry***: The technology to measure the strength of the earth's magnetic field from ships and aircraft (see Vogt 1986b for a historical review) was perhaps the greatest single technological development which later lead to acceptance (by ca. 1968) of seafloor spreading and plate tectonics. Detecting enemy submarines by their magnetic anomaly, as well as prospecting for mineral resources, motivated the development of field magnetometers. In the first years after WWII, fluxgate magnetometers were tried at sea, with the first magnetic field strength profile across the Mid-Atlantic Ridge measured not far south of the Azores in 1948. In later years the sensor of choice for marine work became the proton precession magnetometer, invented by Victor Vacquier in 1939 but classified during World War II. Most of the now vast but still incomplete total field strength magnetic anomaly database (e.g., Luis and Miranda 2008; Miranda et al. 2015) has been collected from research vessels towing sensors far enough behind the ship to reduce anomalies caused by the vessel. Large, nowadays gridded databases are now available to plate kinematicists, much of the data with GPS vessel navigation accuracy. These databases enabled progressively improved plate kinematic reconstructions which include the Azores region (e.g., Pitman and Talwani 1972; Klitgord and Schouten 1986; Srivastava and Tapscott 1986; Roest and Srivastava 1991; Luis et al. 1994; Luis and Miranda 2008; Merkouriev and DeMets 2014a, b, Miranda et al. 2015 and Chapter "The Tectonic Evolution of the Azores Based on Magnetic Data").

5 Some Modern Research Issues

Location of TR and non-TR volcanic islands: Vogt and Jung (2004) nominated the TR as the world's slowest-spreading organised (discrete) spreading axis, and noted that very slow inter-plate extension is more typically associated with broad (diffuse) belts of deformation (see Simkin et al. 2006). One possible explanation for the discrete character of the TR builds on Vink et al. (1984), who suggest typical oceanic mantle lithosphere is stronger, due to the greater strength of ultramafic materials at a given temperature, compared to deep continental crust, and that this difference might explain the preference for new rifts to develop just inside continent-ocean crustal transitions to form microcontinents. We speculate that the thickened AP crust (14 km was measured by Escartin et al. 2001) produced at the MAR axis, i.e., a weaker lithosphere, somewhat resembles continental crust and thus breaks more easily along a discrete axis, versus diffuse rifting. Together with compositionally different (more residual melt inclusions) and possibly warmer AP mantle lithosphere, this might explain why the TR developed only within the pre-existing

Azores Plateau (Fig. 1a, b). The same factors—including a more fertile mantle lithosphere—might also explain why young non-TR volcanic islands (São Jorge, Pico, Faial, Condor Seamount (a guyot; Tempera et al. 2013), as well as Azor and Princess Alice banks grew on the Faial Ridge, while Corvo and Flores rose on the Flores Ridge. The absence of large volcanoes on the intervening younger (largely post-8 Ma) MAR flanks (e.g., Fig. 1e) poses a problem for the model of Neves et al. (2013) unless this crust and associated mantle lithosphere is less vulnerable to transtensional fissuring.

The time of TR origin and the degree of sea-floor spreading (besides rifting of pre-existing AP lithosphere) remain to be quantified, and a dynamic model developed to explain why the Nubia-Eurasia boundary jumped to the TR site.

TR along-axis volcanic relief: If the TR is a spreading axis, what accounts for its great along-axis relief? Western São Miguel volcanoes rise about 4200 m above the adjacent Hirondelle Basin (Fig. 3a), which along with the Povoação Basin SE of São Miguel contains at least 500 m sediment (Weiss et al. 2015b). Copious melt production, combined with very slow export of volcanoes out of the plate boundary zone might help explain such great relief (Vogt and Jung 2004), but simple Pratt isostasy is probably the simplest explanation. Following Vogt (1974), Vogt et al. (2008) showed that maximum along-axis relief of MOR axial volcanic edifices increases systematically with decreasing opening rates (Fig. 4), so the TR volcanoes may represent an end member on this roughly linear distribution of maximum heights versus model axial plate thickness (based on Phipps Morgan and Chen 1993). The slower the spreading, the thicker the axial plate, and the higher the lower-density magma conduit can rise (given sufficient magma) in isostatic balance with adjacent more normal oceanic crust. A thicker TR lithosphere is also consistent with inferred lower degrees of partial melting and a lack of connection between adjacent magma systems (Beier et al. 2008). If it accounts for only ca. 2–3 mm/a of the Eurasia-Nubia separation (Marques et al. 2013),

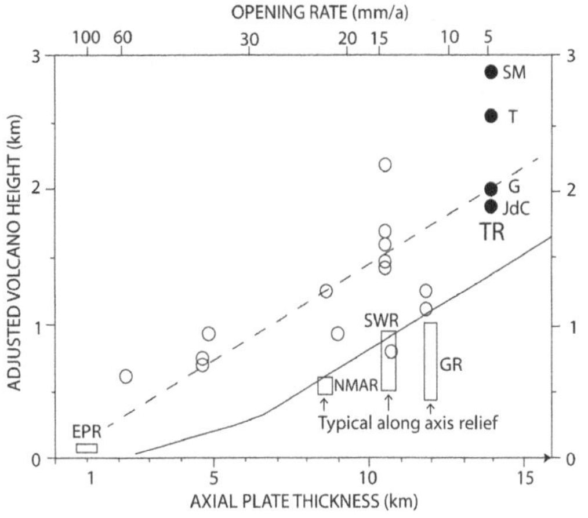

Fig. 4 Maximum (circles) and typical (rectangles) Mid-Oceanic Ridge axial volcano heights (adjusted to remove buoyancy effects of water column) versus opening rates (top) and axial plate thickness (based on Phipps Morgan and Chen 1993). EPR, East Pacific Rise; NMAR, Northern Mid-Atlantic Ridge; SWIR, Southwest Indian Ridge; GR, Gakkel Ridge; TR, Terceira Ridge: Graciosa (G), Terceira (T), João de Castro Bank (Seamount) (JdC) and São Miguel (SM) island points based on highest elevations. Dashed (ultramafic crust) and solid (normal oceanic crust) lines predict maximum volcano heights based on Pratt isostasy, relative to adjacent plate boundary. Adapted from Vogt et al. (2008); see also Vogt (1974)

the TR spreading rate is only half of the 'hyper-slow' rate assumed by Vogt and Jung (2004) and in Fig. 4. Pratt isostasy suggests that most of the TR axial volcanoes are near their final maximum heights, so additional magmas would likely be preferentially injected into flank dikes or form new conduits, versus erupting from summit calderas.

Data on erupted versus intruded magma fluxes versus time are needed to test whether island volcanoes have reached their maximum 'Pratt' heights. The Pratt isostasy approximation needs to be adjusted for the effects of flexural isostasy.

Age of TR and non-TR island volcanism: The TR cannot be younger than the oldest exposed rocks, and radiometric dating in recent years suggests the TR and the volcanoes formed along it are all less than ca. 1 Ma in age: The oldest outcropping lavas, on eastern São Miguel, were erupted 780–880 ka (Johnson et al. 1998). The oldest outcropping lava succession on

Graciosa was dated at 1057 ± 28 ka (Larrea et al. 2014) and probably more accurately at 702 ± 10 ka by Sibrant et al. (2014). On Terceira Island, the oldest measured radiometric age is only 401 ± 6 ka (Hildenbrand et al. 2014; see also Calvert et al. 2006). The oldest known volcanics on non-TR islands were erupted 850 ka on Faial (Hildenbrand et al. 2014), 1.2–1.2 Ma on São Jorge (Hildenbrand et al. 2008) and 5.3–5.7 Ma on Santa Maria (Sibrant et al. 2015).

Although there must obviously be older eruptives and intrusives deeper inside all island edifices, such rocks cannot be much older than the oldest surface materials, provided the average magma flux (F) remained relatively constant: For an assumed conical edifice of slope s and height h (t), $\frac{dh}{dt} = \frac{F}{\pi h}\tan^2 s$. According to this relation, an edifice grows rapidly in height at the beginning and much more slowly later—e.g., at the present elevations of the TR volcanic islands above the adjacent rift valleys, $\frac{dh}{dt}$ would be only 5–10% of what it was at 1 km edifice heights. F of the order 1 km^3/ka would suffice to produce one of the TR islands in about 1 Ma (Vogt and Jung 2004).

By how much does the TR (>1 Ma) predate the oldest island volcanoes? A wide range of ages have been postulated for the TR, but with Sibrant et al. (2015) we consider ages greater than 3 Ma implausible. Miranda et al. (2015) suggested the TR originated when the Nubia-Eurasia plate boundary jumped from the Princess Alice Bank rift (Fig. 1b) to the TR. Sibrant et al. (2015) used tectonic extension measured by normal fault displacements to conclude the TR formed by rifting between 1.4 and 2.7 Ma. However, before sea-floor spreading is discounted entirely, boreholes are needed to date the basement in the apparently rifted belt. Some post-3 Ma spreading from a TR axis appears consistent with the bathymetry in the Formigas area (Fig. 1c). The Formigas islets appear symmetrically located as far from the present TR axis as a paired modest bathymetric high northeast of the axis—somewhat resembling the 'split axial volcanoes' locally present along the MAR (Vogt and Jung 2005) and there formed by an alternate dominance of volcanism and rifting. The lesser East Formigas High (Weiss et al. 2015b) appears also to have a split "twin". These features are hard to explain entirely by tectonics, with no spreading. Moreover, the Big North and North Hirondelle bathymetric highs (Weiss et al. 2015b) might also have formed along or near the earliest TR axis, perhaps with subsequent off-axis volcanism enhancing early asymmetry. Indeed, unless the TR island volcanoes are entirely loaded on older crust, they must themselves represent axial accretion of new crust, balanced by stretching (rifting) in adjacent rift valleys. Topographic asymmetry (higher peaks south of the TR axis vs. on the north side in this case) is common on slow-spreading plate boundaries north of Iceland (Vogt et al. 1982) and thus possible in the Azores. Multibeam bathymetry and seismic reflection profiles across the submerged TR on either side of São Miguel reveal a terrain peppered with volcanic cones (Weiss et al. 2016) and laced with faults typically spaced 1–2 km apart (Weiss et al. 2015b), and thus similar to the axis of a slow-spreading ridge whose rift valley is partially filled with sediments—for example the Knipovich Ridge southwest of Svalbard.

What can have happened during the later Pleistocene and Recent that might explain so much geologically young volcano-tectonic activity, especially along the TR? A geologically recent change in plate motions or plume activity is one possibility. Hinting at such a change is the location of the present (GPS-based) Eurasia-Nubia plate rotation pole some 3000 km south of the MORVEL (3.16 Ma-present average) pole for the same pair of plates (Marques et al. 2013), although the opening rates happen to remain nearly the same in the ATJ area.

Carminati and Doglinoni (2010) have speculated on a different reason for globally increased Pleistocene volcanism: The Mid-Pleistocene Transition (ca. 1.25–0.8 Ma, most recently dated to 0.9 Ma by Elderfield et al. 2012), shifted glacial-interglacial climate cycles from moderate amplitude (in terms of ice sheet volume, hence sea level) 40 ka to high-amplitude 100 ka cycles. This shift increased the amplitude of glacioisostatic loading placed on the mantle by northern ice sheets and to some extent the global ocean, thus forcing a more vigorous upper mantle flow. Could such changes in pressure and flowage have stimulated magma production and migration in mantle regions close to their pressure-melting points distant from the ice sheets? The probable initiation of major edifice construction along the TR is suggestive enough to warrant additional dating and numerical modelling. Relatively short episodes (ca. 850, 700, 400, 200, 120, 50 and 0 ka according to Hildenbrand et al. 2014) of coeval volcanism on several Azores islands exhibit episodicity on time scales comparable to that of major post-Mid Pleistocene glacial cycles. However, the episodicity of Azorean volcanism may simply reflect the time scale of some local upper mantle process, such as Rayleigh-Taylor instabilities in the asthenosphere. Mid-Oceanic Ridge axial volcanism appears locally non-steady state, forming successions of axial volcanoes which are then split to form half-volcanoes on similar time scales of ca. 0.1–1 my (Vogt and Jung 2005).

Azores mantle plume?: Wilson (1963) first postulated that movement of the earth’s crust over localised anomalous melt sources relatively fixed in the cores of mantle convection cells could explain the age progression and trends of Pacific island and seamount chains like the Hawaiian-Emperor chain. The sites where anomalously prolific melts, derived by partial melting of mantle material, rise to the surface to create volcanoes came to be called ‘hotspots’ because it was first assumed only higher mantle temperatures could be the cause. Morgan (1972) modified this concept by proposing that ‘hotspots’—by then including the Azores—actually represent deep mantle convection plumes rising from the core-mantle oundary. Such plumes would be relatively fixed relative to the lithospheric plates moving across them. However, due to the lack of credible evidence, the deep mantle plume concept remains controversial (e.g., Anderson 2005) after nearly a half century of research and numerous publications, many of which modified the plume model, using newer seismic tomography and numerical modelling. For example, Courtillot et al. (2003) classified 49 postulated hotspots into three types: (1) ‘primary’ hotspots may originate at the core-mantle boundary; (2) secondary hotspots originate at the base of the seismic transition zone; and (3) the remainder are the consequence of plate tectonics. Davaille et al. (2005) placed the Azores at the edge of a large deep mantle source domain called the “Indo-Atlantic Box”. Based on numerical convection modelling and seismic tomography, they interpreted Iceland as a narrow deep mantle plume and the Azores as a “secondary plume” rising from a large plume head ponded in the seismic transition zone. Silveira et al. (2006) observed a region of anomalously slow S-waves below the Azores lithosphere down to ca. 250–300 km depths. They suggested this region comprises the remains of a dying plume.

A conservative term for features like the Azores is simply ‘melting anomaly’: Excessive melting is not necessarily- and can often be shown not to be- due to higher mantle temperatures, but rather to anomalous composition or even just to decompression melting by magma rising where the lithosphere is rifted by tectonic forces (Beier et al. 2015). Although given (Anderson 2005) a score of only 2+ (out of 12), and rated only 12th among a list of 60 proposed deep mantle plume sites in terms of observational support, the Azores region is still considered by many the surface expression of at least a shallow or secondary mantle plume, perhaps more ‘wet’ than ‘hot’ (Metrich et al. 2014; Beier et al. 2012). The geochemistry of Azores volcanic rocks suggests heterogeneous, poorly mixed, and not particularly hot mantle sources, ultimately recycled (previously subducted) oceanic crust (Beier et al. 2008, 2015).

How old is the Azores ‘hotspot’? Gente et al. (2003) argue that an Azores hotspot/plume has existed since 85 Ma, but only began to influence the MAR when the latter began to migrate near it. Beier et al. (2015) inferred from a 39 Ma lava dredged on Monaco Bank that the Azores plume already existed below the area at the time, because the lava is rich in radiogenic isotopes, similar to modern São Miguel volcanics. The date—about the crustal age predicted by magnetic lineations—is consistent with an abstract by Campan et al. (1993), who suggested the MAR-Azores plume interaction began 36–39 Ma. The 39 Ma sample may have come from an old NW trending Monaco Bank seamount, constructed near the MAR axis and unrelated to formation of the AP. However, the sample was recovered from depths of 250–500 m (Beier et al., personal communication, 2016), much too shallow for post-39 Ma crustal subsidence of MAR generated crust.

While not discounting the possibility of older antecedents, we propose the origin of the Azores Plateau was 'recorded' (Fig. 1b) by formation of the West Azores Ridge (WAR) on the west flank of the MAR, and by the Northeast Spur—at least partly a belt of small seamounts—on the east flank. Both features also mark a shallowing of the Mid-Atlantic Ridge to form the AP. Magnetic lineations identified by Luis and Miranda (2008) suggest the WAR was formed ca. 18 Ma, slightly postdating anomaly 6 and as implied by its northward divergence from that lineation (Fig. 1b) propagating northwards along the MAR axis. On the east MAR flank, the Northeast Spur appears to coincide with anomaly 6 (19 Ma).

How is formation of the AP related to the southward jump of the triple junction from ca. 45°N (King's Trough) to ca. 39°N (Azores)? The jump ended a period of compressional deformation in the Pyrenees and transferred Iberia from the Africa (Nubia) to the Eurasia plates (Srivastava et al. 1990). This event was dated to anomaly 9 time (Srivastava and Tapscott 1986; Klitgord and Schouten 1986) and later to anomaly 6C time (23 Ma), or at least predating anomaly 6 time (Srivastava et al. 1990; Roest and Srivastava 1991). Most of the change in Eurasia-North America plate motion had happened by anomaly 6C time according to Luis and Miranda (2008). Although the jump and AP development were not instantaneous, the similarity in timing between the AP and the jump is suggestive. However, the AP appears to have developed after the jump, and thus perhaps triggered by it. If the AP developed from a mantle plume below the plates, an improbable coincidence is required for a plume 'happening' to rise into a new triple junction.

Whereas many modern triple junctions (e.g., Fig. 1b inset map) are not associated with 'melting anomalies', and modelling (Adam et al. 2013) does not require such a correlation, Sager (2005) noticed that several Pacific oceanic plateaus also 'happened' to form at triple junction jumps: Were mantle plumes—including secondary plumes Davaille et al. (2005)—responsible for oceanic plateau initiation, this would have been extremely unlikely. A plate tectonic origin might explain the correlation, by decompression melting, at a new triple junction, given the existence of an anomalous asthenosphere susceptible to increased melting under small stress perturbations (Sager 2005). Although this also might explain AP formation soon after the triple junction jump, some underlying asthenosphere anomaly is still required.

The triple junction jump from the King's Trough to the Azores sites coincided, within age-dating uncertainties, with the axis boundary jump from Aegir Ridge in the Norway Basin to the Kolbeinsey Ridge axis just north of Iceland around anomaly 6C time. The oldest clearly recognizable lineation (Vogt 1986b) on the east flank of that ridge is 6B (22 Ma). This seems also the most likely time the Iceland melting anomaly greatly increased its output to begin generating Iceland and its associated platform. Based on the shoaling (over time) of the Reykjanes Ridge, Saemundson (1986) concluded this happened sometime between 19.5 and 35.5 Ma. Luis and Miranda (2008) found a rapid northward motion of the Eurasia-North America plate rotation pole during the C11/12 to C6C interval (31–23 Ma) and continuing after that time. Formation of the Azores triple junction largely by C6C time (23 Ma) was thus an expression of a regional plate kinematic reorganization, and this necessarily also involved at least the upper mantle. This also corresponds to the Neogene-Palaeogene boundary (23.0 Ma, near the end of C6C; Walker et al. 2012), a time long recognised as marking major geological and paleontological changes in the Alpine-Tethys region (Earnes 1970) and a 'major change in global planktonic biogeography' (Kennett 1978). How the plate kinematic changes at C6C time relate to the Neogene-Paleogene boundary is a problem inviting collaboration among geophysicists, geologists and palaeontologists.

Where is the present centre or apex of the Azores melting anomaly, whatever its mantle source may be, and how wide is this centre? The most concentrated off-ridge volcanism would place this Pleistocene-Holocene "volcanic centre" in the area of Pico, Faial, and São Jorge. However, along the MAR axis, the centre corresponds to the shallowest crestal topography, a ca. 80 km long MAR segment also lacking a prominent rift valley (just east of the label 'MAR' in Fig. 1b) and located ca. 150 km WNW of the 'volcanic centre'. The along-axis geochemical anomaly (Schilling 1986) peak corresponds (Fig. 3c) to the segment lacking a rift valley, but crustal thickness there is not particularly thick, compared to the 14 km under Faial Ridge (Escartin et al. 2001). Does the Azores volcanic centre represent a 'second type of hotspot island' (Morgan 1978), supplied by asthenosphere flow from a ridge-centred plume near the MAR axis? If this accounts for the 150 km distance between the Azores 'volcanic centre' and the anomalous MAR axial segment, what explains a similar relationship at ca 45°N, where MAR axial geochemical and topographic anomalies (Schilling 1986) lie 300–350 km west of the tip of the long extinct (ca. 20–25 Ma) King's Trough? Broad regions of anomalous asthenosphere might explain these relationships, or also remnants of dying plumes (e.g., Silveira et al. 2006) but only if such anomalies are elongated in an EW direction. As discussed previously, the Azores 'volcanic centre' simply reflects preferential transtensional fracturing of an anomalously thick and weak lithosphere produced by the MAR ca. 12–8 Ma (the oldest part of the Faial Ridge) as part of the diffuse Nubia-Eurasia plate boundary.

The fixed mantle plume hypothesis predicts an age progression, with the youngest magma generation located over the plume core. Nearly all the known volcanic activity ages associated with the putative plume or hotspot/melting anomaly cluster around ca. 6–4 Ma and again 1.5–0 Ma (e.g., Beier et al. 2015; Hildenbrand et al. 2014; Sibrant et al. 2015), with no clear age progression and thus not supporting a fixed mantle plume model. A widespread flood basalt episode (Beier et al. 2015) would not be expected from the dying plume idea of Silveira et al. (2006). As noted by Hildenbrand et al. (2014), multiple coeval volcanic episodes on several islands is also not explained by the plume model but would require modulation by other processes. However, an age progression would not be easy to detect if the postulated plume head is broad and motion of a plate across the head is slow and poorly constrained (as is the case for the Nubia plate; Gripp and Gordon 2002, and for the Afar triple junction magmatism; Corti 2008). Age dates from within Faial, São Jorge and Terceira (Fig. 2 of Hildenbrand et al. 2014) do suggest age progressions of the order 1–10 cm/a, but younging in the direction of the MAR axis. Such age progression might reflect crustal fracture propagation or plate movement relative to magma sources, but is inconsistent with the motion of the Nubia plate relative to a fixed hotspot or plume (Gripp and Gordon 2002, see below). Moreover, in the MORVEL-based No-Net-Rotation reference frame, the TR region is moving at ca. 20 mm/a in a NE direction (Fig. 1 of Argus et al. 2011), opposite to that predicted by Gripp and Gordon (2002).

Only a few observations support a mantle source (such as Morgan's 1972 deep mantle plume) fixed below the plates in the Azores area. Relative to a set of assumed fixed hotspots (Gripp and Gordon 2002), the TR should be moving ca. 20 mm/a ('absolute' motion, i.e. relative to a relatively fixed mantle below the plates) in the direction S55 W (Fig. 3b of Vogt and Jung 2004), consistent with seismic tomographic evidence for a mantle plume (at least to 400 km depth) now located under or just northeast of the TR (Yang et al. 2006). This direction of absolute motion is consistent with the apparent ca. 150 km NE jump (equivalent to ca. 8 m.y. of 20 mm/a absolute plate motion) of the rift axis from the extinct Princess Alice Rift (Fig. 1b) to the present

TR location. Miranda et al. (2015) also suggested such a jump ca. 3 Ma. The diachronous Faial and Flores ridges (Fig. 1b; Vogt 1979; Cannat et al. 1999) suggest mantle flow away from a fixed upwelling located near the Azores triple junction, but the ridge geometry could alternatively be explained by MAR axis motion across a NE elongated compositional or/and thermal anomaly (Fig. 8B of Vogt and Jung 2005).

A careful search for anomalously high heat flow and/or young volcanism or intrusives on the Eurasia plate just northeast of the TR, particularly around small seamounts north of Terceira (the Northeast Spur in Fig. 3c) and NE of São Miguel, might help determine if the TR and eastern Azores Plateau are being displaced southwestwards relative to a melting anomaly currently below the plates. We consider this possible but unlikely. Detailed magnetic, seismic reflection, multibeam and borehole investigations of the Northeast Spur and the West Azores Ridge are needed to evaluate our suggestion of a TR origin ca. 20 Ma.

We remain agnostic on deep mantle plumes but agree with many other authors that compositionally or thermally anomalous mantle must have existed at least since 20 Ma, and still exist below the AP and surroundings. Needed for future knowledge advances are: (1) improved seismic tomography using broadband OBS and OBH arrays distributed over the seafloor around the Azores Plateau to help resolve what lies at depth (particularly below ca. 300–400 km) in the mantle below the AP (Silveira et al. 2006); (2) detailed mapping and borehole sampling around the western end of King's Trough and along the Northeast Spur and West Azores Ridge to determine more accurately the timing of the triple junction jump and its relation to origin of the AP and to apparently coeval events in the Iceland area; (3) data-based dynamic models predicting why the triple junction and new plate boundary jumped south specifically to the Azores-Gibraltar region.

The Azores Plateau crust: What exactly is it? If the apparently NE-trending, MAR-generated sea-floor spreading magnetic lineations on much of the Azores Plateau (Fig. 1b; Luis and Miranda 2008; Miranda et al., Chapter "The Tectonic Evolution of the Azores Based on Magnetic Data ") have been correctly identified, why has subsequent Azores magmatism (e.g., Weiss et al. 2015b; Beier et al. 2015 and Chapter "Melting and Mantle Sources in the Azores") not obscured the older MAR-derived magnetic lineation signature? If the linear volcanic ridges (including the TR) dominating the modern plateau formed by eruptions controlled by transtensional crustal fractures (Neves et al. 2013), did much of the eastern AP form from extrusive and intrusives emitted along these fissures? The age-dating reported by Beier et al. (2015) suggests much of the elevated AP south of the TR (labeled South Azores Plateau in Fig. 1b to distinguish it from the North Azores Plateau) is underlain by flood basalts erupted during the interval ca. 6–4 Ma, although their dredges likely did not recover the initial stages of this episode. The amount of anomalous AP crustal thickening attributable to such post-MAR igneous 'overprinting' also remains unknown. However, the previously discussed 39 Ma lava from Monaco Bank, compositionally similar to young volcanics on São Miguel (Beier et al. 2015), implies that the overprinting eruptives are at least locally thin.

The existence of the Flores Ridge (Vogt 1979) on the West Azores Plateau, i.e., the west flank of the MAR, capped by two volcanic islands (Flores and Corvo), suggests that anomalously thick crust (Escartin et al. 2001) was already being formed at the MAR axis during the period ca. 12–7 Ma (Fig. 1b), but that this crust and mantle lithosphere was fertile, perhaps also more heterogeneous- to account for geochemical differences among various AP magmas (Beier et al. 2008, 2015 and Chapter "Melting and Mantle Sources in the Azores"), and subsequently more easily fractured. Flores and Corvo are relatively circular islands, located on the North America plate, and thus form a control for the Neves et al. (2013) model, insofar as the Flores Ridge was not subjected to transtensional shear, and therefore island lavas were not erupted from fissures.

Forming the eastern edge of the North Azores Plateau (Fig. 1b), the Northeast Spur, especially Sedlo Seamount, may in part represent off-axis

AP volcanism similar to that forming Corvo and Flores islands. However, the Northeast Spur has a twin (West Azores Ridge, WAR) on the opposite flank of the MAR (Fig. 1b). The WAR must have (like the Flores and Faial ridges) largely formed on or near the MAR axis, but earlier.

The three peaks of Sedlo Seamount (Santos et al. 2010), a guyot, range from 950 to 660 m in summit depth. These volcanoes, located on ca. 20 Ma crust, probably formed as small islands at least partly off-axis—otherwise they would have thermally subsided ca. 1300–1600 m together with the adjoining oceanic MAR crust. If post-volcanic erosion beveled these islets by 5 Ma, subsequent crustal subsidence could account for the present summit depths. Although not yet mapped in detail, much of the WAR was first formed ca. 10 Ma prior to the first Faial and Flores ridges, which based on the locations of magnetic anomalies 5 and 6, formed during the period ca. 12–8 Ma (Fig. 1b). Although Beier et al. (2015) attempted to recover unaltered samples from Northeast Spur, there are no igneous rock ages from this feature, Sedlo Seamount, and the WAR.

In any case, the existence of MAR-parallel magnetic lineations over parts of the South Azores Plateau (Fig. 1b, c; Miranda and Luis 2013; Miranda et al., Chapter "The Tectonic Evolution of the Azores Based on Magnetic Data ") suggests that an original MAR-generated basement topography (e.g., Fig. 2) was buried in lava flows and volcaniclastic debris, much of which must have been shed by the linear, island-capped volcanic ridges. Such randomly oriented debris—even strongly magnetised clasts —would not create a significant thermoremanent (TRM) magnetic anomaly signature and therefore would not obscure the original MAR-generated sea-floor spreading anomalies. Do the flat, acoustically opaque reflector below some sediment ponds (Fig. 2; Weiss et al. 2015b) represent anomalously flat, rapidly erupted MAR flood basalts, or has an original MAR type basement topography been buried and thus obscured under sheet-type lava flows or sills (largely 6–4 Ma; Beier et al. 2015) or by rapidly deposited volcaniclastic debris flows?

Coring is needed to calibrate the reflection seismology of the thick (up to ca. 1 km) sediments in the intermontane valleys (e.g., of Weiss et al. 2015b) and sample, age-date and measure the magnetic polarities of the AP flood basalt sequence down to and including the original MAR-generated basement. Boreholes could map the AP and MAR flank geochemical anomalies over space and time, building on the complex results from DSDP Site 556 (typical MORB recovered from 31 Ma MAR flank crust west of the AP) and DSDP Site 557 (Fig. 1b; LREE-enriched or 'E-type' basalt from 17 Ma AP crust east of the WAR). We suggest an entire ODP drilling leg be planned for the Azores Plateau.

Azores microplate, diffuse plate boundary, or both? The scattered seismicity and volcanism in the ATJ-AP area (Fontiela et al., Chapter "Characterization of Seismicity of the Azores Archipelago: An Overview of Historical Events and a Detailed Analysis for the Period 2000–2012") suggests that the plate boundary—particularly the Nubia-Eurasia one—is somewhat diffuse (distributed) there, as has long been known. GPS technology (e.g., Fernandes et al., Chapter "The Contribution of Space-Geodetic Techniques to the Understanding of the Present-Day Geodynamics of the Azores Triple Junction") has opened the door to measurement of the present Azores inter-island displacement field. Fernandes et al. (2006) reported 1993–2001 displacements among the islands. Although the measured rates are generally consistent with global models for geologically recent plate motions, some of the islands did not move consistent with a simple triple junction, but supported a model of distributed elastic deformation within the Azores Plateau (Bastos et al. 1998). While as expected, Santa Maria moved with the Nubia plate, Graciosa moved with Eurasia, and the others (São Jorge, Pico, Terceira and São Miguel) moved in ways somewhat intermediate between Eurasia and Nubia. However, the displacement of Pico and São Jorge were closer to Nubia, the plate on which they to first order reside. The intermediate displacements of Terceira and São Miguel are to be expected, given their location on the immediate TR plate boundary. The most surprising motion is that of Graciosa, located on the TR but moving entirely with Eurasia. Subsequent work (Marques et al. 2013), based on 35 new GPS stations occupied during 2001–2013, on Terceira, Faial and Pico (the latter two located on same volcanic ridge; Fig. 1b), plus fifteen GPS velocities on São Jorge (Mendes et al. (2013) refined these results: Terceira: Thirteen stations show slow (1–2.8 mm/a) westward motion relative to Eurasia, consistent with location on a very slow-spreading TR plate boundary; São Jorge: Fifteen GPS sites moved WSW at 2.7 ± 0.7 mm/a relative to the Eurasia plate; and Faial and Pico moved WSW at ca. 2–4 mm/a, with average separation of Faial/Pico from São Jorge ca. 1 mm/a. Faial/Pico moved largely, but not entirely, as if fixed to the Nubia plate (Marques et al. 2013).

DeMets et al. (2010) observed that opening rates along the MAR do not change abruptly from south to north anywhere between ca. 38° and 40°N near ATJ, but rather increase gradually from the Nubia-North America rate to the Eurasia-North America rate. They interpreted this gradual variation in rate as evidence for a separate Azores microplate, with its eastern edge presumably in the volcanic and seismically active island group from Graciosa through São Jorge to Pico-Faial. If a microplate exists, there must be three triple plate junctions, not just one. However, GPS stations on Faial-Pico, São Miguel, and Terceira (Marques et al. 2013), the sea-floor structures mapped by multibeam (Miranda et al. 2014), taken together with long-known mid-plate seismicity (Fig. 1e) make a convincing case for a diffuse triple junction, as suggested by Bastos et al. (1998). High resolution multibeam data (Fig. 1d, e; Miranda et al. 2014) do not support the speculative extension of the TR from West Graciosa Basin to the MAR axis suggested by Vogt and Jung (2004).

Presumably the magnetic data can also be interpreted in terms of a systematically deforming region, as suggested by Bastos et al. (1998). DeMets et al. (2010) used the 3.16–0 Ma interval on slow spreading ridges like the MAR to facilitate anomaly picks, with 3.16 Ma being the magnetic anomaly low in the middle of the anomaly 2A sequence. Perhaps a careful re-examination of the 38°–40°N magnetic anomaly data first reported by Luis et al. (1994), plus any newer and higher-resolution data, together with careful magnetic anomaly modelling which includes multibeam-defined seafloor topography, would yield an acceptably accurate rate for just the 0.78–0 Ma interval (0.78 Ma is the older limit of the Brunhes normal chron, recorded by the oceanic crust as anomaly #1). This time interval is closer to the period represented by earliest exposed Terceira Rift island volcanism. This rate might show an opening rate gradient more compatible with the GPS results for the present time and indicate that a change in kinematics has occurred, as in fact might be expected if the TR is only ca. 1 Ma old and was formed because of the change. Alternatively, if no such change has occurred, the northern edge of the microplate (from Graciosa to the MAR) might simply have been locked (seismically inactive) during the WWSSN observation period (i.e., since the early 1960s). Microplates are known at two triple plate junctions in the eastern Pacific (Fig. 1b inset), but are not there clearly associated with excess magmatism.

While the morpho-structural trends mapped by multibeam (Fig. 1c, d, e) and the earthquake epicentres (Fig. 1e) support diffuse deformation, the observed seismicity generally does not correlate with the structures, except for some epicentres clustering around an extension of São Jorge, and around Condor Seamount. Most of the epicentres are actually located on crust with only abyssal hill (MAR-generated) structures (N1, Fig. 1e). Future seafloor OBS and OBH campaigns will be necessary to obtain more accurate hypocentre locations and thus test whether the topographic structures mapped e.g. by Miranda et al. (2014) are tectonically active, perhaps showing that the diffuse triple junction actually comprises a series of smaller, fault-bounded blocks.

Distributed deformation probably also accounts for the misfit (over-rotation) of magnetic lineations from the Azores Triple Junction north ca. 300 km to ca. 42°N prior to ca. 7–8 Ma, i.e., prior to the MAR spreading rate slowdown (Merkouriev and DeMets 2014b) discussed further below. The misfits range from ca. 5 km at C4A to ca. 20 km at C6. 42°N also marks the approximate northern edge of the North Azores Plateau (Fig. 1a, b).

Probably the microplate and the elastic intraplate deformation models are not mutually exclusive, and if the regional tectonic patterns change frequently over human time scales, future geophysicists will have no lack of research material.

5.1 The MAR Spreading Deceleration ca. 8–6 Ma

By careful mapping and back-rotating isochrons represented by magnetic anomalies, Merkouriev and DeMets (2008, 2014b) concluded that Eurasia-North America motion was essentially constant from 19.7 Ma (Chron C6n) to ca. 7 Ma and again from ca. 6 Ma to the present (See Vogt 1986c for a review of differing earlier results). However, during the 7–6 Ma interval the plate rotation rate slowed by ca. 20%, while the rotation pole migrated south ca. 1000 km. Using the same techniques, Merkouriev and DeMets (2014a) showed closely similar kinematic behaviour of Nubia-North America motion: Motion was steady from ca. 20 to 8.2 Ma; a deceleration by 25% happened during the interval 8.2–6.2 Ma, with motion subsequently steady. A deceleration ca. 7 Ma had previously been demonstrated from magnetic surveys along the 28–29°N part of the MAR (Sloan and Patriat 1992). The slowdown was accompanied by an end to the diffuse deformation on the MAR from the present triple junction north to 42° (Merkouriev and DeMets 2014b). Merkouriev and DeMets (2014a) speculate that the close coupling between the two plate motions might have resulted from changes in frictional forces along the Nubia-Eurasia plate boundary.

There are several implications of these results for the Azores Geosyndrome problem: First, Eurasia-Nubia motion—along the Azores plate boundary—must have slowed by similar amounts during this interval. Second, the ‘deceleration crust’ (Fig. 1e) includes most of the N2 and N3 morphostructures mapped by Miranda et al. (2014)—this observation coupled with fairly low historical seismicity (Fig. 1e) leaves open the possibility that the structures were formed as a MAR plate boundary response to past changes in plate motion, and may therefore not all be active parts of the present diffuse boundary. Third, the deceleration crust forms the eastern edge of relatively normal, albeit shallow, oceanic crust. Evidently generation at the MAR axis of very anomalous Azores Plateau crust constituting the Fayal and Flores ridges (Vogt 1979, 1986c) and their cargo of younger volcanoes ceased. Perhaps this change in magmatism was some sort of response to the 8–6 Ma change in plate motion. Alternatively, perhaps the location of the deceleration band is a clue that the deceleration was itself driven by a change (reduction?) in mantle convection associated with a dying (Silveira et al. 2006) Azores plume. Azores. A slower convection and a less melt-rich asthenosphere (i.e., a higher viscosity) might have caused the deceleration. Numerical models of mantle convection its consequences for plate boundary forces can test this speculation.

6 Conclusions

For logistic reasons, the Azores triple plate junction and associated “geosyndrome” are among the best studied such features on the planet. Yet, despite great advances in seismic tomography and numerical modelling of mantle processes, advances have mainly refined earlier (1970s–1980s) interpretations.

Several of the geophysical research methods used to date have reached the point of diminishing marginal returns. A detailed surface ship or aeromagnetic survey of the North Atlantic is unlikely to provide a plate kinematic model greatly improved over (and different from) those of Luis and Miranda (2008) and Merkouriev and DeMets (2008, 2014a, b). GPS data on “present day” inter-island motions (e.g., Fernandes et al. 2006 and Chapter “The Contribution of Space-Geodetic Techniques to the Understanding of the Present-Day Geodynamics of the Azores Triple

Junction"; Marques et al. 2013) and intra-island deformation (e.g., Miranda et al. 2012) cannot detect temporal variations on time scales of centuries to millennia until far in mankind's future. However, 'paleostress' parameters can be recovered by age-dating of faults or dikes, as done for Graciosa (Hipolito et al. 2013), São Miguel (Carmo et al. 2015) and Santa Maria (Sibrant et al. 2015). Where fault or dike orientations changed over time (e.g., between 5.3 and 4.3 Ma on Santa Maria), were these just a local or regional volcano-tectonic effects or were they responses to changes in relative plate motions?

Seismic tomography could in principle resolve mantle anomalies, if any, deeper below the Azores region, particularly below ca. 400 km depths (Yang et al. 2006) and settle the long-standing controversy over the existence of a mantle plume or plume head. However, further progress is limited by current global and regional seismic coverage. A costly network of broadband stations both on the islands and on the wider seafloor region surrounding the AP would be needed (Silveira et al. 2006). Advancing from kinematics to dynamic modelling (e.g., Magde and Sparks 1997; Adam et al. 2013) will remain problematic until the structure, composition, and physical state of the mantle below the Azores are better constrained. This applies not only to the AP region, but also to the apparently vast geographic extent (Fig. 3b) of the Azores Geosyndrome particularly southwest along the MAR from the ATJ. Does partially molten mantle flow southwest, from an upwelling region under the Azores, in the triangular pipe-like region of enhanced melting (hence reduced viscosity) which must exist at depth under the MAR, as proposed by Vogt (1976)? The Faial and Flores ridges (Vogt 1979, 1986c; Cannat et al. 1999; Figs. 1b and 3a), which are not propagating rifts of the sort described by Hey (1977), as well as the SW geochemical gradient (Schilling 1976, 1986; Dosso et al. 1999; Fig. 3a) possibly remain the only expressions of such flow.

Meanwhile, rock sampling and geologic mapping have for practical reasons been almost entirely restricted to the Azores islands and immediate submarine plate boundaries, providing a necessarily biased 'zero age' geochemical/petrologic snapshot of the Azores geosyndrome and its mantle origins (e.g., Beier et al. 2008, 2015, Chapter "Melting and Mantle Sources in the Azores"; Larrea et al., Chapter "Petrology of the Azores Islands", Madureira et al. 2011, 2014; Metrich et al. 2014) and other chapters in the present volume). In terms of magma extraction from the upper mantle, the TR islands themselves probably represent at most 10% of the roughly 0.05 km^3 annual crustal production along the 500 km long TR and 2000 km (conservatively assumed) length of MAR lithosphere produced by the "Azores melting anomaly".

Even along these narrow plate boundary belts, only the most recent eruptive products have been sampled at most sites, while off the axis the igneous crust lies buried under sediment, or where exposed is highly altered. More research boreholes are clearly needed to penetrate the Azores island edifices and, particularly, the surrounding submarine and largely sub-sedimentary AP crust to resolve processes of tectono-magmatic evolution, especially on time scales shorter than those resolved by magnetic lineations generated at low spreading rates (roughly on the order one to a few Ma). Owing to their high costs, boreholes into igneous volcanic islands and AP oceanic crust need to be carefully planned, with detailed 'site surveys'.

Less expensive coring from drilling vessels, penetrating and recovering primarily seafloor sediments, for example the up to ca. 1 km thick (e.g., Weiss et al. 2015b) largely hemipelagic sediment accumulations in Azores Plateau valleys (e.g., Fig. 2), recover not only the paleo-oceanographic history of the AP region, but also the volcano-tectonic history of the islands, as recorded by ash falls from subaerial and submarine explosive eruptions (e.g., Weiss et al. 2015a), volcaniclastic and other debris flows (e.g., Weiss et al. 2016), faulting (e.g., Weiss et al. 2015b) and infrequent but spectacular flank collapses from large edifices (e.g., Sibrant et al. 2014, 2015; Costa et al. 2014). Age dating (i.e., calibrating the seismic stratigraphy) would depend on a combination of sediment biostratigraphy and radiometric dating of volcanogenic clasts from debris flows or ejecta. DSDP Leg 43 sediment cores adjacent to two of the New England Seamounts (DSDP sites 382 and 385) and adjacent to the Bermuda pedestal (DSDP site 387) exemplify this combined dating approach (Tucholke and Vogt 1979).

The demonstration from MAR flank magnetic lineations that both Nubia-North America motion (Merkouriev and DeMets 2014a) and Eurasia-North America (Merkouriev and DeMets 2008, 2014b) simultaneously slowed during the period ca. 8.2 to 6.2 Ma indicates that plate motions are coupled, perhaps controlled by Eurasia-Nubia convergence (Merkouriev and DeMets 2014a, b) or perhaps a regional change in mantle convection associated with the Azores geosyndrome.

The most intriguing, interdisciplinary question concerns the relation between AP formation and the triple junction jump ca. 23 Ma (chron 6C; Walker et al. 2012), closely coeval with plate reorganization in the Iceland area and more widely the 23.0 Ma Paleogene-Neogene boundary, a time of major global environmental change caused by the tectonic opening or closing of an oceanic 'watergate'. The formation of the AP in association with the jump is similar to what Sager (2005) found for other oceanic plateaus in the Cretaceous Pacific, and can scarcely be explained by coincidental rising of deep mantle plumes into newly formed triple junctions. The AP seems to have begun forming AFTER the jump and must thus be a consequence. However, because triple junctions do not REQUIRE excess magmatism (e.g., Georgen and Sankar 2010; Adam et al. 2013), an anomalous source mantle is still required. A secondary plume-type blob present below that part of the MAR would explain many of the observations, especially if the blob or plume 'head' is now slowly dying (Silveira et al. 2006) after creating the Flores and Faial ridges ca. 12–8 Ma. However, the great spatial extent of the MAR axial topographic high and the associated geoid high (e.g., Grevemeyer 1999) suggest that this Azores blob or secondary plume rose from the top of a much larger mantle anomaly (e.g., Davaille et al. 2005).

Acknowledgements We thank Robert I. Tilling and J. Miguel Miranda for thoughtful reviews (2014), and editors Ulrich Kueppers and Christoph Beier for inviting us to contribute this chapter. The chapter was updated in November-December, 2016. Joaquim Luis generously allowed us to reproduce his high-resolution hillshade topography of the triple junction region (Fig. 3d).

References

Abdel-Monem, A. A., Fernandez, L. A., & Boone, G. M. (1975). K-Ar ages for the eastern Azores group (Santa Maria, São Miguel and the Formigas Islands). *Lithos, 8,* 247–254.

Adam, C., Madureira, P., Miranda, J. M., Lourenco, N., Yoshida, M., & Fitzenz, D. (2013). Mantle dynamics and characteristics of the Azores Plateau. *Earth and Planetary Science Letters, 362,* 258–271.

Anderson, D. (2005). Scoring hotspots: The plume and plate paradigms. In G. R. Foulger, J. H. Natland, D. C. Presnall, & D. L. Anderson (Eds.), Plates, plumes, and paradigms. *Geological Society of America Special Paper 388,* 31–54.

Argus, D. F., Gordon, R. G., & DeMets, C. (2011). Geologically current motion of 56 plates relative to the no-net-rotation reference frame. *Geochemistry, Geophysics, Geosystems, 12,* Q11001.

Bastos, L., Osorio, J., Barbeito, A., & Hein, A. (1998). Results from geodetic measurements in the western part of the African-Eurasian plate boundary. *Tectonophysics, 294,* 261–269.

Beier, C., Haase K. M., & Abouchami, W. (2015). Geochemical and geochronological constraints on the evolution of the Azores Plateau. In C. R. Neal, W. Sager, E. Erba, & T. Sano (Eds.), The origin, evolution, and environmental impact of oceanic large igneous provinces. *Geological Society of America Special Paper 511,* 1–29.

Beier, C., Haase, K. M., Abouchami, W., Krienitz, M. S., & Hauff, F. (2008). Magma genesis by rifting of oceanic lithosphere above anomalous mantle: Terceira Rift, Azores. *Geochemistry, Geophysics, Geosystems, 9,* Q12013.

Beier, C., Haase, K. M., & Turner, S. P. (2012). Conditions of melting beneath the Azores. *Lithos, 144–145,* 1–11. https://doi.org/10.1016/j.lithos.2012.02.019.

Bezzeghoud, M., Adam, C., Buforn, E., Borgos, J. F., & Caldeira, B. (2014). Seismicity along the Azores-Gibraltar region and global plate kinematics. *Journal of Seismology, 18,* 205–220.

Borges, J. F., Bezzeghoud, M., Buforn, E., Pro, C., & Fritas, A. (2007). The 1980, 1997 and 1998 Azores earthquakes and some seismo-tectonic implications. *Tectonophysics, 435,* 37–54.

Buforn, E., Udias, A., & Colombas, M. A. (1988). Seismicity source mechanisms and tectonics of the

Azores-Gibraltar plate boundary. *Tectonophysics, 152,* 89–118.
Calvert, A. T., Moore, R. B., McGeehin, J. P., & Rodrigues da Silva, A. M. (2006). Volcanic history and 40Ar/39Ar and 14C geochronology of Terceira Island, Azores, Portugal. *Journal of Volcanology and Geothermal Research, 156,* 103–115.
Campan, A., Royer J.-D., Gente P., Olivet J.-L., & Mueller, R. D. (1993). Evolution of the Azores-Gibraltar plate boundary for the last 36 Ma. Eos [Transactions of the American Geophysical Union], 74, p. 586.
Cannat, M., Briais, A., Deplus, C., Escartin, J., Georgen, J., Lin, J., et al. (1999). Mid-Atlantic Ridge-Azores hotspot interaction: Along-axis migration of a hotspot-derived event of enhanced magmatism, 10 to 4 Ma ago. *Earth and Planetary Science Letters, 173,* 395–406.
Carminati, E., & Doglioni, C. (2010). North Atlantic geoid high, volcanism and glaciations, *Geophysical Research Letters*, L3302.
Carmo, R., Madeira, J., Ferreira, T., Queiros, G., & Hipolito, A. (2015). Chapter 6: Volcano-tectonic structures of São Miguel Island, Azores, *Geological Society, London, Memoirs, 44,* 65–86.
Casalbore, D., Romagnoli, C., Pimentel, A., Quartau, R., Casas, D., Ercilla, G., et al. (2015). Volcanic, tectonic and mass-wasting processes offshore Terceira Island (Azores) revealed by high-resolution seafloor mapping. *Bulletin of Volcanology, 77,* 24.
Catalao, J., Miranda, J. M., & Lourenco, N. (2006). Deformation associated with the Faial (Capelhinhos) 1957-1958 eruption: Inferences from 1937-1997 geodetic measurements. *Journal of Volcanology and Geothermal Research, 155,* 151–163.
Corti, G. (2008). Control of rift obliquity on the evolution and segmentation of the main Ethiopian Rift. *Nature Geoscience, 1*(4), 258–262.
Costa, A. C. G., Marques, F. O., Hildenbrand, A., Sibrant, A. L. R., & Catita, C. M. S. (2014). Large-scale catastrophic flank collapses in a steep volcanic ridge: The Pico-Faial Ridge, Azores Triple Junction. *Journal of Volcanology and Geothermal Research, 272,* 111–125.
Courtillot, V., Davaille, A., Besse, J., & Stock, J. M. (2003). Three types of hotspots in the Earth's mantle. *Earth and Planetary Science Letters, 205,* 295–308.
Dalrymple, G. B. (1991). *The age of the earth* (474p). Stanford, CA: Stanford University Press.
Davaille, A., Stutzmann, E., Silveira, G., Besse, J., & Courtillot, V. (2005). Convective patterns under the Indo-Atlantic. *Earth and Planetary Science Letters, 239,* 233–252.
DeMets, C., Gordon, R. G., & Argus, D. F. (2010). Geologically current plate motions. *Geophysical Journal International, 81,* 1–80.
DeMets, C., Gordon, R. G., Argus, D. F., & Stein, S. (1990). Current plate motions. *Geophysical Journal International, 101,* 425–478.
Dias, N. A., Matias, L., Lourenco, N., Madeira, J., Carrilho, F., & Gaspar, J. L. (2007). Crustal seismic velocity structure near Faial and Pico Islands (AZORES), from local earthquake tomography. *Tectonophysics, 445,* 301–317.
Dietz, R. S. (1961). Continent and ocean basin evolution by spreading of the sea floor. *Nature, 190,* 854–857.
Dosso, L., Bougault, H., Langmuir, C., Bollinger, C., Bonnier, O., & Etoubleau, J. (1999). The age and distribution of mantle heterogeneity along the Mid-Atlantic Ridge (31°-41° N). *Earth and Planetary Science Letters, 170,* 269–286.
Earnes, F. E. (1970). Some thoughts on the Neogene-Palaeogene boundary. *Palaeogeography, Palaeoclimatology, Palaeoecology, 8,* 37–48.
Elderfield, H., Ferratti, P., Greaves, M., Crowhurst, S., McCave, I. N., Hodell, D., et al. (2012). Ocean temperature and ice volume through the Mid-Pleistocene climate transition. *Science, 337,* 704–709.
Escartin, J., Cannat, M., Pouliquen, G., Rabain, A., & Lin, J. (2001). Crustal thickness of V-shaped ridges south of the Azores: Interactions of the Mid-Atlantic Ridge (36°-39° N) with the Azores hot spot. *Journal of Geophysical Research, 106,* 21719–21755.
Ewing, M., & Heezen, B. (1956). Mid-Atlantic seismic belt (Abstract). *Transactions of the American Geophysical Union, 37,* 343.
Fernandes, R. M. S., Bastos, L., Miranda, J. M., Lourenco, N., Ambrosius, B. A. C., Noomen, R., et al. (2006). Defining plate boundaries in the Azores region. *Journal of Volcanology and Geothermal Research, 156,* 1–9.
Forjaz, V. H., Franca, Z. T. M., Tavares, J. M., Almeida, I. M., & Rodrigues, J. A. (2010). *Dos Vulcoes dos Acores (From the Azores Volcanoes)* (2nd ed., 167p), Ponta Delgada: Publicor.
Gaspar, J. L., Queiros, G., Ferreira, T., Amaral, P., Viveiros, F., Marques, R., Silva, C., & Wallenstein, N. (2011). Geological hazards and monitoring at the Azores (Portugal). *Earthzine* (13p), posted 12 April 2011.
Gaspar, J. L., Queiros, G., Ferreira, T., Medeiros, A. R., Goulart, C., & Madeiros, J. (2015). Chapter 4: Earthquakes and volcanic eruptions in the Azores region: Geodynamic implications from major historical events and instrumental seismicity. *Geological Society of London, Memoir, 44,* 33–49.
Gente, P., Dyment, J., Maia, M., & Goslin, J. (2003). Interactions between the Mid-Atlantic Ridge and the Azores hot spot during the last 85 Myr: Emplacement and rifting of the hot spot-derived plateaus. *Geochemistry, Geophysics and Geosystems, 4*(10), 8514.
Georgen, J. E., & Sankar, R. D. (2010). Effects of ridge geometry on mantle dynamics in an oceanic triple junction region: Implications for the Azores Plateau. *Earth and Planetary Science Letters, 298,* 23–34.
Glenn, M. F. (1970). Introducing an operational multi-beam array sonar. *International Hydrographic Review, 47,* 35–40.

Goslin, J., & Party, T. S. (1999). Extent of Azores plume influence on the Mid-Atlantic Ridge north of the hotspot. *Geology, 27,* 991–994.

Grevemeyer, I. (1999). Isostatic geoid anomalies over mid-plate swells in the Central North Atlantic. *Journal of Geodynamics, 28*(1), 41–50.

Grimison, N. I., & Chen, W. P. (1986). The Azores-Gibraltar plate boundary: Focal mechanisms, depths of earthquakes and their tectonic implications. *Journal of Geophysical Research, 91,* 2029–2047.

Grimison, N. I., & Chen, W. P. (1988). Source mechanisms for four recent earthquakes along the Azores-Gibraltar plate boundary. *Geophysical Journal International, 92,* 391–401.

Gripp, A. E., & Gordon, R. G. (2002). Young tracks of hotspots and current plate velocities. *Geophysical Journal International, 150,* 321–361.

Gutenberg, B., & Richter C. F. (1941). Seismicity of the Earth. *Geological Society of America Special Paper 34,* 125p.

Heezen, B. C., Tharp, M., & Ewing, M. (1959). The floors of the oceans, I, the North Atlantic. *Geological Society of America Special Paper 65,* 122p.

Hess, H. H. (1959). Nature of the great oceanic ridges. *International Oceanographic Congress* (Abstract) (pp. 33–34).

Hey, R. N. (1977). A new class of pseudo-faults and their bearing on plate tectonics: A propagating rift model. *Earth and Planetary Science Letters, 37,* 321–325.

Hildenbrand, A., Madureira, P., Marques, F. O., Cruz, I., Henry, B., & Silva, P. (2008). Multistage evolution of a sub-aerial volcanic ridge over the last 1.3 Myr: S. Jorge Island, Azores Triple Junction. *Earth and Planetary Science Letters, 273,* 289–298.

Hildenbrand, A. P., Weis, D., Madureira, P., & Marques, F. O. (2014). Recent plate re-organization at the Azores Triple Junction: Evidence from combined geochemical and geochronological data on Faial, S. Jorge and Terceira Volcanic Islands. *Lithos, 210–211,* 27–39.

Hipolito, A., Madeira, J., Carmo, R., & Gaspar, J. L. (2013). Neotectonics of Graciosa Island (Azores): A contribution to seismic hazard assessment of a volcanic area in a complex geodynamic setting. *Annals of Geophysics, 56.*

Hirn, A., Haessler, A., Hoang Trong, P., Wittlinger, G., & Mendes Victor, L. A. (1980). Aftershock sequence of the January 1st, 1980, earthquake and present-day tectonics in the Azores. *Geophysical Research Letters,* 7(501), 504.

Holmes, A. (1929). Radioactivity and earth movements. *Geological Society of Glasgow Transactions, 18,* 559–606.

Isacks, B. J., Oliver, J., & Sykes, L. R. (1968). Seismology and the new global tectonics. *Journal of Geophysical Research, 73,* 5855–5899.

Johnson, G. J., Wijbrans, J. R., Constable, C. G., Gee, J., Staudigel, H., Tauxe, L., et al. (1998). $^{40}Ar/^{39}Ar$ ages and paleomagnetism of São Miguel lavas, Azores. *Earth and Planetary Science Letters, 160,* 637–649.

Kennett, J. P. (1978). The development of planktonic biogeography during the Cenozoic. *Marine Micropaleontology, 3,* 301–345.

Kidd, R., & Ramsay, A. T. S. (1987). The Geology and formation of the King's Trough Complex in the light of Deep Sea Drilling Project Site 608. In W. F. Ruddiman, R. B. Kidd, & E. Thomas (Eds.), Initial Reports of the Deep Sea Drilling Reports, Leg 94, Part 2, Washington, U.S. Printing Office, pp. 1245–1261.

Klitgord, K. D., & Schouten, H. (1986). Plate kinematics of the central Atlantic. In P. R. Vogt & B. E. Tucholke (Eds.), *The geology of North America, v. M, The Western North Atlantic Region, Geological Society of America* (pp. 351–378).

Knauss, J. (1986). The juridical ocean basin. In P. R. Vogt & B. E. Tucholke (Eds.), *The geology of North America, v. M, The Western North Atlantic Region, Geologic Society of America* (pp. 677–687).

Krause, D. C., & Watkins, N. D. (1970). North Atlantic crustal genesis in the vicinity of the Azores. *Geophysical Journal of the Royal Astronomical Society, 19,* 261–283.

Larrea, P., Wijbrans, P. R., Gale, C., Ubide, T., Lago, M., Franza, Z., et al. (2014). 40Ar/39Ar constraints on the temporal evolution of Graciosa Island, Azores (Portugal). *Bulletin of Volcanology, 76,* 796.

Laughton, A. S., & Whitmarsh, R. B. (1974). The Azores-Gibraltar plate boundary. In L. Kristjannson (Ed.), *Geodynamics of Iceland and the North Atlantic Area* (pp. 63–81). Dordrecht: D. Reidel Publishing Co.

LePichon, X. (1968). Sea-floor spreading and continental drift. *Journal of Geophysical Research, 73,* 3661–3697.

Louden, K. E., Tucholke, B. E., & Oakey, G. N. (2004). Regional anomalies of sediment thickness, basement depth, and isostatic crustal thickness in the North Atlantic Ocean. *Earth and Planetary Science Letters, 224,* 193–211.

Lourenco, N., Miranda, J. M., Luis, J. F., Ribeiro, A., Mendes Victor, L. A., Madeira, J., et al. (1998). Morpho-tectonic analysis of the Azores volcanic plateau from a new bathymetric compilation of the area. *Marine Geophysical Researches, 20*(3), 141–156.

Luis, J. F., & Miranda, J. M. (2008). Reevaluation of magnetic chrons in the North Atlantic between 35°N and 47°N: Implications for the formation of the Azores Triple Junction and associated plateau. *Journal of Geophysical Research, 113,* B10105 (12p).

Luis, J. F., Miranda, J. M., Galdeano, A., & Patriat, P. (1998). Constraints on the structure of the Azores spreading center from gravity data. *Marine Geophysical Researches, 20,* 157–170.

Luis, J. F., Miranda, J. M., Patriat, P., Galdeano, A., Rossignol, J. C., & Mendes Victor, L. A. (1994). Azores triple junction evolution in the last 10 Ma from a new aeromagnetic survey. *Earth and Planetary Science Letters, 125,* 439–459.

Madeira, J., Brum da Silveira, A., Hipolito, A., & Carmo, R. (2015). Chapter 3: Active tectonics in the central and eastern Azores islands along the Eurasia-Nubia plate boundary: A review. *Geological Society, London, Memoirs, 44,* 15–32.

Madureira, P., Mata, J., Mattielli, N., Queiroz, G., & Silva, P. (2011). Mantle source heterogeneity, magma generation and magmatic evolution at Terceira Island (Azores archipelago): Constraints from elemental and isotopic (Sr, Nd, Hf, and Pb) data. *Lithos, 126,* 402–418.

Madureira, P., Moreira, M., Mata, J., Nunes, J. C., Gautheron, C., Lourenco, N., et al. (2014). Helium isotope systematics in the vicinity of the Azores Triple Junction: Constraints on the Azores geodynamics. *Chemical Geology, 372,* 62–71.

Magde, L. S., & Sparks, D. W. (1997). Three-dimensional mantle upwelling, melt generation, and melt migration beneath segmented slow spreading ridges. *Journal of Geophysical Research, 102*(B9), 20571–20583.

Marques, F. O., Catalao, J. C., DeMets, C., Costa, A. C. G., & Hildenbrand, A. (2013). GPS and tectonic evidence for a diffuse plate boundary at the Azores Triple Junction. *Earth and Planetary Science Letters, 381,* 177–187.

Marques, F. O., Catalao, J. C., DeMets, C., Costa, A. C. G., & Hildenbrand, A. (2014). Corrigendium to "GPS and tectonic evidence for a diffuse plate boundary at the Azores Triple Junction". *Earth and Planetary Science Letters, 387,* 1–3.

Matias, L., Dias, N. A., Morais, L., Vales, D., Carrilho, F., Madeira, J., et al. (2007). The 9th of July 1998 Faial Island (Azores, North Atlantic) seismic sequence. *Journal of Seismology, 11,* 275–298.

Maury, M. F. (1855). *The physical geography of the Sea, Harper and Brothers*, New York, 287p.

McKenzie, D. F. (1972). Active tectonics of the Mediterranean region. *Geophysical Journal of the Royal Astronomical Society, 30,* 109–185.

McKenzie, D. F., & Morgan, W. J. (1969). Evolution of triple junctions. *Nature, 224,* 125–133.

Menard, H. W. (1965). Sea floor relief and mantle convection. *Physics and Chemistry of the Earth, 6,* 315–364.

Menard, H. W., & Atwater, T. (1968). Changes in the direction of sea floor spreading. *Nature, 219,* 463–467.

Mendes, V. B., Madeira, J., Brum da Silveira, A., Trota, A., Elosegui, P., & Pagarete, J. (2013). Present-day deformation in São Jorge Island, Azores, from episodic GPS Measurements (2001-2011). *Advances in Space Research, 51,* 1581–1592.

Merkouriev, S., & DeMets, C. (2008). A high-resolution model for Eurasia-North America plate kinematics since 20 Ma. *Geophysical Journal International, 173,* 1064–1083.

Merkouriev, S., & DeMets, C. (2014a). High-resolution estimates of Nubia-North America plate motion: 20 Ma to present. *Geophysical Journal International, 196,* 1281–1298.

Merkouriev, S., & DeMets, C. (2014b). High-resolution Neogene reconstructions of Eurasia-North America Plate motion. *Geophysical Journal International, 198,* 366–384.

Metrich, N., Zanon, V., Creon, L., Hildenbrand, A., Moreira, M., & Marques, F. O. (2014). Is the "Azores Hotspot" a Wetspot? Insights from the Geochemistry of Fluid and Melt Inclusions in Olivine of Pico Basalts. *Journal of Petrology, 55,* 377–393.

Minster, J. B., & Jordan, T. H. (1978). Present day plate motions. *Journal of Geophysical Research, 83*(B11), 5331–5354.

Miranda, J. M., Freire Luis, J., Abreu, A., Mendes Victor, L. A., Galdeano, A., & Rossignol, J. C. (1991). Tectonic framework of the Azores Triple Junction. *Geophysical Research Letters, 18,* 1421–1424.

Miranda, J. M., & Luis, J. F. (2013). *Changes in the Mid-Atlantic Ridge spreading velocity during the last 20 Ma* (unpublished manuscript).

Miranda, J. M., Luis, J. F., Lourenco, N., & Fernandes, R. M. S. (2015). Chapter 2: The structure of the Azores Triple Junction: Implications for São Miguel Island. *Geological Society of London, Memoirs*, 5–13.

Miranda, J. M., Luis, J. F., Lourenco, N., & Goslin, J. (2014). Distributed deformation close to the Azores Triple "Point". *Marine Geology, 355,* 27–35.

Miranda, J. M., Mendes Victor, L. A., Simoes, J., Luis, J., Matias, L., Shimamura, H., et al. (1998). Tectonic setting of the Azores Plateau deduced from a OBS survey. *Marine Geophysical Researches, 20*(3), 171–182.

Miranda, J. M., Navarro, A., Catalao, J., & Fernandes, R. M. S. (2012). Surface displacement field at Terceira island deduced from repeated GPS measurements. *Journal of Volcanology and Geothermal Research, 217–218,* 1–7.

Morgan, W. J. (1968). Rises, trenches, great faults and crustal blocks. *Journal of Geophysical Research, 73,* 1959–1982.

Morgan, W. J. (1972). Deep mantle convection plumes and plate motions. *Bulletin of the American Association of Petroleum Geologists, 56,* 203–312.

Morgan, W. J. (1978). Darwin, Rodriguez, Amsterdam… a second type of hotspot island. *Journal of Geophysical Research, 83,* 5355–5360.

Morley, L. W., & Larochelle, A. (1964). Paleomagnetism as a means to dating geologic events. *Royal Society of Canada, Special Publication, 8,* 512–521.

Navarro, A., Lourenco, N., Chorowicz, J., Miranda, J. M., & Catalao, J. (2009). Analysis of geometry of volcanoes and faults in Terceira Island (Azores): Evidence for reactivation tectonics at the EUR/AFR plate boundary in the Azores Triple Junction. *Tectonophysics, 465,* 98–113.

Neves, M. C., Miranda, J. M., & Luis, J. F. (2013). The role of lithospheric processes on the development of linear volcanic ridges in the Azores. *Tectonophysics, 608,* 376–388.

Phillips, J. D., & Fleming H. F. (1978). Multi-beam sonar study of the Mid-Atlantic Ridge rift valley, 36°-37° N, Map and Chart Series MC-19, Geological Society of America.

Phillips, J. D., Fleming, H. S., Feden, R. H., King, W. E., & Perry, R. K. (1975). Aeromagnetic study of the Mid-Atlantic Ridge near the Oceanographer Fracture Zone. *Geological Society of America Bulletin, 86,* 1348–1357.

Phipps Morgan, J., & Chen, Y. (1993). Dependence of ridge-axis morphology on magma supply and spreading rate. *Nature, 364,* 706–708.

Pitman, W. C., III, & Talwani, M. (1972). Seafloor spreading in the North Atlantic. *Geological Society of America Bulletin, 83,* 619–646.

Rabinowitz, P. D., & Jung, W.-Y. (1986). Gravity anomalies in the Western Atlantic Ocean. In P. R. Vogt & B. E. Tucholke (Eds.), *The geology of North America, v. M, The Western North Atlantic Region, Geological Society of America* (pp. 205–214).

Roest, W. R., & Srivastava, S. P. (1991). Kinematics of the plate boundary between Eurasia, Iberia, and Africa in the North Atlantic from the Late Cretaceous to the present. *Geology, 19,* 613–616.

Runcorn, S. K. (1956). Palaeomagnetism, polar wandering, and continental drift. *Geologie en Mijnbow, 18,* 253–256.

Ryan, W. B. F., Carbotte, S. M., Coplan, J. O., O'Hara, S., Melkonian, A., Arko, R., Weissel, R. A., Ferrini, V., Goodwillie, A., Nitsche, F., Bonczkowski, J., & Zemsky, R. (2009). Global multi-resolution topography synthesis, *Geochemistry, Geophysics, Geosystems, 10*(3). https://doi.org/10.1029/2008gc002332.

Saemundson, K. (1986). Chapter 5: Subaerial volcanism in the Western North Atlantic, In P. R. Vogt & B. E. Tucholke (Eds.), *The geology of North America, v. M, The Western North Atlantic Region, Geological Society of America* (pp. 69–86).

Sager, W. W. (2005). What built the Shatsky Rise, a mantle plume or ridge tectonics? In G. R. Foulger, G. H. Natland, D. C. Presnall, & D. L. Anderson (Eds.), Plates, plumes, and paradigms, *Geological Society of America Special Paper 388,* 721–733.

Santos, R. S., Tempera, P., Menezes, G., Porteiro, F., & Morato, T. (2010). Spotlight 12: Sedlo Seamount. *Oceanography, 23,* 202–203.

Schilling, J. G. (1976). Azores mantle blob: Rare earth evidence. *Earth and Planetary Science Letters, 25,* 103–115.

Schilling, J. G. (1986). Geochemical and isotopic variation along the Mid-Atlantic Ridge axis from 79° N to 0° N. In P. R. Vogt & B. E. Tucholke (Eds.), *The Geology of North America, v. M. The Western North Atlantic Region, Geological Society of America* (pp. 137–156).

Schouten, H., Klitgord, K. D., & Whitehead, J. A. (1985). Segmentation of mid-ocean ridges. *Nature, 317*(6034), 225–229.

Sclater, J. G., Anderson, R. N., & Bell, M. L. (1971). The elevation of ridges and the evolution of the central eastern Pacific. *Journal of Geophysical Research, 76,* 7888–7915.

Searle, R. C. (1980). Tectonic pattern of the Azores spreading center and triple junction. *Earth and Planetary Science Letters, 51,* 415–434.

Sibrant, A. L. R., Hildenbrand, A., Marques, F. O., & Costa, A. C. G. (2015). Volcano-tectonic evolution of the Santa Maria Island (Azores): Implications for paleostress evolution at the western Eurasia-Nubia plate boundary. *Journal of Volcanology and Geothermal Research, 291,* 49–62.

Sibrant, A. L. R., Marques, F. O., & Hildenbrand, A. (2014). Construction and destruction of a volcanic island developed inside an oceanic rift: Graciosa Island, Terceira Rift, Azores. *Journal of Volcanology and Geothermal Research, 284,* 32–45.

Sibrant, A. L. R., Marques, F. O., Hildenbrand, A., Boulesteix, T., Costa, A. C. G., & Catalao, J. (2016). Deformation in a hyperslow oceanic rift: Insights from the tectonics of the São Miguel Island (Terceira Rift, Azores). *Tectonics, 35,* 425–446.

Silveira, G., Stutzmann, E., Davaille, A., Montagner, J.-P., Mendes-Victor, L., & Sebai, A. (2006). Azores hotspot signature in the upper mantle. *Journal of Volcanology and Geothermal Research, 156,* 23–34.

Simkin, T., Tilling, R. I., Vogt, P. R., Kirby, S. H., Kimberly, P., & Stewart, D. B. (2006). This dynamic planet: World map of volcanoes, earthquakes, impact craters, and plate tectonics. *U.S. Geological Survey Geologic Investigations Series Map* I-2800.

Sloan, H., & Patriat, P. (1992). Kinematics of the North American-African plate boundary between 28° N and 29° N during the last 10 Ma: Evolution of the axial geometry and spreading rate and direction. *Earth and Planetary Science Letters, 113,* 323–341.

Smith, A. (1971). Alpine deformation and the oceanic areas of the Tethys, Mediterranean, and Atlantic. *Geological Society of America Bulletin, 82,* 2039–2070.

Smith, W., & Sandwell, D. (1997). Global seafloor bathymetry from satellite altimetry and ship depth soundings. *Science, 277,* 1956–1962.

Srivastava, S. P., & Tapscott, C. R. (1986). Plate kinematics of the North Atlantic. In P. R. Vogt & B. E. Tucholke (Eds.), *The geology of North America, v. M, The Western North Atlantic Region, Geologic Society of America* (pp. 379–404).

Srivastava, S. P., Roest, W. R., Kovacs, L. C., & Oakey, G. (1990). Motion of Iberia since the Late Jurassic: Results from detailed aeromagnetic measurements in the Newfoundland Basin. *Tectonophysics, 84,* 229–260.

Stocks, T., & Wuest, G. (1935). Die Tiefenverhaeltnisse des offenen Atlantischen Ozeans, Deutsche Atlantische Expedition METEOR 1925-1927, *Wissenschaftliche Ergebisse, 3,* Teil 1,1. Lieferung, 31p.

Tams, E. (1927). Die seismischen Verhaeltnisse des offenen Atlantischen Ozeans. *Gerlands Beitraege zur Geophysik, 18,* 319–353.

Tempera, F., Hipolito, A., Madeira, J., Vieira, S., Campos, A. S., & Mitchell, N. C. (2013). Condor

Seamount (Azores, NE Atlantic): A morpho-tectonic interpretation. *Deep-Sea Research II, 98,* 7–23.
Tucholke, B. E., & Vogt, P. R. (Eds.). (1979). Initial Reports of the Deep Sea Drilling Project, v. 43, U.S. Government Printing Office, Washington, D.C., 1115p.
Udias, A., Lopez Arroyo, A., & Mezcua, J. (1976). Seismotectonics of the Azores-Alborean región. *Tectonophysics, 31,* 259–289.
Vine, F. J., & Matthews, D. H. (1963). Magnetic anomalies over oceanic ridges. *Nature, 199,* 947–949.
Vink, G. E., Morgan, W. J., & Zhen, W. I. (1984). Preferential rifting of continents: A source of displaced terranes. *Journal of Geophysical Research, 89,* 10072–10076.
Vogt, P. R. (1974). Volcano height and plate thickness. *Earth and Planetary Science Letters, 23,* 337–348.
Vogt, P. R. (1976). Plumes, subaxial pipe flow, and topography along the Mid-Oceanic Ridge. *Earth and Planetary Science Letters, 29,* 309–325.
Vogt, P. R. (1979). Global magmatic episodes: New evidence and implications for the steady-state mid-oceanic ridge. *Geology, 7,* 93–98.
Vogt, P. R. (1986a). The present plate boundary configuration. In P. R. Vogt & B. E. Tucholke (Eds.), *The Geology of North America, v. M, The Western North Atlantic Region, Geological Society of America* (pp. 189–204).
Vogt, P. R. (1986b). Magnetic anomalies and crustal magnetization. In P. R. Vogt & B. E. Tucholke (Eds.), *The geology of North America, v. M, The Western North Atlantic Region, Geological Society of America* (pp. 229–256).
Vogt, P. R. (1986c). Plate kinematics during the last 20 m.y. and the problem of 'present' motions. In P. R. Vogt & B. E. Tucholke (Eds.), *The Geology of North America, v. M., The Western North Atlantic Region, Geological Society of America* (pp. 405–425).
Vogt, P. R., & Tucholke B. E. (1986). Imaging the ocean floor: History and state of the art. In P. R. Vogt & B. E. Tucholke (Eds.), *The Geology of North America, v. M., The Western North Atlantic Region, Geological Society of America* (pp. 19–44).
Vogt, P. R., & Jung, W.-Y. (2004). The Terceira rift as hyper-slow, hotspot-dominated oblique spreading axis: A comparison with other slow-spreading plate boundaries. *Earth and Planetary Science Letters, 218,* 77–90.
Vogt, P. R., & Jung W.-Y. (2005). Paired basement ridges: Spreading axis migration across mantle heterogeneities? In G. R. Foulger, G. H. Natland, D. C. Presnall, & D. L. Anderson (Eds.), *Plates, plumes, and paradigms, Geological Society of America Special Paper 388,* 555–579.
Vogt, P. R., Michael P. J., White S. M., & Macdonald K. C. (2008). Tall axial edifices (TAEs) on MOR spreading boundaries-characteristics and causes (Abs.), International Geologic Congress.
Vogt, P. R., Kovacs, L. C., Bernero, L. C., & Srivastava, S. P. (1982). Asymmetric geophysical signatures in the Greenland-Norwegian and Southern Labrador Seas and the Eurasia Basin. *Tectonophysics, 89,* 95–160.
Vogt, P. R., & Ostenso, N. A. (1966). Magnetic survey over the Mid-Atlantic Ridge between 42° and 46° N. *Journal of Geophysical Research, 71*(18), 4389–4412.
Walker, J. D., Geissman J. W., Browning J. W., & Babcock L. E. (2012). Compilers, Geologic Time Scale v. 4.0, Geologic Society of America.
Wegener, A. (1966). Die Enstehung der Kontinente und Ozeane, Vieweg und Sohn, Braunschweig, 1929 [The origin of continents and oceans, J. Biram, Trans.]. New York: Dover Publications (246p).
Weiss, B. J., Huebscher, C., & Luedmann, T. (2015a). The tectonic evolution of the southeastern Terceira Rift/São Miguel region (Azores). *Tectonophysics, 654,* 75–95.
Weiss, B. J., Huebscher, C., Luedmann, T., & Serra, N. (2016). Submarine sedimentation processes in the southeastern Terceira Rift/São Miguel region (Azores). *Marine Geology, 374,* 42–58.
Weiss, B. J., Huebscher, C., Wolf, D., & Luedmann, T. (2015b). Submarine explosive volcanism in the southeastern Terceira Rift/São Miguel region (Azores). *Journal of Volcanology and Geothermal Research, 303,* 79–91.
Wilson, J. T. (1963). A possible origin of the Hawaiian Islands. *Canadian Journal of Physics, 41,* 863–870.
Wilson, J. T. (1965). A new class of faults and their bearing on continental drift. *Nature, 207,* 343–347.
Yang, T., Shen, T., van der Lee, S., Solomon, S. C., & Hung, S.-H. (2006). Upper mantle structure beneath the Azores hotspot from finite-frequency seismic tomography. *Earth and Planetary Science Letters, 250,* 11–26.

The Contribution of Space-Geodetic Techniques to the Understanding of the Present-Day Geodynamics of the Azores Triple Junction

Rui M. S. Fernandes, João Catalão and António N. Trota

Abstract

Since the last few decades, Space-Geodetic techniques have provided new observations and understanding of many geodynamics processes. They offer an accuracy in position and spatial coverage that was not available using classical methodologies. Even in regions like the Azores Triple Junction, where the tectonic plates of North America, Eurasia and Nubia meet and that is mostly covered by the ocean, the use of Global Navigation Satellite Systems (GNSS) and Interferometry Synthetic Aperture Radar (InSAR) has undoubtfully contributed to a better understanding of the tectonic and volcanic processes. This chapter focuses on the major contributions obtained with Space-Geodesy in the Azores through a review of results published by different authors that have been actively working in this region since the late eighties. In fact, the first efforts to accurately measure relative crustal deformations between the different islands of the Azores Archipelago date back to 1988, when a network of nine markers (one per island) was observed using the Global Positioning System (GPS). Since then, several networks of GNSS points have been installed and regularly reoccupied in the islands by different research groups, in particular in the Central Group (Graciosa, Terceira, São Jorge, Pico, and Faial) and São Miguel, where the geodynamic processes of this triple junction manifest themselves through recurrent episodes of seismic and volcanic activity. Additionally, and starting in 1999 (Ponta Delgada), continuously operating GNSS (cGNSS) stations have been installed in almost all islands which permits today permanent monitoring even if the existing coverage is still not optimal. The initial results provided by GNSS studies focused on the understanding of the large-scale processes taking place in Azores by modelling inter-island displacements constrained by estimated angular velocities of the three tectonic plates. More recently, using the denser networks installed, several works have been concentrated on intra-island deformations due to tectonic but also volcanic activity. We also discuss recent results published using InSAR data. This technique has been

R. M. S. Fernandes (✉)
Instituto Dom Luiz (IDL), University of Beira Interior (UBI), Covilhã, Portugal
e-mail: rui@segal.ubi.pt

J. Catalão
Instituto Dom Luiz (IDL), Faculdade de Ciências, Universidade de Lisboa, FCUL, Lisbon, Portugal
e-mail: jcfernandes@fc.ul.pt

A. N. Trota
Instituto de Investigação em Vulcanologia e Avaliação de Riscos, University of the Azores, Azores, Portugal

U. Kueppers and C. Beier (eds.), *Volcanoes of the Azores*, Active Volcanoes of the World, https://doi.org/10.1007/978-3-642-32226-6_4

successful applied worldwide for volcanic studies and Azores is no exception. Several works have been published providing relevant information for the understanding of the tectonic dynamics that are taking place at the different islands of the Archipelago.

1 Introduction

GNSS is the most commonly used space-geodetic technique to study geodynamic processes since the early 1990s but it was not the first one. Very Long Baseline Interferometry (VLBI) was the first space-geodetic technique that could estimate distances between stations with sub-centimetre accuracy (e.g., Ryan and Ma 1998). Another space-geodetic technique that has been available since the 1970s is Satellite Laser Ranging (SLR), based on measuring the travel time of a light pulse reflected by a satellite (Degnan 1993).

SLR and VLBI were the first spatial-geodetic techniques contributing to quantify the kinematics of the tectonic plates on a global scale. Dedicated campaigns using mobile VLBI and SLR systems were made during the 1980's decade in some regions of the world with more significant tectonic activity (e.g., Vermaat et al. 1998). In addition, these techniques are also fundamental to materialize the International Terrestrial Reference System (ITRS) since VLBI is able to provide accurate scale and SLR has an important role in the determination of the Earth's geocenter. However, both techniques lack portability, have high costs, and they require complex operational procedures. Consequently, the global distribution of both techniques, VLBI (International VLBI Service for Geodesy and Astrometry—http://ivscc.gsfc.nasa.gov) and SLR (International Laser Ranging Service—http://ilrs.gsfc.nasa.gov) suffer from a bias towards the northern hemisphere (e.g., Bastos et al. 2010).

VLBI and SLR observations were carried out in Azores in the early 1990's. However, the short data-span of the observations did not permit to obtain robust velocities for these stations and they never have been published. The situation might change in the near future with the installation of two VLBI 2010 stations in Santa Maria and Flores in the framework of RAEGE project (RAEGE 2013).

The dissemination of space-geodetic techniques for geodynamic studies occurred in the early eighties with the advent of the GNSS, in particular GPS. The scientific community promptly started to develop and use GPS observations in geodynamic applications since it was able to provide similar, or even better accuracies, as the previously mentioned techniques with a fraction of the (financial and logistics) costs.

Actually, although other GNSS systems are being implemented (e.g., GLONASS), GPS is still the most commonly used space-geodetic technique for geodynamic studies. The interested reader has available immense literature of the principles of the GPS technique (e.g., Hoffman-Wellenhof et al. 1997). Specific reviews on the application of GPS for geodynamics, in particular requirements, technical issues, and developments, can also be consulted in Bastos et al. (2010) and references listed therein.

The other technique discussed in this chapter is the SAR interferometry. It is currently the only geodetic system capable of mapping the deformation (associated with earthquakes, volcanoes, mass movements, or the movement of large ice masses or glaciers.) of vast areas of the Earth's surface with spatial continuity and high accuracy.

The first significant result of the application of differential SAR interferometry was the surface deformation map associated with the 1992 Landers Earthquake presented by Massonnet et al. (1993). It was shown for the first time that the displacement of the surface with centimetre accuracy could be quantified over a wide area using pairs of SAR images and a reference terrain model. Since then, this technique has been applied in many regions to map the deformation caused by major earthquakes: Landers in 1992 (Zebker et al. 1994; Price and Sandwell 1998); Izmit in 1999 (Delouis et al. 2000; Sarti et al. 2000);

Hector Mine in 1999 (Fialko et al. 2001; Zeng 2001) or more recently, Tohoku-Oki in 2011 (Kobayashi et al. 2011; Feng et al. 2012).

The monitoring of volcanic surface deformations using SAR interferometry also dates back to the early stages of this technique with studies focused on active volcanoes like Etna (Massonnet et al. 1995) and Kilauea, Hawaii (Rosen et al. 1996). The use of SAR interferometry for volcanic observing has been also continuously developed with the introduction of the new approaches (e.g., Persistent Scatterers—Hooper et al. 2004).

This chapter presents the current status of the geodetic results in Azores concerning the determination of the present-day plate boundary both at archipelago and island scales (Sect. 2). Afterwards, we discuss in detail the published results focused on São Miguel where dedicated GNSS networks to monitor volcanic deformation has been installed (Sect. 3). Finally, we present the efforts to use SAR interferometry in Azores to detect surface displacements due to earthquake and volcanic activities (Sect. 4).

2 Present-Day Geodynamics of Azores Using GNSS Data

The use of precise space-geodetic techniques to study the present-day displacement field associated with the Azores Triple Junction started in 1988 with the Trans-Atlantic Network for Geodynamics and Oceanography (TANGO) project (Bastos et al. 1998). The initial network consisted of one station per island (cf. Fig. 1) that were re-occupied with an approximate triennial periodicity until 1997.

The TANGO network was densified with new 27 stations in 1999 and 30 stations in 2001, distributed over the islands of the so-called Central group (Terceira, Graciosa, São Jorge, Pico and Faial) in the framework of three projects: Extended-TANGO (Fernandes 2004), STAMINA (Navarro et al. 2003, 2009) and SARAZORES (Catita et al. 2005). This network continued to be re-observed in the framework of several follow-up projects, in particular KARMA (Catalão et al. 2010) and KINEMA (Fernandes et al. 2011).

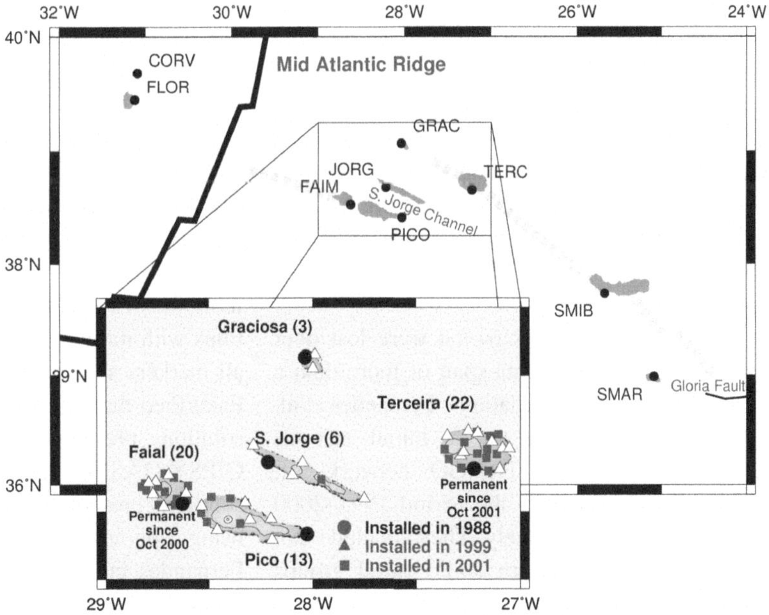

Fig. 1 Episodic GPS sites installed in the framework of several projects: TANGO (black circles), Extended-TANGO (white triangles), STAMINA and SARAZORES (black squares)

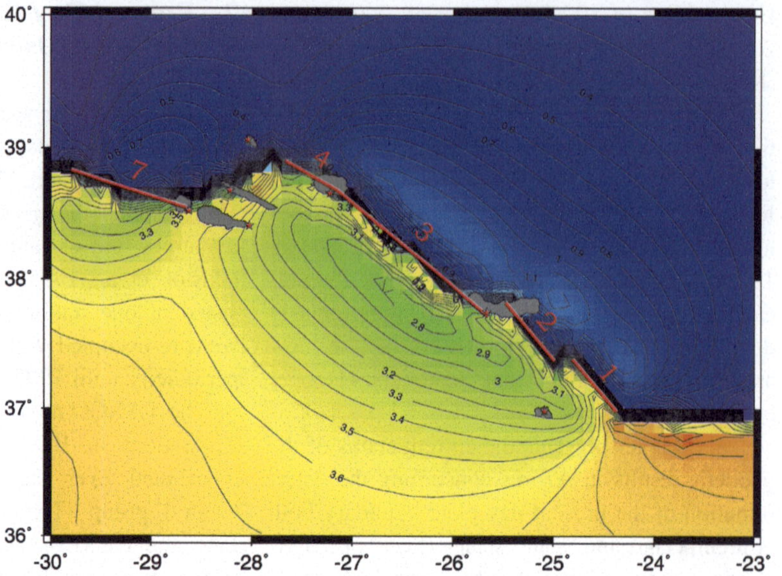

Fig. 2 Horizontal Displacement field—opening rates (represented by the contour lines and background colors; in mm/yr) with respect to the location of the fault segments (red lines) used for elastic modelling presented at Fernandes et al. (2006)

Figure 1 does not depict all episodic stations installed in Azores in the last 15 years. Mendes et al. (2013) refer a network of 59 episodic stations installed in the islands of Faial, Pico, and São Jorge in the framework of the DISPLAZOR project in 2001 that were re-observed three times until 2011.

Furthermore, the geodetic group of Instituto de Investigação em Vulcanologia e Avaliação de Riscos (IVAR), called CVARG before 2017, also installed other GNSS episodic networks during the last decade. Whenever possible, new markers were installed on rocky stable substratum, such as lava flows. However, due to the difficulty to find proper locations, some of the IVAR networks made also use of local geodetic public markers (Trota 2009).

Most of these stations (some were lost over the years) have now a data-span of more than a decade with up to 8 occupations. Fernandes et al. (2004, 2006) analyzed the 9 initial stations (forming the original TANGO network, cf. Fig. 1), corresponding to the period 1993–2000 and 1993–2001, respectively. They conclude that the islands in the Eastern and Central groups present velocities between pure Eurasian (Graciosa) and pure Nubian (Santa Maria) behaviour. The major result of these works was to favor a model (cf. Fig. 2) where two major spreading axes, an eastern one running from western edge of Gloria fault to northwestern of Terceira, indicated by numbers 1–4, and a western one from Faial to the Mid-Atlantic Ridge, number 7 in Fig. 2, were defining the active Nubia-Eurasia plate boundary. Such model was based on an elastic half-space approach with a locking depth of 8 km (Fernandes et al. 2006).

Concerning the present-day intra-island deformations, despite the number of existing stations, robust conclusions are still difficult to draw due to the uncertainties of the used data sets and the small amount of differential displacements expected at each island. This is evident on the velocity solutions (estimated and predicted) shown in Fig. 3. It presents the horizontal motions (w.r.t. ITRF2008) for the episodic stations with more than 6 years of observations (cf. all markers shown in Fig. 1) for a transect from Faial/Pico through São Jorge until Terceira. The solutions presented in Fig. 3, computed using GIPSY-OASIS software package, differ from the solutions presented in Fernandes et al. (2006) by using upgraded models and procedures (e.g., Fernandes et al. 2013) and by including additional observations for some few points carried out in 2010.

Several plate angular velocity models are used here in order to present the variation of predicted

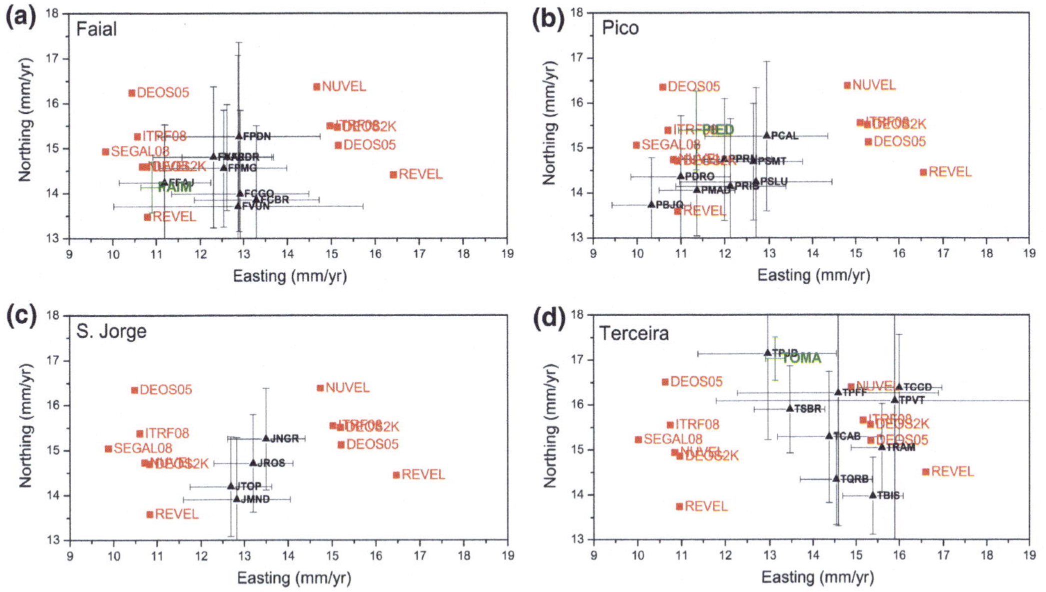

Fig. 3 Observed (black—campaign; green—permanent) and predicted (red) motions for Faial, Pico, São Jorge and Terceira with 95% uncertainty error bars. The predicted motions by different models for Nubia (left) and Eurasia (right) are relative to a point in the middle of each island

velocities for Nubia and Eurasia based on different models: NUVEL-1A (DeMets et al. 1994), REVEL (Sella et al. 2002), DEOS2k (Fernandes et al. 2003), DEOS05 (re-computation with respect to ITRF2005 of the data used in Fernandes et al. 2006), ITRF2008 (Altamimi et al. 2011), and SEGAL08 (Fernandes et al. 2013). In this respect, it is possible to conclude that the uncertainty of these models is nowadays of few mm/yr for both plates in the Azores region since the estimates given by the most recent angular velocity plate models tends to be similar.

The analysis of Figs. 2 and 3 permits to conclude that Graciosa appears to be situated on the Eurasia plate while Santa Maria is on the Nubia plate. All other stations in the Central and East groups appear to be in the deformation region with Faial/Pico closer to Nubia and Terceira closer to Eurasia (cf. Fig. 3).

As observed, the uncertainties of the estimated velocities are still too large to compute robust deformation models at intra-island level. Nevertheless, several intra-island studies have been published in recent years. The research focused in São Miguel, mainly evaluating local volcanic processes, are discussed in the next section. In the Central Group, the majority of the studies focused on kinematic models and derived strain partition. Miranda et al. (2012) analysed 17 stations in Terceira to corroborate the initial results presented by Navarro et al. (2003) that point for lateral compression due to the Terceira Rift. However, the authors were cautious since the rate of strain field obtained after the removal of the rigid-body motion was of the same magnitude as the 95% confidence ellipses concerning the horizontal components. More interestingly, the estimated average rates were close to 10 mm/yr of subsidence for most of the island, which was attributed to the volcanic processes that shaped the Serreta Ridge, NW of Terceira. In this respect, the results presented by Catalão et al. (2010) based on the network installed in Faial and Pico indicate that that some stations were subsiding by 9 mm/yr in those islands. Trota (2009) and Trota et al. (2010) obtained subsidence rates in the order of 10 mm/yr for the period 2000–2007 in Terceira Island and for western part of São Miguel Island, which were independently confirmed in São Miguel by tide-gauge data.

Finally, Mendes et al. (2013) presented results for São Jorge Island based on the analysis of the 2001, 2004, and 2010 GNSS campaigns, which recorded data on 17 stations. Surface velocities estimated at 15 inland locations were interpreted as deformation related to local sub-surficial magmatic/volcanic processes occurring near the island. The authors point out that the intra-island deformation may also be related to the stress field and seafloor spreading occurring in an area situated on the western sector of the Azores Plateau.

cGNSS stations have several advantages when compared with episodic networks: they are much less prone to suffer errors related to installation procedures (normally, only faulty equipment is replaced) and they can sense other signals, in particular seasonal signals, that are not detected or modelled in episodic time-series. In fact, the only limitation of cGNSS when compared with episodic stations is the costs associated with installation and maintenance. Such constraint prevented the fast development of cGNSS network in Azores. Nevertheless, the situation has been continuously improving since the installation of the first cGNSS in Ponta Delgada in 1999. At the time of writing this chapter, only Santa Maria and Corvo still do not have any cGNSS station providing public data to the community (a new station in Santa Maria is planned). The public network of cGNSS stations in Azores is managed by the Regional Government (REPRAA 2013). Although the major goal of this network is to serve technical geo-referencing activities, the highest requirements applied to the monuments permit their use also for geodynamic studies in the region. Currently two stations (PDEL and FLRS, on São Miguel and Flores islands, respectively) are part of the global IGS (International GNSS Service) permanent network (RAEGE 2013). Additionally, and with the specific purpose of geo-hazards monitoring, in particular volcano monitoring, several stations were installed by CVARG. This network presently has 10 cGNSS stations distributed at São Miguel (5), Terceira (2), and Pico (3), respectively.

Most of the cGNSS stations in Azores are relatively recent (2010 onwards). Consequently, their time-series are still too short to derive robust velocity estimates. Until the present, few results have been published based on these stations with the exception of Ponta Delgada station and of two stations installed in previous TANGO markers (FAIM and TOMA) that were operational during the last decade.

Figure 4 shows recent results published by Miranda et al. (2015) using only cGNSS stations. Residual velocities are plotted with respect to stable Nubia (according to SEGAL08, Fernandes et al. 2013). For comparison, Fig. 4 also shows the predicted relative motions for points in the stable Eurasia using SEGAL08 (published in Miranda et al. 2015), GEODVEL (Argus et al. 2010) and MORVEL (DeMets et al. 2010) models. The authors conclude, based on magnetic and this geodetic data, that most of the extension takes place today within the Terceira Rift (cf. Fig. 1) and that São Miguel Island shows significant intra-island extension.

3 Volcanic Studies in Azores Using Geodetic Data

During the last decades, several geodetic and surveying techniques were used to study, monitor and supervise volcanoes around the world. Volcanologists use several observation techniques from inexpensive simple tape measurements on fracture openings to high precision extensometer continuous measurements. However, due to several factors (e.g. harsh landscape, soil use, access, political reasons, price, logistics, precision of technique and deformation rate of the area under study) only some of them can be applied successfully. In Azores, studies aiming at ground deformation due to volcanic activity commonly use the GNSS technique.

The first application of the GNSS technique devoted to volcanic deformation studies was

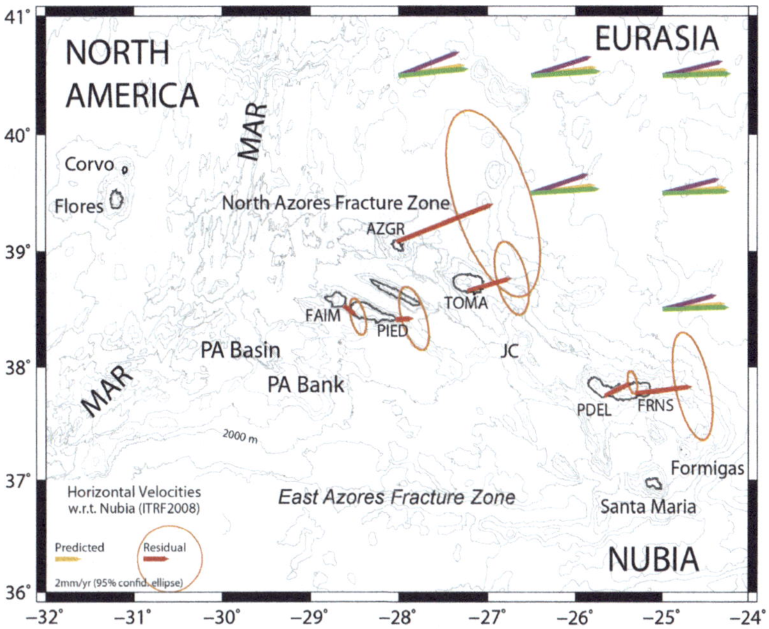

Fig. 4 Velocities relative to stable Nubia. cGNSS stations (red vectors) and predicted velocities for stable Eurasia points as given by MORVEL (violet), GEODVEL (yellow), and SEGAL08 (green)

used at Furnas volcano (in the scope of the European Laboratory Volcanoes, International Decade for Natural Disaster Reduction) on São Miguel. During this experiment, other geodetic techniques were also applied as precise optical levelling (lake, dry tilt, and profile levelling; Sigmundsson et al. 1995; Jónsson et al. 1999). Based on a network of 16 sites, distributed mainly at Furnas volcano and some at Fogo volcano, and three additional surveys carried out in 1993, 1994, and 1997, Jónsson et al. (1999) were able to measure slowly progressing ground deformation of Furnas volcano. Despite the small time interval for the data set and the low deformation rate the authors hypothesized that a source of inflation was located northwest of Furnas caldera or, alternatively, that a combination of two processes is responsible for the observed deformation pattern; i.e. plate divergence between the Eurasian and African plates together with a deflation of the Furnas caldera. Most of the markers used in this study were later on used in other GNSS surveys (e.g., Trota et al. 2006).

From 1999 on, the network installed in the framework of the Furnas Project were augmented to cover all São Miguel Island, and densified in some zones, e.g., in the Fogo Congro area in central São Miguel Island (Trota 2003). Most markers were installed as episodic stations although some are capable (in terms of ancillary systems and security) to be converted into cGNSS stations.

Based on a GNSS data set obtained from 1999 to 2007 an important and complex deformation period was determined in the Fogo Congro area (Trota 2009). The observed deformation (cf. Fig. 5) is now being interpreted as consequence of a basaltic magma intrusion episode, updating the results presented by Trota (2009), who considered them a possible consequence of trachytic magma injection or siliceous magma body intrusion. This new conclusion is based on an updated inversion of the GNSS data for spherical-like and prolate spheroid pressure sources along with reinterpretation of geological, seismic and geochemical results (Trota et al. 2015). The volcanic unrest that took place between 2000 and 2007 in the northeast flank of Fogo stratovolcano was characterized by three different main deformation phases of inflation and deflation process. The maximum surface ground deformation, which was recorded in September 2005, was concomitant with the

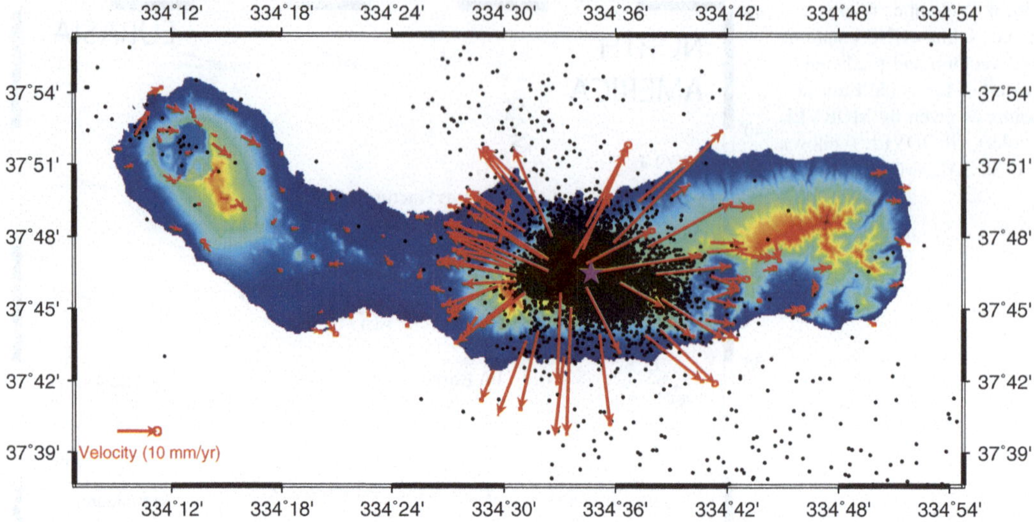

Fig. 5 Representation of the horizontal velocity field (red arrows) for the period 1999–2007 for the Fogo/São Brás/Congro area, São Miguel Island, interpreted as probable basaltic magma intrusion (Trota et al. 2015). The map also shows all the seismic events for the analyzed period, along with plot of location of the modelled deformation centre. The map shows a strong concentration of events plots enclosing the deformation centre (extracted from Trota 2009)

maximum seismic energy released. The new inversion of GNSS data is showing a north-eastward moving source from Fogo lake to São Brás lagune. The pressure source migrated from an initial depth around 3.9 to 2.3 km and then inward moving to 6.1 km, provides indication of a moving pressure source along a probable fault zone trending SW–NE. This trend coincides with local old volcanic vent alignments (craters and cones). The pressure source model volume change was estimated as 0.018 km^3. This volume, estimated for the Phase III of the event (Trota 2009) is one order of magnitude lower than the products emitted by the AD 1563 eruption in Fogo volcano, explaining why the basaltic mass flow intrusion did not turn into an eruption (Trota et al. 2015).

The data obtained in campaign style markers networks installed in one of the most active volcanic systems in Azores along with the cGNSS stations are allowing for a comprehensive investigation of the active volcanoes in Azores and are contributing to a better understanding of the interplay between tectonics and volcanic activity. The large region to monitor, the type of intrusions and eruption styles, and the observed low deformation rates creates additional challenges to the geodetic observations in Azores but will contribute to a better understanding of volcanic processes in this region and finally worldwide.

4 InSAR in Azores

Differential SAR interferometry (DInSAR) was used in 1998 Faial (Azores) earthquake to validate the GPS estimated co-seismic fault rupture model parameters (Fernandes et al. 2002; Catita et al. 2005) and used to discriminate between the sinistral NNW–SSE and the dextral ENE–WSW fault rupture, along with other complementary data (Marques et al. 2013). The displacement associated with a seismic event provides the limits of the co-seismic fault rupture model and can reveal the relations in the adjacent fault system imposing constraints on its model parameters. Examples are the 1992 Landers and 1999 Hector Mine earthquakes in the south of California and 1999 Izmit, Turkey in which

DInSAR has revealed the interaction between the fault rupture and adjacent fault systems.

In the case of the 1998 Faial (Azores) earthquake (Senos et al. 1998), the epicentre was offshore, preventing the imaging of the fault rupture with INSAR or GPS measurements. However, the seismic waves were strongly felt in the northeast Faial causing 8 casualties. Catita et al. (2005) applied the differential InSAR technique to ERS data, acquired between 1992 and 2000, to analyse surface deformation produced by the 9 July 1998 earthquake. Twelve co-seismic interferograms were produced from a small set of 17 ERS images in descending mode, evenly distributed in time (only four SAR images after the earthquake). Even under such limited conditions (temporal and geometric decorrelation) the authors managed to achieve fringe patterns with approximately 3 cm of range change between 1992 and 1998. Although correlation decreases in most areas, the fringe pattern is legible on the NW part of Pico Island.

The fringe pattern detected on the NW of Pico is closely related to the synthetic model computed from the fault parameters of Fernandes et al. (2002), and the authors concluded that the observed interferometric fringes generally agree with the synthetic models and, therefore, are coherent with the available seismological and GPS data. However, it was not possible to decide which fault was responsible for the earthquake mainly due to unfavourable geometric relation between the line of sight and the displacement directions (perpendicular). Besides, Faial is covered with dense vegetation and forest which cause decorrelation with ERS SAR wavelength (5.2 cm).

The study by Catita et al. (2005) was the first attempt to apply differential InSAR to the evaluation of ground displacement in the Azores archipelago. Results obtained are limited if compared with similar studies developed for most of the well-known volcanic systems. For active volcanoes, InSAR has revealed unexpected phenomena such as the magma movement in Isabela and Fernandina Islands in Galápagos (Amelung et al. 2000; Hooper et al. 2007) or the deflation movements in the Etna, Italy (Massonnet et al. 1995). Early detection based on the InSAR technique of the deformation caused by volcanoes can provide a warning indicator of imminent eruptions reducing the loss of lives and mitigating impacts on properties/assets.

The applications of differential interferometry are limited due to geometrical and temporal decorrelation. The first is due to the different acquisition geometries in repeated track interferometry and the second due to variation in the backscattering in the time interval between acquisitions. In the case of Azores, where most of the islands are highly vegetated and with very variable weather conditions, InSAR is limited by several sources of decorrelation (water vapour variability or changes of backscatter characteristics with time, etc.). To overcome these limitations, Ferretti et al. (2001) propose a new technique, the *Permanent Scatterers technique, PS,* (registered trademark). The technique allows the analysis of time series of interferograms of a set of points (pixels) with phase stability in time. The technique was applied in the Azores by Catalão et al. (2010) and Cong et al. (2010) to measure the internal deformation of Faial and Pico Islands and São Miguel Islands, respectively. In the case of São Miguel, the authors have complemented the set of PS with two corner reflectors (CR) installed on Lagoa do Fogo, the summit lake of Fogo volcano. The CRs ensure stability of the signal over a long period of time. A set of SAR stripmap images from TerraSAR-X satellite was used in this study. The authors pointed out the difficulties in the use of CR in Azores, mostly due to extreme weather conditions causing damage or orientation change on the CR.

The episodic stations installed in the framework of several projects in Faial and Pico (cf. Fig. 1), in particular the markers installed by the SARAZORES project (Catita et al. 2005), were used to study the vertical deformation on these islands. Most of the geodetic marks were surveyed 4 times since 2001 until 2013 (Marques et al. 2014). The results point to a time consistent and spatial coherent deformation rate between most of the points. Nonetheless, the GPS network is only the expression of a spatial sampling of the surface deformation and better spatial coverage is required. For that, a set of ASAR (Advanced SAR, on board of the ENVISAT satellite) images (ascending and descending passes), acquired between 2006 and 2009, were interferometrically processed and the line-of-sight deformation was derived using the Persistent Scatterers approach (Hooper et al. 2004). Because of different geometry acquisitions, ascending and descending passes produce different sets of Persistent Scatterers, with complementary spatial coverage, and different line-of-sight velocities. Besides, the estimated velocities are relative to the master image (different from ascending and descending) and must be referred to an absolute velocity (in the sense of referred to a geodetic reference frame). The strategy proposed by the authors to overcome aforementioned problems was based on the combination of sparse GPS 3D-velocities with two sets of Persistent Scatterers determined from ascending and descending passes.

The result of the integration of both ascending and descending persistent scatterers with GPS velocity data, is shown in Fig. 6 (Catalão et al. 2011). A large subsiding area on the west of Pico in the Madalena area can be identified, corresponding mostly to creep movement and the subsidence of the summit crater of Pico Island. On the west of Faial, on the flank of Capelinhos eruption, a large continuous area of subsidence is observed. The integrated geodetic approach, by merging GPS and PS-InSAR velocities, allowed the authors to draw several conclusions: (a) Faial, as a whole, behaves like a rigid body whose absolute motion is similar to Nubia motion (Marques et al. 2013); (b) intra-island deformation, given by the estimated strain rates, reveals north-south areal contraction at the western part of the island (Catita 2005); (c) the north-western flank of western Faial is subsiding, with a maximum subsidence rate of 89 mm/yr (Catalão et al. 2010); (d) The subsidence of the north west Faial or related with Capelinhos eruption, due to magma cooling and migration at depth and may be precursor of large-scale mass wasting events (Catalão et al. 2006).

Large scale mass wasting has been reported by Hildenbrand et al. (2012) for Pico Island. Due to its steep topography, Pico is particularly sensitive to flank instability. The southern flank of the ridge close to Lajes village shows several curved structures concave toward the ocean, previously interpreted as reflecting early caldera development, faulting or ancient lateral collapse(s). Hildenbrand et al. (2012) show from high-resolution digital elevation model (DEM), fieldwork, GPS and InSAR data that part of the flank in the analysed period (2001–2006) displacing toward the ocean, accommodated by the motion of large blocks that may eventually detach and have catastrophic consequences.

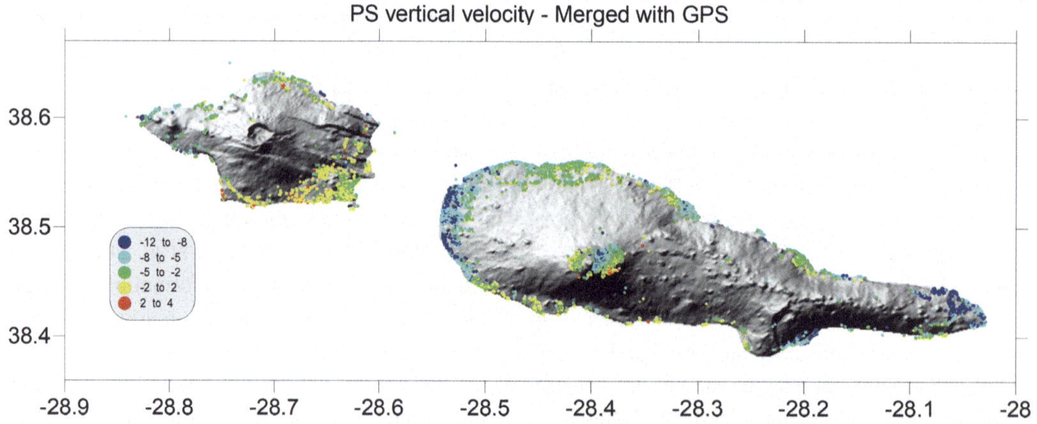

Fig. 6 Vertical velocity map resulting from the integration of ascending and descending PS and GPS data. The velocity is shown in mm/yr (Catalão et al. 2010)

SAR interferometry has shown great potential in measuring surface deformation with ability to map natural (tectonic, volcanic) and anthropogenic processes with particular focus on uplift/subsidence, due to its acquisition geometry. As already mentioned, this technique is strongly limited in the Azores by the density of vegetation with severe effects on the coherence of interferograms and the consequent reduction of the precision of the measurement phase. Nevertheless, the evolution of SAR sensors with higher spatial resolution, continuous SAR imaging of the earth surface by the new ESA Sentinel-1, JAXA ALOS2 and NASA DESDynI missions, as well as the ongoing evolution of algorithms for image processing are foreseen to improve the reliability and accuracy of the InSAR results even in such in such highly vegetated place as the Azores.

References

Altamimi, Z., Collilieux, X., & Métivier, L. (2011). ITRF2008: An improved solution of the international terrestrial reference frame. *Journal of Geodesy, 85*(8), 457–473. https://doi.org/10.1007/s00190-011-0444-4.

Amelung, F., Jónsson, S., Zebker, H., & Segall, P. (2000). Widespread uplift and trap door faulting on Galapagos volcanoes. *Nature, 407*(993–996), 2000.

Argus, D. F., Gordon, R. G., Heflin, M. B., Ma, C., Eanes, R. J., Willis, P., et al. (2010). The angular velocities of the plates and the velocity of Earth's center from space geodesy. *Geophysical Journal International, 180,* 913–960. https://doi.org/10.1111/j.1365-246X.2009.04463.x.

Bastos, L., Bos, M. S., & Fernandes, R. M. S. (2010). Deformation and tectonics: Contribution of GPS measurements to plate tectonics—Overview and recent developments. In G. Xu (Ed.), *Sciences of geodesy—I, advances and future directions* (pp. 155–184). Heidelberg: Springer.

Bastos, L., Osório, J., Barbeito, A., & Hein, G. (1998). Results from geodetic measurements in the western part of the African-Eurasian plate boundary. *Tectonophysics, 294,* 261–269.

Catalão, J., Miranda, J. M., & Lourenço, N. (2006). Deformation associated with the Faial (Capelinhos) 1957–1958 eruption: Inferences from 1937–1997 geodetic measurements. *Journal of Volcanology and Geothermal Research, 155,* 151–163.

Catalão, J., Nico, G., Hanssen, R., & Catita, C. (2010). Integration of InSAR and GPS for vertical deformation monitoring: A case study in Faial and Pico Islands. FRINGE 2009—Advances in the science and applications of SAR interferometry. ESA ESRIN, 30th November–4th December 2009, Frascati, Italy. ESA as Special Publication SP-677.

Catalão, J., Nico, G., Hanssen, R., & Catita, C. (2011). Merging GPS and atmospherically corrected InSAR data to map 3D terrain displacement velocity. *IEEE Transactions on Geoscience and Remote Sensing, 49* (6), Part: 2, https://doi.org/10.1109/tgrs.2010.2091963.

Catita, C., Feigl, K. L., Catalão, J., Miranda, J. M., & Victor, L. M. (2005, July 10). Time series analysis of SAR data applied to the study of the co-seismic deformation of the 9th July 1998 Pico-Faial (Azores) earthquake. *International Journal of Remote Sensing, 26*(13), 2715–2729.

Cong, X., Minet, C., & Eineder, M. (2010). Deformation monitoring using Sar interferometry on the Azores within Exupéry Project, Proceedings of 'Fringe 2009 Workshop', Frascati, Italy, 30 November–4 December 2009 (Esa Sp-677, March 2010).

Degnan, J. J. (1993). Millimeter accuracy satellite laser ranging: A review. In D. E. Smith & D. L. Turcotte (Eds.), *Contributions of Space Geodesy to Geodynamics: Technology*, Washington, D. C.: American Geophysical Union. https://doi.org/10.1029/gd025p0133.

Delouis, B., Lundgren, P., Salichon, J., et al. (2000). Joint inversion of InSAR and teleseismic data for the slip history of the 1999 Izmit (Turkey) earthquake. *Geophysical Research Letters, 27*(20), 3389–3392. https://doi.org/10.1029/2000gl011511.

DeMets, D. C., Gordon, R. G., & Argus, D. F. (2010). Geologically recent plate motions. *Geophysical Journal International, 181,* 1–80. https://doi.org/10.1111/j.1365-246X.2010.04491.x.

DeMets, C., Gordon, R. G., Argus, D. F., & Stein, S. (1994). Effect of recent revisions to the geomagnetic reversal timescale on estimates of current plate motions. *Geophysical Research Letters, 21*(20), 2191.

Feng, G., Ding, X., Li, Z., et al. (2012). Calibration of an InSAR-derived coseimic deformation map associated with the 2011 Mw-9.0 Tohoku-Oki earthquake. *IEEE Geoscience and Remote Sensing Letters, 9*(2), 302–306. https://doi.org/10.1109/lgrs.2011.2168191.

Fernandes, R. M. S. (2004). *Present-day kinematics at the Azores-Gibraltar plate boundary as derived from GPS observations*. Delft: DUP. ISBN 90-407-2557-8.

Fernandes, R. M. S., Ambrosius, B. A. C., Noomen, R., Bastos, L., Wortel, M. J. R., Spakman, W., et al. (2003). The relative motion between Africa and Eurasia as derived from ITRF2000 and GPS data. *Geophysical Reseach Letters, 30*(16), 1828. https://doi.org/10.1029/2003GL017089.

Fernandes, R. M. S., Bastos, L., Ambrosius, B. A. C., Noomen, R., Matheussen, S., & Baptista, P. (2004). Recent geodetic results in the Azores Triple Junction region. *Pure and Applied Geophysics, 161*(3), 683–699. https://doi.org/10.1007/s00024-003-2469-y.

Fernandes, R. M. S., Bastos, L., Miranda, J. M., Lourenço, N., Ambrosius, B. A. C., Noomen, R., et al. (2006). Defining the Plate Boundaries in the Azores Region. *Journal of Volcanology and Geothermal Research, 156,* 1–9. https://doi.org/10.1016/j.jvolgeores.2006.03.0192006.

Fernandes, R. M. S., Miranda, J. M., Catalão, J., Luis, J. F., Bastos, L., & Ambrosius, B. A. C. (2002). Coseismic displacements of the $M_w = 6.1$, July 9, 1998, Faial earthquake (Azores, North Atlantic). *Geophysical Research Letters, 29*(16), 21-1, 21-4. https://doi.org/10.1029/2001gl014415.

Fernandes, R. M. S., Miranda, J. M., Delvaux, D., Stamps, D. S., & Saria, E. (2013). Re-evaluation of the kinematics of Victoria Block using continuous GNSS data. *Geophysical Journal International, 193,* 1–10. https://doi.org/10.1093/gji/ggs071.

Fernandes, R. M., Miranda, J. M., & Luis, J. M. (2011). Strain Distribution across the Terceira (Azores) Rift from 13 years of GPS data. *AGU Fall Meeting Abstracts, 1,* 0839.

Ferretti, A., Prati, C., & Rocca, F. (2001). Permanent scatterers in SAR interferometry. *IEEE Transactions on Geoscience and Remote Sensing, 39*(1).

Fialko, Y., Simons, M., & Agnew, D. (2001). The complete (3-D) surface displacement field in the epicentral area of the 1999 M(w)7.1 Hector Mine earthquake, California, from space geodetic observations. *Geophysical Research Letters, 28*(16), 3063–3066. https://doi.org/10.1029/2001gl013174.

Hildenbrand, A., Ornelas, F., Catalão, J., Catita, C., & Costa, C. (2012, October). Large-scale active slump of the southeastern flank of Pico Island, Azores. *Geology, 40,* 939–942 (First published on August 9, 2012). https://doi.org/10.1130/g33303.1.

Hoffman-Wellenhof, B., Lichteneger, H., & Collins, J. (1997). *Global positioning system: Theory and practice* (4th ed.). Berlin: Springer.

Hooper, A., Segall, P., & Zebker, H. (2007). Persistent scatterer InSAR for crustal deformation analysis, with application to Volcán Alcedo, Galápagos. *Journal of Geophysical Research, 112,* B07407. https://doi.org/10.1029/2006JB004763.

Hooper, A., Zebker, H., Segall, P., & Kampes, B. (2004). A new method for measuring deformation on volcanoes and other natural terrains using InSAR persistent scatterers. *Geophysical Reseach Letters, 31,* L23611. https://doi.org/10.1029/2004GL021737.

Jónsson, S., Alves, M. M., & Sigmundsson, F. (1999). Low rates of deformation of the Furnas Volcanoes, São Miguel, Azores, observed with the Global Positioning System, 1993–1997. *Journal of Volcanology and Geothermal Research, 92,* 83–94.

Kobayashi, T., Tobita, M., Nishimura, T., et al. (2011). Crustal deformation map for the 2011 off the Pacific coast of Tohoku Earthquake, detected by InSAR analysis combined with GEONET data. *Earth Planets and Space, 63*(7), 621–625 https://doi.org/10.5047/eps.2011.06.043.

Marques, F. O., Catalão, J., DeMets, C., Costa, A. C. G., & Hildenbrand, A. (2013). GPS and tectonic evidence for a diffuse plate boundary at the Azores triple Junction. *Earth and Planetary Science Letters, 381,* 177–187.

Marques, F. O., Catalão, J., Hildenbrand, A., Costa, A. C. G., & Dias, N. A. (2014). The 1998 Faial earthquake, Azores: Evidence for a transform fault associated with the Nubia-Eurasia plate boundary? *Tectonophysics, 633*(2014), 115–125. https://doi.org/10.1016/j.tecto.2014.06.024.

Marques, F. O., Catalão, J., Hildenbrand, A., & Madureira, P. (2015). Groundmotion and tectonics in the Terceira Island: Tectonomagmatic interactions in an oceanic rift (Terceira Rift, Azores Triple Junction). *Tectonophysics, 651–652*(2015), 19–34. https://doi.org/10.1016/j.tecto.2015.02.026.

Massonnet, D., Briole, P., & Arnaud, A. (1995). Deflation of Mount Etna monitored by spaceborne radar interferometry. *Nature, 375*(6532), 567–570. https://doi.org/10.1038/375567a0.

Massonnet, D., Rossi, M., Carmona, C., Adragna, F., Peltzer, G., Feigl, K., et al. (1993). The displacement field of the Landers earthquake mapped by radar interferometry. *Nature, 364,* 138–142.

Mendes, V., Madeira, J., Brum da Silveira, A., Trota, A., Elosegui, P., & Pagarete, J. (2013). Present-day deformation in São Jorge Island, Azores, from episodic GPS measurements (2001–2011). *Advances in Space Research, 51*(8), 1581–1592.

Miranda, J. M., Luis, J. F., Fernandes, R. M. S., & Lourenço, N. (2015). The structure of the Azores triple junction: Implications for S. Miguel Island. In *Volcanic Geology of S. Miguel Island (Azores Archipelago)* (Vol. 44), Geological Society, London, Memoirs series.

Miranda, J. M., Navarro, A., Catalão, J., & Fernandes, R. M. S. (2012). Surface displacement field at Terceira Island deduced from repeated GPS measurements. *Journal of Volcanology and Geothermal Research.* https://doi.org/10.1016/j.jvolgeores.2011.10.009.

Navarro, A., Catalão, J., Miranda, J. M., & Fernandes, R. M. S. (2003). Estimation of the Terceira Island (Azores) main strain rates from GPS data. *Earths, Planets and Space, 55,* 637642.

Navarro, A., Lourenço, N., Chorowicz, J., Miranda, J. M., & Catalão, J. (2009). Analysis of geometry of volcanoes and faults in Terceira Island (Azores): Evidence for reactivation tectonics at the EUR/AFR plate boundary in the Azores triple junction. *Tectonophysics, 465*(1–4), 98–113.

Price, E. J., & Sandwell, D. T. (1998). Small-scale deformations associated with the 1992 Landers, California, earthquake mapped by synthetic aperture radar interferometry phase gradients. *Journal of Geophysical Research, 103*(B11), 27001–27016. https://doi.org/10.1029/98JB01821.

RAEGE. (2013). http://www.raege.net, viewed September 16, 2013.

REPRAA. (2013). http://www.repraa.azores.gov.pt, viewed September 16, 2013.

Rosen, P. A., Hensley, S., Zebker, H. A., Webb, F. H., & Fielding, E. J. (1996). Surface deformation and coherence measurements of Kilauea Volcano, Hawaii, from SIR C radar interferometry. *Journal Geophysical Research, 101*(E10), 23109–23126. https://doi.org/10.1029/96JE01459.

Ryan, J. W., & Ma, C. (1998). NASA-GSFC's geodetic VLBI program: a twenty-year retrospective. *Physics and Chemistry of the Earth, 23*(9–10), 1041–1052. https://doi.org/10.1016/S0079-1946(98)00144-X.

Sarti, F., Adragna, F., Pourthie, N. et al. (2000). InSAR analysis of the effects of the Izmit earthquake with ERS data. In R. A. Harris & L. Ouwehand (Eds.), *CEOS SAR* (Vol. 450, pp. 663–669), Toulouse, France, October 26–29, 1999. European Space Agcy; CNES; Comm Earth Observat Satellites. CEOS SAR Workshop, ESA Special Publications.

Sella, G. F., Dixon, T. H., & Mao, A. (2002). REVEL: A model for recent plate velocities from space geodesy. *Journal Geophysical Research, 107*(B4), 2081. https://doi.org/10.1029/2000JB000033.

Senos, M. L., Gaspar, J. L., Cruz, J., Ferreira, T., Nunes, J. C., Pacheco, J. M., et al. (1998). O terramoto do Faial de 9 de Julho de 1998. 1º Simpósio de Materorologia e Geofisica da APMG, Lagos, pp. 61–67.

Sigmundsson, F., Tryggvason, E., Alves, M. M., Alves, J. L., Pálsson, K., & Olafsson, H. (1995). Slow inflation of the Furnas volcano, São Miguel, Azores, suggested from initial levelling and Global Positioning System measurements. *Geophysical Reseach Letters, 22,* 1681–1684.

Trota, A. (2003). "Aplicação da técnica de GPS à monitorização de vulcões activos na ilha de S. Miguel (Açores)." Tese de Mestrado em Vulcanologia e Avaliação de Riscos Geológicos, Departamento de Geociências, Universidade dos Açores, Ponta Delgada, 152p (in Portuguese).

Trota, A. (2009). "Estudos de deformação crustal nas ilhas de S. Miguel e Terceira (Açores). Avaliação da actividade vulcânica na área do Fogo/Confro (S. Miguel)". Tese de doutoramento no ramo de Geologia, especialidade de Geodesia. Universidade dos Açores. 281p (in Portuguese).

Trota, A., Houlié, N., Briole, P., Gaspar, J. L., Sigmundsson, F., & Feigl, K. L. (2006). Deformation studies at Furnas and Sete Cidades Volcanoes (São Miguel Island, Azores). Velocities and further investigations. *Geophysical Journal International, 166*(2), 952–956. https://doi.org/10.1111/j.1365-246X.2006.03039.x.

Trota, A. N., Mendes, V., & Amaral, P. (2010). Subsidência generalizada nas ilhas de São Miguel e Terceira (Açores) detetada por GPS. In Actas da VI conferência nacional de cartografia e geodesia (pp. 330–335), 7 a 8 de maio 2009. Ed. Lidel, Lisboa, Portugal.

Trota, A., Mendes, V. B., Amaral., P., Rodrigues, R., Gaspar, J. L., Houlié, N., et al. (2015). Crustal deformation modelling of the Fogo/S. Brás/Congro volcanic unrest, São Miguel, Azores (in preparation).

Vermaat, E., Degnan, J. J., Dunn, P., Noomen, R., & Sinclair, A. (1998). Satellite laser ranging, status and impact for Wegener. *Journal of Geodynamics, 25*, 195–212. Deformation and Tectonics: Contribution of GPS Measurements to Plate Tectonics—Overview and Recent Developments.

Zebker, H. A., Rosen, P. A., Goldstein, R. M., Gabriel, A., & Werner, C. L. (1994). On the derivation of coseismic displacement fields using differential radar interferometry: The Landers earthquake. *Journal of Geophysical Research, 99,* 19617–19634.

Zeng, Y. H. (2001). Viscoelastic stress-triggering of the 1999 Hector Mine earthquake by the 1992 Landers earthquake. *Geophysical Research Letters, 28*(15), 3007–3010. https://doi.org/10.1029/2000gl012806.

Crust and Mantle Structure Beneath the Azores Hotspot—Evidence from Geophysics

Craig O'Neill and Karin Sigloch

Abstract

The Azores hotspot marks the triple junction between the North American, Eurasian, and African plates, and is responsible for the ~20 Ma Azores plateau, and ongoing, off-axis volcanism today. The dynamics of the interaction between the Azores hotspot and the slow-spreading North Atlantic ridge has led to short wavelength V-shaped bathymetric and geochemical anomalies along the mid-ocean ridge, suggesting variations in the flow of mantle plume material towards the southwest. The depth extent of the Azores plume is unclear, or indeed whether it constitutes a traditional plume at all. Surface-wave models have suggested that the "plume" is confined to the upper 250–300 km of the mantle, suggesting either a shallow origin to the Azores hotspot, or that the plume is waning. In contrast, recent finite-frequency body-wave tomography has suggested that the Azores conduit may extend to the core-mantle boundary, and that the Azores, Canary, and Cape Verde hotspots may have a common origin under West Africa. Here we assess geophysical constraints on crustal and mantle structure beneath the Azores hotspot. Geochemical constraints and body-wave tomography results argue for a deep origin of the Azores hotspot. Radial anisotropy suggests significant vertical flow in the vicinity of the hotspot, and this is consistent with the geoid and gravity field. Calculations of plume conduit dynamics in simulations of the global mantle flow field suggest that the present conduit tilts towards West Africa, as observed in the body-wave tomography, and support a common origin for the Azores, Canary, and Cape Verde plumes.

1 Introduction

The Azores archipelago marks the triple junction of African, European/Iberian and North American plates (Gente et al. 2003). The archipelago itself consists of three island groupings (Fig. 1): (1) The Western Islands, consisting of Flores and Corvo, located west of the Mid-Atlantic ridge; (2) the Central Islands east of the ridge, which include Faial, Pico, Graciosa, Terceira, and São Jorge; and (3) the Eastern group, consisting of São Miguel and Santa Maria.

C. O'Neill (✉)
Department of Earth and Planetary Science, CCFS ARC Centre of Excellence, Macquarie University, Sydney, Australia
e-mail: craig.oneill@mq.edu.au

K. Sigloch
Department of Earth Sciences, University of Oxford, Oxford, UK
e-mail: karin.sigloch@earth.ox.ac.uk

U. Kueppers and C. Beier (eds.), *Volcanoes of the Azores*, Active Volcanoes of the World, https://doi.org/10.1007/978-3-642-32226-6_5

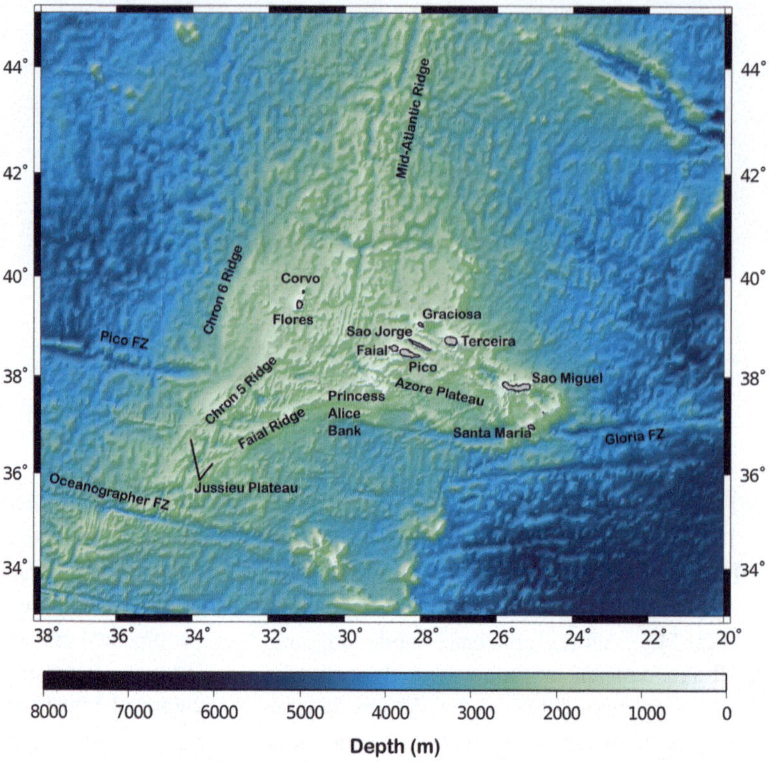

Fig. 1 Major bathymetric features of Azores archipelago and tectonic surrounds

The shallow bathymetry in the Azores region is largely a result of thickened crust. Seismological studies have estimated crustal thicknesses varying between 8 and 20 km across the archipelago. The thick volcanic units making up the Azores plateau have been suggested to be due to the interaction of an upwelling plume beneath the Azores, with the Mid-Atlantic spreading ridge (e.g., Gente et al. 2003). However, a plume origin for the Azores plateau and archipelago is not uncontested. Luis and Neves (2006) were unconvinced by the lack of a definitive bathymetry/geoid anomaly in the region, and suggested the volcanism was due to a lossy plate boundary/triple junction effect. King (2005) suggests that the origin of volcanism in the Azores and Iceland may be the result of upper-mantle edge-driven convection cells that initiated with breakup of Eurasia and North America. Bonatti (1990) and Asimow and Langmuir (2003) both suggest that melting in the Azores is primarily a function of enhanced volatile content, with little excess temperature contributing to melting.

In contrast, a number of lines of evidence support an active (thermal) upwelling in the Azores. The alkaline volcanics of the Azores plateau suggest a very isotopically heterogeneous source, with high volatile contents (Schilling et al. 1983; Bonatti, 1990, see Larrea et al., Chapter "Petrology of the Azores Islands", Beier et al., Chapter "Melting and Mantle Sources in the Azores", Moreira et al., Chapter "Noble Gas Constraints on the Origin of the Azores Hotspot"). The geochemistry of mid-ocean ridge basalts in the vicinity of the Azores document a long-wavelength, extended, asymmetric anomaly, similar in extent to the bathymetry anomaly, and consistent with the inferred influence of an Azores plume (Dosso et al. 1999).

Bourdon et al. (1996) suggest an excess temperature of 100 °C beneath the Azores, based on

U-series disequilibria. An excess temperature of 75 °C is estimated by Cochran and Talwani (1978) using gravity data, and Schilling (1991) used ridge bathymetry to estimate the temperature excess at 198 °C. While there are large differences in these estimates owing to non-uniqueness (e.g., gravity), sensitivity, depth-averaging, and source effects, together these temperature estimates are consistent with a moderate thermal anomaly beneath the Azores archipelago.

Beier et al. (2012) constrain melting conditions within the Azores by demonstrating similar degrees of partial melting within the central island group, despite variable Sr–Nd–Pb isotope ratios and incompatible trace element ratios (Yu et al. 1997), the latter suggesting a heterogeneous source. They infer variations in SiO_2, FeO and TiO_2 then are primarily related to temperature and pressure of melting, constrained to be between ~1400 and <1480 °C for the majority of lavas, and between ~2.5 and 3.7 GPa. The shallowest melting is documented at Faial and Graciosa (and partly beneath Pico), and the deepest beneath São Jorge, which also shows the highest temperatures at 1490 °C. This is broadly consistent with increasing melting pressures away from the Mid-Atlantic ridge (Beier et al. 2012). Beier et al. (2012) also note that their low melting temperatures, and high pressures, are below that of a dry peridotite solidus, requiring at least 200 ppm of water in melting source region (see also Beier et al., Chapter "Melting and Mantle Sources in the Azores").

Bourdon et al. (2005) suggested the plume was centred on the island of Terceira, on the basis of U–Th–Pa disequilibria, which suggest the fastest upwelling velocities beneath the island (3–4 cm/yr, compared to ~1 cm/yr on the periphery). They suggest that, São Miguel aside, the $^{230}Th/^{238}U$ and $^{231}Pa/^{235}U$ variations indicate that the plume centre is characterised by deeper initial melting, and faster melting rates, than the periphery. They also note that the lack of correlation between $^{230}Th/^{238}U$ is probably due to the effect of water in the source.

Moreira et al. (1999) measured the $^{4}He/^{3}He$ ratio from a number of samples from the Azores archipelago (see Moreira et al., Chapter "Noble Gas Constraints on the Origin of the Azores Hotspot"). The results are highly variable, ranging from very primitive, to highly radiogenic helium ($^{4}He/^{3}He$ > 140,000)—suggesting a heterogeneous mantle source with perhaps a contribution from relic continental lithosphere from the opening of the Atlantic. In contrast, the $^{4}He/^{3}He$ for Terceira and Pico islands—while variable—contains very primitive contributions ($^{4}He/^{3}He$ of 64,000 for Terceira, and 70,000 for Pico)—strongly suggesting a primitive, lower mantle source (Madureira et al. 2005).

Schaefer et al. (2002) measured $^{187}Os/^{188}Os$ across the archipelago, and suggested the low ratios observed across the centre of the plume may represent sampling of a deep, recycled (perhaps Archaean) depleted harzburgitic lithosphere. This again suggests a significant lower mantle contribution to the Azores upwelling.

To summarize, there is evidence for active upwelling beneath the Azores archipelago, centred at Terceira (e.g., from U–Th–Pa disequilibria, Bourdon et al. 2005). The involvement of a deep lower mantle source is suggested by the primitive $^{4}He/^{3}He$ ratios for Terceira and Pico (Moreira et al. 1999), and osmium data (Schaefer et al. 2002). However, there is strong evidence for the involvement of water in the melt source (e.g., Beier et al. 2012), consistent with high volatile content of Azores volcanics (e.g., Bonatti 1990). The ultimate question is whether the Azores represents the impingement at a plate triple junction of a typical (though hydrous) thermal plume rising from the core-mantle boundary, or whether the volcanic activity can plausibly be ascribed to another mechanism (Foulger 2007). Recent advances in deep geophysical imaging of the Azores region can shed light on the mantle state and structure of the hotspot at depth, and discern between these end-members. We summarize the relevant observations in the following sections.

2 Tectonic History

Construction of the broad Azores plateau began around 20 Ma (Chron 6; Gente et al. 2003), at the Mid Atlantic Ridge axis. Between 20 and 10 Ma, plateau construction extended along the ridge axis, reaching its most northerly extent by around ~10 Ma (Chron 5, Gente et al. 2003). Gente et al. (2003) suggest volcanism crossed the Pico transform at approximately 15 Ma, and the most southerly volcanism, on the Jussieu plateau and Faial ridge, erupted 6–7 Ma (Cannat et al. 1999). The eruption of the Jussieu plateau was associated with a redistribution of plate boundaries around the triple junction, with the previous Eurasian(/Iberian)-African boundary, the Gloria Fracture zone, being supplanted by the Azores Spreading Centre or Terceira Ridge (Fig. 1), which was orientated ~N123°E at that time.

Rifting of the Azores plateau began at about 8–9 Ma at the northern end, and about 4 Ma (Chron 3) to the south (Gente et al. 2003), with the ridge moving away from the hotspot volcanic source (Gripp and Gordon 2002) (see also Vogt and Jung, Chapter "The "Azores Geosyndrome" and Plate Tectonics: Research History, Synthesis, and Unsolved Puzzles"). The rifting propagated, in discrete intervals, from the north at about ~150 km/Myr. Currently the Mid Atlantic rift separating the plateau is 175 km in width to the north, and ~74 km wide in the south (Gente et al. 2003). Since about 4 Ma, the Mid Atlantic ridge in the proximity to the Azores has operated as a normal slow-spreading ridge segment (i.e. distal from the Azores hotspot). The spreading rate has been approximately constant for the last 40 Myr (Cande et al. 1985), and increases from 20 mm/yr in the south (35°N) to 22 mm/yr in the north (40°N; De Mets et al. 1990). The present-day Eurasian/Iberian—African plate margin is complex, incorporating not only the Azores region, which acts as a transtensional domain, but also the Gloria transform fault, and the Horseshoe seamounts off the Iberian margin, the latter of which are compressional (Gente et al. 2003; Madeira and Ribeiro 1990). Global plate kinematics suggest the Azores region is accommodating right lateral transtensional motion, the extensional component being around 3–4 km/Myr (Minster and Jordan 1978; De Mets et al. 1990), and evidence for this is borne out by detailed bathymetry of the region (Lourenco et al. 1998; Miranda et al. 1998).

3 Potential Field and Bathymetry

A large bathymetric anomaly exists over the Azores region (Figs. 1 and 2) (Thibaud et al. 1997). In addition to the mid-ocean ridge bathymetric rise, there exists a distributed swell to the west, in the vicinity of Corvo and Flores, and also to the east along the Azores Archipelago, along the Azores plateau (Terceira Ridge). The bathymetric anomaly extends further south than north, along the Faial ridge on the eastern side of the Mid-Atlantic ridge, and along the Chron 5 ridge to the west. Gente et al. (2003) removed the seafloor subsidence trend to create a residual bathymetry map of the area. The plateau is split into two components, on either side of the mid-Atlantic ridge, reflecting its volcanic history, and the residual depth is generally less than ~1000 m. Gente et al. (2003) suggest the topographic rise of the plateau is largely the result of thickened volcanic crust in the region. Sleep (1990) used the volcanic ridge geometry and plate speed relative to the hotspot to calculate the mass flux of the Azores at 1100 kg/s—a factor of 8 lower than Hawaii, and about 78% the flux of Iceland.

While gravity data are unable to resolve the actual Azores plume, they are sensitive to crustal structure and mantle density anomalies. Cochran and Talwani (1978) used gravity data to argue for a diffuse, low-density anomaly within the mantle in the vicinity of the Azores plume. The gravity anomaly they identified may be related to an anomalously warm mantle beneath the Azores archipelago—approximately 75 °C above ambient mantle temperatures, for their estimated source depth and extent—though this estimate is non-unique. Gente et al. (2003) used an FFT approach to calculate the mantle Bouguer gravity anomaly, and show a large (−300 mGal) negative Bouguer anomaly over the Azores region (see Fig. 2). The region is asymmetric, extends

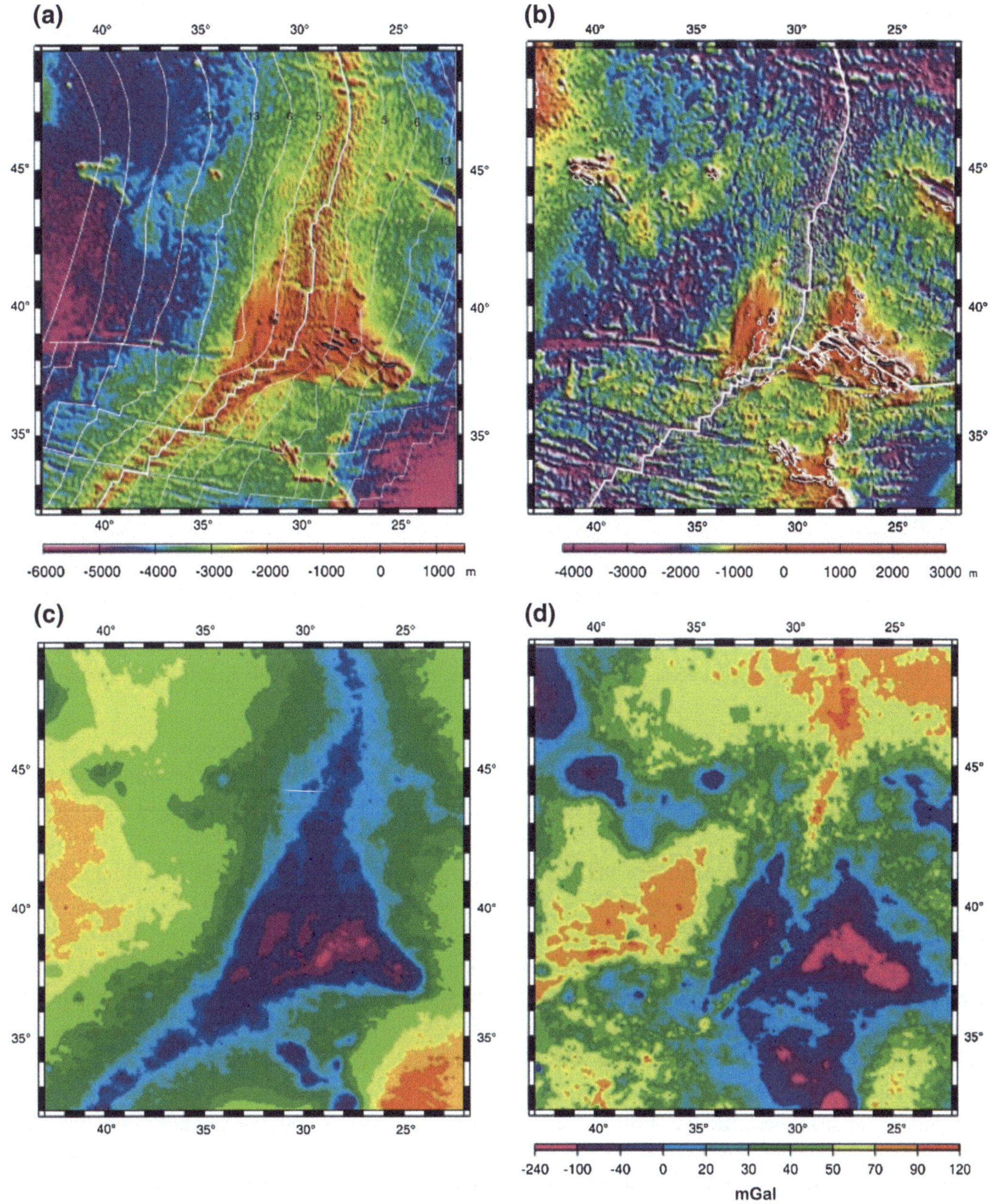

Fig. 2 Bathymetry and Bouguer gravity of the Azores region, from Gente et al. (2003). **a** Bathymetry of the Azores region, with white isochrons plotted. **b** Residual bathymetry after the effect of age-related seafloor subsidence is removed. **c** Bouguer gravity over the Azores archipelago. **d** Residual Bouguer anomaly after the ridge and seafloor subsidence anomalies are removed

further south than it does north, and also extends along the Terceira Ridge. They also created a residual Bouguer anomaly map, which removed the gravitational effect of plate thickening, and which correlates well with observed bathymetry. Gente et al. (2003) argue that the residual Bouguer anomaly is up to 60% contribution from the crustal thickness, and perhaps 40% from deeper

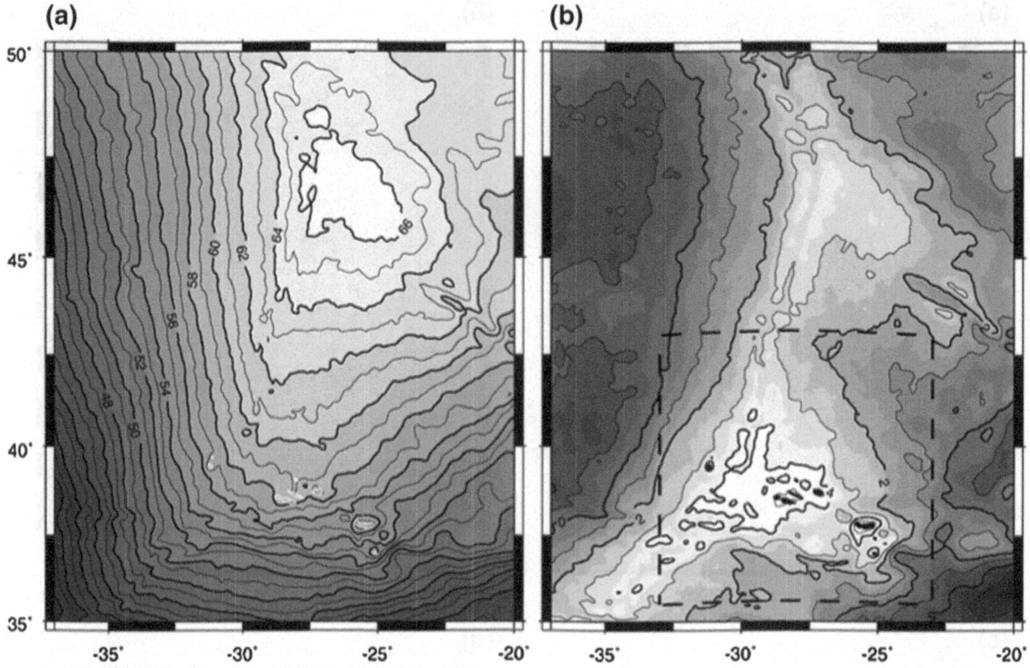

Fig. 3 Total geoid (**a**) and filtered geoid (**b**) to remove spherical harmonic degrees less than 10—preserving mid to shorter wavelength anomalies. From Luis and Neves (2006)

mantle anomalies. They created a crustal thickness map on the assumption that the entire residual anomaly is due to the crust – this resulted in crustal thickness estimates up to 12–15 km in places. Independent crustal thickness estimates from seismology can help resolve the relative contributions of crust versus lithospheric mantle to the gravity anomaly. Searle (1976) estimated the crustal thickness of the Azores platform to be ~8 km from Rayleigh waves, which would imply a sizable mantle contribution to the observed Bouguer gravity. However, Dias et al. (2007) used local earthquake tomography to estimate the crustal velocity structure near Faial and Pico islands, and estimated the crust surrounding the islands to be of the order of 14 km thick—requiring little mantle contribution to the Bouguer gravity signal.

Most hotspot swells demonstrate small, positive geoid anomalies of the order of ~2–10 m. One complication with the Azores' geoid signature is that the North Atlantic is dominated by the North Atlantic geoid anomaly, a ~60 m positive anomaly Detrick et al. (1995) centred around 47° N, which extends from Iceland down to north of the Azores (Fig. 3). This is vastly above the expected geoid anomaly of a plume, which should be of the order of 1–5 m (e.g., Sleep 1990), and most analyses remove the long-wavelength signal to resolve the geoid swell from the hotspot. Bowin et al. (1984) examined the residual geoid over the Atlantic Ocean, and concluded the residual geoid high centred on the Azores region was due to an upwelling plume. They filtered out long-wavelength contributions (less than degree 10), and observed a positive anomaly of ~6 m over the Azores region—which is a contrast of over ~10 m from the adjacent abyssal plains, also significantly greater than ridge-axis geoid anomalies distal from hotspot influence (<~2–3 m). Bowin et al. (1984) conclude that a combination of crustal thickness variations and thermal influence of the plume may explain the residual geoid. Goslin et al. (1998), Goslin and Party (1999) using the model of Lerch et al. (1994), showed a positive,

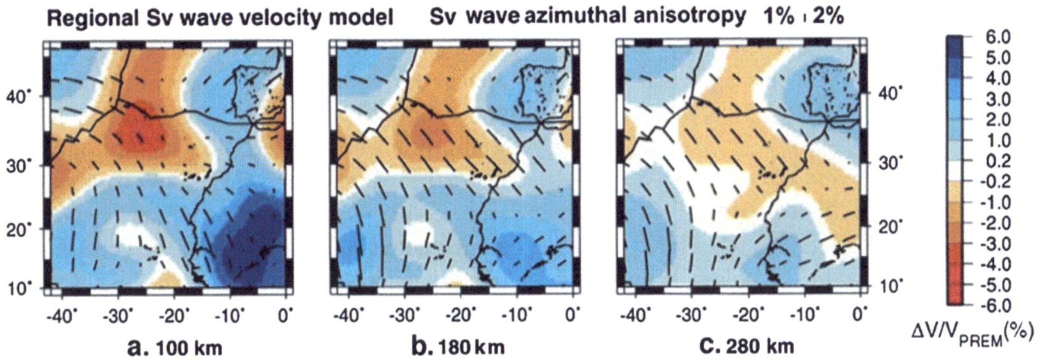

Fig. 4 Regional S-wave velocity of Silveira et al. (2006), relative to PREM, with azimuthal anisotropy directions plotted on top

along-ridge anomaly of the order of 6–8 m in the Azores, which extends further south than north. Silveira et al. (2006) argued that the positive, asymmetric geoid anomaly associated with the Azores is generally coincident with geochemical variations observed along-ridge (Fig. 4).

Luis and Neves (2006, see Fig. 3) looked at both the admittance functions of geoid (EGM96) (Lemoine et al. 1998) to bathymetry, and gravity (both free air and Bouguer) to bathymetry. They filtered out the long-wavelength signal of the North Atlantic geoid anomaly, and considered only short wavelength (degree 10 and higher) contributions to the local geoid of the Azores. They argue the low geoid to bathymetry ratios (admittance $\sim$1.6 m/km at long wavelengths) suggest Airy compensation, rather than a form of dynamic uplift, which would require admittance ratios >6 m/km (Sandwell and Renkin 1988). Luis and Neves (2006) conclude that the Azores plateau is largely isostatically compensated, thickened crust—which is consistent with its volcanic history. They also suggest, based on gravity-bathymetry admittance, that the Moho depth is $\sim$12 km, and the elastic lithospheric thickness is around 3–6 km. However, they also note about a third of the buoyant load is due to bottom loading—which they ascribe to buoyant material at the Moho, perhaps as a result of underplating, though the origin of this bottom-load is not fully understood.

This argument is rather at odds with the conclusions of Monnereau and Cazenave (1990), who group the Azores in a worldwide compilation of hotspots they consider to have dynamic support. They calculate a bathymetric swell height of 1100–1200 m, and a geoid anomaly of 3–4 m. Monnereau and Cazenave (1990) consider the age of the ocean floor onto which their compiled swells were formed, and note that the Azores (along with Cape Verde and Crozet) have larger bathymetric swells, and larger geoid anomalies, than the global trend. This they attribute to relative immobility of the hotspot relative to the overlying plate, which has enabled significant convective thinning in these cases. The thick volcanic plateau also amplifies the bathymetric anomaly of the hotspot swell. Their geoid-to-depth ratio for the Azores, at 2.5 m/km, is not dissimilar to Luis and Neves's (2006) estimate—but Monnereau and Cazenave (1990) demonstrate this lies directly on the global trend for hotspot swells on different age plates. So the Azores geoid anomaly is consistent, if a bit larger, with that of other hotspot geoid anomalies, and the geoid-to-swell ratio ($\sim$2.5 m/km) is also consistent with the global compilation.

While both mantle Bouguer gravity and residual geoid have been used to argue for a deep plume contribution to the Azores (Bowin et al. 1984; Cochran and Talwani 1978; Monnereau and Cazenave 1990), it has been also shown that the magnitude of the contribution can equally be explained by the crustal structure of the Azores plateau (e.g., Gente et al. 2003; Luis and Neves 2006). This does not mean that there is no mantle contribution, but rather that the contribution is both non-unique, and at the limit of signal-to-noise ratio for this area, and therefore not definitive.

4 Seismological Imaging

The main challenge to seismological exploration of the Azores is their ocean island setting. Seismograms from islands are more afflicted by (ocean-generated) noise than comparable recordings on continents, reducing the quantity and quality of usable data. Seismological instrumentation of the seafloor is expensive, requires research ships, and is currently not sustainable for more than a year or two. Only a few of the oceanic hotspots suspected to be fuelled by deep mantle plumes have seen such longer-term deployments of their surrounding ocean-bottom with broadband seismometers, including French Polynesia/Tahiti (Suetsugu et al. 2005), Hawaii (Wolfe et al. 2009), Tristan da Cunha (Geissler 2013), and La Réunion (Barruol and Sigloch 2013)—though not the Azores.

The instrumentation gaps dictated by the distribution of islands in an archipelago puts fundamental limits on the spatial resolution of seismological imaging. In interpreting any imaging study, assessing its resolution limits is as important as evaluating the (most likely) model itself. Hence, we include resolution considerations in the discussion of individual studies but emphasize that brief verbal summaries cannot replace the (graphical) resolution tests of the original studies. In the absence of resolution tests, the robustness and relevance of a tomographic model cannot be evaluated. Hence we do not review global-scale tomographies unless they present resolution tests specifically for the Azores region. The spatial resolution of any imaging technique is a complex interplay between the density and arrangement of seismic sensors, the number and distribution of seismic sources (earthquakes), the seismic wave type, and the recording quality.

In general, body-wave tomography promises the highest 3-D resolution at all depths, but can realize this potential only when instrumentation is relatively dense (e.g., beneath a well-instrumented island, or within an archipelago). The size of the smallest resolved subsurface feature is correlated with the density of seismometers. The depth extent of good resolution is commensurate with the lateral width of the sensor array (e.g., the diameter an evenly covered island). A dense seismometer array may also be used for measurements of point-like character in the tens or hundreds of kilometres directly beneath it (e.g., receiver functions of the Moho and transition zone discontinuities, or SKS splitting). In the absence of dense receiver coverage (e.g., across ocean basins), surface-wave tomography may yield more relevant information than body waves, in the form of laterally averaged properties of the upper few hundred kilometres of the mantle.

The islands of the Azores are instrumented with six permanent broadband stations (the type of instrumentation useful for the crust and mantle studies discussed here). Most tomographic models that we review could rely only on the two oldest of these permanent broadband stations (on Faial and São Miguel), in addition to using dozens or hundreds of broadband stations on the continents rimming the Atlantic. Ocean-bottom seismometers have not been deployed in the ocean regions immediately surrounding the Azores. From October 2000 to September 2002, a dedicated Portuguese/French/American/Swiss experiment named COSEA ("Coordinated Seismic Experiment in the Azores," Silveira et al. 2002) deployed seismometers throughout the Azores archipelago as a whole. COSEA deployed five temporary broadband seismometers in addition to the two then-existing permanent stations, one on each of the major islands. These data turned out to be challenging due to very high ambient noise levels, i.e., microseismic noise generated by frequent high winds and wave action in this part of Atlantic Ocean (Webb 1998; Schutt et al. 2002; Yang et al. 2006). The limited data harvest of COSEA was used in the dedicated tomographic body-wave study by Yang et al. (2006), and to investigate mantle discontinuities (Schutt et al. 2002; Silveira et al. 2010).

Compared to oceanic regions globally, seismic illumination of the mantle under the Central/North Atlantic (which includes the Azores) is generally decent, in terms of distance and azimuthal coverage from reliable earthquake sources. Hence large-scale tomographic mantle models, which rely mostly on teleseismic receivers on nearby continents, are fairly resolving for this oceanic setting. Our discussion will proceed spatially downward, from seismological constraints on the crust, to those on upper mantle structure, the transition zone, and the lower mantle.

4.1 Crustal Structure

Seismological estimates of crustal thickness in the Azores vary between 8 and 20 km, as briefly mentioned in the Introduction and in the context of gravity measurements (Zandomeneghi et al. 2008). Searle (1976) found a crustal thickness of ~8 km from the analysis of surface-wave dispersion. Marillier and Mueller (1982) arrived at a similar estimate from regional body waves. Luis et al. (1998) used elastic plate model estimates to calculate a mean crustal thickness between 9 and 12 km thick. Dias et al. (2007) found a thickness of 14 km for the region of Faial/Pico from local tomography, and Silveira et al. (2010) estimated crustal thicknesses of 20 km from receiver functions from six COSEA island stations.

Hence the estimates of crustal thickness have been revised upward over time, but note that the variable values are not unexpected on account of the different spatial sampling inherent to the different seismological techniques. The thicker estimates (Dias et al. 2007; Silveira et al. 2010) derive from sampling direct beneath islands, whereas the shallower results effectively seem to represent averages over islands and the oceanic plateau crust between them (Senos and Nunes 1993).

4.2 Upper Mantle Structure

The average upper mantle structure beneath the North/Central Atlantic basin was first studied using surface-wave dispersion measurements (Weidner 1974; Searle 1976: Canas and Mitchell 1981; Marillier and Mueller 1982). This early work could distinguish the seismically slow Mid-Atlantic spreading ridge from the remainder of the basin. Searle (1976) showed that the crust of the Azores Plateau is 60% thicker than that of the surrounding ocean, and that upper-mantle seismic velocities are anomalously low. Marillier and Muller (1982) found a clear velocity asymmetry between of the oceanic upper mantle across the Azores-Gibraltar plate boundary: mantle profiles to the north of the ridge are faster than to the south.

With increasing station coverage and computing power, seismic tomography became possible. Honda and Tanimoto (1987) could localize a low-velocity zone beneath the Azores triple junction, using surface waves with a lateral resolution of 1000–2000 km. The S-wave model by Grand (1994) showed a clear low-velocity anomaly in the upper mantle beneath the Azores triple junction. In hindsight, it is worth noting that the Azores anomaly was robustly present in these early models, whereas many other aspects of velocity structure along and off of the MAR have since been revised.

Subsequent regional and global, surface- and body-wave studies reproduced this strong low-velocity anomaly under the Azores in increasing detail (e.g., Silveira et al. 1998; Ritsema et al. 1999, 2004; Pilidou et al. 2004, 2005; Silveira et al. 2002, 2006; Lebedev and van der Hilst 2008; Montelli et al. 2004, 2006; Van der Hilst and de Hoop 2005). Surface-wave studies have suggested that this anomalously slow (and thus presumably hot) mantle volume extends no deeper than 300 km (Silveira et al. 2006) or 400 km (Pilidou et al. 2004, 2005). Note however that surface waves have only very little sensitivity to mantle structure below 300–400 km depth, and may well not detect deeper slow material even if it were present.

In the uppermost mantle, this Azores anomaly is elongated about 2200 km in north-south direction (25°N to 45°N), tracking the Mid Atlantic Ridge (Silveira et al. 2006; Pilidou et al. 2005; Steinmetz et al. 1976, 1977). These most recent surface-wave tomography models of the North Atlantic region estimate lateral image resolution to around 300–500 km. Elongation of the anomaly is asymmetric: relative to the Azores archipelago, slow seismic anomalies extend further to the south than to the north. This has been interpreted as manifestation of hotspot-ridge interaction through asthenospheric flow from the heat source (plume) to the sink (spreading ridge) (Pilidou et al. 2005; Yang et al. 2006). From the relatively northerly location of the Azores, plume material should flow southward to join the Mid Atlantic Ridge (Yang et al. 2006). This scenario would also explain the observed extents and north-south asymmetries of basalt geochemistry (Schilling et al. 1983), ocean spreading (Gente et al. 2003), and geoid anomalies (e.g., Luis and Neves 2006) along the ridge.

Among slow upper-mantle anomalies underlying Atlantic hotspots, the Azores anomaly is second only to Iceland in size and strength (Pilidou et al. 2004; Silveira et al. 2002). It seems to extend less deep, but comparison is difficult because the mantle under Iceland is confidently imaged to greater depths, thanks to the large size and good instrumentation of the island (Bjarnason et al. 1996; Shen et al. 1998; Foulger et al. 2001). For Iceland as well, there are strong indications of hotspot-ridge interaction.

Arguably the most highly resolving study of upper-mantle structure under the Azores archipelago is the P-wave tomography by Yang et al. (2006), the only tomography to use the full set of teleseismic data from the dedicated COSEA experiment (Silveira et al. 2002). Oceanic noise levels in the Azores proved to be high, so that the two-year deployment yielded only a relatively modest number of 41 usable earthquakes at 6 island stations. Lateral imaging resolution is limited by inter-island distances (no seismometers deployed in the ocean), and yet this multiple-island coverage can be expected to provide significantly better image resolution of the upper mantle than the surface-wave tomography models discussed—an expectation borne out by the resolution tests of Yang et al. (2006). (None of the surface wave tomography models include COSEA data.) Yang et al. (2006) analysed the COSEA data using a modern waveform inversion method, finite-frequency tomography, which is effective in avoiding noise in the worst affected frequency bands. Their model (Fig. 5c) shows a strong slow anomaly confined to the uppermost 200 km along the Azores Islands and beneath the centre of the Azores Plateau (38.5°N, 28.5°W). This is underlain by a moderate slow anomaly from 250 km to at least 400 km depth, resembling a plume conduit of 200 km diameter, and dipping toward the northeast. They argue a northeasterly anchoring region for a plume conduit (at 400 km depth, the slow anomaly is located north of Terceira). Such a laterally sheared plume would account for a variety of observations attributed hotspot-ridge interactions: geochemical and thermal asymmetries along strike of the MAR (discussed above), and repeated northward migrations of the triple junction due to ridge attraction by the heat source.

Observations of seismic anisotropy are considered an informative diagnostic of sheared mantle fabric, and thus of present or past flow directions. The reliable detection of anisotropy in data-limited settings is challenging. Kuo and Forsyth (1992) found no SKS splitting at station PDA, and thus diagnosed the absence of systematic anisotropy beneath São Miguel. Using more and higher-quality SKS splitting measurements at station CMLA, Silveira et al. (2010) found a similar lack of anisotropic signature beneath São Miguel. Silveira et al. (2006) argue for significant negative radial anisotropy in the larger vicinity of the Azores hotspot, i.e., S-waves traveling faster in the vertical than in the horizontal direction, which would be consistent with upwelling mantle (vertical flow) in that area. They note that the radial anisotropy is less well resolved and probably associated with an error of 3–4%. Their map shows a radial anisotropy

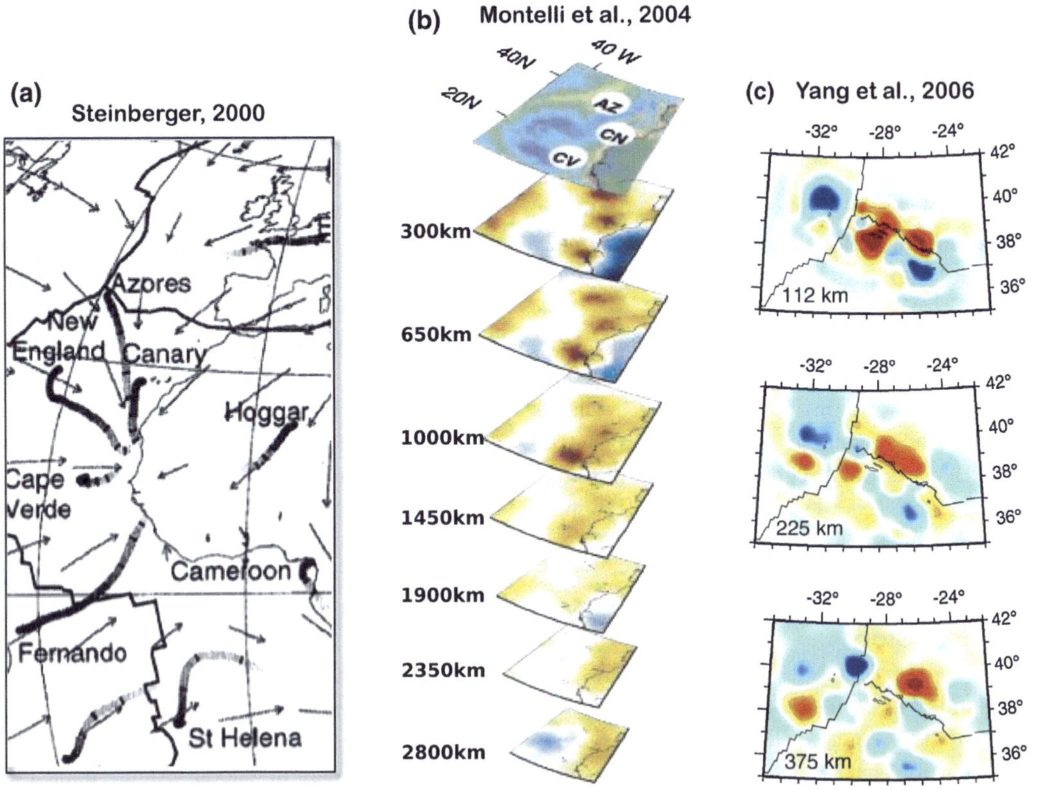

Fig. 5 Comparison of plume conduit tilts. **a** Calculated tilts in a mantle flow field from Steinberger (2000) based on his model 3—which used Grand's et al. (1997) tomographic model. **b** Imaged conduits for the Azores, Canary, and Cape Verde plumes from Montelli et al. (2004). **c** The shallow finite-frequency model of Yang et al. (2006)

anomaly to the west of the Azores, near the mid-Atlantic ridge, rather than coincident with the inferred location of the plume.

4.3 Transition Zone and Lower Mantle

Information about the transition zone (410–670 km) beneath the Azores is limited. Computing receiver functions from the COSEA data, Schutt et al. (2002) report a thinning of the transition zone by 20 km, consistent with hot upwelling in a localised plume conduit. By contrast Li et al. (2003), using the permanent station CMLA, detected no thinning of the transition zone, averaged over a diameter of about 450 km beneath São Miguel. Computing P- and S-receiver functions from CMLA and COSEA stations, Silveira et al. (2010) detect normal transition zone thickness outside the Azores plateau, but they report a lack of wave coverage beneath the plateau itself. They also infer severely slow S-velocities at 460–500 km depth immediately beneath the plateau but not outside of it, and interpret this as evidence for high water content in the transition zone, possibly representing the source region for the hotspot's wet lavas.

The global, finite-frequency P and S-wave tomographies of Montelli et al. (2004, 2006) report resolving a plume beneath the Azores from the upper mantle down to the core-mantle boundary (Fig. 5b). The radius of its conduit in the lower mantle must be at least 300 km according to P-wave tomography, and at least 100 km according to S-wave tomography. The plume conduit seems to merge with those of the

Canary and Cape Verde hotspots at depths greater than ~1450 km, suggesting a common origin for this group. While Montelli et al. (2004) caution that the apparent merger with the Canary plume may not be well resolved, overall they consider the trio Azores/Canary/Cape Verde among their best and most continuously resolved plume conduits globally. The merged anomaly continues into the lowermost mantle, connecting to the eastern edge of the African Large Low Shear Velocity Province (Fig. 5). Structure above 300 km should not be interpreted, since near-vertical wave incidence at isolated islands stations tends to cause vertical smearing artefacts. Hence a direct comparison to the finding by Yang et al. (2006) of a northeastward dipping conduit in the upper 400 km is not possible.

5 Discussion

Resolution of crustal and mantle structure beneath the Azores is limited by the relative signal, in for instance geoid studies, being near the noise range (here due to unknown crustal variations) for the region. In the case of the geoid, there is a clear ~6 m residual geoid anomaly in the vicinity of the Azores swell (e.g., Bowin et al. 1984), in line with that expected for a plume. However, this magnitude of geoid anomaly could equally be explained by isostatic support of thickened crust (Luis and Neves 2006). Estimates for crustal thickness of the Azores plateau vary significantly, from as little as ~8 km from Rayleigh waves (Searle 1976), up to ~14 km for recent tomography models (Dias et al. 2007). Knowledge of the exact structure of the crust is critical for delineating the geoid effects of the crust, as opposed to the deeper upwelling mantle. Luis and Neves (2006) argue that the local geoid gradients suggest the plateau is primarily isostatically compensated. Given the crustal thickness, this is not surprising, but is a poor constraint on the geoid contribution of a weak plume in an area of unknown crustal thickness variations. In fact, the need for bottom loading in their model (similar to the Bouguer gravity analysis of Gente et al. 2003) is suggestive of basal support—either by some sort of underplating, as suggested by Luis and Neves (2006), or perhaps dynamically by an upwelling conduit.

The Azores have been branded as an example of hotspot-ridge interaction, but the postulated flow processes are difficult to resolve seismically. The flow might generate a strong signature, but if the hotspot sits away from the ridge, then instrumentation of the ocean-bottom atop the flow channel would be needed. If the hotspot sits atop the ridge, hotspot and ridge processes are mixed and hard to delineate, and the advantage conferred by relative ease of instrumentation of the Azores islands is lost.

Recent regional surface-wave studies agree with regard to the depth extent of the slow upper-mantle anomaly: mostly to around 300 km, and no deeper than 400 km, according to Silveira and Stutzmann (2002), Silveira et al. (2006), Pilidou et al. (2004, 2005). The body-wave tomography by Yang et al. (2006), which should be the most highly resolving study both in depth and in lateral extent, finds that anomalously slow material continues to at least 400 km depth (the lower limit of their resolved domain), but they also report that this columnar anomaly gets narrower and weaker in magnitude below 200 km. At an estimated radius of 100 km [Yang et al. (2006), consistent with Montelli et al. (2006)], this conduit is probably simply not resolved by surface-wave tomography, the lateral resolution of which is estimated around 300–500 km (Pilidou et al. 2004; Silveira et al. 2006).

Montelli et al. (2004) suggests a plume conduit down to the lowermost mantle (and hence through the transition zone). Their global body-wave data set, while informative for the lower mantle, has poor vertical resolution in the upper 400 km, so that a vertical smearing artefact at those depths are likely, and hence not directly comparable to the upper-mantle geometry of Yang et al. (2006). While numerous other global-scale tomography studies have used data and methods similar to Montelli et al. (2004, 2006), they do not appear to discuss mantle structure and imaging resolution beneath the Azores in particular.

The Montelli et al. (2004) finite-frequency tomography model illustrates an interesting aspect of plume dynamics, in that the conduits for the Azores, Canary and Cape Verde plumes seem to converge at depth, at around ~1450 km, and originate from a common source at D", beneath western Africa. Conduit breakup, and splitting into multiple conduits, particularly under shear, is rather common in fluid dynamics experiments (e.g., Whitehead 1982), and raises the possibility of the Azores/Canary/Cape Verde representing a single system, i.e., a plume rising from the edge of the African superwell, which splits into three branching conduits between 1500 and 1900 km depth. This also raises another interesting question: is the observed tilt of the conduits consistent with the recent mantle flow field in the region? Yang et al. (2006) also constrain a tilted plume conduit within the upper 400 km of the mantle—however, their tilt is towards the northeast—compared to southeast tilting lower-mantle conduit of Montelli et al. (2004). Yang et al. (2006) suggest that a northeast tilt may be consistent with a southwesterly flow of the asthenosphere. Light can be shed on this discrepancy by dynamic modelling of plume conduits. Steinberger (2000) calculated plume conduit tilts for a global selection of hotspots, based on a mantle flow field derived from past plate motions and present-day tomography. The results are sensitive to the tomography model assumed, viscosity structure of the mantle, and conduit viscosity. Interestingly, for Steinberger's (2000) mantle model 3—the one based on the global tomography model of Grand (1994), Grand et al. (1997)—the calculated plume structure of the Azores, Canary, and Cape Verde all converge at depth, in the vicinity of the coast of west Africa. The reason is illustrated in the calculated present-day mantle flow, which within D" is converging on the point of plume origin—a result of global three-dimensional flow due to plate motion history and mantle density anomalies. The conduits need not have formed together in this model, they may have formed individually, and geographically separated, and been swept together by flow at the base of the mantle. Nevertheless, the tomography structures of Montelli et al. (2004), and the flow calculations of Steinberger (2000), are two completely independent approaches to deriving plume conduit structure. The similarity of the inferred Azores plume structure in both cases is a strong argument for its deep origin. Many of the plume conduit tilts calculated by Steinberger (2000) exhibit changes in tilt between the upper mantle and lower mantle as a result of a change in the dominant flow regime (plate return for the upper mantle, broad return flow/upwelling for the lower mantle). Neither of the cases modelled by Steinberger (2000) show a NE tilt in the upper mantle however, as suggested by Yang et al. (2006), and this tilt is also hard to reconcile with Montelli's observed conduit tilt. However, Steinberger's (2000) flow calculations show slow transition zone velocities in the vicinity of the Azores, with velocities either weakly to the south—southwest, or northwest—west, depending on the model. This is reasonably consistent with Yang et al.'s (2006) inferred asthenospheric flow. A number of slow anomalies extend through the upper mantle in Yang et al.'s (2006) model, and their geometry is complex, so it is possible the interconnecting conduit has subtle signature.

How does this bear on the question of a deep mantle plume? It is very likely that a plume in the transition zone and lower mantle would have a weaker seismic expression than in the upper mantle. Global tomography models consistently show much stronger (slow and fast) anomalies in the uppermost mantle than at deeper depths, probably mostly due to variable depth dependence of seismic velocities on temperature anomalies, and the potentially strong effects of partial melting in the upper 150 km. In addition, vertically rising plume material needs to divert laterally as it approaches the surface, and would be likely to pond in the vicinity of the plume conduit. Such a widened puddle is much more readily detectable by surface waves than the narrow conduit at deeper depths. Thus surface wave studies, which gradually lose sensitivity beneath 200–300 km depth, are probably detecting only the ponding material, but cannot resolve the weaker and thinner conduit at depth.

This is illustrated by the resolution tests of Pilidou et al. (2004).

6 Conclusions

Surface- and body-wave studies agree that the uppermost 200–300 km of mantle beneath the Azores plateau are anomalously slow, as is a >2000 km long segment beneath the adjacent mid-Atlantic ridge (Pilidou et al. 2004; Silveira et al. 2006). The regional body-wave tomography by Yang et al. (2006) resolves a conduit-shaped downward continuation to at least 400 km depth under the central to northwestern part of the archipelago. P- and S-receiver functions indicate anomalously slow S-wave velocities at 460–500 km depth beneath the Azores plateau, but not outside of it (Silveira et al. 2010). The global P- and S-wave tomographies by Montelli et al. (2004, 2006) indicate a conduit down to the lowermost mantle. Pieced together, these observations point towards a classical, deep mantle plume as hypothesised by Morgan (1971).

However, seismic illumination especially of the transition zone and lower mantle remains incomplete and sparse, so that no consensus has been reached about the true depth extent and feeding mechanism of the Azores hotspot, and of comparable oceanic hotspots globally. Extensive seismological instrumentation of the oceans will be needed in order to make progress. The Azores would be interesting targets because their lower-mantle anomaly has emerged as one of the most robust among oceanic hotspots worldwide, even without dedicated instrumentation (Montelli et al. 2004, 2006). However, the COSEA experiment has shown that ambient noise is high in this region of the Atlantic, and that therefore the Azores will not be easy targets for future island/ocean-bottom deployments.

To date, seismic tomography has been used to argue for (e.g., Montelli et al. 2004, 2006) or against (Silveira et al. 2006) a deep plume, and due to uncertainties in crustal structure, the geoid response and bathymetric swell are inconclusive. However, a number of lines of evidence are suggestive. These include: (1) U-series disequilibria, indicating upwelling velocities of 3–4 cm/yr beneath the Azores (Bourdon et al. 2005); (2) primitive $^4He/^3He$ ratios at Terceira and Pico; (3) Osmium isotope evidence suggesting a deep recycled component (Schaefer et al. 2002); (4) radial anisotropy in the vicinity of the Azores region (Silveira et al. 2006), consistent with local upwelling; (5) the consistency between deep plume structure and mantle flow modelling results (Montelli et al. 2004, 2006; Steinberger 2000). Taken together, a solid argument can be made for a deep mantle origin for the Azores plume, but a number of question remain, including the low excess temperatures of the hotspot, its high volatile content, and the potential involvement of the transition zone, e.g., as a volatile repository.

References

Asimow, P. D., & Langmuir, C. H. (2003). The importance of water to oceanic mantle melting regimes. *Nature, 421*(6925), 815–820.

Barruol, G., & Sigloch, K. (2013). Investigating La Réunion Hot Spot From Crust to Core. *Eos, Transactions American Geophysical Union, 94*(23), 205–207.

Beier, C., Haase, K. M., & Turner, S. P. (2012). Conditions of melting beneath the Azores. *Lithos, 144*, 1–11.

Bjarnason, I. T., Wolfe, C. J., Solomon, S. C., & Gudmundson, G. (1996). Initial results from the ICEMELT Experiment: Body-wave delay times and shear-wave splitting across Iceland. *Geophysical Research Letters, 23*(5), 459–462.

Bonatti, E. (1990). Not so hot "hot spots" in the oceanic mantle. *Science, 250*(4977), 107–111.

Bourdon, B., Langmuir, C. H., & Zindler, A. (1996). Ridge-hotspot interaction along the Mid-Atlantic Ridge between 37° 30′ and 40° 30′ N: The U Th disequilibrium evidence. *Earth and Planetary Science Letters, 142*(1), 175–189.

Bourdon, B., Turner, S. P., & Ribe, N. M. (2005). Partial melting and upwelling rates beneath the Azores from a U-series isotope perspective. *Earth and Planetary Science Letters, 239*(1), 42–56.

Bowin, C., Thompson, G., & Schilling, J. G. (1984). Residual geoid anomalies in Atlantic Ocean Basin: Relationship to mantle plumes. *Journal of Geophysical Research, 89*(B12), 9905–9918.

Canas, J. A., & Mitchell, B. J. (1981). Rayleigh wave attenuation and its variation across the Atlantic Ocean. *Geophysical Journal International, 67*, 159–176.

Cande, S. C., & Kent, D. V. (1992). A new geomagnetic polarity time scale for the Late Cretaceous and Cenozoic. *Journal of Geophysical Research: Solid Earth, 97*(B10), 13917–13951.

Cannat, M., Briais, A., Deplus, C., Escartín, J., Georgen, J., Lin, J., da Silva, P. (1999). Mid-Atlantic Ridge–Azores hotspot interactions: Along-axis migration of a hotspot-derived event of enhanced magmatism 10 to 4 Ma ago. *Earth and Planetary Science Letters, 173*(3), 257–269.

Cochran, J. R., & Talwani, M. (1978). Gravity anomalies, regional elevation, and the deep structure of the North Atlantic. *Journal of Geophysical Research, 83*(B10), 4907–4924.

De Mets, C., Gordon, R. G., Argus, D. F., & Stein, S. (1990). Current plate motions. *Geophysical Journal International, 101*(2), 425–478.

Detrick, R. S., Needham, H. D., & Renard, V. (1995). Gravity anomalies and crustal thickness variations along the Mid-Atlantic Ridge between 33 N and 40 N. *Journal of Geophysical Research, 100*(B3), 3767–3787.

Dias, N. A., Matias, L., Lourenço, N., Madeira, J., Carrilho, F., & Gaspar, J. L. (2007). Crustal seismic velocity structure near Faial and Pico Islands (AZORES), from local earthquake tomography. *Tectonophysics, 445*(3), 301–317.

Dosso, L., Bougault, H., Langmuir, C., Bollinger, C., Bonnier, O., & Etoubleau, J. (1999). The age and distribution of mantle heterogeneity along the Mid-Atlantic Ridge (31–41N). *Earth and Planetary Science Letters, 170*(3), 269–286.

Foulger, G. R. (2007). The "plate" model for the genesis of melting anomalies. In G. R. Foulger & D. M. Jurdy (Eds.), *Plates, plumes, and planetary processes: Geological Society of America. Special Paper 430* (pp. 1–28).

Foulger, G. R., Pritchard, M. J., Julian, B. R., Evans, J. R., Allen, R. M., Nolet, G., et al. (2001). Seismic tomography shows that upwelling beneath Iceland is confined to the upper mantle. *Geophysical Journal International, 146*(2), 504–530.

Geissler, W. (2013). Electromagnetic, gravimetric and seismic measurements to investigate the Tristan da Cunha hot spot and its role in the opening of the South Atlantic Ocean (MARKE)—Cruise No. MSM24—December 27, 2012–January 21, 2013—Walvis Bay (Namibia)—Cape Town (South Africa). In MARIA S. MERIAN-Berichte (cruise reports); MSM24, pp. 1–56. https://doi.org/10.2312/cr_msm24, https://getinfo.de/app/details?id=awi:doi ~ 10.2312%252Fcr_msm24.

Gente, P., Dyment, J., Maia, M., & Goslin, J. (2003). Interaction between the Mid-Atlantic Ridge and the Azores hot spot during the last 85 Myr: Emplacement and rifting of the hot spot-derived plateaus. *Geochemistry, Geophysics, Geosystems, 4*(10), 8514. https://doi.org/10.1029/2003GC000527.

Goslin, J., & Party, T. S. (1999). Extent of Azores plume influence on the Mid-Atlantic Ridge north of the hotspot. *Geology, 27,* 991–994.

Goslin, J., Thirot, J.-L., Noel, O., & Francheteau, J. (1998). Slow-ridge/hotspot interactions from global gravity, seismic tomography and 87Sr/86Sr isotope data. *Geophysical Journal International, 135,* 700–710.

Grand, S. P. (1994). Mantle shear structure beneath the Americas and surrounding oceans. *Journal of Geophysical Research: Solid Earth, 99*(B6), 11591–11621.

Grand, S. P., van der Hilst, R. D., & Widiyantoro, S. (1997). High resolution global tomography: A snapshot of convection in the Earth. *Geological Society of America Today, 7*(4).

Gripp, A. E., & Gordon, R. G. (2002). Young tracks of hotspots and current plate velocities. *Geophysical Journal International, 150,* 321–361.

Honda, S., & Tanimoto, T. (1987). Regional 3D heterogeneities by waveform inversion-application to the Atlantic area. *Geophysical Journal International, 94,* 737–757.

King, S. D. (2005). North Atlantic topographic and geoid anomalies: The result of a narrow ocean basin and cratonic roots? *Geological Society of America Special Papers, 388,* 653–664.

Kuo, B. Y., & Forsyth, D. W. (1992). A search for split SKS waveforms in the North Atlantic. *Geophysical Journal International, 108*(2), 557–574.

Lebedev, S., & van der Hilst, R. D. (2008). Global upper-mantle tomography with the automated multimode inversion of surface and S-wave forms. *Geophysical Journal International, 173,* 505–518.

Lemoine, F. G., Kenyon, S. C., Factor, J. K., Trimmer, R. G., Pavlis, N. K., Chinn, D. S., et al. (1998). The development of the joint NASA GSFC and the National Imagery and Mapping Agency (NIMA) geopotential model EGM96. Report no. NASA/TP-1998-206861. National Aeronautics and Space Administration.

Lerch, F. O., Nerem, R. S., Putney, B. H., Felsentreger, T. L., Sanchez, B. V., Marshall, J. A., ... & Pavlis, E. C. (1994). A geopotential model from satellite tracking, altimeter, and surface gravity data: GEM-T3. *Journal of Geophysical Research, 99,* 2815–2815.

Li, X., Kind, R., & Yuan, X. (2003). Seismic study of upper mantle and transition zone beneath hotspots. *Physics of the Earth and Planetary Interiors, 136,* 79–92.

Lourenco, N. L. J. F., Miranda, J. M., Luis, J. F., Ribeiro, A., Victor, L. M., Madeira, J., & Needham, H. D. (1998). Morpho-tectonic analysis of the Azores Volcanic Plateau from a new bathymetric compilation of the area. *Marine Geophysical Researches, 20*(3), 141–156.

Luis, J. F., Miranda, J. M., Galdeano, A., & Patriat, P. (1998). Constraints on the structure of the Azores spreading center from gravity data. *Marine Geophysical Researches, 20*(3), 157–170.

Luis, J. F., & Neves, M. C. (2006). The isostatic compensation of the Azores Plateau: A 3D admittance and coherence analysis. *Journal of Volcanology and Geothermal Research, 156*(1), 10–22.

Madeira, J., & Ribeiro, A. (1990). Geodynamic models for the Azores triple junction: A contribution from tectonics. *Tectonophysics, 184*(3), 405–415.

Madureira, P., Moreira, M., Mata, J., & Allègre, C. J. (2005). Primitive neon isotopes in Terceira Island (Azores archipelago). *Earth and Planetary Science Letters, 233,* 429–440.

Marillier, F., & Mueller, S. (1982). Structure of the upper mantle in the Northeastern Atlantic close to the azores-gibraltar ridge from surface-wave and body-wave observations. *Tectonophysics, 90,* 195–213.

Minster, J. B., & Jordan, T. H. (1978). Present-day plate motions. *Journal of Geophysical Research: Solid Earth, 83*(B11), 5331–5354.

Miranda, J. M., Victor, L. M., Simoes, J. Z., Luis, J. F., Matias, L., Shimamura, H., ... & Lepine, J. C. (1998). Tectonic setting of the Azores Plateau deduced from an OBS survey. *Marine Geophysical Researches, 20* (3), 171–182.

Monnereau, M., & Cazenave, A. (1990). Depth and geoid anomalies over oceanic hotspot swells: A global survey. *Journal of Geophysical Research: Solid Earth* (1978–2012), *95*(B10), 15429–15438.

Montelli, R., Nolet, G., Dahlen, F. A., & Masters, G. (2006). A catalogue of deep mantle plumes: New results from finite-frequency tomography. *Geochemistry, Geophysics, Geosystems, 7,* Q11007. https://doi.org/10.1029/2006GC001248.

Montelli, R., Nolet, G., Dahlen, F. A., Masters, G., Engdahl, E. R., & Hung, S.-H. (2004). Finite-frequency tomography reveals a variety of plumes in the mantle. *Science, 303,* 338–343.

Moreira, M., Doucelance, R., Kurz, M. D., Dupré, B., & Allègre, C. J. (1999). Helium and lead isotope geochemistry of the Azores Archipelago. *Earth and Planetary Science Letters, 169*(1), 189–205.

Morgan, W. J. (1971). Convection plumes in the lower mantle. *Nature, 230,* 42–43.

Pilidou, S., Priestley, K., Gudmundsson, Ó., & Debayle, E. (2004). Upper mantle S-wave speed heterogeneity and anisotropy beneath the North Atlantic from regional surface wave tomography: The Iceland and Azores plumes. *Geophysical Journal International, 159,* 1057–1076.

Pilidou, S., Priestley, K., Debayle, E., Gudmundsson, O. (2005). Rayleigh wave tomography in the North Atlantic: High resolution images of the Iceland, Azores and Eifel mantle plumes. *Lithos, 79,* 453–474.

Ritsema, J., van Heijst, H. J., & Woodhouse, J. H. (1999). Complex shear wave velocity structure imaged beneath Africa and Iceland. *Science, 286,* 1925–1928.

Ritsema, J., van Heijst, H. J., & Woodhouse, J. H. (2004). Global transition zone tomography. *Journal Geophysical Research, 109,* 14p.

Schaefer, B. F., Turner, S., Parkinson, I., Rogers, N., & Hawkesworth, C. (2002). Evidence for recycled Archaean oceanic mantle lithosphere in the Azores plume. *Nature, 420*(6913), 304–307.

Schilling, J.-G. (1991). Fluxes and excess temperature of mantle plumes inferred from their interaction with migrating mid-ocean ridges. *Nature, 352,* 397–403.

Schilling, J.-G., Zajac, M., Evans, R., Johnston, T., White, W., Devine, J. D., et al. (1983). Petrologic and geochemical variations along the Mid-Atlantic Ridge from 29°N to 73°. *American Journal of Science, 283,* 510–586.

Schutt, D. L., Solomon, S. C. Burkett, P. G. (2002). Evidence for a Thinned Mantle Transition Zone beneath the Azores Hotspot. *American Geophysical Union, Fall Meeting Abstract* #S11A-1125

Searle, R. C. (1976). Lithospheric structure of the Azores Plateau from Rayleigh- wave dispersion. *Geophysical Journal International, 44,* 537–546.

Senos, M. L., & Nunes, J. C. (1993). *Constituição da litosfera Atlântica na região das ilhas dos Açores.* I.N. M.G, Lisboa.

Senos, M. L., Nunes, J. C., & Moreira, V. S. (1980). *Estudos da estrutura da crosta e manto superior nos Açores.* I.N.M.G, Lisboa.

Shen, Y., Solomon, S. C., Bjarnason, I. T., & Wolfe, C. J. (1998). Seismic evidence for a lower-mantle origin of the Iceland plume. *Nature, 395*(6697), 62–65.

Silveira, G., & Stutzmann, E. (2002). Anisotropic tomography of the Atlantic Ocean. *Physics of the Earth and Planetary Interiors, 132,* 237–248.

Silveira, G., Stutzmann, E., Davaille, A., Montagner, J. P., Mendes-Victor, L., & Sebai, A. (2006). Azores hotspot signature in the upper mantle. *Journal of Volcanology and Geothermal Research, 156*(1), 23–34.

Silveira, G., Stutzmann, E., Griot, D., Montagner, J. P., & Mendes, V. L. (1998). Anisotropic tomography of the Atlantic Ocean from Rayleigh surface waves. *Physics of the Earth and Planetary Interiors, 106,* 257–273.

Silveira, G., van der Lee, S., Stutzmann, E., Matias, L., James, D., Burkett, P., et al. (2002). Coordinated seismic experiment in the Azores. *ORFEUS Newsletter, 4*(2), 10.

Silveira, G., Vinnik, L., Stutzmann, E., Farra, V., Kiselev, S., & Morais, I. (2010). Stratification of the Earth beneath the Azores from P and S receiver functions. *Earth and Planetary Science Letters, 299,* 91–103.

Sleep, N. H. (1990). Hotspots and mantle plumes: Some henomenology. *Journal of Geophysical Research: Solid Earth* (1978–2012), *95*(B5), 6715–6736.

Steinberger, B. (2000). Plumes in a convecting mantle: Models and observations for individual hotspots. *Journal of Geophysical Research: Solid Earth, 105* (B5), 11127–11152.

Steinmetz, L., Hirn, A., Sapin, M., Whitmarsh, R., & Moreira, V. (1976). Zone of P wave attenuation beneath the crest of Mid-Atlantic ridge. *Bulletin de la Société Géologique de France, 4*(7), 931–936.

Steinmetz, L., Whitmarsh, R., & Moreira, V. (1977). Upper mantle structure beneath mid- Atlantic ridge north of Azores based on observations of compressional waves. *Geophysical Journal International, 50* (2), 353–380.

Suetsugu, D., Shiobara, H., Sugioka, H., Barruol, G., & Reymond, D. (2005). Probing South Pacific mantle plumes with ocean bottom seismographs. *Eos, 86*(44).

Thibaud, R., Gente, P. & Maia, M. (1997). A systematic analysis of the Mid-Atlantic Ridge morphology and gravity between 15°N and 40°N, *Journal of Geophysical Research, 103*(B10), 24 223–24 243.

Van der Hilst, R. D., & de Hoop, M. V. (2005). Banana-doughnut kernels and mantle tomography. *Geophysical Journal International, 163,* 956–961.

Webb, S. C. (1998). Broadband seismology and noise under the ocean. *Reviews of Geophysics, 36,* 105–142.

Weidner, D. J. (1974). Rayleigh wave phase velocities in the Atlantic Ocean. *Geophysical Journal International, 36,* 105–139.

Whitehead, J. A. (1982). Instabilities of fluid conduits in a flowing earth—are plates lubricated by the asthenosphere? *Geophysical Journal International, 70*(2), 415–433.

Wolfe, C. J., Solomon, S. C., Laske, G., Collins, J. A., Detrick, R. S., Orcutt, J. A., et al. (2009). Mantle shear-wave velocity structure beneath the Hawaiian hot spot. *Science, 326*(5958), 1388–1390.

Yang, T., Shen, Y., van der Lee, S., Solomon, S. C., & Hung, S.-H. (2006). Upper mantle structure beneath the Azores hotspot from finite frequency seismic tomography. *Earth and Planetary Science Letters, 250,* 11–26.

Yu, D., Fontignie, D., & Schilling, J.-G. (1997). Mantle plume—ridge interactions in the Central North Atlantic: a Nd isotope study of Mid-Atlantic Ridge basalts from 30°N to 50°. *Earth and Planetary Science Letters, 146,* 259–272.

Zandomeneghi, D., Almendros, J., Ibáñez, J. M., & Saccorotti, G. (2008). Seismic tomography of Central São Miguel, Azores. *Physics of the Earth and Planetary Interiors, 167,* 8–18.

The Tectonic Evolution of the Azores Based on Magnetic Data

J. Miguel Miranda, J. Freire Luis and Nuno Lourenço

Abstract

The Azores attracted the interest of geoscientists since the beginning of the XX century. In the late 60s, when plate tectonics was established as the basic geodynamic paradigm, the peculiar morphology of the Azores Islands and the surrounding plateau, located close to the Mid-Atlantic ridge, were early interpreted as the result of the separation between the Eurasian and the North-America plates. Nevertheless, a number of particular geological features were targeted for explanation: (i) the long active fault going from Gibraltar to the Azores (now called Gloria Fault), (ii) the existence of a large but inactive fracture on the North-American plate, offset tenths of kilometres to the north with respect to Gloria Fault, (iii) the curvilinear succession of islands marked by pervasive volcanic and seismic activity, (iv) the development of a plateau, partially split by the Mid-Atlantic Ridge. These questions remained elusive for a long time, despite the large amount of geological and geophysical data available, as most of the conventional approaches were not as fruitful as expected, and new identified features raise new unknowns or revealed uncommon geological environments. Here, we present a review of the progress made in the understanding of the tectonic evolution of the Azores, mainly based on the interpretation of magnetic and morphological data and we present an updated interpretative scheme for the genesis and evolution of the Azores triple junction.

J. M. Miranda (✉) · N. Lourenço
Instituto Português do Mar e da Atmosfera, IDL, Rua C Aeroporto de Lisboa, 1749-077 Lisbon, Portugal
e-mail: miguel.miranda@ipma.pt

J. F. Luis · N. Lourenço
University of Algarve, IDL, Campus de Gambelas, 8000 Faro, Portugal

1 Triggering of the Azores Triple Junction

The early identification of the Azores morphological plateau can be traced up to 1855 (Vogt and Jung, Chapter "The "Azores Geosyndrome" and Plate Tectonics: Research History, Synthesis, and Unsolved Puzzles"). Early works by Azores researchers include the pioneering studies made by Agostinho (1927) and Machado (1959).

The use of magnetic data to derive the first kinematic interpretation of the Azores was attempted by Krause and Watkins (1970), based on the pioneering works of Heirtzler et al. (1968) on marine magnetic anomalies, and the first sea-floor spreading models (Le Pichon 1968). They were particularly puzzled by a number of "unique" features: (i) the seismically active East Azores Fracture Zone (EAFZ) extending from Gibraltar to the Mid-Atlantic Ridge; (ii) the

U. Kueppers and C. Beier (eds.), *Volcanoes of the Azores*, Active Volcanoes of the World, https://doi.org/10.1007/978-3-642-32226-6_6

seismically inactive West Azores Fracture Zone (WAFZ, often misleadingly called "Pico Fracture Zone") offset northwards from the trend of EAFZ; (iii) the transverse island chain of the Azores, almost linear, but oblique to the EAFZ; (iv) the change in direction of the Mid-Atlantic Ridge (MAR) from northeast-southwest to north-south across the Azores; and (v) the broadening of the Mid-Atlantic Ridge (Krause and Watkins 1970; Vogt and Jung, Chapter "The "Azores Geosyndrome" and Plate Tectonics: Research History, Synthesis, and Unsolved Puzzles").

Some of these questions were early explained by Krause and Watkins (1970) as consequences of differences between plate spreading directions and velocities, north and south of the East Azores Fracture Zone after the establishment of a new plate boundary new plate boundary. Using simple geometric considerations, Krause and Watkins (1970) concluded that the onset of velocity contrast is older than 45 Ma. This would have triggered the development of a secondary spreading axis, within a "leaky transform" environment, and creating a triple junction between the Eurasian, African and North America plates. The change in spreading directions was also interpreted as the responsible for the latitudinal offset between the EAFZ and the WAFZ. Krause and Watkins (1970) considered EAFZ as an active feature, in contrast to WAFZ, considered a fossil structure within the single North-American plate. Using 13.3 mm/year as the half-spreading rate between the Eurasian and North American plates immediately north of the Azores, and 11.0 mm/year for the African and North American plate pair, south of the Azores, they predicted a 2.5 mm/year of half-spreading rate for Terceira Rift and partially checked this computation with the direct modelling of three magnetic profiles across almost perpendicular to the new rift. Additional geophysical surveys (Schilling and Krause 1970; Krause and Schilling 1970) were made in the following years to test their interpretation, but new interpretations were never presented afterwards.

The rough estimation made by Krause and Watkins (1970) for the triggering of the Azores triple junction was refined by Srivastava et al. (1990) based on the analysis of a large compilation of magnetic data for the North Atlantic and on a hypothesis put forward by Schouten et al. (1984) that the Iberian plate has either been attached to Africa or to Eurasia at various times. They show that (i) Iberia was attached to Africa from the late Cretaceous to middle Eocene; (ii) between the middle Eocene to late Oligocene it behaved as an independent plate with slight motion relative to Africa, while most of the deformation was concentrated along its northern border, which corresponds now to the King's Trough and Azores-Biscay Rise; (iii) Iberia became part of the Eurasia plate since the late Oligocene (Srivastava et al. 1990), ultimately leading to the development of a plate boundary south of the Iberian plate along the Azores-Gibraltar Fracture Zone (see Fig. 1 for locations).

The structure and geometry of this plate boundary was studied in the early seventies by Laughton and Whitmarsh (1974) and Laughton et al. (1975). They made bathymetric and side-scan sonar surveys between Gibraltar and the Mid-Atlantic Ridge, mapping a major scarp now called Gloria Fault, where several large magnitude strike-slip earthquakes were known to occur. The Gloria Fault was interpreted as a segment of the plate boundary separating Eurasia from Núbia and it was confirmed that close to the Azores, west of 25°W, there is no continuity between Gloria Fault and a marked topographic linear depression that corresponds to the now called East Azores Fracture Zone (EAFZ), earlier interpreted also as part of the plate boundary (see Krause and Watkins 1970). East of the Gloria fault, the geometry of the plate boundary up to the Mediterranean Sea is still controversial.

Luis and Miranda (2008) improved the magnetic compilation of the North Atlantic and re-interpreted the magnetic chrons of the Eurasia-North America plate pair. They confirmed most of the interpretation made by previous studies (e.g. Srivastava et al. 1990) and concluded that the establishment of the Azores triple junction occurred between chrons C6c (ca. 24 Ma) and C11-C12 (ca. 30 Ma) following the

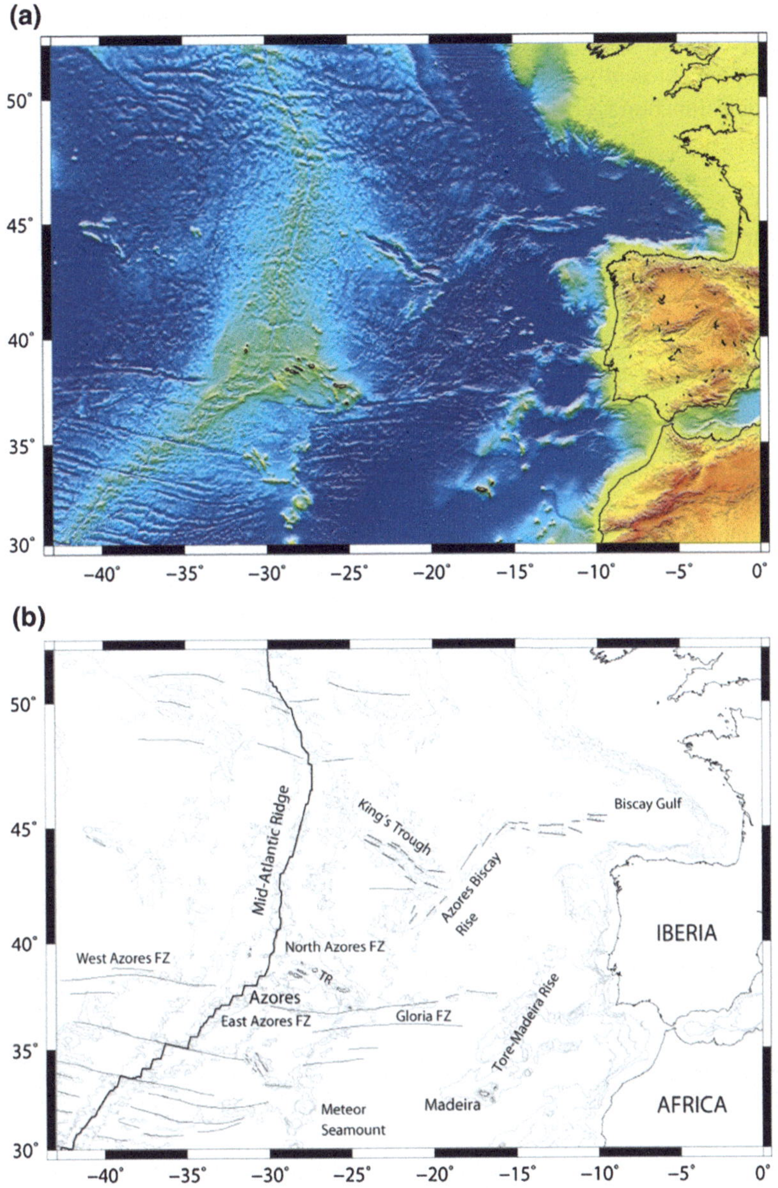

Fig. 1 General framework of the Azores triple junction. Main morphological structures are sketched below: FZ: Fracture Zone. The plate boundary between Eurasia and Nubia is supposed to presently follow Terceira Rift (TR in the figure), Gloria Fault up to Tore-Madeira Rise. Close to Iberia the plate boundary becomes more complex and its geometry within the Southwest Iberian transpressive area is controversial

welding of Iberia to Eurasia, at ~27 Ma ago. Main stages can be simply summarised as:

(i) The initial opening of the North Atlantic at the latitude of Iberia took place after chron M0 (~120 Ma); Spreading continued during Cretaceous times, leading to the opening of Biscay Gulf and the development of a plate boundary along the Azores-Biscay Rise and King's Trough;

(ii) At ~27 Ma Iberia became welded to Eurasia and its northern plate boundary became inactive. Extension developed south of the Iberian plate leading to the onset of a new triple junction, presently known as the Azores Triple Junction (ATJ).

2 Northward Migration of the Azores Triple Junction

Searle (1980) improved the bathymetric and the magnetic compilations and acquired a large set of side-scan sonar data between the Azores islands and the Mid-Atlantic Ridge (MAR) to understand the geometry of the triple junction. He confirmed several of the Krause and Watkins (1970) interpretations, and proposed a final configuration change corresponding to a northward jump of the Eurasia-Africa plate boundary. This jump took place from the latitude of the East Azores Fracture Zones, to the vicinity of the North Azores Fracture Zone, which would connect the western tip of Terceira Rift to the MAR (Searle 1980) and would correspond to its present-day location. This interpretation was challenged by Luis et al. (1994) based on a detailed Azores Aeromagnetic Survey covering mostly the area close to the MAR, up to chron C5 ($\sim$10 Ma) on both plates. They concluded that the Eurasia-Nubia-North America triple junction was located north of the East Azores Fracture Zone between chron C4 and chron C3a, approximately at 38°20′N, 30°15′W (Eurasia fixed co-ordinates) and proposed a present-day location close to 38°55′N, 30°00′W (Eurasia fixed co-ordinates), after chron C2a, approximately 2.45 Ma ago, and not to the North-Azores Fracture Zone as predicted by Searle (1980).

The Azores plateau is interpreted by Schilling (1975) as a hotspot derived feature (see also O'Neill and Sigloch, Chapter "Crust and Mantle Structure Beneath the Azores Hotspot—Evidence from Geophysics"). Geochemical studies (e.g. Dosso et al. 1999) detect a long wavelength geochemical anomaly from 33°N to 41°N and seismic tomography (Yang et al. 2006) confirms the existence of an anomaly on body wave seismic velocity matching the plateau (O'Neill and Sigloch, Chapter "Crust and Mantle Structure Beneath the Azores Hotspot—Evidence from Geophysics"). Gente et al. (2003) studied the role of mantle processes in the time evolution of the Azores triple junction. They hypothesize that a plume was formed at $\sim$85 Ma ago, SW of the Azores and that it is responsible for the early formation of the Azores plateau between 20 and 7 Ma. Gente et al. (2003) invoke a "magmatic phase" followed by a "tectonic phase", splitting the western from the eastern plateaus.

Numerical modelling shows that the topographic excess associated with the plateau can be partially attributed to the geometry of the triple junction (Georgen and Sankar 2010) but most of it is of dynamic origin (Adam et al. 2013). The shift between the magmatic and the tectonic phases corresponds to chron C4a, and matches a sudden global $\sim$25% decrease in spreading velocity occurring between $\sim$8.2 and $\sim$6.2 Ma simultaneously in both plate pairs (Merkouriev and DeMets 2008, 2014a, b). Despite the relevance of the mantle processes in the buildup of the Azores domain, abyssal hill morphology is not destroyed (see Miranda et al. 2014) thus implying that most of the plateau construction is sub-crustal.

Recent data acquired in the framework of the project for the extension of the Portuguese continental shelf allow a better description of magnetic anomalies over the entire plateau and provide reliable magnetic chrons (see Miranda et al. 2014), which confirm some of the abovementioned hypotheses (see Fig. 2):

(i) Between the Mid-Atlantic Ridge and chrons C4a-C5 ocean floor magnetic anomalies are well developed and spreading is mostly regular since chron C4a, coherent with the split of the pre-existing plateau into an "eastern" and a "western" units.
(ii) Magnetic anomalies are continuous between the North Azores Fracture Zone

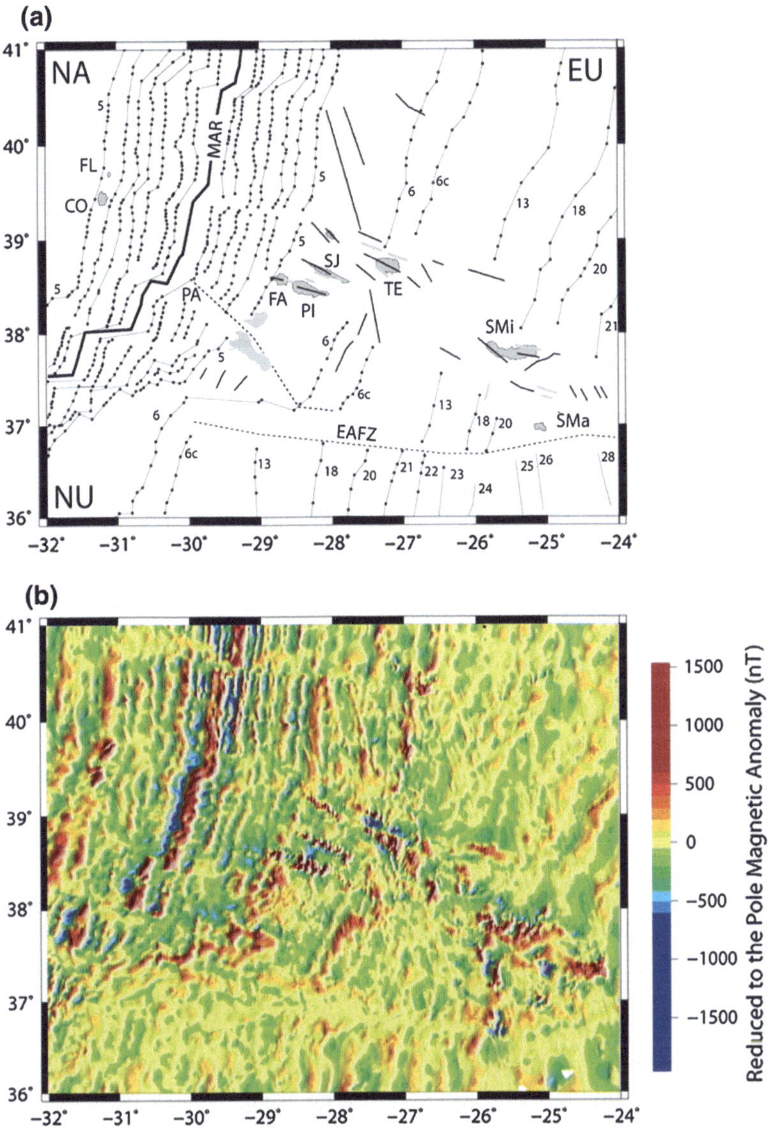

Fig. 2 **a** Morphotectonic sketch and main magnetic chrons. Magnetic chrons 5, 6, 13, 18, 20 and 21 are interpreted according to Luis and Miranda (2008). Younger chrons 2, 2A, 3, 3A, 4 and 4A are also interpreted close to the Mid-Atlantic Ridge. Along Terceira spreading axis magnetic lineations corresponding to the Brunhes epoch (black) and Matuyama (gray) are also plotted. NA: North America; EU: Eurasia; NU: Nubia; EAFZ: East Azores Fracture Zone; MAR: Mid-Atlantic Ridge; FL: Flores Island; CO: Corvo Island; FA: Faial Island; PI: Pico Island; SJ: São Jorge Island; TE: Terceira Island; SMi: São Miguel Island; SMa: Santa Maria Island. Below **b** Reduced to the Pole Magnetic anomaly map, with an horizontal resolution of 30"

and Princess Alice Basin South of 38°30′ N, with no sign of a discrete plate boundary.

(iii) There is evidence of a jump of the ridge axis to the west in the segment 38°N–38° 30′N at chron C3.

3 Rifting of the Azores Domain

The extension in the Azores domain can be quantified using the poles determined for the whole plates (Miranda et al. 2014; Merkouriev

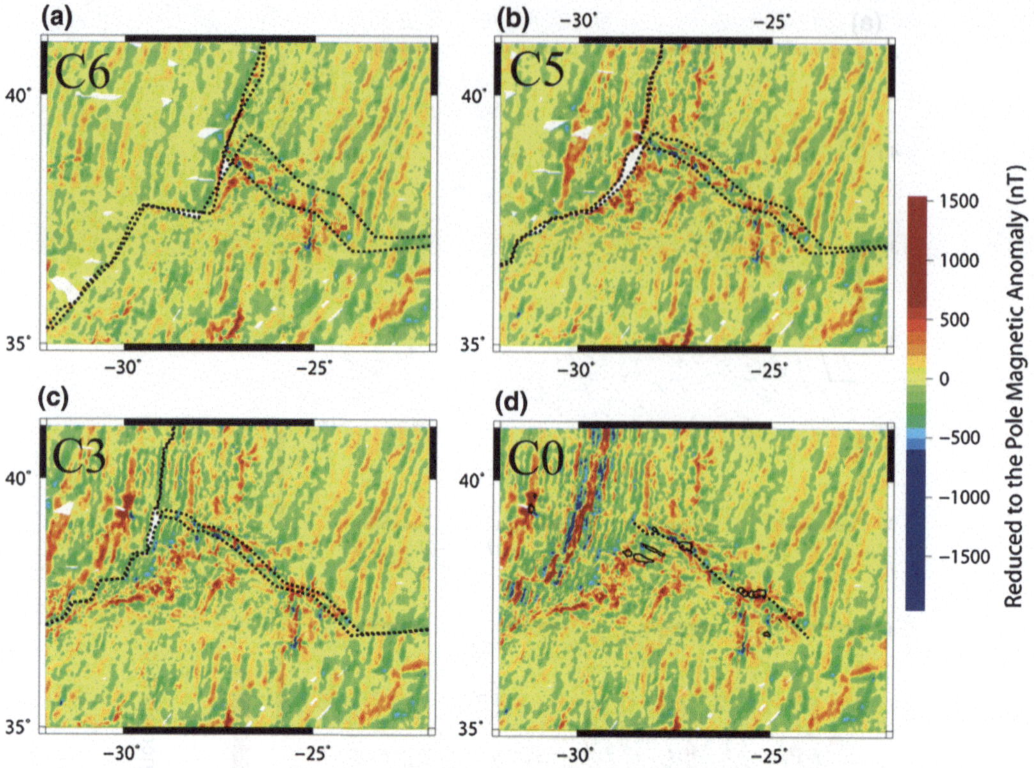

Fig. 3 Magnetic reconstruction of Eurasia, Iberia and Nubia at the time of chrons C6 (**a**), C5 (**b**), C3 (**c**) and C0 (**d**). Euler rotation parameters are taken from Miranda et al. (2014). Dashed lines delimit the lithospheric blocks based on the interpretation of magnetic anomalies. In the case of Terceira Rift they follow the shoulders of the basins as depicted in the swath bathymetry data. The Azores domain was rotated as Nubia and so, the reconstitution at C6 times shows an overlap of Nubia and Eurasia that quantifies the integrated extension that took place south of Terceira rift; the angular mismatch in the MAR segment that corresponds to the Azores plateau quantifies the integrated shear deformation that is applied to the Azores domain by the difference in velocity north and south of that segment

and DeMets 2014a, b) and reconstructing the previous locations of the "lithospheric blocks". In this case, significant overlaps or gaps between the rotated lithospheric blocks correspond to extensional or compressional processes respectively. In Fig. 3 we present the reconstructions for chrons C6 (early phase of the development of the ATJ), C5, C3 and the present configuration using Eurasian-fixed coordinates. The delimitation of the lithospheric blocks (dashed lines in Fig. 3) is based on the interpretation of magnetic anomalies and transform faults. In the case of Terceira Rift the lithospheric blocks follow the axis of the basin deduced from the location of the rift shoulders, as depicted by the swath bathymetry data. For the reconstructions corresponding to C6 when the Azores domain is rotated using Nubian poles with respect to the North American plate, there is an overlap in the north of the domain and an angular mismatch in the west. This results from the internal deformation in the Azores domain during the last ~20 Ma.

While the amount of total extension can be estimated from the magnetic reconstructions, the location of the extensional features can be inferred from morphological data. From the morpho-structural point of view two major rift features can be found in the plateau which can correspond to stable configurations of the triple junction: Terceira Rift that corresponds to the present-day configuration (see Vogt and Jung, Chapter "The "Azores Geosyndrome" and Plate

Tectonics: Research History, Synthesis, and Unsolved Puzzles"), and Princess Alice basin and bank, now located in the Nubian plate (Luis et al. 1994). The Princess Alice Basin disrupts the magnetic pattern generated by the Mid-Atlantic Ridge spreading. Its western limit follows approximately the chron 2A ($\sim$3 Ma), close to the Mid-Atlantic Ridge transform zone of 38°30′N. We interpret it as a previous location of the rift (see Fig. 2).

It is now relatively straightforward to infer the amount and probable evolution of rifting within the Azores domain (see Fig. 3):

(i) There is an overlap between Nubia and Eurasia in the magnetic reconstructions for C6 times, when the Azores domain south of Terceira Rift is rotated as Nubia. This quantifies the extension in the Azores domain that took place in the last $\sim$20 Ma ($\sim$60 km in ENE direction).

(ii) The comparison between the total extension and the width of Terceira Rift allows to conclude that Terceira Rift accommodated most, but not all, of this extension (compare Fig. 3a, d).

(iii) The first phase of extension in the Azores triple junction was probably focused on the formerly continuous WAFZ-EAFZ transform fault, within a "leaky transform" environment. The morphology of the western segment of the EAFZ reflects this extensional phase, as well as the gap that is found between the magnetic anomalies north and south of EAFZ (see Fig. 2).

(iv) Terceira Rift is the main structure presently active, and Princess Alice basin and bank are likely remnants of a previous triple junction configuration which predated the development of Terceira Rift.

4 Spreading at Terceira Rift

The magnetic map used by Krause and Watkins (1970) to locate oceanic spreading related to Terceira rift (see also Vogt and Jung, Chapter "The "Azores Geosyndrome" and Plate Tectonics: Research History, Synthesis, and Unsolved Puzzles") was based on a few profiles where chron picking could not be robustly accomplished. Miranda et al. (1991) and Luis et al. (1994) analyses of the Azores Aeromagnetic Survey show no stable magnetic anomaly pattern parallel to Terceira rift (see Fig. 3), in disagreement with the original interpretation of Krause and Watkins (1970). They also conclude that a few high-amplitude magnetic lineaments, which could be interpreted to reflect more recent chrons, are not strictly confined to the Terceira Rift: São Jorge and the Pico-Faial Islands also show large positive magnetic anomalies with the same trend and linear prolongation on the plateau. The linear magnetic anomalies that match the volcanic highs of São Jorge, Pico and Faial islands with strikes $\sim$N110°–N120° do not reflect topographic effects as shown by inversion (Luis 1996). Miranda et al. (1991) also show that the abovementioned young chrons could be interpreted as Brunhes and Matuyama, because their on-shore prolongation into the Faial island was confirmed by both a refined aeromagnetic survey (Miranda et al. 1991) and by radiometric dating (Féraud et al. 1980, 1981). This was interpreted as a demonstration that the Azores islands are mostly young and do not show any age progression either as a function of the distance to the MAR or a function of the distance to Terceira Rift (Miranda et al. 1991).

Hildenbrand et al. (2008) confirmed this assumption with a new set of radiometric ages of volcanic outcrops in the main units of Faial, arguing for widespread volcanism older than 800 ka, but within the Matuyama chron. The interpretation of a recent aeromagnetic survey covering the island of São Miguel (Miranda et al. 2015) shows magnetization lows associated with hydrothermal alteration above the main volcanic systems, and evidence of the Bruhnes-Matuyama transition on-shore, matching the radiometric ages determined by Johnson et al. (1998) in surface outcrops. A similar situation was found on the island of São Jorge (Silva et al. 2012) supporting the interpretation that the Matuyama-Brunhes transition is found almost everywhere in the Azores islands and that Azorean magnetic anomalies associated with the islands are younger than ~2 Ma. The exception is Santa Maria Island, believed to have formed between 5.2 and 4.6 Ma (Féraud et al. 1981), and the only island of the Azores archipelago where fossiliferous sediments of Zanclean age (5.3–3.6 Ma) were found (Abdel-Monem et al. 1975; Janssen et al. 2008).

The only place where magnetic lineations seem better developed is the East Graciosa Basin close to Terceira Island, where a succession of normal and reversed anomalies can be identified. The two anomalies with normal magnetic polarity that follow the southern and northern flanks of the rift basin are the oldest magnetic signature associated with the development of the Terceira Rift. Princess Alice basin also disrupts the MAR-related magnetic pattern and its western limit follows chron C2a (~3 Ma) at 38°30′N.

The thermal regime of Terceira rift, deduced from the analysis of gravity data also favors the interpretation that it is a young feature, with a small thermal signature which could be associated with a stable spreading regime (Luis et al. 1998).

The main conclusions can be summarised as:

(i) The Terceira rift is a relatively young feature, and the same occurs with the main volcanic islands that separate the individual basins: Graciosa, Terceira and São Miguel (see Fig. 3 for locations). Available magnetic and radiometric data point to ~3 Ma as an upper limit;
(ii) Spreading is discontinuous, with several linear magnetic anomalies that can be interpreted as the signature of linear neovolcanic ridges, sub-parallel to the rift direction;
(iii) Lithospheric rifting is focused on Terceira Rift at least after C3. So, the whole EAFZ and Princess Alice basin are now fossil. The western Eurasia-Nubia plate boundary follows now Terceira Rift and Gloria Fault. The northward migration of the plate boundary to Terceira Rift led to the subsidence of Princess Alice island, now a submerged bank (Fig. 4).

5 The Development of Linear Volcanic Ridges

The relative motion between Eurasia and Nubia after the establishment of the Azores Triple Junction was mostly accommodated by Terceira Rift, Princess Alice Rift and, in the early phases, close to the EAFZ. However, there is a significant amount of deformation along a series of Linear Volcanic Ridges (LVR) located on the Azores plateau. Islands like São Jorge, Pico and Faial are extreme cases of LVR development. Lourenço et al. (1998) mapped fault scarps, and elongated seamounts, to conclude that the two most frequent strikes are N110E-N120E, at west, and N140E-N150E, at east, in agreement with the arcuate form of the Terceira axis itself. Lourenço et al. (1998) interpreted this pattern as the result of the prevalence of co-axial oblique extension, focalised within the Terceira axis, and a stress field with minimum compressive axis sub-parallel to the opening directions predicted by geological and geodetic kinematic models.

The mechanism beyond the development of LVR in the Azores plateau was also discussed by Vogt and Jung (2003 and Chapter "The "Azores Geosyndrome" and Plate Tectonics: Research History, Synthesis, and Unsolved Puzzles"), interpreting them as traces of failed rifts either spreading obliquely (west of Terceira island) or normal (east of Terceira island) to the relative motion between the Eurasian and the Nubian plates. They also suggested that the arcuate shape of the Terceira Ridge may result from the relocation of active spreading in response to the emplacement of massive volcanic loads or tectonic piles. Such a mechanism seems valid at a broad scale and coherent with the northward migration of the Triple Junction. However, the age progression it implies for the LVR, progressively younger towards NE, is not supported by magnetic or geochronology data.

Neves et al. (2013) suggest an alternative explanation for the development of LVRs. They assumed that the crustal deformation is driven by plate boundary forces applied at the edges, as describe by global plate kinematic models.

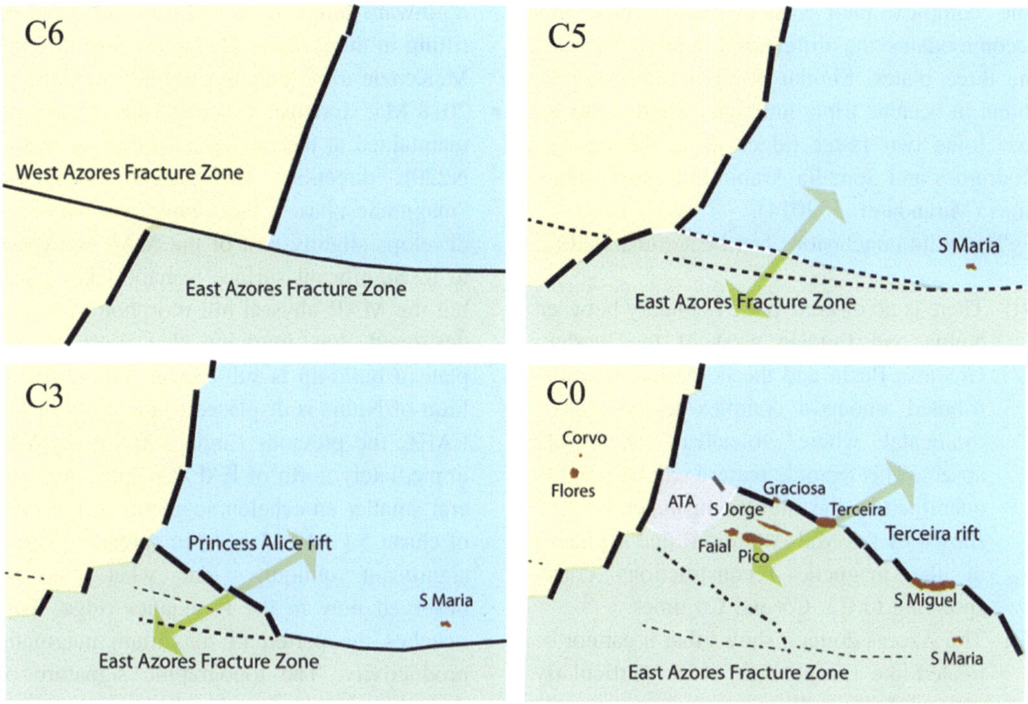

Fig. 4 Sketch of the main phases of the evolution of the Azores domain. At the time of chron C6 the western and the eastern Azores fracture zones form a single transform boundary; at the time of chron C5 the extensional component of the motion between Eurasia and Nubia is mainly accommodates as transtension close to the East Azores Fracture Zone; at the time of chron C3 there is a new RRR triple junction with an active rift incorporating the Princess Alice basin; after chron C3, the rift moves to the NE, leading to the development of Terceira rift, ~100 km away from the Mid-Atlantic Ridge. The area between Terceira rift and the MAR (ATA, Azores triple area) is marked by significant internal deformation

Using a 3D numerical model where the brittle layer is described by an elasto-plastic rheology and the mantle underneath is modeled as a viscoelastic layer. Assuming also that fractures are analogous to localised shear bands, they show that lithospheric processes alone can justify the spatial distribution of most of the linear volcanic ridges.

6 Distributed Deformation at the Azores Domain

The spatial distribution of the extensional processes associated with the Nubian-Eurasian plate boundary along the Azores was systematically addressed by Vogt and Jung (2003) in what concerns the problem of distributed versus non-distributed deformation. Miranda et al. (2014) tried to understand the tectonic mechanism associated with the internal deformation of the Azores domain combining very high resolution bathymetric data with detailed magnetic chron identification. They conclude that there is no discrete plate boundary between the Mid-Atlantic Ridge and Terceira rift as thought by early interpretations (e.g. Searle 1980 or Luis et al. 1994). The triple junction area is marked by an apparent northward increase in the spreading velocity as measured by magnetic chrons, between a "pure" Nubian velocity at the latitude of Faial Island and a "pure" Eurasian velocity at the latitude of Graciosa Island. This defining a "triple junction area" with approximately 90 km × 100 km characterised by brittle faulting where volcanism is barren or absent.

The complex fault system within this area accommodates the differential motion between the three plates. Similar configurations can be found in oceanic triple junctions where a slower axis joins two faster ridges as is the case of Rodrigues and Somalia-Arabia-India triple junctions (Miranda et al. 2014).

The main conclusions can be summarised as:

(i) There is no discrete plate boundary between Nubia and Eurasia west of the western Graciosa Basin and the deformation is distributed among a complex tectonic environmental where volcanism is mostly absent. This tectonic pattern can be roughly quantified by the angular mismatch ($\sim 12°$) shown by the Mid-Atlantic Ridge segments in the magnetic reconstructions corresponding to C3, C5 and C6 times.

(ii) The Azores domain shows that it cannot be treated like a single rigid block, particularly west of Terceira Rift.

7 Main Steps of ATJ Evolution

After more than four decades of research, a few questions have been clarified concerning the different phases of development of the Azores triple junction and allowing the design of the following interpretation sketch:

- ~ 27 Ma ago there was a major rearrangement of tectonic plates and microplates in the Atlantic amalgamating Iberia to Eurasia and developing a new plate boundary along the Azores-Gibraltar Fracture Zone, here comprising the Gloria Fault and the East Azores Fracture Zone. This was early realised by Krause and Watkins (1970), and confirmed by later kinematic studies covering the evolution of the North Atlantic since the Cretaceous (e.g. Srivastava et al. 1990; Luis and Miranda 2008).
- 20 Ma ago the East Azores Fracture Zone ceased to act either as a plate boundary or a transform fault. This corresponded to a northward jump of the ATJ and to onset of rifting in the Azores. This is predicted by the McKenzie triple junction stability model.
- 20–8 Ma: Extension across the Azores is maintained at the rate of 3.7 mm/year in the N220E direction. The plateau is in the "magmatic phase": the topographic anomaly develops slightly east of the MAR but close to it. Significant surface reshape takes place but the MAR abyssal hill morphology is not destroyed thus implying that most of the plateau build-up is sub-crustal. The northern limit of Nubia is displaced to the north of the EAFZ: the previous single 200 km segment immediately north of EAFZ is split into several smaller en-echelon segments at the time of chron 5 (~ 10 Ma), when spreading had a significant obliquity like what can be observed now in the Reykjanes ridge. This matches the period of maximum magmatic productivity. The topographic signature of this period can be found both east and west of the MAR and even south of EAFZ.
- 8 Ma: Major rearrangement in the Atlantic corresponding to a sudden decrease of spreading velocity in the MAR and acceleration of the stretching rate across the Azores (Merkouriev and DeMets 2008, 2014a, b; Miranda et al. 2014), with a change of azimuth of the stretching direction; this corresponds also to the end of the magmatic phase of the plateau, and its split by the MAR (Gente et al. 2003) into the "eastern" and the "western" plateaus.
- 8–3 Ma: Extension at the rate 4.5 mm/year in the N240E direction (Miranda et al. 2014), corresponding to the rifting phase of the plateau. In this phase Santa Maria Island developed. Rifting started probably at Princess Alice Rift into a RRR configuration.
- 3 Ma: Abandon of Princess Alice Rift and rifting starting at Terceira Rift close to the northern limb of the plateau. Rifting is controlled by boundary kinematic conditions (McKenzie 1972) and mantle heterogeneities (Vogt and Jung 2003; Yang et al. 2006; Adam et al. 2013), associated with the two well know families of surface faults (Miranda

et al. 1998): N120E associated with magmatic processes and N150E associated with brittle tectonics.

- 3–0 Ma: Extension at the rate 4.5 mm/year in the N240E direction. Opening of all TR basins, and accommodation of distributed extension into the ATA (Azores triple area), with the development of mesoscale blocks, almost deprived of volcanism. Development of most of the Azores islands in both pre-existing Azores plateaus, by a combination of buoyancy and tectonics; island volcanos drifted away of the TR to the Eurasia or the Nubia plates, except for São Miguel Island internally stretched by TR (Miranda et al. 2015), and Flores and Corvo trapped in the North-America plate (see Fig. 3 for locations and Fernandes et al., Chapter "The Contribution of Space-Geodetic Techniques to the Understanding of the Present-Day Geodynamics of the Azores Triple Junction "); development of off-ridge extension controlled by mantle dynamics.

Acknowledgements We thank the helpful reviews made by Dr. Madureira and one anonymous reviewer that significantly improved the paper. This is a contribution from MAREKH, PTDC/MAR/108142/2008.

References

Abdel-Monem, A. A., Fernandez, L. A., & Boone, G. M. (1975). K-Ar ages from the eastern Azores group (Santa Maria, São Miguel and the Formigas islands). *Lithos, 8*(4), 247–254.

Adam, C., Madureira, P., Miranda, J. M., Lourenço, N., Yoshida, M., & Fitzenz, D. (2013). Mantle dynamics and characteristics of the Azores plateau. *Earth and Planetary Science Letters, 362,* 258–271.

Agostinho, J. (1927). The earthquake in the Azores islands on 31st August 1926. *Vulkanologie, 10,* 268–272.

Dosso, L., Bougault, H., Lagmuir, C., Bollinger, C., Bonnier, O., & Etoubleau, J. (1999). The age and distribution of mantle heterogeneity along the mid-atlantic ridge (31–41 N). *Earth and Planetary Science Letter, 170,* 269–286.

Féraud, G., Kaneoka, I., & Allegre, C. J. (1980). K-Ar ages and stress pattern in the Azores: geodynamic implications. *Earth and Planetary Science Letters, 46,* 275–286.

Féraud, G., Schminke, H. U., Lietz, J., Gastaud, J., Pritchard, G., & Bleil, U. (1981). New K-Ar ages, chemical analyses and magnetic data of rocks from the islands of Santa Maria (Azores), Porto Santo and Madeira (Madeira Archipelago) and Gran Canaria (Canary Islands). *Bulletin of Volcanology, 44,* 3.

Fernandes, R. M. S., Catalão, J., & Trota, A. (2017). The contribution of Space-Geodetic techniques to the understanding of present-day geodynamics of the Azores Triple Junction, this issue.

Gente, P., Dyment, J., Maia, M., & Goslin, J. (2003). Interaction between the Mid-Atlantic Ridge and the Azores hot spot during the last 85 Ma: Emplacement and rifting of the hot spot-derived plateaus. *Geochemistry, Geophysics, Geosystems, 4,* 8514.

Georgen, J. E., & Sankar, R. D. (2010). Effects of ridge geometry on mantle dynamics in an oceanic triple junction region: Implications for the Azores Plateau. *Earth and Planetary Science Letters, 298*(1), 23–34.

Heirtzler, J. R., Dickson, G. O., Herron, E. M., Pitman, W. C., & Le Pichon, X. (1968). Marine magnetic anomalies, geomagnetic field reversals and motions of the ocean floor and continents. *Journal Geophysical Research, 73*(6), 2119–2136.

Hildenbrand, A., Madureira, P., Marques, F. O., Cruz, I., Henry, B., & Silva, P. (2008). Multi-stage evolution of a sub-aerial volcanic ridge over the last 1.3 Myr: S. Jorge Island, Azores Triple Junction. *Earth and Planetary Science Letters, 273,* 289–298.

Janssen, A. W., Kroh, A., & Avila, S. P. (2008). Early Pliocene heteropods and pteropods (Mollusca, Gastropoda) from Santa Maria Island (Azores, Portugal): systematics and biostratigraphic implications. *Acta Geologica Polonica, 58,* 355–369.

Johnson, C. L., Wijbrans, J. R., Constable, C. G., Gee, J., Staudigel, H., Tauxe, L., et al. (1998). $^{40}Ar/^{39}Ar$ ages and paleomagnetism of São Miguel lavas, Azores. *Earth and Planetary Science Letters, 160,* 637–649.

Krause, D. G., & Schilling, J.-G. (1970). Cruise Report R/V Trident, Cruise TR-86, Terceira I. University of Rhode Island, Kingston.

Krause, D. G., & Watkins, R. B. (1970). North Atlantic crustal genesis in the vicinity of the Azores. *Geophysical Journal Royal Astronomical Society, 19,* 261–263.

Laughton, A. S., Roberts, D. G., & Graves, R. (1975). Bathymetry of the Northeast Atlantic: Mid-Atlantic ridge to South-west Europe. *Deep-Sea Research, 22,* 791–810.

Laughton, A. S., & Whitmarsh, R. B. (1974). The Azores-Gibraltar plate boundary. In *Geodynamics of Iceland and the North Atlantic area* (pp. 63–81). Springer Netherlands.

Le Pichon, X. (1968). Sea-floor spreading and continental drift. *Journal of Geophysical Research, 73,* 3661–3697.

Lourenço, N., Miranda, J. M., Luis, J. F., Ribeiro, A., Mendes-Victor, L. A., Madeira, J., et al. (1998). Morpho-tectonic analysis of the Azores Volcanic

Plateau from a new bathymetric compilation of the area. *Marine Geophysical Researches, 20,* 141–156.
Luis, J. F. (1996). Le Plateau des Açores et le Point Triple Associé: Analyse Géophysique et Evolution, Ph.D. Thesis, Paris VII, Paris (in French).
Luis, J. F., & Miranda, J. M. (2008). Re-evaluation of Magnetic Chrons in the North-Atlantic between 35 N and 47 N: Implications for the formation of the Azores triple junction and associated plateau. *Journal Geophysical Research, 113,* B10105.
Luis, J. F., Miranda, J. M., Galdeano, A., & Patriat, P. (1998). Constraints on the structure of the Azores spreading center from gravity data. *Marine Geophysical Researches, 20*(3), 157–170.
Luis, J. F., Miranda, J. M., Patriat, P., Galdeano, A., Rossignol, J. C., & Mendes-Victor, L. (1994). Açores triple junction evolution in the last 10 Ma from a new aeromagnetic survey. *Earth and Planetary Science Letters, 125,* 439–459.
Machado, F. (1959). Submarine pits of the Azores plateau. *Bulletin of Volcanology, II,* XXI.
McKenzie, D. P. (1972). Active tectonics of the Mediterranean region. *Geophysical Journal Royal Astronomical Society, 30,* 109–185.
Merkouriev, S., & DeMets, C. (2008). A high-resolution model for Eurasia-North America plate kinematics since 20 Ma. *Geophysical Journal International, 2008* (173), 1064–1083. https://doi.org/10.1111/j.1365-246X.2008.03761.x.
Merkouriev, S., & DeMets, C. (2014a). High-resolution Neogene reconstructions of Eurasia-North America Plate motion. *Geophysical Journal International, 198*(1), 366–384.
Merkouriev, S., & DeMets, C. (2014b). High-resolution estimates of Nubia-North America plate motion: 20 Ma to present. *Geophysical Journal International, 196* (3), 1281–1298.
Miranda, J. M., Luis, J. F., Abreu, I., & Victor, L. A. M. (1991). Tectonic framework of the Azores triple junction. *Geophysical Research Letters, 18*(8), 1421–1424.
Miranda, J. M., Luis, J. F., Lourenço, N., & Fernandes, R. M. S. (2015). The structure of the Azores triple junction: Implications for Sao Miguel Island. *Geological Society, London, Memoirs, 44*(1), 5–13.
Miranda, J. M., Luis, J. F., Lourenço, N., & Goslin, J. (2014). Distributed deformation close to the Azores Triple point. *Marine Geology, 355,* 27–35.
Miranda, J. M., Mendes Victor, L. A., Simões, J. Z., Luis, J. F., Matias, L., Shimamura, H. H., et al. (1998). Tectonic setting of the Azores Plateau deduced from a OBS survey. *Marine Geophysical Researches, 20*(3), 171–182.
Neves, M. C., Miranda, J. M., & Luis, J. F. (2013). The role of lithospheric processes on the development of linear volcanic ridges in the Azores. *Tectonophysics, 608,* 376–388.
Schilling, J. G. (1975). Azores mantle blob: Rare-earth evidence. *Earth and Planetary Science Letters, 25*(2), 103–115.
Schilling, J.-G., & Krause, D. G. (1970). Cruise Report R/V Trident, Cruise TR-89, Terceira II. University of Rhode Island, Kingston.
Schouten, H., Srivastava, S. P., & Klitgord, K. D. (1984). Iberian plate kinematics: Jumping plate boundaries, an alternative to ball-bearing plate tectonics. *EOS Transactions, AGU, 65,* 190.
Searle, R. C. (1980). Tectonic pattern of the Azores spreading centre and triple junction. *Earth and Planetary Science Letters, 51,* 415–434.
Silva, P. F., Henry, B., Marques, F. O., Hildenbrand, A., Madureira, P., Mériaux, C., et al. (2012). Palaeomagnetic study of a sub-aerial volcanic ridge (São Jorge Island, Azores) for the past 1.3 Myr: Evidence for the Cobb Mountain Subchron, volcano flank instability and tectono-magmatic implications. *Geophysical Journal International, 188*(3), 959–978.
Srivastava, S. P., Schouten, H., Roest, W. R., Klitgord, K. D., Kovaks, L. C., Verhoef, J., et al. (1990). Iberian Plate Kinematics: A jumping plate boundary between Eurasia and Africa. *Nature, 344,* 756–759.
Vogt, P. R., & Jung, W. Y. (2003). The Terceira Rift as hyper-slow, hotspot-dominated oblique spreading axis: A comparison with other slow-spreading plate boundaries. *Earth and Planetary Science Letters, 218,* 77–90.
Vogt, P. R., & Jung W. Y. (2017). The "Azores Geosyndrome" and Plate Tectonics: Research History, present understanding and unsolved puzzles (Chapter 3).
Yang, T., Shen, Y., van der Lee, S., Solomon, S. C., & Hung, S. (2006). Upper mantle structure beneath the Azores hotspot from finite-frequency seismic tomography. *Earth and Planetary Science Letters, 250,* 11–26.

Volcanism in the Azores: A Marine Geophysical Perspective

Neil C. Mitchell, Rachelle Stretch, Fernando Tempera and Marco Ligi

Abstract

In contrast to oceanic islands produced by mantle melting anomalies or "hotspots" lying in mid-plate settings, the central and eastern Azores islands are distributed on the Nubia-Eurasia tectonic plate boundary and experience frequent tectonic earthquakes. As they lie in an extensional to trans-tensional tectonic environment, volcanism is organised into submarine and subaerial ridges oriented perpendicular or oblique to the direction of plate separation. Much of the marine geophysical work undertaken to study these features and the submarine flanks of the islands has involved seabed mapping with sonars of various kinds, beginning with the GLORIA long-range sidescan sonar deployed in the late 1970s and continuing more recently with the deeply towed sidescan sonar TOBI and multibeam bathymetric sonars in dedicated expeditions and ship transits. These datasets give us a view of the topographic structure of the ridges and of the morphologies of the volcanic and tectonic features comprising them. Sonars also have been used to investigate the incidence of large- and small-scale landsliding around the islands, as well as the morphologies of lava flows originating from land and entering the sea. The overall picture emerging is one of landslides, cones, terraces and ridges analogous to those of Hawaii, Canaries and other oceanic island groups but with notable differences. For example, large-scale landslides appear to be rarer in the Azores and submarine cones tend to be steep-sided and pointed, not flat-topped or cratered as they are in some parts of the Hawaiian Islands. Faults are also more common in the Azores tectonic environment.

N. C. Mitchell (✉)
School of Earth and Environmental Sciences, University of Manchester, Williamson Building, Oxford Road, Manchester M13 9PL, UK
e-mail: Neil.Mitchell@manchester.ac.uk

R. Stretch
Department of Geography, University of Cambridge, Downing Place, Cambridge CB2 3EN, UK

F. Tempera
Departamento de Oceanografia e Pescas, Universidade dos Açores, Rua Prof. Dr. Frederico Machado, 4, Horta, Açores, Portugal

M. Ligi
Istituto Scienze Marine, Consiglio Nazionale delle Ricerche, Via Gobetti 101, 40129 Bologna, Italy

1 Introduction

Whereas the islands of Corvo and Flores lie in a stable tectonic environment west of the Mid-Atlantic Ridge on the North America plate, the central and eastern Azores islands are distributed on and around the Nubia-Eurasia diffuse plate boundary (Fig. 1). Submarine volcanism

U. Kueppers and C. Beier (eds.), *Volcanoes of the Azores*, Active Volcanoes of the World, https://doi.org/10.1007/978-3-642-32226-6_7

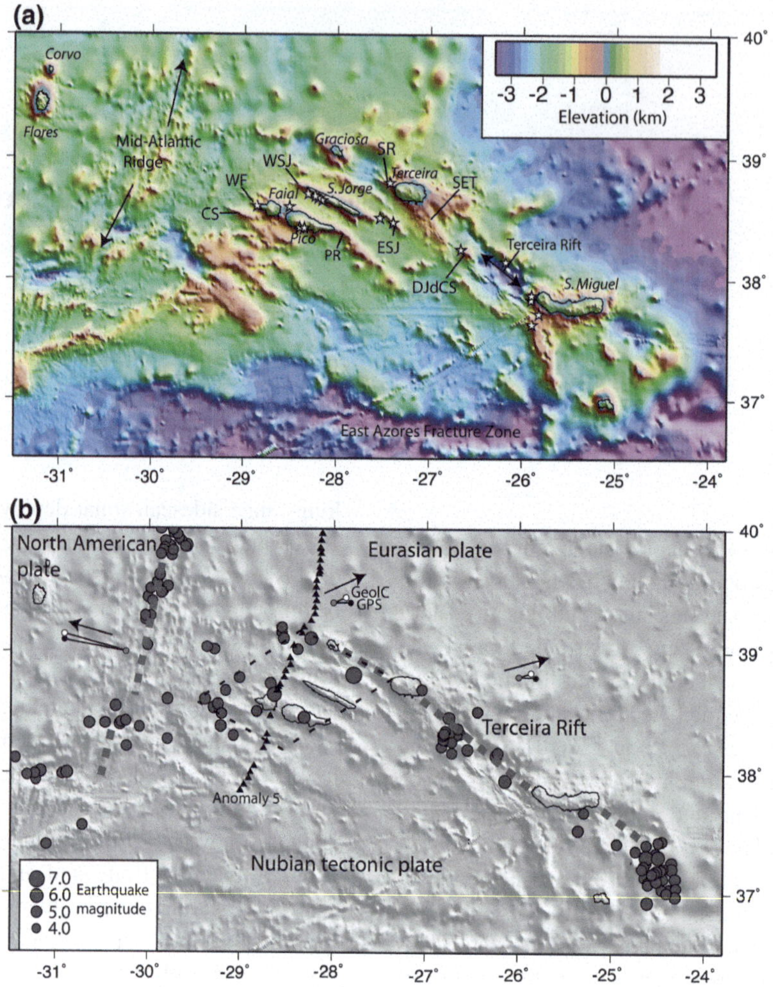

Fig. 1 Overview of the Azores arquipelago. **a** Bathymetry and topography data from the Global Multi-Resolution Topography Synthesis (Ryan et al. 2009). Annotation DJdCS: Dom João de Castro Seamount, and submarine volcanic ridges: CS: Condor Seamount, PR: Pico Ridge, SR: Serreta Ridge, WF: West Faial, WSJ: West São Jorge, SET: Southeast Terceira and ESJ: East São Jorge. Island names are in itallics. Open star symbols locate reported submarine eruptions prior to 1964 from Weston (1964). **b** Tectonics of the Azores. Grey shading is based on the elevation and bathymetry data in (**a**). Heavy dashed line east of 29°W is the Terceira Rift, which is commonly regarded as the Nubia-Eurasia plate boundary. Grey circles are earthquake magnitudes from the US Geological Survey (scale in key, lower-left). Grey-filled triangles are identifications of magnetic anomaly 5 from Freire Luis et al. (1994). Small white and solid circles connected to grey-filled circles represent the effect of 3 m.y. of displacement of Eurasia and North America (Nubia fixed) based on the plate rotation poles GeolC and GPS of Calais et al. (2003), i.e., the displacement that would occur if these rates persisted for 3 m.y. into the future. Large arrows are drawn parallel to the long-term motion (GeolC pole prediction)

has not been much studied around Corvo and Flores, but areas around the central and eastern islands have now been extensively surveyed, providing an opportunity to study the products of submarine volcanism in an extensional to transtensional tectonic environment [an area undergoing extension and transcurrent motion, i.e., where the displacement vector is oblique to the deforming zone (McCoss 1986)]. Volcanism arising from the Azores mantle melting anomaly has produced many submarine ridges (Ligi et al. 1999; Lourenço et al. 1998, Miranda et al.,

Chapter "The Tectonic Evolution of the Azores Based on Magnetic Data") as well as elongated subaerial ridges on some islands, such as eastern Pico and São Jorge (Walker 1999; Hildenbrand et al. 2008; Fig. 1a). Earthquake seismic tomography has revealed a low-velocity anomaly at 112 km depth under Pico and São Jorge islands (Yang et al. 2006; O'Neill and Sigloch, Chapter "Crust and Mantle Structure Beneath the Azores Hotspot—Evidence from Geophysics"), which probably represents the mantle plume or other anomaly underlying the Azores that has caused the excess volcanism generating the islands (O'Neill and Sigloch, Chapter "Crust and Mantle Structure Beneath the Azores Hotspot—Evidence from Geophysics"; Beier et al., Chapter "Melting and Mantle Sources in the Azores"). These ridges can be considered distinct from those produced by "normal" mid-ocean ridge volcanism or from volcanism around oceanic islands without plate-tectonic extension. Most oceanic islands that have grown near to mid-ocean ridges are isolated from the influence of extension associated with the spreading processes (for example, Galapagos, Bouvet, Ascension, Tristan da Cunha, Easter, etc.). Examples of islands and seamounts that have apparently grown in extensional environments producing major elongated structures include Reykjanes Peninsula and Ridge in SW Iceland (Parson et al. 1993; Saemundsson 1986), Jan Mayen island (Fitch 1964), Spiess Ridge near Bouvet Island (Ligi et al. 1997; Mitchell and Livermore 1998), Guadalupe Island of Mexico (Batiza 1977), Marsili Seamount of Italy (Marani and Trua 2002), Axial Seamount on the Juan de Fuca Ridge (Embley et al. 1990), the region around St Paul and Amsterdam Islands in the Indian Ocean (Conder et al. 2000), volcanic ridges in the Gulf of California of Mexico (Fabriol et al. 1999) and in back-arc basins such as near the Antarctic Peninsula (Gracia et al. 1996). Elongated seamounts have also been found in the continental margin of California (Davis et al. 2002).

Ridges radiate from large oceanic guyots and seamounts probably because of processes related to volcanic rifting or spreading (Mitchell 2001; Vogt and Smoot 1984), such as the processes actively generating the submarine Puna Ridge (Kilauea volcano) (Fiske and Jackson 1972; Dieterich 1988) and southerly ridges of La Palma and El Hierro in the Canaries (Gee et al. 2001a; Mitchell et al. 2002). The volcanic rift zones and the volcanic ridges that are commonly associated with them in the Azores lie perpendicular or oblique to the Nubia-Eurasia plate motion vectors (Fig. 1b) so they are more likely to have arisen from the plate tectonics than volcanic edifice spreading (Fiske and Jackson 1972), also because the short 1–3 km heights of the edifices may be insufficient to create much gravitational stress. The Azores provide an opportunity to study such submarine volcanic ridges in a shallow marine environment where high-resolution geophysical data can be acquired relatively easily using surface vessels. Resolution or acoustic "footprint" of a sonar is proportional to water depth, so swath systems with typically 1°–3° beam widths produce high quality data due to the proximity of the seabed to sensors on surface vessels.

Although many historical eruptions have been reported in the Azores, they are spread among the islands and building rates on individual islands tend to be modest. For example, the building rate on São Miguel was about 1 km^3 per thousand years for the last 50,000 years (Booth et al. 1978) and was probably about an order of magnitude lower previously (Moore 1990). The rate of Booth et al. (1978) is an order of magnitude smaller than historical and geological rates for Kilaeua (Clague and Dalrymple 1989, their Table 5) and the earlier rate (Moore 1990) is two orders of magnitude smaller. Volcanism in the Azores seems to be occurring at modest rates on individual islands, while the islands are subjected to frequent earthquakes (Fig. 1b), hence the islands are characterised by modest volcanism but frequent plate-tectonic seismicity.

The Azores region comprises an elevated plateau, roughly corresponding with areas shallower than 2000 m depth (Fig. 1a). From a grid of aeromagnetic data, seafloor spreading magnetic anomalies have been mapped across the plateau (Miranda et al., Chapter "The Tectonic Evolution of the Azores Based on Magnetic Data

"); the anomaly 5 locations in Fig. 1b are from Freire Luis et al. (1994). Their trend parallel with the Mid-Atlantic Ridge demonstrates that the plateau formed by excess volcanism at the Mid-Atlantic Ridge (similar to the modern Reykjanes Ridge and Peninsula), probably over a period beginning at 20 Ma (Gente et al. 2003). Dispersion of Rayleigh waves crossing the plateau and recorded at Ponta Delgada in São Miguel suggest an oceanic crustal structure approximately 60% thicker than normal (Searle 1976). Comparing with the Icelandic plateau, the differing rates of magma production in the two plateaux have led to contrasting gross morphologies and volcanic processes. Whereas the Icelandic plateau is mostly above sea level and includes central shield volcanoes, the Azorean plateau is largely below sea level and edifices are mostly stratovolcanoes (Saemundsson 1986), although there are also some shield volcanoes, e.g., on Faial and Terceira islands.

In this contribution, marine geophysical data are described to show how they can reveal aspects of the structure and morphology of the submarine volcanic features to suggest how they have developed by volcanism associated with the Azores mantle hotspot (O'Neill and Sigloch, Chapter "Crust and Mantle Structure Beneath the Azores Hotspot—Evidence from Geophysics"). Research by biologists characterising substrates and seabed morphology carried out as part of marine habitat mapping is also described where relevant to volcanology. The review concentrates on the seafloor around the central and eastern Azores Islands. Although not much marine geophysical work has been reported to date for the western islands, signs of volcanism within the last 2 m.y. on Flores and Corvo islands (Azevedo and Portugal Ferreira 2009; França et al. 2006; Larrea et al. 2013) suggest that interesting submarine volcanic structures may surround them also.

The main source of information on submarine volcanic structures has originated from seabed mapping sonars rather than data from other types of geophysical instruments. These data show the gross morphologies of ridges, the shapes of cones superimposed on them, how those cones vary in shape with water depth suggesting influences of shallow marine eruptions, the structures of lava flows in shallow water originating from subaerial eruptions and landslides around the slopes of the islands. Each of these aspects is described in turn, after first outlining the main datasets that have been collected.

2 Marine Geophysical Datasets Collected Amongst the Azores Islands

The first systematic marine geophysical survey in the Azores was carried out on RRS *Discovery* using the Geological LOng-Range Inclined Asdic (GLORIA) reconnaissance sidescan sonar (Searle 1980). The focus of that study was mainly the area west of the central Azores islands, in order to decipher its tectonic history, but two lines were run among the islands, also collecting single-channel seismic and other geophysical data (track marked by solid bold line in Fig. 2). With a lateral range set to 30 km, the system was capable of imaging the submarine slopes and ridges between the islands, outlining broad structures. The data have too low a resolution to reveal much detail; resolution is ~20–30 m cross-track from the effective pulse duration of the sonar but along-track resolution deteriorates considerably with acoustic spreading of the 2.7° beam (Mitchell 1991), e.g., 500 m at 10 km range.

The ports in the Azores have provided convenient locations for changing crews of research vessels working in the Atlantic and many lines of multibeam bathymetric sonar data have been collected opportunistically. Lourenço et al. (1998) compiled a bathymetric map from transits amongst the islands and some systematic surveys on the Azores plateau. More recently, Miranda et al. (2015) updated the compilation with data collected by the Task Group for the Extension of the Portuguese Continental Shelf, which includes the most extensive swath survey around the Azores so far. Although most of that survey work occurred 200 nautical miles beyond the island coastlines, sizeable areas within the Portuguese

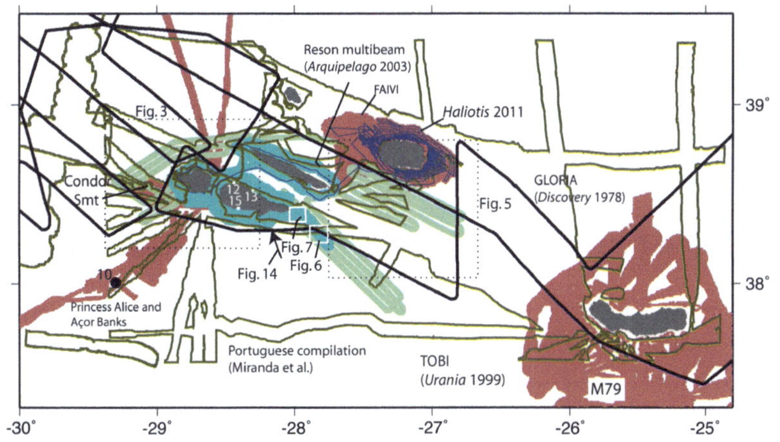

Fig. 2 Locations of some of the marine geophysical datasets collected among the central and east Azores islands. Numbers 10–13, 15 locate Figs. 10, 11, 12, 13 and 15. See main text for explanation

Exclusive Economic Zone were covered including parts of the Mid-Atlantic Ridge, major seamounts and other areas around the islands (Dias Correia 2010). The extent of this combined dataset is outlined by the grey lines in Fig. 2 (marked "Portuguese compilation").

In 1999, a survey was carried out on the Italian RV *Urania* with the Towed Ocean Bottom Instrument (TOBI) sidescan sonar (Ligi et al. 1999; Stretch et al. 2006). This system was towed nearer to the seabed along the tracks marked by wide dark grey lines marked "TOBI" in Fig. 2. It carried a 30 kHz sidescan sonar and a high-frequency sediment profiler, providing a higher resolution view (somewhat analogous to aerial photographs on land) than is typically obtained by either GLORIA or the many multibeam sonars.

A portable Reson Seabat 8160 multibeam sonar (50 kHz) was installed on the University of Azores RV *Arquipelago* in 2003 (Mitchell et al. 2008). The bathymetric coverage of that expedition is shown by the solid black areas in Fig. 2 around Faial, Pico and São Jorge islands (marked "Reson multibeam"). As the crew of the vessel were familiar with the local seas, surveying was allowed into shallow water, which as shown later was important for imaging fine lava structures. The 1.5° beam widths of the system provide an acoustic resolution of 0.26 and 2.6 m at 10 and 100 m depth, respectively. A subsequent deployment of a University of St Andrews phase-difference Submetrix bathymetric sonar allowed some inshore areas to be surveyed (Tempera 2008). Inshore surveys with chirp and boomer seismic systems around the coasts of Faial, Pico, São Miguel and Flores (Bates 2005; Quartau et al. 2003, 2005) have also been useful for resolving shallow marine structures (Quartau et al. 2012, 2015a; Tempera et al. 2012a, b).

Some additional surveys have been carried out recently. Hübscher (2009) led a research cruise on the RV *Meteor*, collecting multibeam and seismic data around São Miguel island in 2009 (area of light grey marked M79 in Fig. 2). Those results are now appearing in the literature (Weiss et al. 2015a, b; Sibrant et al. 2015; Weiss et al. 2016). Results of two further cruises of RV *Meteor* in 2015 (Hübscher 2015) and RV *Meteor* in 2016 (Beier 2016) will extend that work. In 2010, a selection of seamounts to the SW of Faial were mapped by one of us (FT) for habitat mapping purposes including Condor Seamount and parts of Açor and Princess Alice banks (light grey areas emanating from Faial Island in Fig. 2). A shallow water multibeam survey of the Formigas Bank Marine Protected Area southeast of São Miguel Island in 2011 (also by FT) has imaged the headscarps of shallow landslides around the seamount and an intricate pattern of faults on the flat summit of the bank.

Two other surveys in 2011 (project FAIVI) covered a broad area around Terceira Island (Chiocci et al. 2013; Quartau et al. 2014; Casalbore et al. 2015). One of the surveys on RV *l'Atalante* mapped the submarine slopes and ridges around the island (light grey area marked FAIVI in Fig. 2), while a survey with the inshore launch *Haliotis* (marked in Fig. 2 with continuous lines) acquired dense chirp and bathymetric data.

3 Plate-Tectonic Setting

The movements of the tectonic plates create the lithospheric stress conditions under which dykes are intruded and lava is erupted to build the submarine ridges in the Azores. The pattern of tectonic strain is therefore important for understanding the orientations of the ridges. McKenzie (1972) was one of the first researchers to speculate on the nature of the tectonics in the Azores region involving separation of the Eurasian and Nubian plates in the form of a "leaky" transform fault. Such a pattern of strain involving both shear and extension within a deforming zone has more recently been described as transtension (McCoss 1986; Madeira and Ribeiro 1990). The Terceira Rift, an elongate depression filled by São Miguel, Terceira and Graciosa islands (dashed line in Fig. 1b), is commonly considered the main Nubia-Eurasia plate boundary (e.g., Vogt and Jung 2004; Vogt and Jung, Chapter "The "Azores Geosyndrome" and Plate Tectonics: Research History, Synthesis, and Unsolved Puzzles"). However, the seismicity shown in Fig. 1b becomes more widely distributed among the central islands (dashed parallelogram in Fig. 1b) and a variety of other geophysical and geodetic data now suggest that the deformation occurs broadly across the region encompassed by Faial, Graciosa, Terceira and Pico islands and not solely along the Terceira Rift (Madeira and Ribeiro 1990). For example, modelling of aeromagnetic anomalies suggests that the area between the islands and the Mid-Atlantic Ridge has moved independently of Nubia at times in the past (Freire Luis et al. 1994) rather than with Nubia as would be expected if the plate boundary followed the trend of the Terceira Rift to the ridge. This is confirmed by numerous faults and volcanic ridges cutting the older abyssal hill fabric of the Mid-Atlantic Ridge in this area (Lourenço et al. 1998; Searle 1980) and by more recent studies of the magnetic anomalies (Luis and Miranda 2008; Miranda et al., Chapter "The Tectonic Evolution of the Azores Based on Magnetic Data").

Fault-plane solutions of earthquakes among the central islands show both strike-slip and normal fault motions (Hirn et al. 1993) as expected within a transtensional deformation zone. A recent compilation including earthquakes to 1998 (Borges et al. 2007) shows that the T (extensional) axes of the solutions are mostly oriented NNE-SSW, perpendicular to the Terceira Rift, with the central islands dominated by strike-slip mechanisms (with conjugate NE-SW and NW-SE nodal planes) and the rift east of Terciera Island (including São Miguel Island) characterised by more nearly purely normal fault mechanisms. Generally, seismic strain appears to be oriented perpendicular to the direction of elongation of the central islands and submarine volcanic ridges.

GPS measurements at sites around the Azores show deformation occurring broadly throughout the central islands rather than in a narrow plate boundary zone (Fernandes et al., Chapter "The Contribution of Space-Geodetic Techniques to the Understanding of the Present-Day Geodynamics of the Azores Triple Junction"). For example, data collected in six campaigns over an eight-year period show sites on Pico, São Jorge and western Terceira moving at intermediate rates between Eurasia and Nubia (Fernandes et al. 2006), rather than being clearly attached to either of the two plates (Fernandes et al., Chapter "The Contribution of Space-Geodetic Techniques to the Understanding of the Present-Day Geodynamics of the Azores Triple Junction").

Calais et al. (2003) have characterized the motions of the major plates both from geologic (magnetic anomalies and transform faults) and GPS data. The rotation poles computed with these different data types are significantly different in terms of pole locations and rotation rates, implying a change in motion since 3 Ma. The open and solid circles in Fig. 1b show 3 m.y. of forward movement of the plates with respect to fixed Nubia predicted by their GeolC and GPS poles. (Arrows highlight the trends in the reconstructions from the GeolC.) If the components of movement are resolved parallel to N025°E (perpendicular to São Jorge and Pico islands), the GPS results suggest 26% less movement than that in the GeolC reconstructions. This is interesting because the islands of São Jorge and Pico both have active strike-slip faults along their axial ridges revealed by sheared cinder cones (Madeira and da Silveira 2003). The Calais et al. (2003) results may provide an explanation for this observation, with a geologically recent change in motion (reduced extensional component) leaving a more important strike-slip motion.

4 General Morphologic Structure of the Azores Volcanic Ridges

Figure 3 shows an overview of the mosaicked TOBI data from west of Faial and São Jorge islands together with descriptions of samples collected by C. Beier and co-workers (Stretch 2007). Figure 4 shows two enlargements of those data. In Figs. 3 and 4, dark areas represent low acoustic backscattering (usually mud or acoustic shadows) and white areas represent high backscattering (usually fault scarps or exposed rocks such as lavas, rock talus or cold water corals). These data show a variety of irregular and smooth textures on submarine cones,

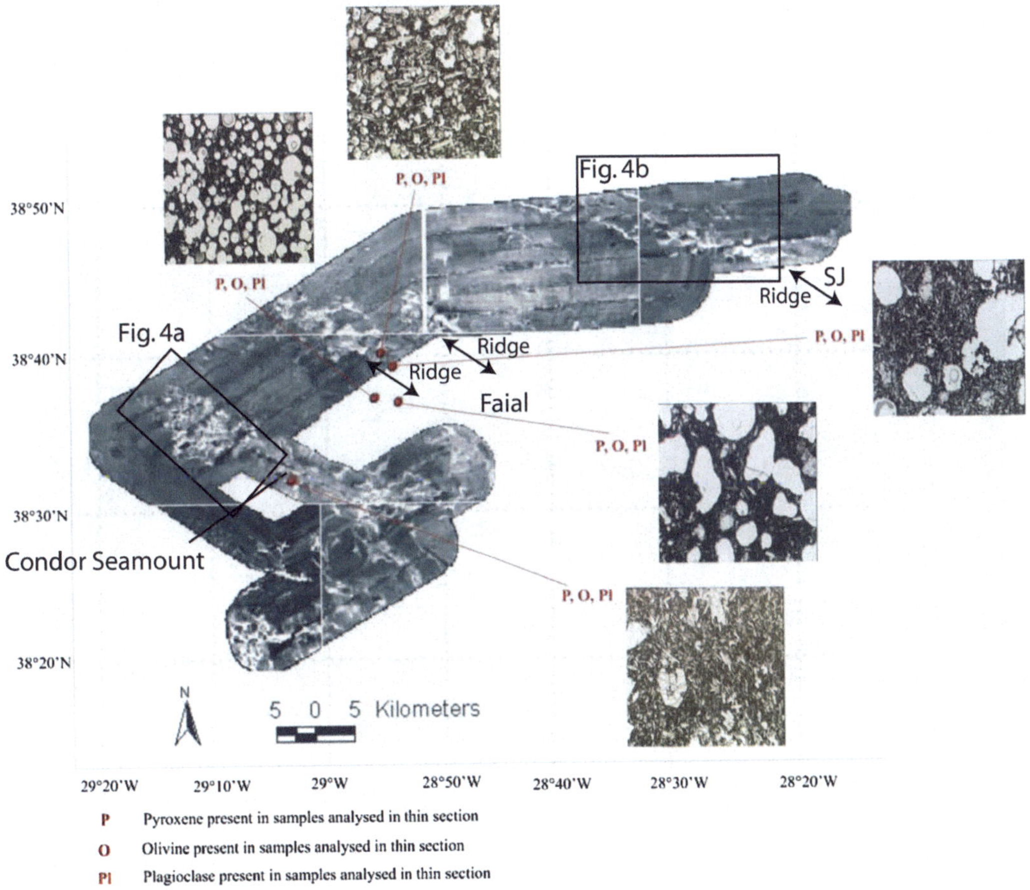

Fig. 3 TOBI deeply towed sidescan sonar data collected immediately west of Faial and São Jorge islands (white represents high acoustic backscattering), shown with some thin section images and minerals present in dredged rock samples (Stretch 2007). Annotation P, O and Pl represents the presence of pyroxene, olivine and plagioclase. "SJ" represents the western extension of São Jorge. (See Fig. 2 for location of tracks relative to the islands.)

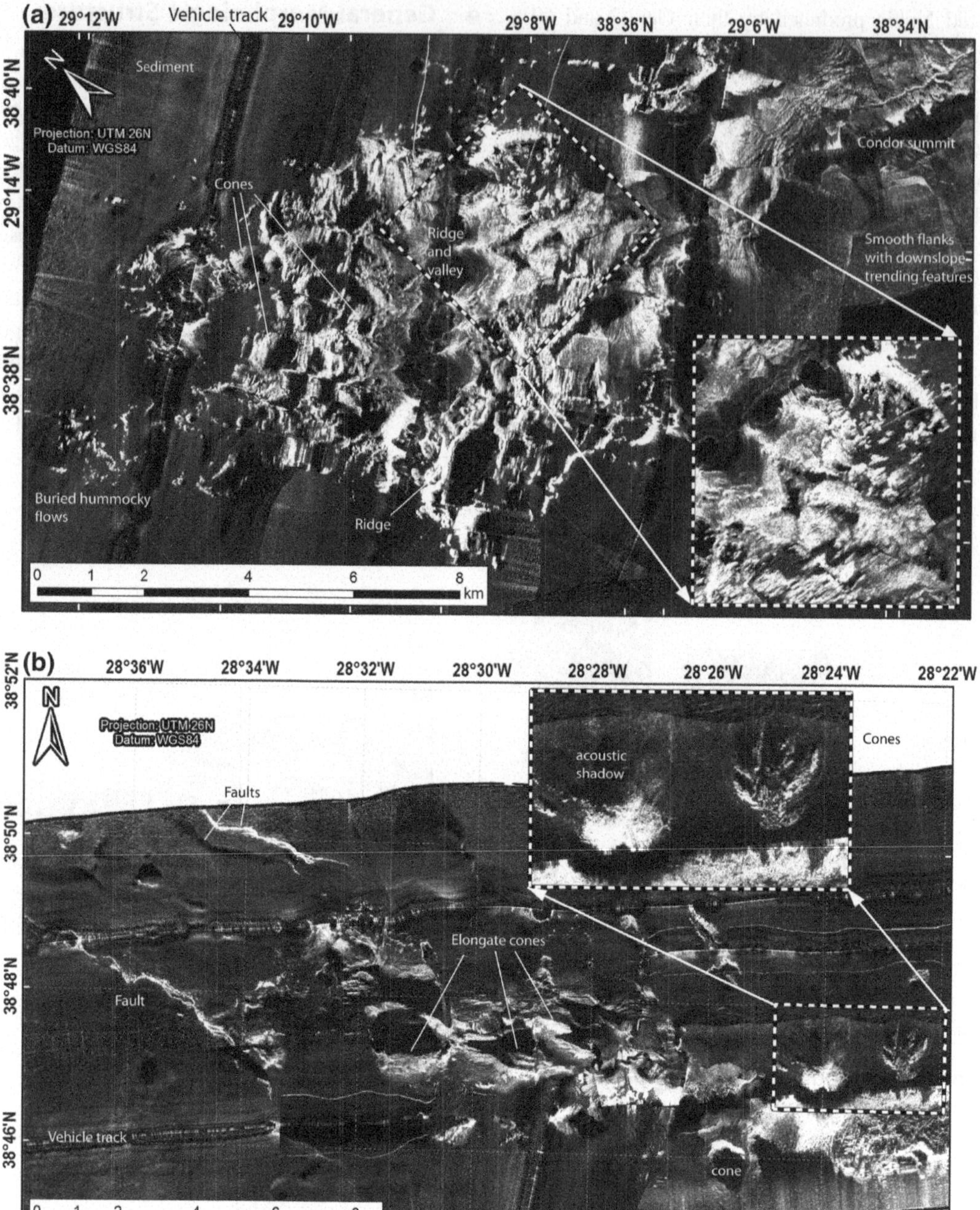

Fig. 4 Selection of the TOBI deeply towed sidescan sonar data from west of **a** Faial and **b** São Jorge. White represents high backscatter. Data are located in Fig. 3

probably representing different lava types, similar to features in the TOBI data collected over the submarine ridge extending SE of Pico island (Stretch et al. 2006). In Fig. 3, a generally rugged seabed of cones and other outcrops forms three main volcanic ridges—Condor Seamount at 38° 30′N, the western extension of Faial Island (off the Capelo peninsula) at 38°40′N and, at the far

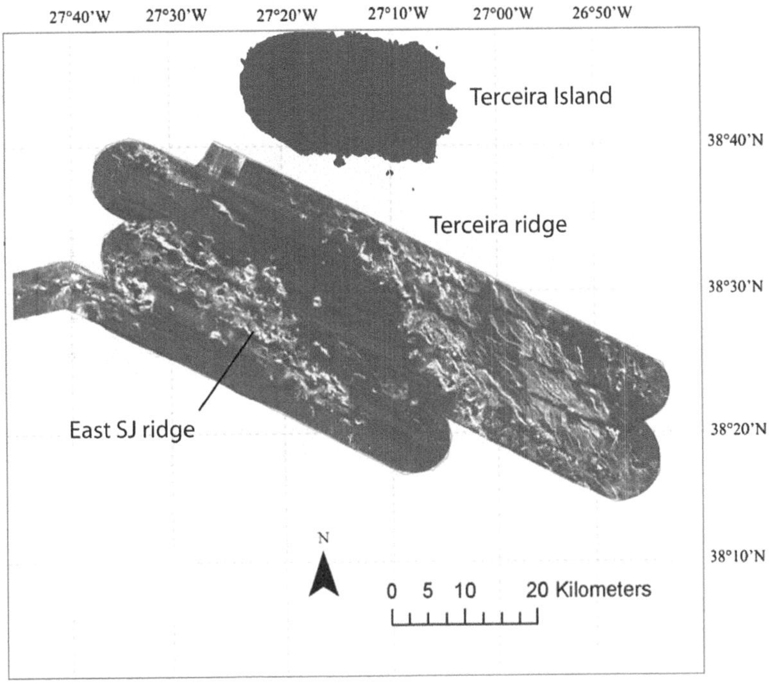

Fig. 5 TOBI sidescan sonar data collected east of São Jorge Island and south of Terceira Island adapted from Stretch (2007). The image reveals volcanic cones and other features on a ridge east of São Jorge and a tectonized ridge south-southeast of Terceira Island. White represents high backscatter

right at 38°45′N, the western extension of São Jorge island. Another small ridge runs parallel to these three immediately northwest of the "Faial" annotation in Fig. 3.

In Fig. 4a, a variety of image textures can be observed. For example, a "speckled" appearance can be seen in the far west of the data, most likely due to hummocky flows that are partly buried by sediment. Elsewhere, surfaces are smooth, as expected for volcaniclastic cones (Mitchell et al. 2012a; Stretch et al. 2006), which are a few hundred metres to 1–2 km across here. Where the sonar track runs up the page slightly left of centre in the figure, the beam pattern of the sonar projected onto the topography leads to a widening and narrowing of the near-nadir dark zone (Mitchell 1991). That pattern suggests a series of ridges and valleys running roughly parallel to the broader topography of Condor Seamount ("ridge and valley" in Fig. 4a, also see inset lower-right). A further highly elongate cone or ridge is highlighted in the lower part of the map. Little detail of the summit of Condor Seamount was obtained as it was obscured beneath the sonar track, but a series of downslope-trending bands in the backscatter are most likely sedimentary features (channels) on the seamount flanks (upper-right in Fig. 4a).

Figure 4b from beyond the western end of São Jorge shows cones that are relatively circular in plan-view but also a series of elongate cones up to ~2 km long and oriented parallel to the main ridge trend. Furthermore, two systems of faults (revealed by highly backscattering bands and acoustic shadows in the west of the map) also run parallel with this trend. The elongate cone morphologies may indicate eruption over dykes (Mitchell et al. 2012a) oriented parallel with the island and these faults.

In further TOBI data shown in Fig. 5, a volcanic ridge east of São Jorge island is also associated with elongate and circular features, probably small ridges and cones, respectively. In contrast, the seabed south of Terceira island, which forms a broad ridge in the regional bathymetry (Fig. 1), is widely disrupted by faults and volcanic features are difficult to identify.

The Reson multibeam multibeam bathymetry collected in 2003 allows a closer look at the morphology of the ridge southeast of Pico Island (Mitchell et al. 2012a) and to a lesser extent ridges west of Faial and São Jorge islands near to

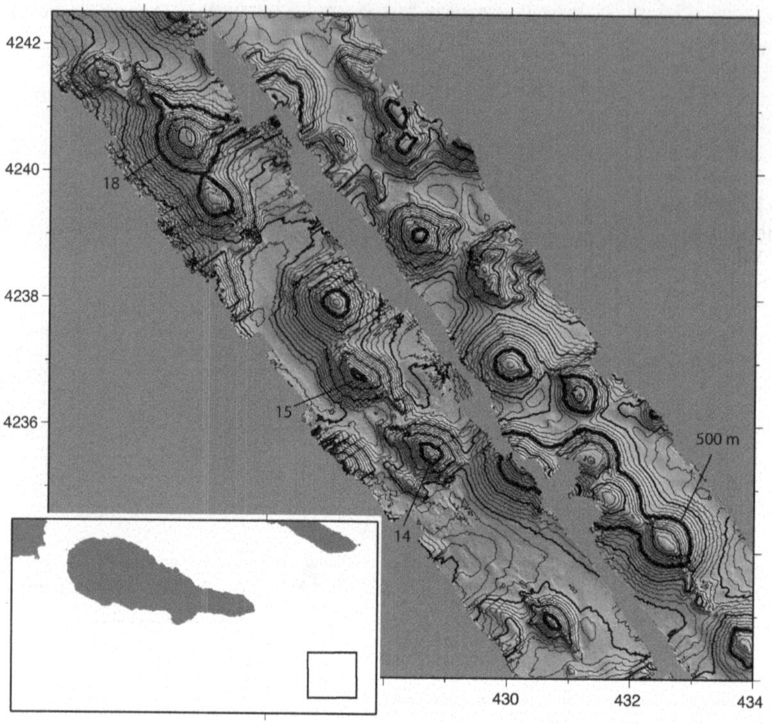

Fig. 6 Bathymetry of deeper section of the submarine ridge extending SSE from the eastern end of Pico Island. Coordinates are Universal Transverse Mercator (UTM) projection (zone 26) distances in km. Bathymetry data (from the Reson 8160 survey) have been gridded at 25 m and contoured at 50 m (500 m intervals are shown in bold). The bathymetry is shaded with an artificial sun to the upper-right. Lower-left inset shows location relative to Pico Island. Values linked to cones refer to cone numbering scheme in Mitchell et al. (2012a). Note the relatively uniform spacing of contours on cone flanks consistent with volcaniclastic deposits at a uniform angle of repose. Small ridges oriented parallel to the main ridge trend are evidence for effusive eruptions over dykes intruded with this orientation

their coastlines. The crest of the SE Pico ridge comprises many small ridges and cones (Figs. 6 and 7), which are described in more detail in a separate section below. The morphology confirms the inference from TOBI images (Stretch et al. 2006) that the large-scale ridges are produced by eruption of many small cones and ridges over dykes. The greater apparent abundance of cones under water compared with scoria cones on the subaerial Pico Ridge probably arises because erosion of cones and their burial by lava flows are less effective underwater than on land (Stretch et al. 2006).

Condor Seamount was also surveyed with multibeam sonar as part of habitat mapping efforts and has been investigated with remotely operated vehicles (ROVs) and "drop-down" cameras to provide "ground truth" (Tempera et al. 2012a). The multibeam data (Fig. 8) show smooth slopes on the flanks of the ridge in its central part (corresponding with the sedimentary features in the upper-right of Fig. 4a mentioned previously), whereas the distal ends are covered with many small cones. This contrast is similar to that observed at Kilauea volcano, Hawai'i, where the submarine slope of the island below the East Rift Zone has a smooth morphology of clastic material produced from fragmentation of lavas crossing the coastline, contrasting with the more rugged Puna Ridge where lavas have erupted underwater (Moore and Chadwick 1995). Over the smooth areas, the flanks of Condor seamount

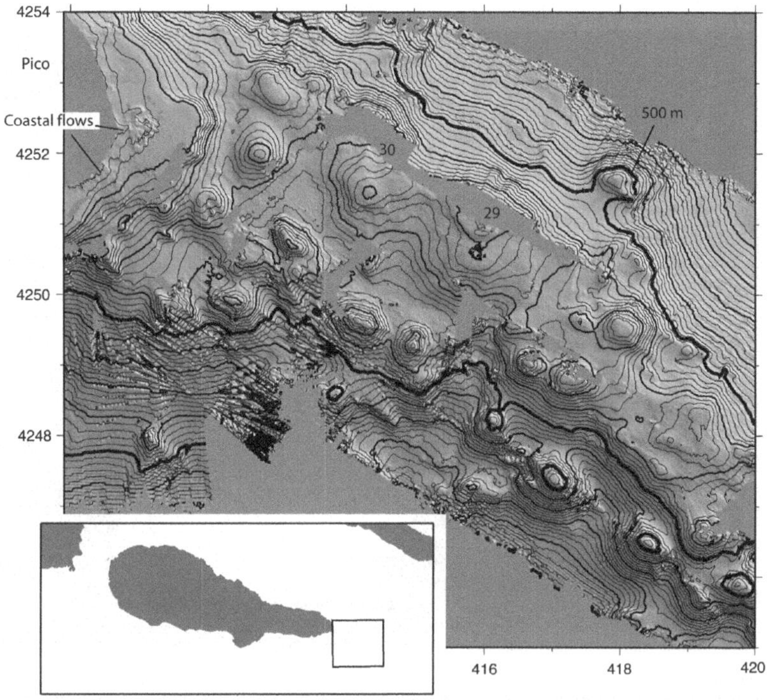

Fig. 7 Bathymetry of shallower section of the submarine ridge extending SSE of Pico Island from Mitchell et al. (2012a), as Fig. 6 and also collected with the Reson 8160 multibeam sonar. Bathymetry shaded with an artificial sun to the upper-right. Lower-left inset shows location relative to Pico Island. Note the subdued relief of cones 29 and 30, speculated to have arisen from forced spreading of the eruption column by the air-sea density interface

observed in camera and ROV imagery are mostly covered with unconsolidated sediments that appear to overlie a volcaniclastic substrate. These smooth slopes are suspected to have been produced by a similar mechanism to that described for Kilauea, though perhaps with shallow marine eruptions creating the clastic material, rather than subaerial lavas reaching coastlines. The summit is highly backscattering acoustically and is covered with large rocky outcrops and some rounded boulders and gravels, characteristic of a paleo-beach. As the shallowest point of the summit at 184 m is deep for the Pleistocene sea level low stands, the seamount has probably subsided since it formed.

The Serreta Ridge west of Terceira Island (Fig. 1) was the site of a submarine eruption in 1999 (Freire Luis et al. 1999), which was observed from sea by some of us (NCM and ML) on RV *Urania*. During the eruption, porous balloon-like blocks of lava were observed reaching the sea surface, where they floated for ~15 min before sinking (Gaspar et al. 2003; Kueppers et al. 2012). The area has since been surveyed with multibeam sonars on RVs *Knorr* (DK Smith chief scientist) and *l'Atalante* (Chiocci et al. 2013; Casalbore et al. 2015). The ridge morphology viewable online in Geomapapp software (Ryan et al. 2009) is similar to that of Condor Seamount, with a smooth central section (where the eruption occurred) and rugged deeper section with superimposed cones.

5 Cone Morphologies

Mitchell et al. (2012a), Weiss et al. (2015b) and Casalbore et al. (2015) have described morphometric studies of Azorean submarine volcanic cones. Figure 6 shows a selection of the 2003 Reson multibeam bathymetry data from a deeper part of the ridge, revealing cones generally a few hundred metres to 2 km in basal width and up to 400 m in height. Their summits generally lack collapse pits or craters. The spacings of contours on the cone flanks are regular, indicating they are smooth with a common gradient, which is close to the 28°–30° angle of repose of submarine talus (Mitchell et al. 2000). Some indentations of the contours may indicate small superficial collapses,

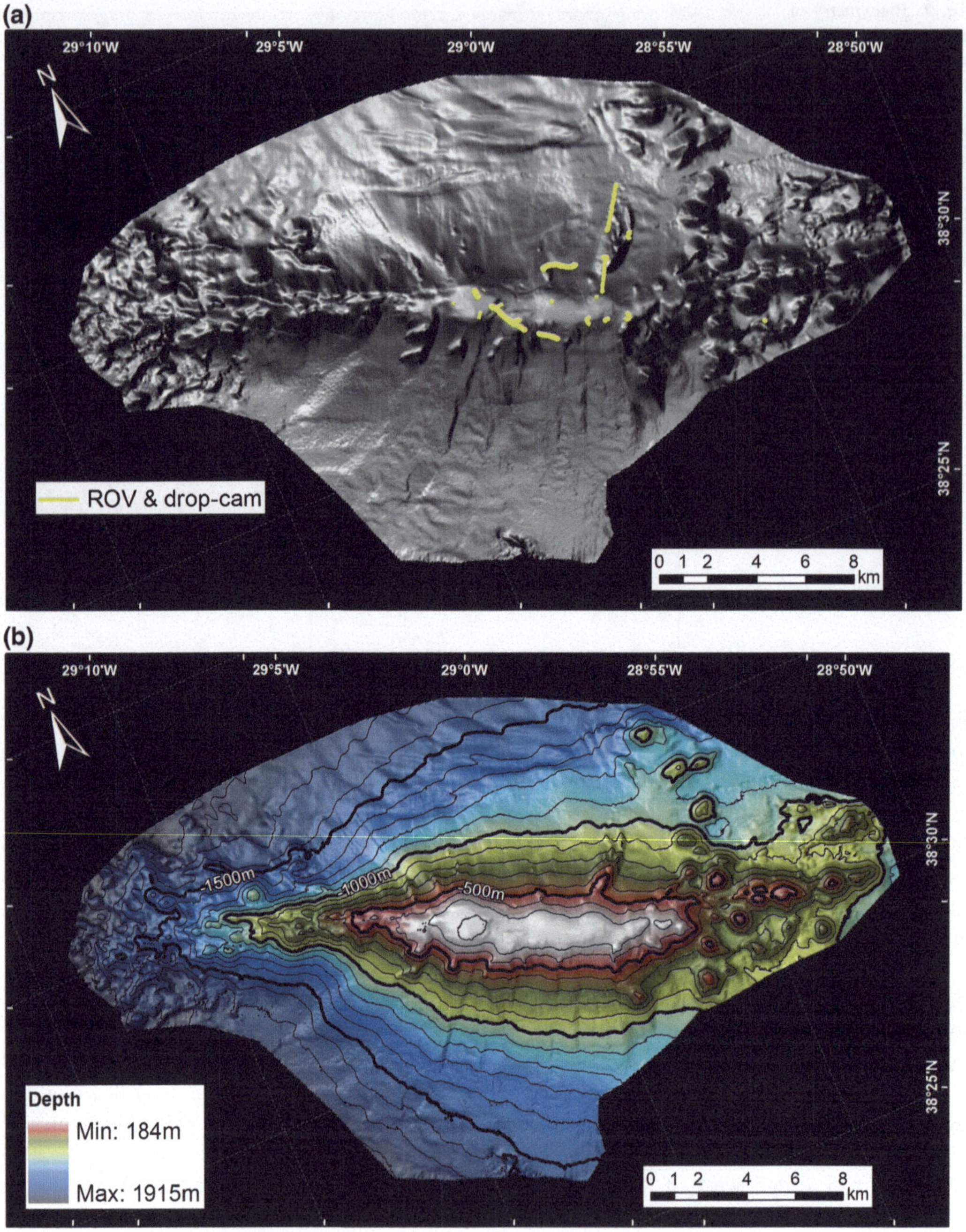

Fig. 8 **a** Shaded relief bathymetry (artificial sun from left of map) and **b** colour-coded bathymetry (10 m contours) of Condor Seamount. Figures adapted from Tempera et al. (2012a) based on data collected with a Kongsberg-Simrad EM120 multibeam sonar and gridded at 50 m. Note the contrasting morphology of the flanks, which is smooth in the central section (probably slopes of volcaniclastic deposits) and rugged at both ends of the seamount (intact volcanic cones and lavas)

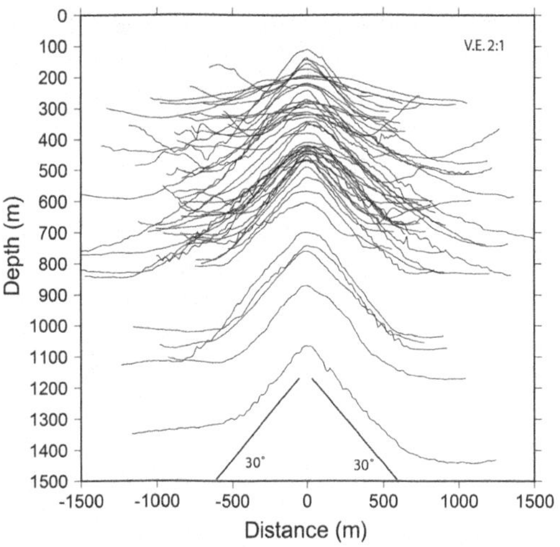

Fig. 9 Bathymetry profiles of the submarine cones from Mitchell et al. (2012a). Vertical exaggeration is 2:1. Whereas the gradients of cone flanks are typically ~30°, the angle of repose of granular particles, some cones in shallow water have flatter profiles with shallower flank gradients. Two straight lines at base are at 30° to the horizontal

which serve to redistribute material from the summit to the flanks of the cones (Chadwick et al. 2008). Several of the cones in Fig. 6 are associated with small ridges oriented parallel with the submarine Pico Ridge. They have been interpreted as produced by effusive eruption over dykes intruded parallel to Pico Ridge. Clague et al. (2000) have described low-relief flat-topped (coin-like) cones below a depth of 650 m around Hawaiʻi, which they interpreted as caused by eruption of degassed lava forming submarine lava-lakes. Such features have not been observed in the Azores. The common pointed shapes of cones may indicate that gas-rich eruptions dominate in the Azores and have produced mostly fragmented material.

Figure 7, in contrast, shows some low-relief (mostly <200 m high), rounded cones in shallower water (e.g., cones 29 and 30) with summit depths of 200–300 m. These flatter cones are also evident in the stacked cone cross-sections in Fig. 9, where low-relief rounded profiles are more common for summits shallower than 400 m and steeper cones are more common in the deeper water. These flat cones have been interpreted as comprising volcaniclastic particles because their acoustic backscattering suggests a smooth surface texture, unlike the rugged morphology of effusive lavas. Their flat profiles may have been produced by eruption into shallow water, which led to forced spreading of the eruption column by the air-sea density contrast (Mitchell et al. 2012a), as predicted by Cashman and Fiske (1991).

The multibeam datasets also reveal a number of possible shallow-water Surtseyan cones. Figure 10 shows an oblique 3D image of bathymetry data from the shallowest part of Princess Alice Bank to the southwest of Faial Island and Condor Seamount. Its summit (50–100 m depth) includes low-relief sub-concentric furrows and ridges. These features are suspected to have arisen from episodic eruption into shallow water (potentially intermittently subaerial) and subsequent wave erosion as observed at Surtla near Iceland (Kokelaar and Durant 1983). Erosion of separate volcaniclastic layers with varying cohesion have left the concentric furrows and ridges. The small mound in the shallowest part of the summit may represent intrusive rocks of the central conduit of the cone that remain after erosion. The small circular features around the margins may be littoral cones.

In 1720 AD, a volcanic eruption between Terceira and São Miguel islands within the Terceira Rift created an island reaching 150 m above sea level, which was subsequently eroded by waves to below sea level, forming Dom João de Castro Seamount (DJdCS, see Fig. 1 for location) (Cardigos et al. 2005; Santos et al. 2010). Bathymetric mapping has revealed that the summit now comprises a 300 by 600 m elongate crater (Cardigos et al. 2005). The shallowest part of the edifice was last reported at 13 m below sea level. In contrast, the summit of volcaniclastic materials on Surtla (Iceland) was eroded down to 45 m depth in the 18 years following its eruption (Kokelaar and Durant 1983). This difference may either reflect different degrees of cementation of the substrate or that bed shear stresses are larger during storms at Surtla associated with longer periods and larger wave heights [100-year significant wave heights are estimated to be 30% higher at Surtla (Sterl and Caires 2005)]. The DJdCS edifice has active

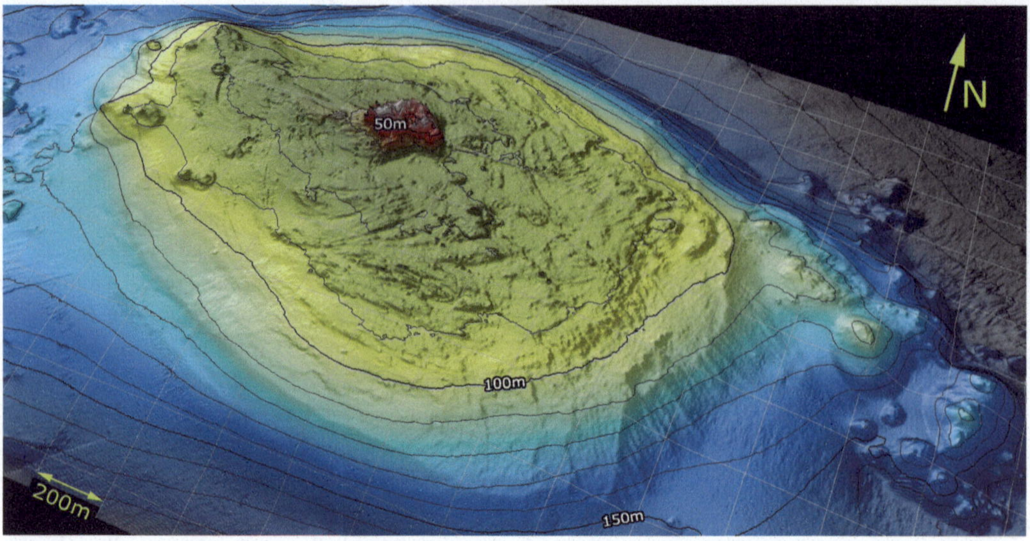

Fig. 10 Possible Surtseyan cone surveyed on the Princess Alice Bank to the southwest of Faial Island and Condor Seamount (oblique view of multibeam sonar data collected with a Kongsberg-Simrad EM120 multibeam sonar and gridded at 2 m). Note the concentric-like pattern of ridges and troughs on the summit, which are possible evidence for different layers of volcaniclastic particles left after wave erosion. Located by "10" in Fig. 2

hydrothermal vents as shallow as 20 m depth, which emit primarily CO_2 but also other volcanic gases (Cardigos et al. 2005).

Figure 11a shows an unusual near-shore cone that was previously published in electronic supplementary material to Mitchell et al. (2008). It shows two smooth surfaces dipping towards a central 20–30 m central high mound (best seen in the cross-sections in Fig. 11b). This inward-dipping morphology was also confirmed by the sidescan sonar display of the corresponding backscatter data shown in Fig. 11c collected roughly along line 'p'. Those data are shown without slant-range correction so that the shape of the seabed can be observed (i.e., the water column has not been removed). Various features can be correlated with the bathymetry map in Fig. 11a, including a linear depression (apparently filled with a highly backscattering, rugged material casting small acoustic shadows), the central mound and the inward-dipping surfaces. In the grey-shaded image in Fig. 11a and where marked "Circular trends" in the sidescan data in Fig. 11c, the inward-dipping surfaces contain concentric furrows and ridges as in Fig. 10. The central mound is highly backscattering and rugged, casting acoustic shadows. The presence of acoustic shadows close to the track implies very steep outward-facing surfaces (Mitchell 1991). This feature is suggested to have formed by eruption of a littoral cone. The inward-dipping surfaces forming a roughly inverted cone shape probably were created by excavation of central material by hydromagmatic explosions (Mattox and Mangan 1997; Moore 1985), a similar though much smaller structure to that originally produced by the 1957–1958 Capelinhos eruption (Cole et al. 2001). Varied cohesion of clastic materials has left the concentric trends.

6 Submarine Lava Flows

Subaerial lava flows have been observed entering water directly only rarely, for example, around Hawai'i by SCUBA divers (Moore et al. 1973). Other observations have been on already emplaced lavas, for example, deeper lavas mapped with sonar data and submersible dives and sampled for geochemistry around Hawai'i (e.g., Wanless et al. 2006) and with sonar data in a lake

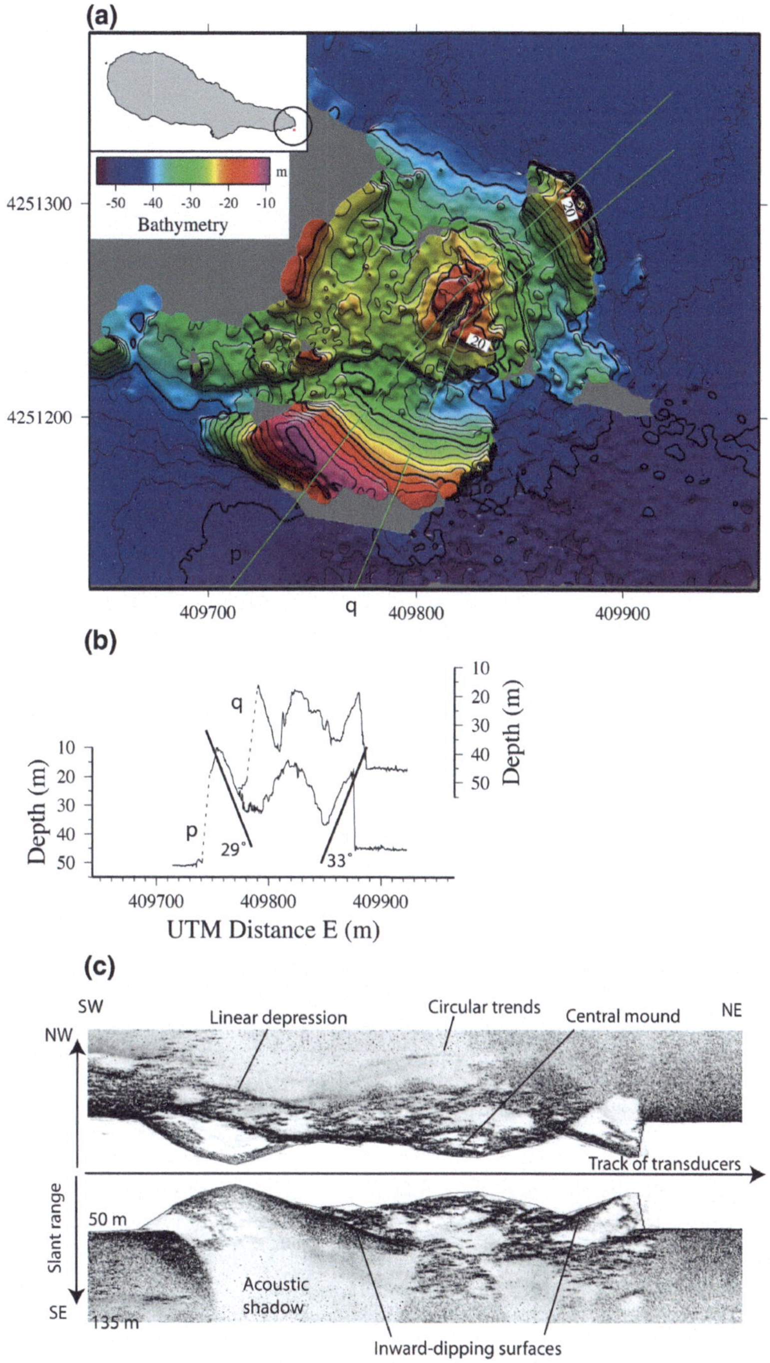

Fig. 11 **a** Multibeam bathymetry data showing an unusual littoral cone off the east coast of Pico Island (from the 2003 Reson 8160 survey). Coordinates are UTM metres and artificial sun is from the northeast. The data are gridded at 0.25 m and contoured at 2 m intervals with every 10 m in bold. **b** Depth profiles along two separate lines located in (**a**). Straight lines against profiles and associated numbers are representative gradients (corrected for directions of the lines). Profile q has been offset vertically for clarity. Map is located by "11" annotation in Fig. 2. **c** Sidescan sonar display collected with the Reson 8160 multibeam approximately along line p in (**a**). Black represents high backscattering and white represents low backscattering or acoustic shadows cast by seabed features. In order to show the shape of the seabed profile, the image data have not been corrected for slant-range distortions. Distances from the central track are acoustic slant ranges from the transducers (values in the lower-left; seabed echo for example is at 50 m depth)

on Iceland (Stevenson et al. 2012). One remarkable LiDAR dataset has been collected by Japanese scientists of a lava flow that entered a lake near Mount Fuji and is now exposed following draining of the lake (Obata and Umino 1999). Systematic mapping of such features around the nearshore parts of volcanic islands has been rare, however. The 2003 multibeam survey in the Azores was effective because the small draft of the RV *Arquipelago* permitted

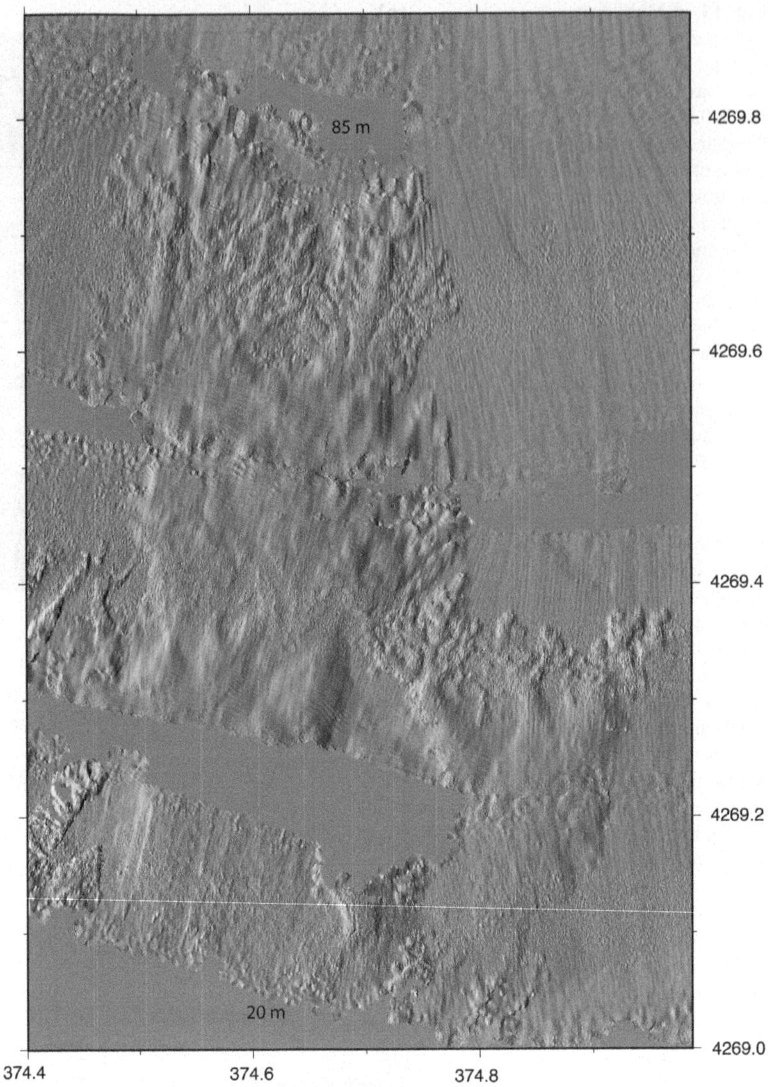

Fig. 12 Lava flow on the shelf immediately north of where the historical flow produced by the 1718 AD eruption on Pico entered the ocean revealed in multibeam bathymetry data collected during the 2003 Reson 8160 survey (located by small box above "12" in Fig. 2). The depth data have been gridded at 0.25 m and are shown in shaded-relief form with an artificial sun from the NW. Coordinates are UTM distance in km. Depth variation is indicated by "20 m" and "85 m" annotation [colour depth-coded version was shown in Mitchell et al. (2008)]

surveying close to coasts and the crew were familiar with the working environment.

Mitchell et al. (2008) described many off-shore lava structures revealed in the 2003 Reson 8160 multibeam bathymetric data, which they argued were mostly Holocene in age. Figure 12 shows a shaded relief image of data collected immediately seaward of where an a'a' lava flow entered the sea in 1718 AD. The lava on land immediately adjacent to this image (underlying the airport of Pico Island) has the blocky morphology of a'a' lava. However, the underwater flows in Fig. 12 show a remarkable series of branching flow lobes, forming a dendritic pattern, quite unlike the rounded fronts of 'a'a lava typically found on land. Individual flow lobes in Fig. 12 are 5–10 m in vertical relief, while the central body of the flow is about 20 m in relief. The flow front was shown to be a statistical fractal within the limits of the data resolution, with a fractal dimension more similar to those of pahoehoe than a'a' flow fronts.

Many examples of these dendritic flow morphologies were found around Pico Island and some examples have also been mapped around Faial (Quartau et al. 2012). Pahoehoe lavas on

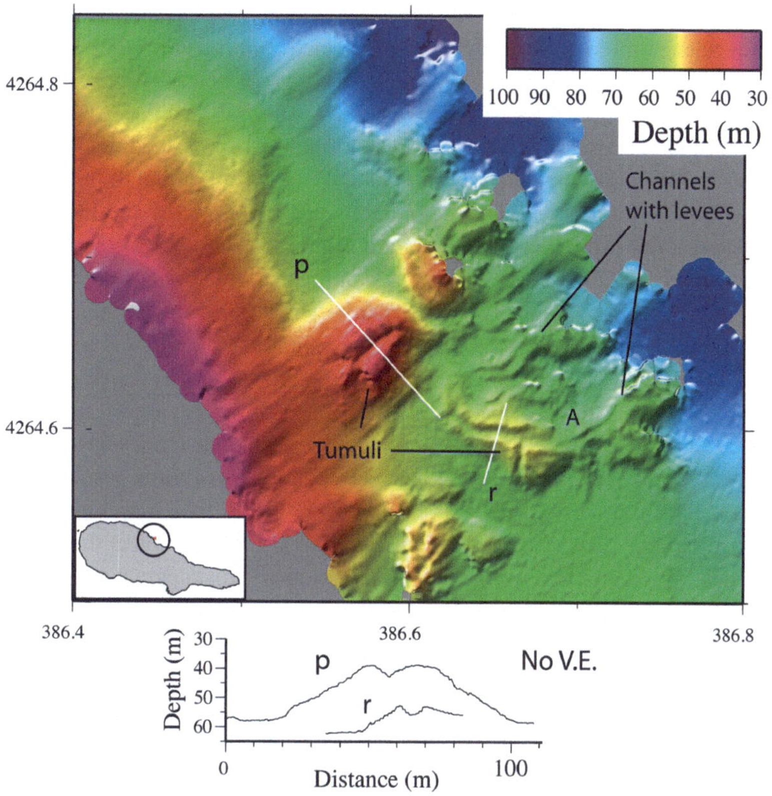

Fig. 13 Shallow-water bathymetry (2003 Reson 8160 survey) showing two tumulis and flows with levées on the central north shelf of Pico Island. The depth data have been gridded at 0.25 m and are shown with shading from the NW. Coordinates are UTM kilometres

land typically develop a viscoelastic layer as they cool, forming the strongest part of the lava surface (Hon et al. 1994). Pahoehoe lava fields evolve by successively inflating lobes under this viscoelastic layer like oblate baloons until the visco-elastic layer ruptures allowing a new lobe to form at a breakout. Given the similar branching patterns of these submarine lavas to pahoehoe, in Mitchell et al. (2008) the branching behaviour was speculated to have arisen from a similar development. Enhanced cooling by seawater entering cracks on the lava surface was suggested to thin the visco-elastic layer of the lava locally, reducing its strength and rendering it susceptible to breakouts. Thus, whereas cooling might be expected to toughen lava (and solidification will clearly reduce mobility ultimately), initial irregular cooling may actually promote mobility. The fractal-like branched patterns of the lava then occur because of episodicity in lava pressure arising from flow discharge variations and from the distribution of open cracks in the lava surface controlling the locations of breakouts. Measurements of heat loss on actively flowing lava along with its evolving morphology are needed to test whether this occurs in practice.

Figure 13 shows further examples of lava flows, in this case revealing the typical clefts of 10–20-m-high tumuli and flow channels with levées in front of a coastal lava delta. The former is evidence that pressure varied within the flow beneath the crust during emplacement, as typically occurs during emplacement of subaerial lavas (Hon et al. 1994).

7 Gross Morphology and Structure of Island Flanks

The submarine slopes of volcanic ocean islands can be broadly classified into two types on the basis of gross morphology in profile—smooth and concave, or rugged and steep. Debris avalanches (produced by catastrophic collapses

(Moore et al. 1989) or smaller downslope movements of volcaniclastic sediments) leave smooth, low-gradient slopes with exponential-like profiles, whereas slopes constructed of lava or affected by deep-seated slumps tend to be more rugged (varied gradient) and steeper (Gee et al. 2001b). In submarine slopes offshore from lava deltas, gradients are also typically steep, reaching 28°–30°, the angle of repose of rock talus in tectonically active marine environments (Mitchell et al. 2000). Unlike in the Canary Islands where submarine channels commonly lie immediately below onshore valleys (Krastel et al. 2001; Mitchell et al. 2003), the submarine slopes of the Azores islands contain few deep submarine channels or canyons.

Graphs of the statistical characteristics of seabed gradients for sectors of island slopes plotted versus elevation can be useful to characterise the slope profiles semi-objectively. For example, the median average of gradient magnitudes and their inter-quartile range (a measure of variability somewhat similar to standard deviation but less affected by outliers) can be computed in given depth intervals. Plotting those gradient statistics versus elevation for Pico and Faial Islands (Mitchell et al. 2008; Quartau and Mitchell 2013) reveal that they are most like those of constructional flanks or those containing debris below lava deltas (Quartau et al. 2015b) and unlike those of debris avalanches, supporting the contention below that the islands have experienced few major catastrophic landslides. In both cases, the median gradient increases from around 5° or less around the deep base of the islands to 23°–28° just below the shelf break.

Low sulphur contents of lava emplaced on the deep submarine flanks of Hawai'i were speculated by Garcia and Davis (2001) to be due to the lava having originally degassed on land and then emplaced in the submarine environment, potentially via lava tubes. The submarine parts of islands may thus not necessarily be formed by direct eruption of lavas. Unfortunately, the smooth morphology of Kilauea down-slope from where lava enters the sea suggests that the lava usually breaks up to form a steep ramp of clastic material (Moore and Chadwick 1995) so details are commonly obscured. The submarine parts of volcanic islands are much larger by volume than their subaerial parts, so resolving which of these alternatives (direct eruption or material provided from land) is more common in the Azores is important to understanding the growth of islands. Most of the lava flows mapped in 2003 in the Azores (Mitchell et al. 2008) terminated on the relatively flat shelf, though some, such as those in Figs. 12 and 13 reached the outer shelf. One example was traced to nearly 400 m depth on the southwest side of the island. Unfortunately, these observations still cannot resolve which alternative applies best to the Azores because the lack of lava flow structures on the steeper submarine slopes may either represent a problem of sonar resolution at those depths or that lava may simply disintegrate to a breccia quickly after emplacement on steep slopes (which reach the ~30° angle of repose immediately below the shelf break) as has been observed around Kilauea (Sansone and Smith 2006).

8 Landslides

With systematic mapping with sonars around many oceanic islands, the early observations of widespread landslides (Holcomb and Searle 1991; Moore et al. 1989) have been generally confirmed (Keating and McGuire 2000). However, the Azores islands are unusual in that only two large landslides are revealed by the subaerial geomorphology (Mitchell 2003). One lies in the southeast side of Pico Island [Topo volcano, Fig. 14, Costa et al. (2015)]. A smaller embayment on the south subaerial flank of Pico volcano was dismissed as unlikely in Mitchell (2003) but probably also is a landslide based on the presence of an embayment of the volcano topography on land typical of a landslide headwall. More recently, Costa et al. (2014) have interpreted two further landslides using multibeam data from the Pico-São Jorge channel. Such landslides, if they were to occur today, could produce unusually large tsunami impacts on the adjacent island coasts because the channel may effectively form a reflecting waveguide (Omira et al. 2016).

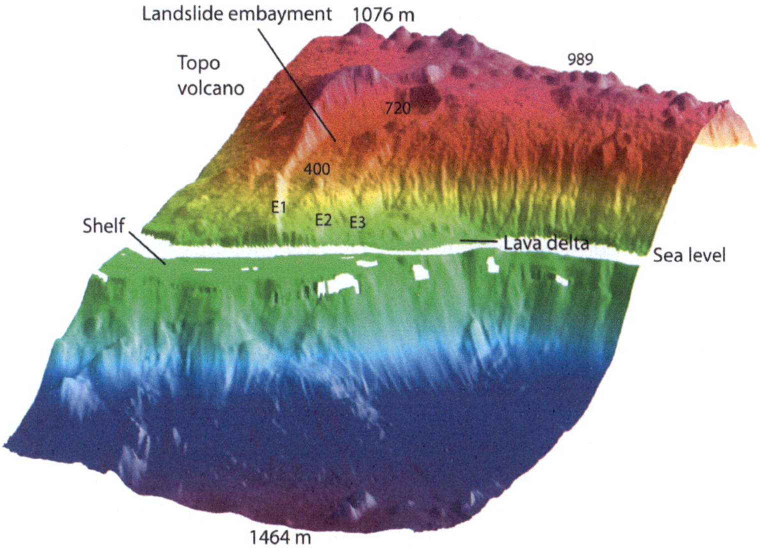

Fig. 14 3D oblique view of topography and bathymetry data from the south coast of Pico Island showing slump structure on land (Mitchell et al. 2012b) (view location shown by the arrow in Fig. 2). Numbers 400–989 are altitudes (m) of volcanic cones identified as potential sources of lava forming the Holocene lava delta marked on the figure. Geophysical data collected on the shelf just below sea level here show little evidence of relief on faults, suggesting little movement of the slump has occurred during the Holocene

The conditions at the time of failure are unknown for major pre-historic landslides, hence many reasons for failure have been suggested (Keating and McGuire 2000). However, the lack of major landsliding in the Azores is consistent with the observation that landslides or sector collapses are common in volcanic ocean islands and seamounts taller than 2500 m but rare in those shorter than 2500 m (Mitchell 2003) as the Azores Islands lie on an elevated plateau (Fig. 1a). In oceanic seamounts and guyots, this transition also coincides with a transition in the shapes of seamounts, whereby the taller edifices are more star-like in plan-view, with radiating volcanic ridges (Mitchell 2001). Landslides commonly form embayments between the ridges, thus contributing to the star-like morphology. The fact that volcanic rift zones occur in the Azores (where they appear in shorter edifices than in the global pattern) thus reflects their plate-tectonic rather than volcano-tectonic origin.

Some of the explanations for deep-seated slumping that have been suggested for the Hawaiian islands probably do not apply here. For example, the Hawaiian islands are built on a layer of impermeable hemipelagic sediment which forms a basal décollement (Moore et al. 1994), a ductile layer of thick olivine cumulates may induce movement (Clague and Denlinger 1994) and over-pressuring of pore fluids during intrusions may temporarily elevate fluid pressures (Elsworth and Voight 1995). The presence of an igneous basement in the Azores may imply higher permeability than beneath the Hawaiian islands and there is likely to have been less compaction of volcanic materials leading to permeability reduction given the shorter elevations of the islands, so pore fluids are potentially less pressurized. The lower volcanic building rate suggests that major bodies of hot olivine cumulate are less likely, although a density anomaly suggesting a possible intrusion has been identified in the SE of Pico Island (Nunes et al. 2006) beneath the slump described below.

Despite the general absence of major slumping, there is potential evidence that one of the slumps is active, a rare feature of volcanic ocean islands and so worthy of further study. Mitchell et al. (2012b) studied the Topo landslide using boomer seismic and chirp sediment profiler data collected on the shelf as well as coincident multibeam data. The island shelves were last modified by surf erosion during sea level transgression after the last glacial maximum (Quartau et al. 2010) so the presence or absence of relief on faults can help reveal whether they have been active in the Holocene after they became isolated from surf. Where the three onshore escarpments E1-E3 observed in Fig. 14 continue offshore, little evidence was found of fault relief on the shelf. Furthermore, no offsets are observed in the lavas on land within the east and north sides of the landslide embayment (Madeira 1998). In contrast, Hildenbrand et al. (2012) have published GPS and InSAR data, which suggest that parts of the coast here within the landslide embayment are presently mobile and some evidence from a separate GPS campaign (Madeira, personal communication, 2012) also suggest the feature is mobile at the coast. The apparent discrepancy between these results might indicate that the landslide has only recently begun moving (Mitchell et al. 2012c).

In the uppermost submarine slopes of the islands, smaller, superficial landslides are common (Mitchell et al. 2008; Quartau et al. 2012). For example, the multibeam data in Fig. 15, from the south coast of Pico Island show a number of embayments of the uppermost slope forming discrete depressions of the seabed. The depressions, which are several 10 s of metres deep, probably represent failure of material that is cohesive, because steep head- and side-walls remain (steeper than the ~30° angle of repose of granular material). Whereas the hazards from major landslides have received attention as potentially causing ocean-traversing tsunamis

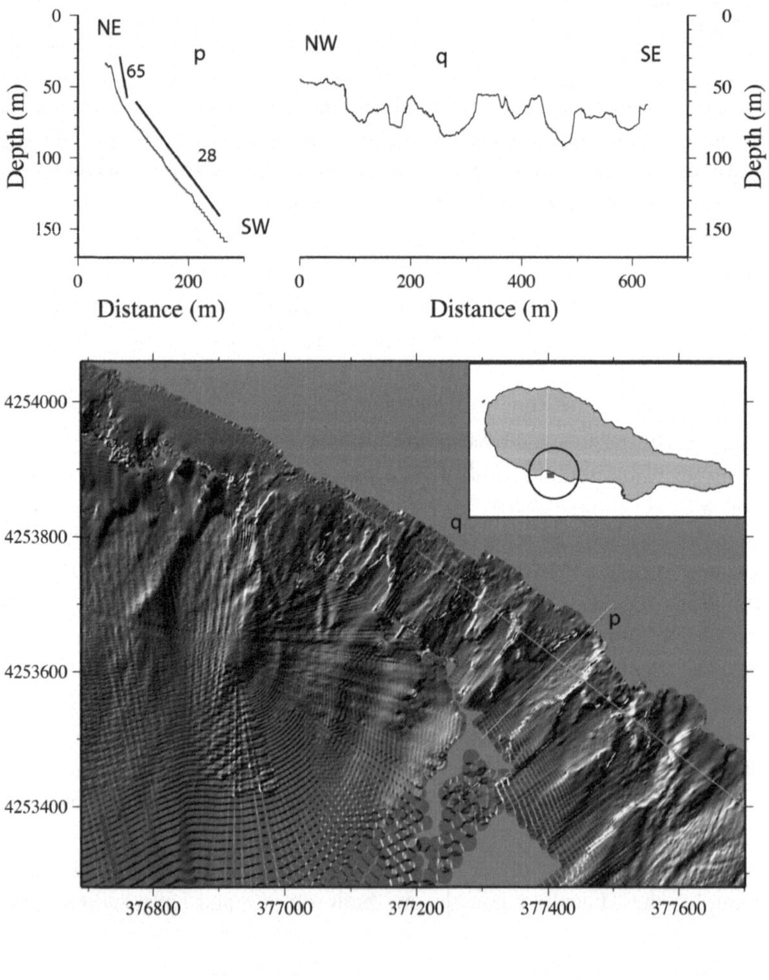

Fig. 15 Examples of surficial landslides in the uppermost slope of south Pico Island in shaded relief imagery of the 2003 *Arquipelago* Reson 8160 multibeam sonar data (shaded from the upper-left). The data have been gridded at 0.25 m. Fan-shaped pattern on left is an artifact from where the vessel was turning. Lines locate the cross-sections shown in upper panels. Values above cross-sections are gradients in degrees (the steep headwall gradients above the typical angle of repose of clastic material ~30° suggest that the failed material has finite cohesion). Coordinates are UTM metres

(Ward and Day 2001), these smaller landslides may present a threat to local populations by creating smaller but nevertheless locally significant water surges. Such a surge was caused by a landslide entering the sea at Stromboli island, Italy (Tinti et al. 2005). Similar upper submarine slope landslides may explain some of the local tsunamis in the Azores that have not been associated with specific earthquakes (Andrade et al. 2006). Although much smaller than the large trans-ocean tsunamis, such small landslide events are probably more frequent.

9 Future Submarine Research in the Azores

Further work on the morphology of volcanic structures can be expected to be published over the coming years, based on the new datasets. The islands of Flores and Corvo remain to be surveyed but evidently will likely reveal interesting features if surveyed also. The potentially active slump in south Pico should be subject to detailed monitoring. Given the number of lava flows and cones now well characterised in the various kinds of sonar data, there is ample scope for more detailed work based on samples, photographic observations and relating their structures to lavas on land (requiring perhaps scientific diving). However, although there is certainly much to be learned though "forensic" earth science, a remaining challenge will be to observe active processes with the level of detail that is common in subaerial volcanology (Mitchell 2012d). For example, the branching behaviours of these submarine lava flows may reflect an interesting effect of cooling by seawater on their visco-elastic layers, an inference that is difficult to test with further observations of already emplaced flows. In the ocean, we are unable to measure temperatures with infra-red emissions as is possible on land but heat flow might instead be measurable from heat advection above actively developing submarine lavas and could be combined with morphologic and other data from autonomous vehicles.

Acknowledgements The research teams and crews of research vessels involved in collecting the datasets shown here are thanked for their work, in particular, on RVs *Arquipélago* and *Urania*. Rui Quartau (UGM-LNEG, Portugal) is acknowledged for many discussions and help with these efforts, as also are researchers of the ISMAR, Bologna who processed the TOBI data. Christoph Beier and colleagues provided the samples and descriptions shown in Fig. 3. Funding was provided by a large number of agencies listed in the cited publications. Reviews by Sebastian Krastel and David Chadwell were very helpful for improving this chapter.

References

Andrade, C., Borges, P., & Freitas, M. C. (2006). Historical tsunami in the Azores archipelago (Portugal). *Journal of Volcanology and Geothermal Research, 156,* 172–185.

Azevedo, J. M. M., & Portugal Ferreira, M. R. (2009). The volcanotectonic evolution of Flores Island, Azores (Portugal). *Journal of Volcanology and Geothermal Research, 156,* 90–102.

Bates, C. R. (2005). Sub-bottom profiling for sand resource assessment, Flores island, Azores. TEAM Internal Report 05-04, 21 pp.

Batiza, R. (1977). Petrology and chemistry of Guadalupe Island: An alkalic seamount on a fossil ridge crest. *Geology, 5,* 760–764.

Beier, C. (2016). Short Cruise Report R/V METEOR – M128, Ponta Delgada - Ponta Delgada 2. July 2016–27. July 2016. In Friedrich-Alexander Universität Erlangen-Nürnberg, Erlangen, Germany (p. 13).

Borges, J. F., Bezzeghoud, M., Buforn, E., Pro, C., & Fitas, A. (2007). The 1980, 1997 and 1998 Azores earthquakes and some seismo-tectonic implications. *Tectonophysics, 435,* 37–54.

Booth, B., Croasdale, R., & Walker, G. P. L. (1978). A quantitative study of five thousand years of volcanism on Sao Miguel, Azores. *Philosophical Transactions of the Royal Society of London A, A288,* 271–319.

Calais, E., DeMets, C., & Nocquet, J.-M. (2003). Evidence for a post-3.16-Ma change in Nubia-Eurasia-North America plate motions? *Earth and Planetary Science Letters, 216,* 81–92.

Cardigos, F., Colaço, A., Dando, P. R., Ávila, S. P., Sarradin, P.-M., Tempera, F., et al. (2005). Shallow water hydrothermal vent field fluids and communities of the D. João de Castro Seamount (Azores). *Chemical Geology, 224,* 153–168.

Cashman, K. V., & Fiske, R. S. (1991). Fallout of pyroclastic debris from submarine volcanic eruptions. *Science, 253,* 275–280.

Casalbore, D., Romagnoli, C., Pimentel, A., Quartau, R., Casas, D., Ercilla, G., et al. (2015). Volcanic, tectonic and mass-wasting processes offshore Terceira Island

(Azores) revealed by high-resolution seafloor mapping. *Bulletin of Volcanology, 77,* 24.

Chadwick, W. W., Jr., Wright, I. C., Schwartz-Schampera, U., Hyvernaud, O., Reymond, D., & de Ronde, C. E. J. (2008). Cyclic eruptions and sector collapses at Monowai submarine volcano, Kermadec arc: 1998–2007. *Geochemistry, Geophysics, Geosystems, 9,* Q10014. https://doi.org/10010.11029/12008GC002113.

Chiocci, F. L., Romagnoli, C., Casalbore, D., Sposato, A., Martorelli, E., Alonso, B., et al. (2013). Bathy-morphological setting of Terceira Island (Azores) after the FAIVI cruise. *Journal of Maps, 9,* 590–595.

Clague, D. A., & Dalrymple, G. B. (1989). Tectonics, geochronology, and origin of the Hawaiian-Emperor volcanic chain. In E. L. Winterer, D. M. Hussong, & R. W. Decker (Eds.), *The geology of North America, Volume N: The Eastern Pacific Ocean and Hawaii* (pp. 188–217). Boulder, CO: The Geological Society of America.

Clague, D. A., & Denlinger, R. P. (1994). Role of olivine cumulates in destabilizing the flanks of Hawaiian volcanoes. *Bulletin of Volcanology, 56,* 425–434.

Clague, D. A., Moore, J. G., & Reynolds, J. R. (2000). Formation of flat-topped volcanic cones in Hawai'i. *Bulletin of Volcanology, 62,* 214–233.

Cole, P. D., Guest, J. E., Duncan, A. M., & Pacheco, J.-M. (2001). Capelinhos 1957–1958, Faial, Azores: Deposits formed by an emergent surtseyan eruption. *Bulletin of Volcanology, 63,* 204–220.

Conder, J. A., Scheirer, D. S., & Forsyth, D. W. (2000). Seafloor spreading on the Amsterdam-St. Paul hotspot plateau. *Journal of Geophysical Research, 105,* 8263–8277.

Costa, A. C. G., Hildenbrand, A., Marques, F. O., Sibrant, A. L. R., & Santos de Campos, A. (2015). Catastrophic flank collapses and slumping in Pico Island during the last 130 kyr (Pico-Faial ridge, Azores Triple Junction). *Journal of Volcanology and Geothermal Research, 302,* 33–46.

Costa, A. C. G., Marques, F. O., Hildenbrand, A., Sibrant, A. L. R., & Catita, C. M. S. (2014). Large-scale flank collapses in a steep volcanic ridge: Pico-Faial Ridge, Azores Triple Junction. *Journal of Volcanology and Geothermal Research, 272,* 111–125.

Davis, A. S., Clague, D. A., Bohrson, W. A., Dalrymple, G. B., & Greene, H. G. (2002). Seamounts at the continental margin of California: A different kind of oceanic intraplate volcanism. *Geological Society of America Bulletin, 114,* 316–333.

Dias Correia, A. J. (2010). O Mar no Século XXI. Fedrave.

Dieterich, J. H. (1988). Growth and persistence of Hawaiian volcanic rift zones. *Journal Geophysical Research, 93*(B5), 4258–4270.

Elsworth, D., & Voight, B. (1995). Dike intrusion as a trigger for large earthquakes and the failure of volcano flanks. *Journal of Geophysical Research, 100,* 6005–6024.

Embley, R. W., Murphy, K. M., & Fox, C. G. (1990). High-resolution studies of the summit of axial volcano. *Journal of Geophysical Research, 95,* 12785–12812.

Fabriol, H., Delgado-Argote, L. A., Dañobeitia, J. J., Córdoba, D., González, A., García-Abdeslem, J., et al. (1999). Backscatering and geophysical features of volcanic ridges offshore Santa Rosalia, Baja California Sur, Gulf of California, Mexico. *Journal of Volcanology and Geothermal Research, 93,* 75–92.

Fernandes, R. M. S., Bastos, L., Miranda, J. M., Lourenço, N., Amrosius, B. A. C., Noomen, R., et al. (2006). Defining the plate boundaries in the Azores region. *Journal of Volcanology and Geothermal Research, 156,* 1–9.

Fiske, R. S., & Jackson, E. D. (1972). Orientation and growth of Hawaiian volcanic rifts: The effect of regional structure and gravitational stresses. *Proceedings of the Royal Society of London, Series A, 329,* 299–326.

Fitch, F. J. (1964). The development of the Beerenberg volcano, Jan Mayen. *Proceedings of the Geologists' Association, 75,* 133–165.

França, Z., Lago, M., Nunes, J. C., Galé, C., Forjaz, V. H., Pueyo, O., et al. (2006). Geochemistry of alkaline basalts of Corvo Island (Azores, Portugal): Preliminary data. *Geogaceta, 40,* 87–90.

Freire Luis, J., Lourenco, N., Miranda, J. M., Gaspar, J. L., & Queiroz, G. (1999). A submarine eruption west of Terceira Island (Azores Archipelago). *InterRidge News, 8*(1), 13–14.

Freire Luis, J., Miranda, J. M., Galdeano, A., Patriat, P., Rossignol, J. C., & Victor, L. A. M. (1994). The Azores triple junction evolution since 10 Ma from an aeromagnetic survey of the Mid-Atlantic Ridge. *Earth and Planetary Science Letters, 125,* 439–459.

Garcia, M. O., & Davis, M. G. (2001). Submarine growth and internal structure of ocean island volcanoes based on submarine observations of Mauna Loa volcano, Hawaii. *Geology, 29,* 163–166.

Gaspar, J. L., Queiroz, G., Pacheco, J. A., Ferreira, T., Wallenstein, N., Almeida, M. H., & Coutinho, R. (2003). Basaltic lava balloons produced during the 1998–2001 Serreta Submarine Ridge eruption (Azores). In J. D. L. White, J. L. Smellie, & D. A. Clague (Eds.), *Subaqueous explosive volcanism: Vol. 140. American Geophysical Union Geophysical Monograph* (pp. 205–212). Washington, DC: American Geophysical Union.

Gee, M. J. R., Masson, D. G., Watts, A. B., & Mitchell, N. C. (2001a). Offshore continuation of volcanic rift zones, El Hierro, Canary Islands. *Journal of Volcanology and Geothermal Research, 105,* 107–119.

Gee, M. J. R., Watts, A. B., Masson, D. G., & Mitchell, N. C. (2001b). Landslides and the evolution of El Hierro in the Canary Islands. *Marine Geology, 177,* 271–293.

Gente, P., Dyment, J., Maia, M., & Goslin, J. (2003). Interaction between the Mid-Atlantic Ridge and the Azores hot spot during the last 85 Myr: Emplacement and rifting of the hot spot-derived plateaus. *Geochemistry, Geophysics, Geosystems, 4,* 8514. https://doi.org/8510.1029/2003GC000527.

Gracia, E., Canals, M., Li Farran, M., Preito, M. J., Sorribas, J., & Team, G. (1996). Morphostructure and evolution of the central and eastern Bransfield basins (Nw Antarctic Peninsula). *Marine Geophysical Researches, 18,* 429–448.

Hildenbrand, A., Madureira, P., Marques, F. O., Cruz, I., Henry, B., & Silva, P. (2008). Multi-stage evolution of a sub-aerial volcanic ridge over the last 1.3 Myr: S. Jorge Island, Azores Triple Junction. *Earth and Planetary Science Letters, 273,* 289–298.

Hildenbrand, A., Marques, F. O., Catalão, J., Catita, C. M. S., & Costa, A. C. G. (2012). Large-scale active slump of the southeastern flank of Pico Island, Azores. *Geology, 10,* 939–942.

Hirn, A., Lépine, J.-C., & Sapin, M. (1993). Triple junction and ridge hotspots: Earthquake, faults, and volcanism in Afar, the Azores, and Iceland. *Journal of Geophysical Research, 98,* 11995–12001.

Holcomb, R. T., & Searle, R. C. (1991). Large landslides from oceanic volcanoes. *Marine Geotechnology, 10,* 19–32.

Hon, K., Kauahikaua, J., Denlinger, R., & Mackay, K. (1994). Emplacement and inflation of pahoehoe sheet flows: Observations and measurements of active lava flows on Kilauea Volcano, Hawaii. *Geological Society of America Bulletin, 106,* 351–370.

Hübscher, C. (2009). Short cruise report: RV Meteor cruise M79/2, São Miguel/Azores to Las Palmas/Gran Canaria/Spain, 26 August–21 September 2009. In University of Hamburg, Hamburg, p. 10.

Hübscher, C., et al. (2015). Cruise report: Azores Plateau, Cruise No. M113/1 December 29, 2014–January 22, 2015, Ponta Delgada (Portugal) - Ponta Delgada (Portugal). Ozeanographie D-Sf (ed), Bremen, p. 31.

Keating, B. H., & McGuire, W. J. (2000). Island edifice failure and associated tsunami hazards. *Pure and Applied Geophysics, 157,* 899–955.

Kokelaar, B. P., & Durant, G. P. (1983). The submarine eruption and erosion of Surtla (Surtsey), Iceland. *Journal of Volcanology and Geothermal Research, 19,* 239–246.

Krastel, S., Schmincke, H.-U., & Jacobs, C. L. (2001). Formation of submarine canyons on the flanks of the Canary Islands. *Geo-Marine Letters, 20,* 160–167.

Kueppers, U., Nichols, A. R. L., Zanon, V., Potuzak, M., & Pacheco, J. M. R. (2012). Lava balloons—Peculiar products of basaltic, submarine eruptions. *Bulletin of Volcanology, 74,* 1379–1393.

Larrea, P., França, Z., Lago, M., Widom, E., Galé, C., & Ubide, T. (2013). Magmatic processes and the role of antecrysts in the genesis of Corvo Island (Azores archipelago, Portugal). *Journal of Petrology, 54,* 769–793.

Ligi, M., Bonatti, E., Bortoluzzi, G., Carrara, G., Fabretti, P., Penitenti, D., et al. (1997). Death and transfiguration of a triple junction in the South Atlantic. *Science, 276,* 243–245.

Ligi, M., Mitchell, N. C., Marani, M., Gamberi, F., Penitenti, D., Carrara, G., et al. (1999). Giant volcanic ridges amongst the Azores islands. EOS Trans. AGU 80, Fall Meet. Suppl.:F913.

Lourenço, N., Miranda, J. M., Luis, J. F., Ribeiro, A., Victor, L. A. M., Madeira, J., et al. (1998). Morpho-tectonic analysis of the Azores volcanic plateau from a new bathymetric compilation of the area. *Marine Geophysical Researches, 20,* 141–156.

Luis, J. F., & Miranda, J. M. (2008). Reevaluation of magnetic chrons in the North Atlantic between 35°N and 47°N: Implications for the formation of the Azores Triple Junction and associated plateau. *Journal of Geophysical Research, 113.* https://doi.org/10.1029/2007jb005573.

Madeira, J. (1998). Estudos de neotectónica nas ilhas do Faial, Pico e S. Jorge: uma contribuição para o conhecimento geodinâmico da junção tripla dos Açores. Lisbon University, Lisbon, Portugal, p. 481.

Madeira, J., & Ribeiro, A. (1990). Geodynamic models for the Azores triple junction: A contribution from tectonics. *Tectonophysics, 184,* 405–415.

Madeira, J., da Silveira, A. B. (2003). Active tectonics and first paleoseismological results in Faial, Pico and S. Jorge islands (Azores, Portugal). *Annals of Geophysics, 46,* 733–761.

Marani, M. P., & Trua, T. (2002). Thermal constriction and slab tearing at the origin of a superinflated spreading ridge: Marsili volcano (Tyrrhenian Sea). *Journal Geophysical Research, 107,* 2188. https://doi.org/2110.1029/2001JB000285.

Mattox, T. N., & Mangan, M. T. (1997). Littoral hydrovolcanic explosions: A case study of lava-seawater interaction at Kilauea Volcano. *Journal of Volcanology and Geothermal Research, 75,* 1–17.

McCoss, A. M. (1986). Simple constructions for deformation in transpression/transtension zones. *Journal of Structural Geology, 8,* 715–718.

McKenzie, D. P. (1972). Active tectonics of the Mediterranean region. *Geophysical Journal Royal Astronomical Society, 30,* 109–185.

Miranda, J. M., Luis, J. F., Lourenço, N., & Fernandes, R. M. S. (2015). The structure of the Azores triple junction: Implications for São Miguel Island. *Geological Society of London Memoirs, 44,* 5–13.

Miranda, J. M., Luis, J. F., & Lourenço, N. (this volume) The tectonic evolution of the Azores based on magnetic data. In: U. Kueppers & C. Beier (Eds.), *Volcanoes of the Azores.* Springer series "Active volcanoes".

Mitchell, N. C. (1991). Improving GLORIA images using Sea Beam data. *Journal of Geophysical Research, 96* (B1), 337–351.

Mitchell, N. C. (2001). The transition from circular to stellate forms of submarine volcanoes. *Journal of Geophysical Research, 106,* 1987–2003.

Mitchell, N. C. (2003). Susceptibility of mid-ocean ridge volcanic islands and seamounts to large-scale landsliding. *Journal of Geophysical Research, 108.* https://doi.org/10.1029/2002jb001997.

Mitchell, N. C. (2012). Hot, cracking rocks deep down. *Nature Geoscience, 5*, 444–445.

Mitchell, N. C., Beier, C., Rosin, P., Quartau, R., & Tempera, F. (2008). Lava penetrating water: Submarine lava flows around the coasts of Pico Island, Azores. *Geochemistry, Geophysics, Geosystems, 9*, Q03024. https://doi.org/03010.01029/02007GC001725.

Mitchell, N. C., Dade, W. B., & Masson, D. G. (2003). Erosion of the submarine flanks of the Canary Islands. *Journal of Geophysical Research, 108*. https://doi.org/10.1029/2002jf000003.

Mitchell, N. C., & Livermore, R. A. (1998). Spiess Ridge: An axial high on the slow-spreading Southwest Indian Ridge. *Journal of Geophysical Research, 103*, 15457–15471.

Mitchell, N. C., Masson, D. G., Watts, A. B., Gee, M. J. R., & Urgeles, R. (2002). The morphology of the flanks of volcanic ocean islands: A comparative study of the Canary and Hawaiian hotspot islands. *Journal of Volcanology and Geothermal Research, 115*, 83–107.

Mitchell, N. C., Stretch, R., Oppenheimer, C., Kay, D., & Beier, C. (2012a). Cone morphologies associated with shallow marine eruptions: East Pico Island, Azores. *Bulletin of Volcanology, 74*, 2289–2300.

Mitchell, N. C., Quartau, R., & Madeira, J. (2012b). Assessing landslide movements in volcanic islands using near-shore marine geophysical data: South Pico island, Azores. *Bulletin of Volcanology, 74*, 483–496.

Mitchell, N. C., Quartau, R., & Madeira, J. (2012c). Comment on: "Large-scale active slump of the southeastern flank of Pico Island, Azores" by A. Hildenbrand et al. *Geology, 41*, e301.

Mitchell, N. C. (2012d). Aspects of marine geoscience: a review and thoughts on potential for observing active processes and progress through collaboration between the ocean sciences. *Philosophical Transactions of the Royal Society A, 370*, 5567–5612.

Mitchell, N. C., Tivey, M. A., & Gente, P. (2000). Slopes of mid-ocean ridge fault scarps from submersible observations. *Earth and Planetary Science Letters, 183*, 543–555.

Moore, J. G. (1985). Structure and eruptive mechanisms at Surtsey Volcano, Iceland. *Geological Magazine, 122*, 649–661.

Moore, J. G., & Chadwick, W. W. (1995). Offshore geology of Mauna Loa and adjacent areas, Hawaii. In J. M. Rhodes & J. P. Lockwood (Eds.), *Mauna Loa Revealed: Structure, composition, history and hazards* (pp. 21–44). Washington DC: American Geophysical Union.

Moore, J. G., Clague, D. A., Holcomb, R. T., Lipman, P. W., Normark, W. R., & Torresan, M. E. (1989). Prodigious submarine landslides on the Hawaiian Ridge. *Journal of Geophysical Research, 94*, 17465–17484.

Moore, J. G., Philips, R. L., Grigg, R. W., Peterson, D. W., & Swanson, D. A. (1973). Flow of lava into the sea, 1969–1971, Kilauea volcano, Hawaii. *Geological Society of America Bulletin, 84*, 537–546.

Moore, J. G., Normark, W. R., & Holcomb, R. T. (1994). Giant Hawaiian landslides. *Annual Reviews of Earth and Planetary Sciences, 22*, 119–144.

Moore, R. B. (1990). Volcanic geology and eruption frequency, São Miguel, Azores. *Bulletin of Volcanology, 52*, 602–614.

Nunes, J. C., Camacho, A., França, Z., Montesinos, F. G., Alves, M., Vieira, R., et al. (2006). Gravity anomalies and crustal signature of volcano-tectonic structures of Pico Island (Azores). *Journal of Volcanological and geothermal Research, 156*, 55–70.

Obata, S., & Umino, S. (1999). Morphology of the A.D. 864 Aokigahara lava flow exposed on the coast of Motosuku Lake, Fiji Volcano (in Japanese). *Bulletin of the Volcanological Society of Japan, 44*, 201–216.

Omira, R., Quartau, R., Ramalho, I., Baptista, M. A., & Mitchell, N. C. (2016). The tsunami effects of a collapse of a volcanic island on a semienclosed basin: The Pico-São Jorge Channel in the Azores Archipelago. In J. C. Duarte & W. P. Schellart (Eds.), *Plate boundaries and hazards: Vol. 219. Geophysical Monograph* (pp. 271–287). American Geophysical Union and Wiley.

Parson, L. M., Murton, B. J., Searle, R. C., Booth, D., Evans, J., Field, P., et al. (1993). En echelon axial volcanic ridges at the Reykjanes Ridge: A life cycle of volcanism and tectonics. *Earth and Planetary Science Letters, 117*, 73–87.

Quartau, R., Curado, F., Bouriak, S., Monteiro, J. H., & Pinheiro, L. (2003). Projecto Gemas – Localização e distribuição de areias em redor da ilha do Pico, Relatório Técnico INGMARDEP 16/2003. In Dept. de Geologia Marinha, IGM, Lisbon, Portugal, p. 49.

Quartau, R., Duarte, H., & Brito, P. (2005). Projecto Gemas – Relatório da campanha de amostragem de sedimentos (FAPI-3) realizada na plataforma e na orla costeira das ilhas do Faial e do Pico. Relatório Técnico INGMARDEP 2/2005. In Dept. de Geologia Marinha e Costeira, INETI, I.P., Lisbon, Portugal, p. 17.

Quartau, R., Hipólito, A., Romagnoli, C., Casalbore, D., Madeira, J., Tempera, F., et al. (2014). The morphology of insular shelves as a key for understanding the geological evolution of volcanic islands: Insights from Terceira Island (Azores). *Geochemistry, Geophysics, Geosystems, 15*, 1801–1826.

Quartau, R., Madeira, J., Mitchell, N. C., Tempera, F., Silva, P. F., & Brandão, F. (2015a). The insular shelves of the Faial-Pico Ridge (Azores archipelago): A morphological record of its evolution. *Geochemistry, Geophysics, Geosystems, 16*, 1401–1420.

Quartau, R., Madeira, J., Mitchell, N. C., Tempera, F., Silva, P. F., & Brandão, F. (2015b). Reply to comment by Marques et al. on ""The insular shelves of the Faial-Pico Ridge (Azores archipelago): A morphological record of its evolution"". *Geochemistry, Geophysics, Geosystems, 17*. https://doi.org/10.1002/2015gc006180.

Quartau, R., & Mitchell, N. C. (2013). Comment on "Reconstructing the architectural evolution of volcanic islands from combined K/Ar, morphologic, tectonic, and magnetic data: The Faial Island example (Azores)" by Hildenbrand et al. (2012). *Journal of Volcanology and Geothermal Research, 255,* 124–126.

Quartau, R., Tempera, F., Mitchell, N. C., Pinheiro, L. M., Duarte, H., Brito, P. O., et al. (2012). Morphology of Faial Island's shelf: The results of volcanic, erosional, depositional and mass-wasting processes. *Geochemistry, Geophysics, Geosystems, 13,* Q04012. https://doi.org/04010.01029/02011GC003987.

Quartau, R., Trenhaile, A. S., Mitchell, N. C., & Tempera, F. (2010). Development of volcanic insular shelves: Insights from observations and modelling of Faial Island in the Azores Archipelago. *Marine Geology, 275,* 66–83.

Ryan, W. B. F., Carbotte, S. M., Coplan, J. O., O'Hara, S., Melkonian, A., Arko, R., et al. (2009). Global multi-resolution topography synthesis. *Geochemistry, Geophysics, Geosystems, 10,* Q03014.

Saemundsson, K. (1986). Subaerial volcanism in the western North Atlantic. In P. R. Vogt & B. E. Tucholke (Eds.), *The geology of North America, Volume M, The western North Atlantic region* (pp. 69–86). Geological Society of America.

Sansone, F. J., & Smith, J. R. (2006). Rapid mass wasting following nearshore submarine volcanism on Kilauea volcano, Hawaii. *Journal of Volcanology and Geothermal Research, 151,* 133–139.

Santos, R. S., Tempera, F., Colaço, A., Cardigos, F., & Morato, T. (2010). Dom João de Castro Seamount. *Oceanography, 23,* 200–201.

Searle, R. C. (1976). Lithospheric structure of the Azores Plateau from Rayleigh-wave dispersion. *Geophysical Journal Royal Astronomical Society, 44,* 537–546.

Searle, R. C. (1980). Tectonic pattern of the Azores spreading centre and triple junction. *Earth and Planetary Science Letters, 51,* 415–434.

Sibrant, A. L. R., Hildenbrand, A., Marques, F. O., Weiss, B., Boulesteix, T., Hübscher, C., et al. (2015). Morpho-structural evolution of a volcanic island developed inside an active oceanic rift: S. Miguel Island (Terceira Rift, Azores). *Journal of Volcanological and Geothermal Research, 301,* 90–106.

Sterl, A., & Caires, S. (2005). Climatology, variability and extrema of ocean waves: The web-based KNMI/ERA-40 wave atlas. *International Journal of Climatology, 25,* 963–977.

Stevenson, J. A., Mitchell, N. C., Mochrie, F., Cassidy, M., & Pinkerton, H. (2012). Lava penetrating water: The different behaviours of pahoehoe and 'a'ā at the Nesjahraun, Pingvellir, Iceland. *Bulletin of Volcanology, 74,* 33–46.

Stretch, R. (2007). *A morphometric and textural analysis of the submarine volcanic ridges of the Azores Plateau.* Ph.D. thesis, Faculty of Earth Sciences and Geography. University of Cambridge, Cambridge, p. 360.

Stretch, R., Mitchell, N. C., & Portaro, R. A. (2006). A morphometric analysis of the submarine volcanic ridge of Pico Island. *Journal of Volcanology and Geothermal Research, 156,* 35–54.

Tempera, F. (2008). *Benthic habitats of the extended Faial Island shelf and their relationship to geologic, oceanographic and infralittoral biological features.* Ph.D. thesis, University of St Andrews, St Andrews, UK.

Tempera, F., Giacomello, E., Mitchell, N. C., Campos, A. S., Henriques, A. B., Martins, A., et al. (2012a). Mapping the Condor seamount seafloor environment and associated biological assemblages (Azores, NE Atlantic). In P. T. Harris et al. (Eds.), *Seafloor geomorphology as benthic habitat: Geohab atlas of seafloor geomorphic features and benthic habitats* (pp. 807–818). London: Elsevier.

Tempera, F., MacKenzie, M., Bashmachnikov, I., Puotinen, M., Santos, R. S., & Bates, R. (2012b). Predictive modeling of dominant macroalgae abundance on temperate island shelves (Azores, Northeast Atlantic). In P. T. Harris et al. (Eds.), *Seafloor geomorphology as benthic habitat: Geohab atlas of seafloor geomorphic features and benthic habitats* (pp. 169–184). London: Elsevier.

Tinti, S., Manucci, A., Pagnoni, G., Armigliato, A., & Zaniboni, F. (2005). The 30 December 2002 landslide-induced tsunamis in Stromboli: Sequence of events reconstructed from the eyewitness accounts. *Natural Hazards and Earth System Science, 5,* 763–775.

Vogt, P. R., & Jung, W. Y. (2004). The Terceira Rift as a hyper-slow, hotspot-dominated oblique spreading axis: A comparison with other slow-spreading plate boundaries. *Earth and Planetary Science Letters, 218,* 77–90.

Vogt, P. R., & Smoot, N. C. (1984). The Geisha Guyots: Multi-beam bathymetry and morphometric interpretation. *Journal of Geophysical Research, 89,* 11085–11107.

Walker, G. P. L. (1999). Volcanic rift zones and their intrusion swarms. *Journal of Volcanology and Geothermal Research, 94,* 21–34.

Wanless, V. D., Garcia, M. O., Trusdell, F. A., Rhodes, J. M., Norman, M. D., Weis, D., et al. (2006) Submarine radial vents on Mauna Loa Volcano, Hawai'i. *Geochemistry, Geophysics, Geosystems, 7,* Article Q05001. doi:05010.01029/02005GC001086.

Ward, S. N., & Day, S. (2001). Cumbre Vieja Volcano – Potential collapse and tsunami at La Palma, Canary Islands. *Geophysical Research Letters, 28,* 3397–3400.

Weiss, B. J., Hübscher, C., & Lüdmann, T. (2015a). The tectonic evolution of the southeastern Terceira Rift/São Miguel region (Azores). *Tectonophysics, 654,* 75–95.

Weiss, B. J., Hübscher, C., Lüdmann, T., & Serra, N. (2016). Submarine sedimentation processes in the southeastern Terceira Rift/São Miguel region (Azores). *Marine Geology, 374,* 42–58.

Weiss, B. J., Hübscher, C., Wolf, D., & Lüdmann, T. (2015b). Submarine explosive volcanism in the southeastern Terceira Rift/São Miguel region (Azores). *Journal of Volcanology and Geothermal Research, 303,* 79–91.

Weston, F. S. (1964). List of recorded volcanic eruptions in the Azores with brief reports. *Boletim do Museu e Laboratorio Mineralogico e Geologico da Faculdade de Ciencias (Lisbon), 10*(1), 3–20.

Yang, T., Shen, Y., van der Lee, S., Solomon, S. C., & Hung, S.-H. (2006). Upper mantle structure beneath the Azores hotspot from finite-frequency seismic tomography. *Earth and Planetary Science Letters, 250,* 11–26.

Characterisation of Seismicity of the Azores Archipelago: An Overview of Historical Events and a Detailed Analysis for the Period 2000–2012

João Fontiela, Carlos Sousa Oliveira and Philippe Rosset

Abstract

The Azores Archipelago is located in the Middle Atlantic Ridge, at the Triple Junction formed by the contact of the Euro-Asiatic, the Nubia (African) and the American plates (see also Vogt and Jung, Chapter "The "Azores Geosyndrome" and Plate Tectonics: Research History, Synthesis, and Unsolved Puzzles"). Its seismicity rate is very high with earthquakes with relatively low magnitude, defining quite well the contact regions. This chapter gives an overview of the existing historical and instrumental catalogues, describes the seismicity of the region essentially since early 1915, and analyses in more detail the characteristics of the recorded data in the period 2000–2012. The spatial variations of the minimum magnitude of completeness (Mc) as well as the b-value is studied for this period within the stripe of observed seismicity which contains the alignment of the Archipelago islands. A preliminary interpretation of the Mc and b-values is made keeping in mind the geological transition between the Gloria fault to the East and the Mid-Atlantic Region to the West.

Keywords

Azores tectonic structures · Earthquake catalogue · Seismic zoning · Minimum magnitude of completeness · b-value

Electronic supplementary material
The online version of this chapter (https://doi.org/10.1007/978-3-642-32226-6_8) contains supplementary material, which is available to authorized users.

J. Fontiela (✉)
Instituto Ciências da Terra (ICT), Escola de Ciências e Tecnologia (ECT), Universidade de Évora, Colégio Luís Verney, Rua Romão Ramalho 59, 7000-671 Évora, Portugal
e-mail: jfontiela@uevora.pt

C. Sousa Oliveira
Instituto Superior Técnico, CEris, Universidade de Lisboa, Lisbon, Portugal

P. Rosset
ICES, International Centre for Earth Simulation Foundation, Geneva, Switzerland

1 Introduction

The Azores Archipelago is composed of nine islands divided into three groups; from west to east, the Western group includes Flores and

U. Kueppers and C. Beier (eds.), *Volcanoes of the Azores*, Active Volcanoes of the World, https://doi.org/10.1007/978-3-642-32226-6_8

Corvo, the Central group is composed of Graciosa, Faial, Pico, São Jorge and Terceira and the Eastern group comprises São Miguel and Santa Maria. The Archipelago is located at the Triple Junction of the Eurasia, Nubia and, the North America lithospheric plates. The Central and Eastern groups are eastward of the Middle Atlantic Ridge (MAR) and at the boundary of the Eurasian and Nubian plates. The Western group is located to the west of MAR and lies on the American plate. The Azores Islands lie on a plateau formed by tectonic and/or magmatic activity. Various theories have been proposed (e.g., a melting anomaly, Schilling 1975); Vogt and Jung (Chapter "The "Azores Geosyndrome" and Plate Tectonics: Research History, Synthesis, and Unsolved Puzzles") have compiled an overview of the models explaining the presence of the volcanic islands of the Azores. The eastern side of the Azores region has a roughly triangular shape delimited by three tectonic discontinuities: the East Fracture Zone (EFZ) in the south, the MAR in the west and the Terceira Rift (TR) in the north. The EFZ is oriented E-W with a seismic activity decreasing to the East, and jumped from the EFZ to a point further north (Luís et al. 1994). MAR is a pure extensional structure, seismically active and, divided into several segments. At the 38°50′ latitude North, the direction is N10°E with a low inflection to the south around N20°E. While the origin, morphology and geodynamic features of these two tectonics limits are consensual, it is not the case for the Terceira Rift which is still debated. TR has a general WNW-ESE orientation and is characterised by a sequence of basins, seamounts, and Islands. According to Udías (1980) and Buforn et al. (1988), it is an extensional zone normal to the MAR or an oblique extension for Searle (1980). Madeira and Ribeiro (1990) state that TR is a leaky transform. According to them, the compressive stress axis (σ_1) is horizontally rotating from N-S out of the boundaries limit to NW-SE in the boundary limits. The maximum tensional stress (σ_3) is horizontal with a NE-SW orientation. Due to the stress deviation near the boundary, the dextral faults change from NNW-SSE to WNW-ESE and sinistral faults from NNE-SSW to NNW-SSE. Nevertheless, the wideness of the disturbed zone given by the direction of the maximum and minimum compression and by the morphological features indicates that Azores domain is a diffuse plate boundary with an oblique ultra-slow spreading centre that accommodates the shear movements between the Nubia and Eurasia plate (Lourenço et al. 1998). According to Vogt and Jung (2004), the TR is an ultra-slow ridge that comprises the segment formed by the Graciosa, Terceira and São Miguel Islands.

Seismicity encompasses a stripe formed by the Graciosa, São Jorge, Faial, Pico, Terceira, São Miguel and Santa Maria Islands and goes to east along the Gloria Fault. Figure 1 depicts instrumental seismicity between 1915 and 2012. Most of the seismicity in this period has $M \leq 4$ and the maximum magnitude recorded in the instrumental period is M_S 7.1. Focal mechanisms of the Azores region calculated and revised by several authors are compiled in the Fig. 2 (McKenzie 1972; Arroyo and Udías 1972; Udías et al. 1976; Grimison and Chen 1986, Buforn et al. 1988; Borges et al. 2007; Bezzeghoud et al. 2014). The deformation process in TR is dominated by earthquakes with right-lateral strike-slip faults or normal faults oriented N120°E or N150°E (Grimison and Chen 1986; Buforn et al. 1988). Both faulting systems are under horizontal tensions with mean direction N25°E normal to TR (Buforn et al. 1988).

Several authors divide TR into two zones with distinct seismicity. Borges et al. (2007) and Bezzeghoud et al. (2014) distinguish them from the total seismic moment tensor whereas Fontiela et al. (2014) consider the frequency magnitude relation. The first seismic zone comprises the area between the MAR up to the Terceira Island and the second one, from the Terceira Island up to the transition of the Azores domain to the western end of the Gloria fault.

Testimonies of the seismicity of the Archipelago begin in the middle of the 15th century with the first Portuguese settlements in the Archipelago. One of the largest events ever known in the Azores occurred in 1522 causing heavy damage in Vila Franca do Campo, the first capital city of São Miguel Island.

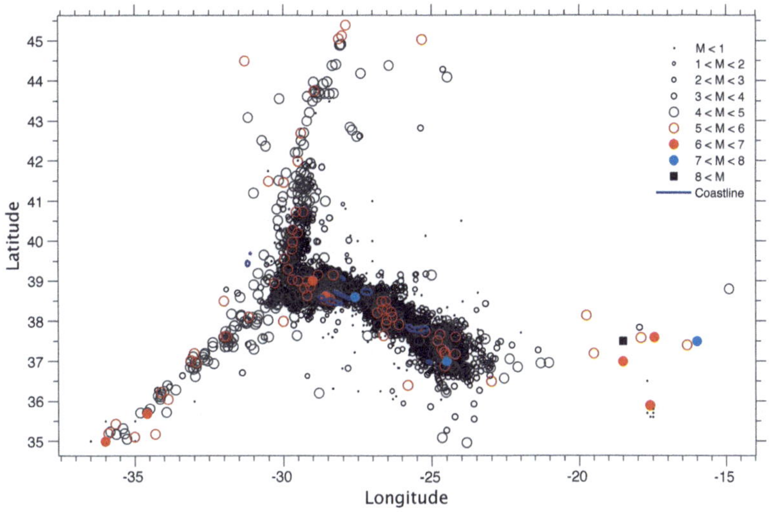

Fig. 1 Earthquake catalogue for the period 1915–2012 (EC2012). EC2012 includes 34.874 events

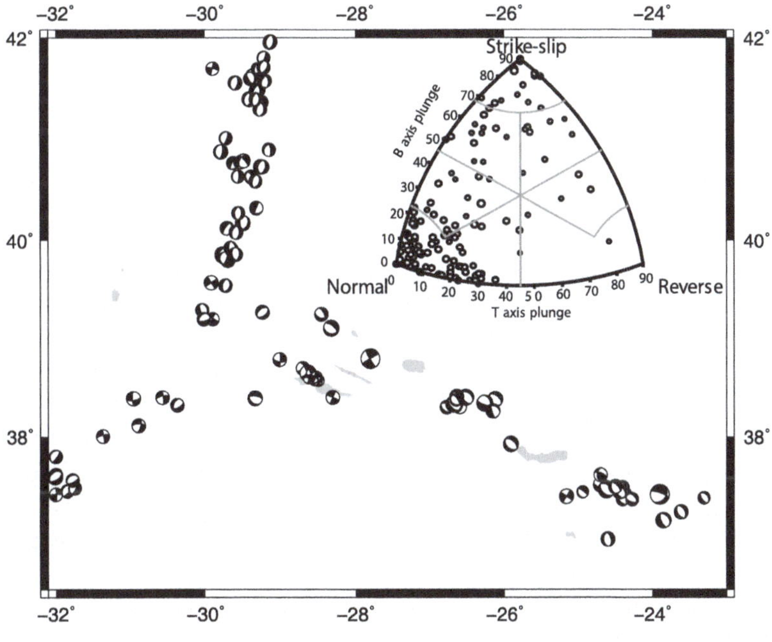

Fig. 2 Focal mechanisms map of the Azores region and triangle diagram to classify focal mechanism (Álvarez-Gómez 2014). Data represents time span 1939–2012 (data sources: Grimison and Chen 1986; Buforn et al. 1988; Borges et al. 2001; Dias 2005; Borges et al. 2007; Matias et al. 2007; Bezzeghoud et al. 2014; ISC—International Seismological Centre)

Among the several authors publishing on the historical seismicity of the Azores, Costa Nunes (1986) is the one who edited the first earthquake catalogue for the period 1444–1980. Later, another catalogue covering the period 1850–1998 was published by Nunes et al. (2004), correcting several events and adding new ones.

The present chapter gives an overview of the main sources of information and data used in these catalogues. It analyses the main characteristics of the evolution of seismicity in the period 1915–2012. For the period 2000–2012, a detailed analysis of the data completeness (i.e. minimum magnitude of completeness, Mc) and of a- and b-values of the Gutenberg-Richter magnitude law is made. Then, the variability of these values for the different zones is discussed.

2 Earthquake Catalogue

As in many other seismic areas of the world, the seismicity in the Azores is described in two periods: the historical and instrumental ones. The historical period concerns data reported by witnesses, its uncertainties depending on the perception and capabilities of the observers to describe the effects (see also Beier and Kramer, Chapter "A Portrait of the Azores: From Natural Forces to Cultural Identity"). For the instrumental period, data are collected on paper or digitally. The increasing number of operating seismic stations associated with an increasing quality of the seismological observatory practices have permitted rapid progress in the detection of small events. As a consequence, the number of events recorded and the quality of the determination of seismic source parameters have increased quite significantly in the last decades.

The 1522 earthquake was the first historical event referred in the Azores. Gaspar Frutuoso (*1522, †1591) described in detail the effects of this earthquake that triggered a massive landslide burying the city of Vila Franca do Campo. Other authors contributed with earthquake reports such as Drumond (1859), Macedo (1871), Junior et al. (1983), and Bessone (1932). Agostinho (*1888, †1978) and Machado (*1918, †2000) also actively contributed to a better understanding of the earthquake phenomena in the Azores. They studied the most severe earthquakes that struck the Islands. For further readings, one could refer to Machado (1948, 1949, 1966) and Agostinho (1927a, b, 1955a, b). Other valuable sources of historical seismicity are the Arquivo dos Açores (1981–1986) and local newspapers.

The first seismometer was installed in São Miguel Island in 1902, followed by the installation of a second station in Terceira in 1932 and a third one in Faial in 1957. Until 1980, the Azores seismic network was composed by these three stations only. Due to the small number of sensors during the period 1957–1980, the minimum magnitude threshold was very high, restraining the detection of events of small magnitude. The Terceira earthquake of the 1st January 1980 is a milestone in the seismology studies of the Azores. Few days after the main shock, the Universidade dos Açores (UAz) deployed a temporary seismic network that was operating on the geothermal project in São Miguel Island. Few months later, the seismic network operated by the UAz had six seismic stations in Terceira Island and eight in São Miguel Island. This network was able to detect tens of micro-earthquakes (Nunes 1991). After Terceira earthquake, and during many others seismic swarms, temporary seismic networks were deployed, but considering the aims and specificity of temporary seismic networks, they will not be referred herein. In 1996, IRIS, within the Global Seismic Network Project (IRIS/IDA), installed one surface and one borehole seismometer in São Miguel Island. Before 1997, two seismic networks operated independently of each other; one operated by the UAz and the other one by the Instituto de Meteorologia (IM). In 1997, these two institutions created the SIVISA (*Sistema de Vigilância Sismológica dos Açores*) Consortium which joins the two networks in a single one (Fig. 3a, b). With the Consortium creation, the number of seismic stations increased producing a decrease in the inland network gaps and improving the determination of the focal parameters. Ten years later, in March 2007, the SIVISA Consortium ended, splitting the network in two separate networks as it was prior to 1997. The seismic network operated by IM with 21 stations was updated with two broadband seismometers installed in Corvo and Flores Islands under the CTBTO (Comprehensive Nuclear-Test-Ban Treaty) framework (Fig. 3a, b). Figure 3a shows the spatial distribution of the SIVISA stations and later IPMA (former IM) while Fig. 3b shows the seismic network evolution as well as the seismic stations operating in the Azores between 2001 and 2012.

As referred previously, Costa Nunes (1986) edited the first earthquake catalogue of the Azores for the period 1444–1980. For this period, earthquake dataset was divided into three groups according to the data quality. The first group includes data collected before 1900. The number of events is small and concerned mainly strong earthquakes. The second group covers the

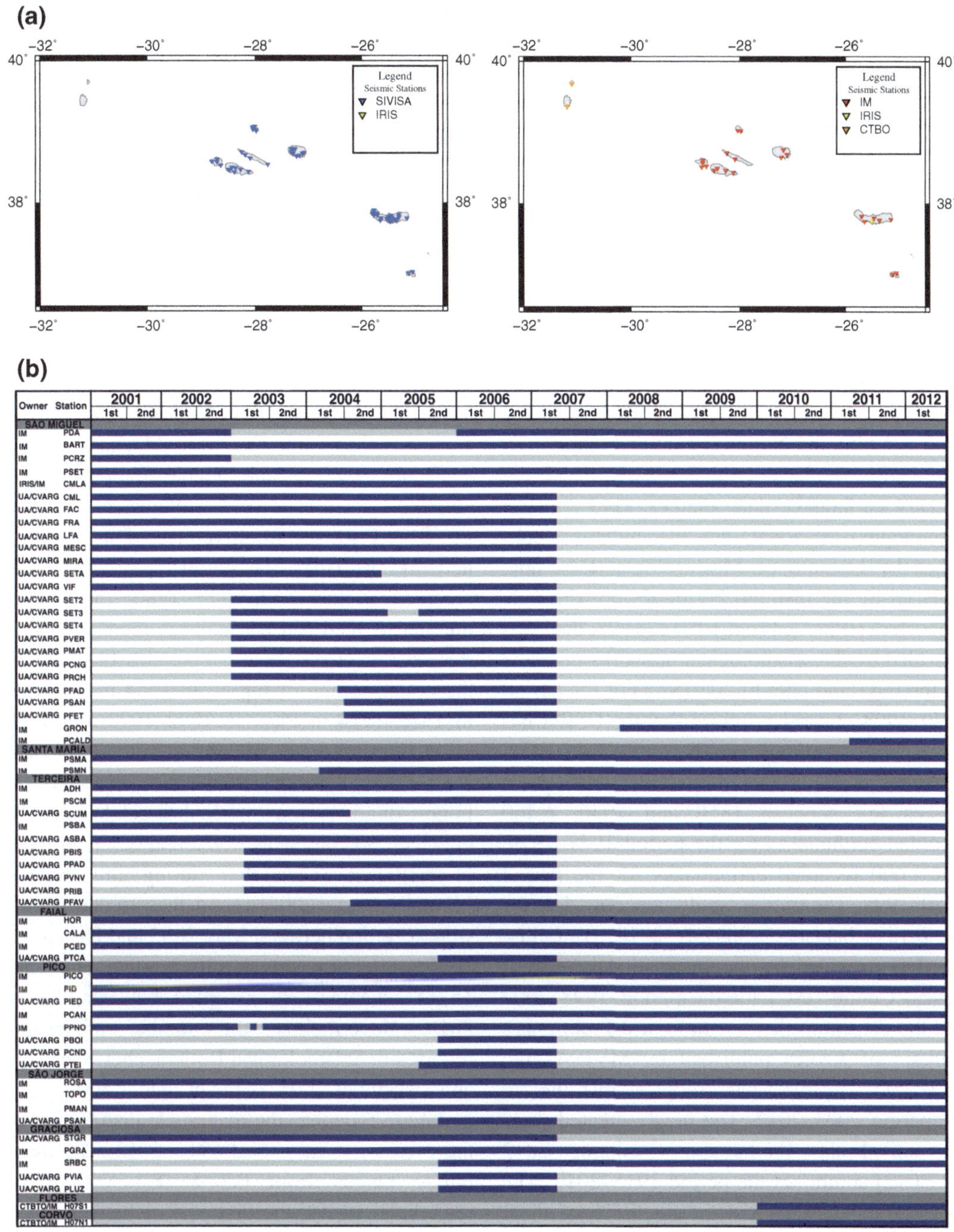

Fig. 3 Evolution of the seismic network in Azores for 2001–2012. **a** the left and right Figures show the seismic network operated by SIVISA and IPMA (former IM), respectively up to the end of 2014. **b** Depicts the seismic stations operating by SIVISA Consortium and after the end in April 2007. The time scale is in semesters. Operating stations are in blue, unknown or not-operating stations are in light grey and unknown status is marked in dark (*Source* Boletim Sismológico Preliminar dos Açores, published by SIVISA in monthly basis between 2001 and 2012. In the period 2007–2012 published by IM)

period 1900–1975 with more detailed and quality improved data. The number of events still remains relatively small. The third group, in the period 1975–1980, concerns mainly instrumental data provided by the seismic stations in the region.

The Earthquake Catalogue for the Azores Region (CSRA) published by Nunes et al. (2004) comprises the period 1850–1998. The CSRA combines the Azores Earthquake Database (BDSA) from Nunes (1991), the earthquake catalogue of Costa Nunes (1986) and data from worldwide earthquake catalogues. Nunes et al. (2004) reviewed the events reported in Costa Nunes (1986) and Nunes (1991) and added new information whenever available. After 1947, the primary earthquake data sources were Anuário Sismológico Nacional and the bulletins published by Instituto de Meteorologia (IM), now replaced by the Instituto Português do Mar e da Atmosfera (IPMA). From 1980 to 1998, data was provided by the seismological networks operated by the UAz and later by SIVISA. For the period 1970–1998, it can be noticed that the CSRA has some periods without records.

Nunes et al. (2004) claim about 900 earthquakes felt between 1850 and 1946, 6,460 recorded earthquakes for the period 1947–1979, and 2,600 recorded earthquakes for the period 1980–1998. The present study concerns data for the period January 2000–July 2012, compiling the data provided by the International Seismological Centre (ISC) for the period 1998–2002 and the data from the Seismological Preliminary Bulletin of Azores published by SIVISA and, since 2007, by the Meteorological Institute for the period 2003–2012. The newly created dataset, called the Earthquake Catalogue 2012 (EC2012) comprises the period 1915–2012. The EC2012 includes 34,874 entries, 75% of them recorded during the period 1998–July 2012. Magnitude scale reported on EC2012 are M_L, M_d, M_b and M_S. M_L and M_d scales are the most common ones with 60 and 23% of the cases, respectively. About 10% of the data have no reported magnitude.

The map locating the epicentres of the EC2012 (Fig. 1) shows seismicity distribution along a stripe that accommodates a complex system of tectonic structures, reinforcing that the Azores domain is a complex area. One could distinguish clusters of events with magnitude greater than five such as the one in the southeast of São Miguel Island, between São Miguel-Terceira, Faial-Pico Islands, west of Faial and Graciosa Islands, and on the MAR. The strongest earthquakes ($M \geq 6$) are located directly on the MAR, in the Central Group (Terceira, Graciosa, São Jorge, Pico, and Faial) and to NW and SE of São Miguel Island. The strongest earthquake ever felt in the Azores struck São Jorge Island on July 9th, 1757 with an estimated magnitude M of 7.4 (Machado 1949).

Figure 4 shows the annual frequency and cumulative number of earthquakes of the EC2012. Prior the improvement of the seismic network, in 1980, the number of earthquakes recorded is low and, after 1980, the annual number of recorded earthquakes increases. Several seismic crises (swarms) are clearly shown in the bottom graph of Fig. 4: in 1964 in S. Jorge, in 1973–1974 in Pico, in 1992–1993 and 1998 crises around Faial Island, and in 1989 and 2005–2006 seismic crisis that occurred in Fogo-Congro region of São Miguel Island. For the latter, no consensus was found on the number of earthquakes recorded during this period. Marques et al. (2007) report that 46,000 events were recorded between May 2005 and December 2005 while Silva et al. (2012) state that only 15,000 events were recorded between 2002–2010. Nevertheless, the number of events reported in Seismological Bulletins between May and December 2005, is lower than 7,500.

Figure 5 depicts the magnitude distribution of the events for the period 1915–2012. The magnitude ranges from 0.7 to 8.1 with three remarkable peaks; the first peak is between 0.8–1.2 and associated to seismic swarms characterised by a high number of events with smaller magnitudes. The second peak referring to magnitudes range between 1.5 and 2.3 corresponds to the largest number of recorded earthquakes in the Azores. The last peak, in the interval 2.5–2.8, is associated with seismic sequences like the one of the Faial in 1998. In general, the Azores

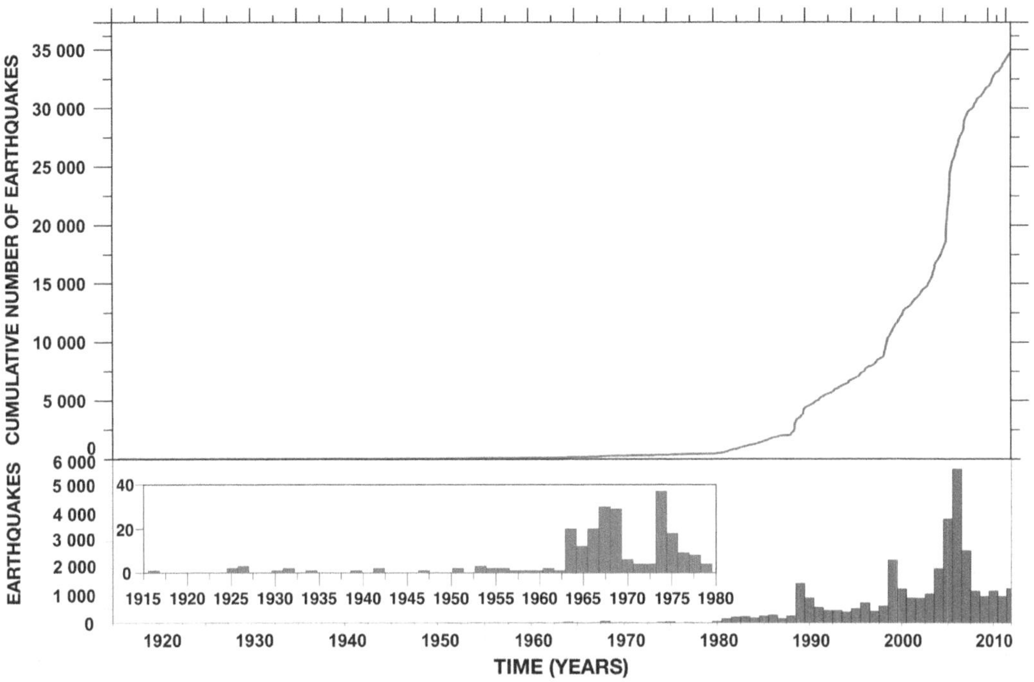

Fig. 4 Annual distribution of earthquake in the EC2012 (1915–mid2012). The cumulative number of earthquakes (above) and the annual number of events (below). The inset shows in detail the annual distribution between 1915 and 1979

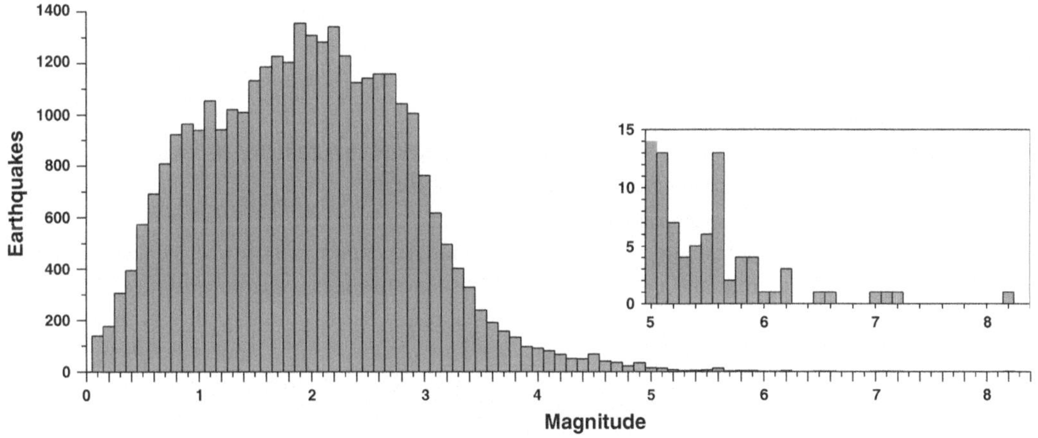

Fig. 5 Frequency magnitude distribution of the EC2012 data. The inset shows in detail the frequency magnitude distribution for magnitude higher or equal than 5. In both cases, bins have 0.1 unit of magnitude

seismicity is characterised by a high number of events of relatively small magnitude.

Most of the earthquakes in the Azores have a shallow depth, less than 10 km, as illustrated by the hypocentres distribution of Fig. 6. According to Kearey et al. (2009), the thickness of continental crust is around 40 km while the oceanic one is around 7 km. Nevertheless, the crust in the Azores is not a typical one. Searle (1976) estimated the thickness of oceanic crust around 8 km

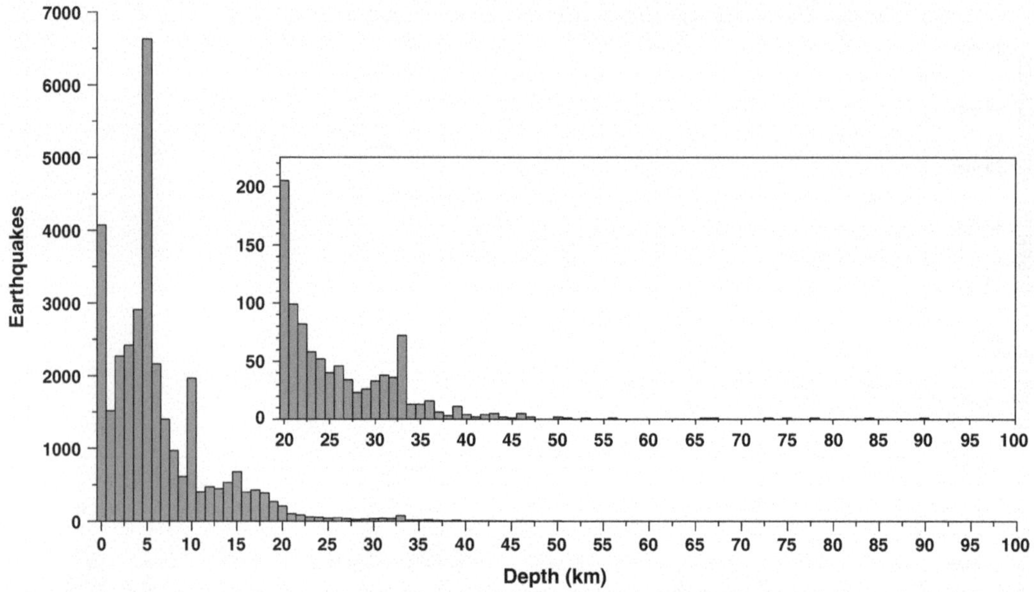

Fig. 6 Depth distribution of events in the of EC2012. 0, 5, 10 and 15 km depths corresponds to the automatic determination. The inset show in detail the seismicity for depths equal or higher than 30 km. In both cases, bins have 1 km depth interval

using the Rayleigh-Wave dispersion method; Steinmetz et al. (1977) in the MAR identified two reflectors at 9 and 30 km depth; Luís et al. (1998) estimated an elastic plate with 7–8 km thickness while Luís and Neves (2006) estimated in 4 km. Later, Matias et al. (2007) estimated the oceanic crust in 14 km in the area that comprises Faial, Pico and São Jorge Islands. Silveira et al. (2010) established a thick crust around 20–30 km. Figure 7 shows hypocentres in two cross-sections along the major tectonic structures of Azores. Three main earthquake clusters at a depth greater than 10 km are identified on cross-section A–B (Fig. 7b) beneath Graciosa Island, to the east of Terceira and beneath the central part of São Miguel Island which comprises the region beneath Fogo-Furnas volcanoes. Concerning cross-section C–D (Fig. 7c) is identifiable a seismic cluster that is related to the Faial earthquake (1998), where the foci are deeper than 25 km. Dias (2005) studied in detail the aftershocks sequence of Faial earthquake and calculated using inversion models that the Moho discontinuity is around 11–14 km. Another cluster is West of the Graciosa Island where earthquake depth reaches 40 km. The last cluster concerns the SE of the Terceira Island on a small area of a submarine section of the Lajes fault where deepest events attain 50 km depth (Lourenço et al. 1998).

3 Characterization of Seismic Activity

Earthquake catalogues are the primary source of information to study seismo-tectonics, seismicity, and hazard. Seismic data quality is the main criteria to obtain reliable and accurate results and provide pertinent statistical analysis. Unfortunately, data is not homogeneous since they come from the analysis of seismic waves recorded by different instruments with different operational practices and procedures, varying in space and time. Several authors (e.g. Habermann 1987, 1991; Habermann and Creamer 1994; Zuniga and Wiemer 1999) gave special attention to the heterogeneities in earthquake catalogues, introduced by network's limitations and man made changes. These heterogeneities in space and time

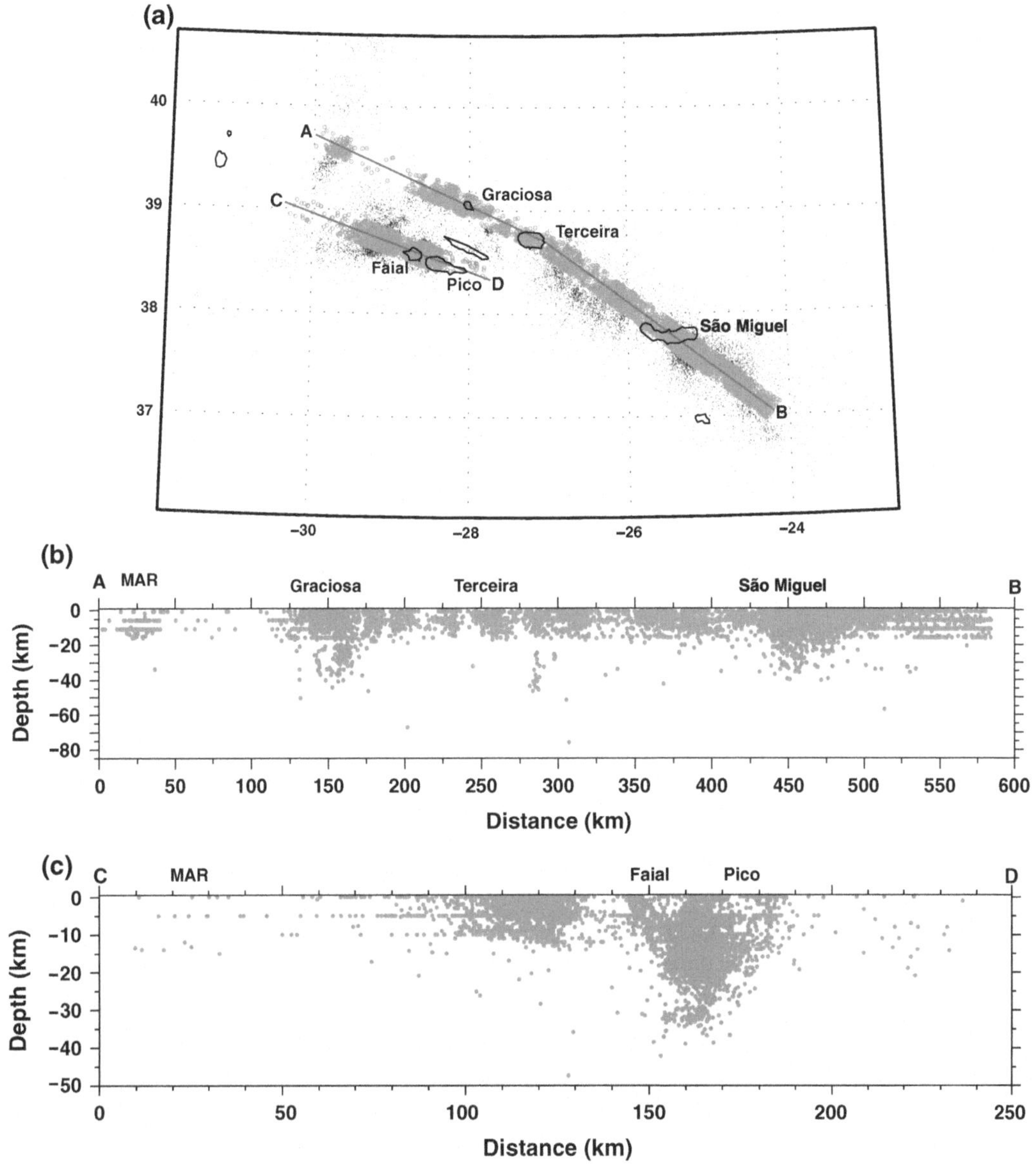

Fig. 7 **a** Location of cross sections along the major tectonic structures of Azores. The empty grey circles are earthquakes selected to draw the cross sections and black dots are earthquakes which are not represented on cross-sections. **b** Cross section A–B starts in the Middle Atlantic Ridge and crosses Terceira Rift until Gloria fault. **c** Cross section C–D starts in the Middle Atlantic Ridge and crosses Faial-Pico volcanic ridge

affect the completeness of the catalogue, and they need to be avoided. A parameter useful to verify the quality of an earthquake catalogue is the minimum magnitude of completeness (*Mc*). *Mc* is the lowest magnitude for which all earthquakes in a space volume and time period can be detected (Rydelek and Sacks 1989; Taylor et al. 1990; Wiemer and Wyss 2000, 2002). To estimate the minimum magnitude of completeness, a simple power-law can approximate the frequency magnitude distribution (FMD) for a given volume. The relationship between frequency of occurrence and earthquakes magnitude (Ishimoto and Iida 1939; Gutenberg and Richter 1944) is as followed:

$$\log_{10} N(M) = a - b\mathrm{M} \quad (1)$$

where N is the number of earthquakes having a magnitude larger than M, a and b are constants. The a-value describes seismic activity while b-value relates to tectonic stress (Mogi 1967; Scholz 1968). The b-value is near 1 in most seismically active regions on Earth (Frohlich and Davis 1993). However, b-value can be disrupted by an increase of material heterogeneity that produces high values (Mogi 1962) or by an increase of shear stress (Scholz 1968), and effective stress (Wyss 1973) that produces lower values. It is common in volcanic areas to have high b-values, up to 2, due to the high temperature gradient (Warren and Latham 1970) or during swarms. Several authors (e.g. Wyss et al. 1997; Wiemer and McNutt 1997; Wiemer et al. 1998; Murru et al. 1999) found high b-values limited to small volumes corresponding to active magma chambers and conduits.

To estimate *Mc*, we choose events with M_L due to their higher number of events in EC2012 for the period 2000–2012 (20,880 events). On the same time span, only 7,970 events were reported with duration magnitude (M_d). We reject M_d due to the dependence of signal-to-noise ratio of this magnitude scale and to the low number of events reported. The choice of having a single magnitude scale reduces the errors in the estimation of b-value (Wyss 1991; Zuniga and Wiemer 1999). Wiemer and Wyss (2000) proposed to combine Mc95 with Mc90 estimates and the maximum curvature method to estimate *Mc*. The Mc90 and Mc95 are the magnitudes at which 90 or 95% of the observed data are modelled by power law fit, respectively. The maximum curvature method estimates *Mc* using the point of maximum curvature of the FMD.

Large earthquakes produce hundreds to thousands of aftershocks that can reduce the statistical significance or introduce bias on seismicity rate. To remove aftershocks clusters, we apply the Reasenberg (1985) method to the earthquake catalogue. From 20,880 earthquakes, 9,625 were in clusters, representing 54% of the dataset (Fig. 8). Minimum magnitude of completeness Mc and b-value are determined using the ZMAP software (Wiemer 2001).

Figure 9 shows the frequency magnitude distribution and the power law fitting for all events declustered in the span 2000–2012. The best fit of the Eq. (1) is found with Mc = 2.5 and $\boldsymbol{b} = 0.9$. Estimated error on Mc and b-value is ±0.4 and ±0.12, respectively.

To qualify the earthquake catalogue regarding *Mc*, b-value (for the period 2000–2012), and identify heterogeneities and spatial variations we create a regular grid of 2 by 2 degree. We obtained ten rectangular regions (Fig. 10), called cells. The grid has a superimposition window of 1° in latitude and longitude in the region of the Middle Atlantic Ridge (MAR) (cells 1, 2 and 4), and 1.5° by 1° in the Terceira Rift (TR) (cells 3, 5, 6, 7, 8, 9, and 10). The review of the seismological studies developed in the Azores Islands as well as the historical seismicity, with consequences in the islands, are mentioned in cells 6 and 9 to give a general view of seismicity of both regions. The following gives an overview of the seismic activity per cell. For all cells (but CELLs 6 and 8), the electronic supplement (S1–S9) show the cumulative annual distribution, the annual number of events and the frequency magnitude distribution.

CELL 1 includes the Northern part of the Azores plateau and a segment of the Middle Atlantic Ridge (MAR). Its seismic activity is scattered and irregular on time. Before 2003, no events were reported in this section of the MAR. Largest event is ML4.6, however most of the events have M_L between 2.8 and 3.8. The FMD fits very well all events above Mc 3.1 with an estimated error for Mc of ±0.2. The a- and b-value are 5.2 and 1.0 ± 0.2, respectively.

CELL 2 shows a cluster of earthquakes centres along a NE-SW trend that coincides with the intersection of MAR and North Azores Fracture Zone (NAFZ). The number of events is much higher than in the other sections of MAR. Around 700 earthquakes were recorded between 2000 and 2012, half of them belonging to the mentioned cluster. Most of the events reported

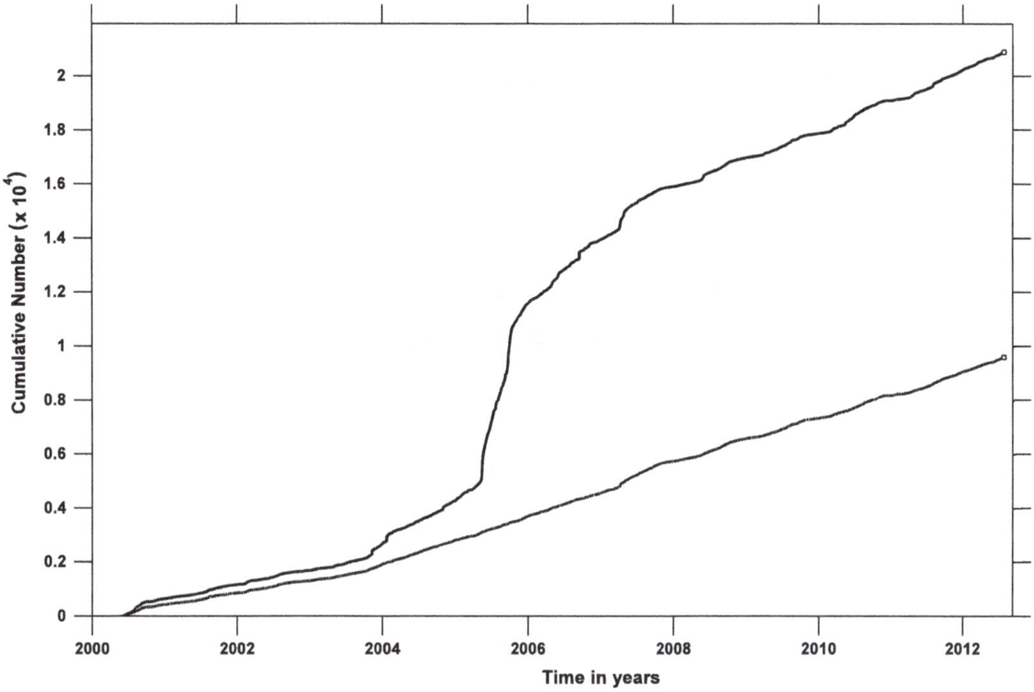

Fig. 8 Temporal distribution of earthquakes in the EC2012 catalogue. Cumulative number is in dark grey and declustered catalogue in light grey. Around 54% of the events on earthquake catalogue were located in clusters

Fig. 9 Frequency magnitude distribution of the EC2012. The error in magnitude completeness is ±0.04 and b-value ±0.12. The red line is the fitted power law beginning at the minimum magnitude completeness (triangle)

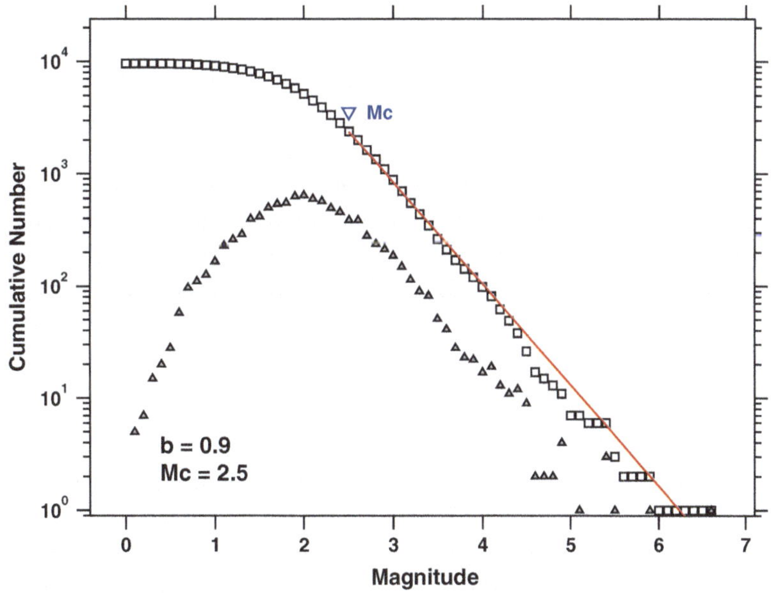

have a magnitude between 2 and 4.5. Since 2005, the number of recorded earthquakes have increased partly (or wholly) due to the decrease of the magnitude threshold to 1.5. Mc is 3.2 ± 0.2 while a is 6.7 and b is 1.4 ± 0.3. The relatively high value of Mc in the cells 1 and 2 is due to the difficulty for seismic networks to record low magnitude offshore earthquakes.

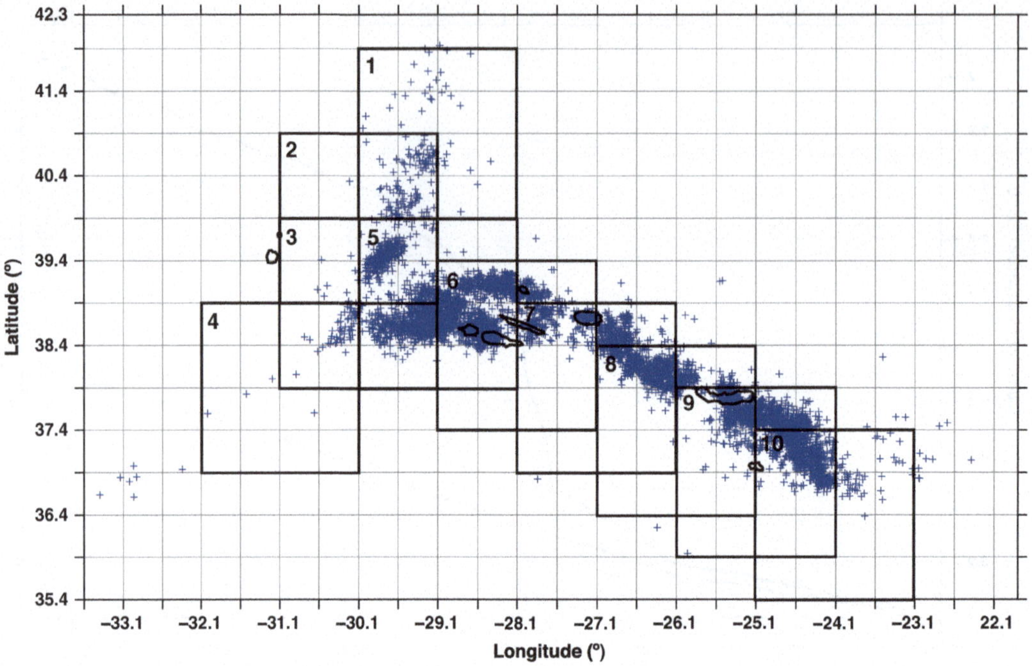

Fig. 10 Declustered earthquake catalogue EC2012 in a regular grid of 2° by 2°. The grid in the Mid Atlantic Ridge has superimposed windows of 1° in latitude and longitude (cells 1, 2 and 4) while in Terceira Rift, the superimposition is 1.5° × 1° (cells 3, 5, 6, 7, 8, 9 and 10)

CELL 3 includes a part of MAR and the west of Terceira Rift. The distribution of the seismic activity shows quiescence periods alternating with intense seismic activity. The seismogenic sources of the cell 3 produce more earthquakes (~2,000) than in cells 1 and 2. The cumulative number of earthquakes softly increase until 2006, then sharply increase up to 2012. Two peaks at magnitudes $M_L = 2.2$ and 2.6 are identified which are related to the seismic activity to the West of Faial Island and with the activity in the MAR, respectively. Before 2005, the magnitude range is 2–3.5 (M_L) with few events of magnitude higher than 3.5. After 2005, the range of magnitude is 1.5–4. The magnitude-frequency distribution follows the power law model from *Mc* to M_L 4.1. In this case, Mc = 2.8 ± 0.2 whilst a = 5.9 and b = 1.2 ± 0.2.

CELL 4 is fairly aseismic. Most of the seismicity is limited to an area of 0.5° × 0.5°. In this area only 55 earthquakes were recorded, and magnitude range was 2.2 to 4.4. Due to the small number of events, the power law does not fit well the FMD. In this case, Mc = 2.8 ± 0.2, a = 4.1 and b = 0.9 ± 0.3.

CELL 5 covers the MAR and extends over the Islands of Faial, Pico and partially over São Jorge. Seismic activity occurs in a broad area between the Islands of Faial-Pico and the W of Graciosa Island. This cell shows a cluster with a general NE-SW trend and rough elliptical shape of 24 × 12 km at 23 km of the western Faial. Several authors (Luís et al. 1994; Miranda et al. 1998; Lourenço 2007) relate this seismic activity to the extension of the Faial Fracture Zone (FFZ).

In this cell, we notice two periods where seismicity increases; the first one on April–May 2007 when 550 earthquakes occurred with a maximum magnitude (M_L) of 4. Fourteen of them were felt by the population with the highest intensity IV (MM) in Faial Island. The second one in March–June 2010 with more than 400 earthquakes recorded. The maximum magnitude (M_L) was 4.5 and intensity reached V (MM) in Faial. Most of the events have a magnitude in the range 1.2–2.4 with a maximum at 4.5. At this stage, the seismic network can detect small events with M_L lower than 0.5. For this cell, the Mc = 2.8 ± 0.16; a = 6.5 and b = 1.3 ± 0.22.

CELL 6 includes the islands of the Central Group, namely Graciosa, São Jorge, Faial, Pico and the western part of Terceira. A high value of seismic activity can be seen, except in São Jorge Island. It could be explained by the release of energy which occurred with the 1757 earthquake (M7.4, Machado 1949). All the Islands in this cell felt the effects of strong earthquakes at least once on the historical period (Table 1).

Since 1900, Faial Island had to handle two earthquakes that caused heavy damage and deaths and a volcanic eruption. In chronological order the first one was:

- The earthquake of August 31st, 1926 is inserted in a seismic sequence started at 5 April 1926 and lasted up the end of September. Agostinho (1927a, b) estimated the epicentre near Faial Island at the channel between Faial-Pico Islands. The maximum intensity X (MSK) and estimated magnitude (M_b) 5.3–5.6 (Nunes et al. 2001). The main event caused severe damage in the main town of the Island and on nearby villages.
- The second one is the Capelinhos volcanic eruption which begun in September 1957 and finished in October 1958. Initially, the eruption was of surtseyan type and in April 58 decrease gradually to give place to strombolian and hawaiian eruption types. Suddenly, during the early evening of May 12th, 1958 a seismic crisis started and continued through next day. During the first night, the population felt around 450 earthquakes, two of them with maximum intensity X (MMI) and one of VIII (Machado et al. 1962). One maximum intensity was located in southwest of the caldeira and the other one in northwest at half-distance between the caldeira and the active crater whilst. The intensity VIII was in the northeast side of the Island at a focal depth of 1 km.
- On 9 July 1998 Faial Island was hit by a strong earthquake (Mw6.2), 10 km NE of Faial which affected mostly Faial and Pico Islands. The result was eight deaths, more than one hundred injuries and heavy damage on building stock. According to Senos et al. (2008),

Table 1 Historical earthquakes in cell 6 with Imax > VII (modified from Nunes et al. 2001)

Date	Intensity	Death toll	Region most affected
17/05/1547	VII/VIII (MMI)	>3	North zone of Terceira Island
24/05/1614	IX (MMI)	>200	Praia da Vitória, Terceira Island
13/06/1730	VIII/IX (MMI)		Luz, Graciosa Island
09/07/1757	XI (MMI)	1,046	Calheta, São Jorge Island
24/06/1800	VII/VIII (MMI)		Praia da Vitória, Terceira Island
26/01/1801	VIII (MMI)	2	São Sebastião, Terceira Island
21/01/1837	IX (MMI)	3	Guadalupe e Santa Cruz, Graciosa Island
15/06/1841	IX (MMI)		Praia da Vitória, Terceira Island
11/06/1912	VII/VIII (MMI)		Praia da Vitória, Terceira Island
31/08/1926	X (MMI)	9	Horta, Faial Island
27/12/1946	VII/VIII (MMI)		Serreta, Terceira Island
13/05/1958	VIII/IX (MMI)		Praia do Norte, Faial Island
21/02/1964	VIII (MMI)		Rosais, São Jorge Island
23/11/1973	VII/VIII (MMI)		Bandeiras, Pico Island
01/01/1980	VIII/IX (MMI)	61	Doze Ribeiras, Terceira Island
09/07/1998	VIII/IX (MMI)	8	Ribeirinha, Faial Island

around 7,600 aftershocks were recorded during the first month and more than 15,000 up the end of 2003, extending the seismic annual rate up to early 2005 (Dias 2005).

The study of the seismic sequence of Faial 98 earthquake by Matias et al. (2007) reveals an unusual high *p*-value of 1.40 using the Omori modified law, a value typical of mid-oceanic ridges. The authors calculated a 1D seismic velocity model fitted to the local conditions where the Vp/Vs ratio range between 1.77 and 1.91. The hypocentre relocation disclose two alignments. One with direction NNW-SSE and the other one with a rough orientation ENE–WSW. The analysis of the focal mechanisms of the mainshock and aftershocks indicate that the rupture is compatible with strike-slip mechanism; and few events with normal/reverse mechanism. From the epicentres and focal mechanism, the authors deduced that the alignment NNW-SSE are associated to left-lateral strike-slip fault solution and the WSW-ENE exhibit right-lateral component.

The tomographic study conducted by Dias et al. (2007) in the area of Faial, São Jorge, and Pico Islands with data recorded after the Faial 98 earthquake reveal that the shallow layers are consistent with tomographic signal on Faial Island; probably the plumbing system or even a magma chamber beneath Caldeira with low Vp (<6.0 km/s); a region of high Vp (>6.0 km/s) beneath NE Faial; the NE slopes of Faial – Pico ridge reveal a gradient in Vp and Vp/Vs models, and a tectonic segmentation of Faial Island.

Borges et al. (2007) conducted a study on the seismic source of the 1998 earthquake using the inversion of seismic body waves. They found that the rupture with the sub events was similar to the one of Terceira 1980 earthquake. Most of the energy was released on the first sub-event at 8 km depth with left-lateral strike-slip rupture. The location of the second sub-event was at the same horizontal position but at 7 km depth and less energetic than the first one. The scalar seismic moment of the first sub-event was 1.1×10^{18} Nm and second one 0.3×10^{18} Nm; the duration of the source time function was 2.5 s.

In Pico Island, earthquakes are located in the western part of the island around the volcanic edifice named Pico Mountain (Nunes et al. 1997). From October 1973 to March 1974, 802 earthquakes were recorded, 490 of them were felt by the population (Nunes et al. 1997). On November 23, and December 11, 1973 two moderate earthquakes occurred with M_d 5.8 and 5.6 and intensity VII–VIII and VI (MMI), respectively.

On the 9 July of 1757, a great earthquake struck São Jorge Island collapsing all buildings of the eastern region and causing more than 1,000 fatalities. The historical accounts refer that cracks were formed, some of them with more than 2.2 m depth, and changes on the topography of the island were observed. Machado (1949) assigned maximum intensity XI (MI 31) and estimated magnitude 7.4 for the event. From paleoseismological studies, Madeira (1998) estimated that the magnitude ranges from 6.4 to 6.8.

Concerning the volcanic eruption of 1963/64 near São Jorge, the first signs started towards the end of 1963 with the Horta seismic station recording continuous seismic tremor that lasted up to January of 1964 (Zbyszewski et al. 1977). Early in the morning of February 15th, 1964 a seismic crisis started and, up to midnight, 179 earthquakes were recorded; and 125 the next day. On the first days, the epicentres were located near the historical eruptions of Manadas in 1580 and Urzelina in 1808. According to the authors, the seismic tremor started around noon the 15th of February and ended the 22nd of February and were felt strongly in Manadas and Urzelina settlements. The strongest events, with maximum intensity VIII, occurred the following days, the 18th, 20th and 21st of February. The seismic crisis due to the eruption ended in September 1964 and more than 500 earthquakes were felt by the population.

In Graciosa Island, many old chronicles referred to damaging events without accurate details on the date and damage level. In 1730 and 1837, two earthquakes of maximum estimated intensity of VIII–IX and IX, respectively, destroyed most of the dwellings located in the southern part of the Island (Nunes et al. 2001).

In the past, strong earthquakes (intensity $\geq$ VII) inflicted economic and social losses to the inhabitants of Terceira Island (Table 1). According to this table, the island suffered the effects of seven earthquakes with maximum intensity $\geq$ VII/VIII MMI. The eastern and south-eastern areas were most severely affected. The largest events occurred in 1614, 1800, 1801 and 1841. With the exception of 1801, the other earthquakes occurred in the eastern part, more precisely in the Lajes Graben which extends to the seafloor, SE of Terceira (Nunes et al. 2001).

On January 1st, 1980 Terceira Island was affected by a strong earthquake Ms7.2 (Hirn et al. 1980) and maximum intensity IX MMI. The islands of São Jorge and Graciosa felt the earthquake with maximum intensity VII/VIII and VI/VII, respectively. The earthquake caused heavy damage in some areas of Terceira, especially in the city of Angra do Heroísmo where around 30% of the building stock collapsed, and caused a death toll of 61. The data referring to seven days of aftershocks made it possible to locate the epicentres within a ribbon of 40 km long by 6 km wide, and the strike N150°E (Hirn et al. 1980). These authors inferred from the aftershock sequence that the mainshock fault mechanism was pure sinistral shear along a N150°E fault plane, the same plane being found by the composite solution for aftershocks. The body wave inversion suggests the source time function had two peaks separated by 10 s (Grimison and Chen 1988). The authors state that the first sub-event is consistent with a pure strike-slip mechanism, while the second sub-event exhibits a large component of thrust faulting at depths of 12 and 20 km. These depths are not in agreement with the ones constrained by Hirn et al. (1980) nor consistent with an oceanic type of crust. Borges et al. (2007) obtained two sub-events with left-lateral strike-slip mechanisms. The second sub-event was located 25 km NNE of the main rupture plane, and was 12 s later than the first. The total scalar seismic moment obtained by Buforn et al. (1988) (2.0×10^{19} Nm) from long-period Rayleigh-wave inversion, is similar to the one obtained by Borges et al. (2007) (1.9×10^{19} Nm) from body wave inversion; nevertheless, Grimison and Chen (1988) obtained a higher value (3×10^{19} Nm).

The last volcanic eruption in the Azores was between 1998 and 2001, and took place at sea about 10 km west of Serreta (Terceira Island, Gaspar et al. (2003), Kueppers et al. (2012)). According to Gaspar et al. (2003) the number of earthquakes increased above background values on November 23rd, 1998. On November 27th, the number of earthquakes rose to 200, and two days later reached 400. Then the seismic swarm decreased to background values. On December 18th, the first sign of an ongoing eruption was observed.

Seismic hazard analysis by Carvalho et al. (2001) estimates a PGA value of around 2.5–2.75 m/s^2 (exceedance probability of 10% in 50 years) for this cell mainly in the west of Faial and in the Terceira Rift segment between the islands of Terceira and São Miguel.

The seismic temporal evolution is like cell 5, with magnitudes ranging from 0.2 to 4.5 (Fig. 11a, b). Earthquakes with magnitude (M_L) larger than 3.5 are isolated (e.g. in 2003 to the first quarter of 2004, and in 2010). Periods with low radiated energy (2005–2007 and 2008–2009) are followed by periods of increasing radiated energy (2000–2004, 2007 and 2010).

The FMD in cell 6 is well constrained by the power law for events larger than 2.4 (M_L). Mc is fixed at 2.4 ± 0.2; $a = 5.9$ and $b = 1.2 \pm 0.1$ (Fig. 11c). Despite the generally improved azimuthal coverage along the Archipelago, some seismicity has been analysed in regions with poor azimuthal coverage, such as those to the west of both Graciosa and Faial Islands.

CELL 7 lies in a section of the Terceira Rift that includes Terceira Island, half of São Jorge Island and a small western section of Pico Island. Temporal evolution of seismicity shows a low annual rate from 2000 to 2003, a slight increase until 2007 and then an average rate. Several periods have reduced or no recorded earthquakes (e.g. 2000–2002, 2007 and 2009). Magnitudes (M_L) range between 0.2 and 4.4, with few events with a magnitude higher than 3. The

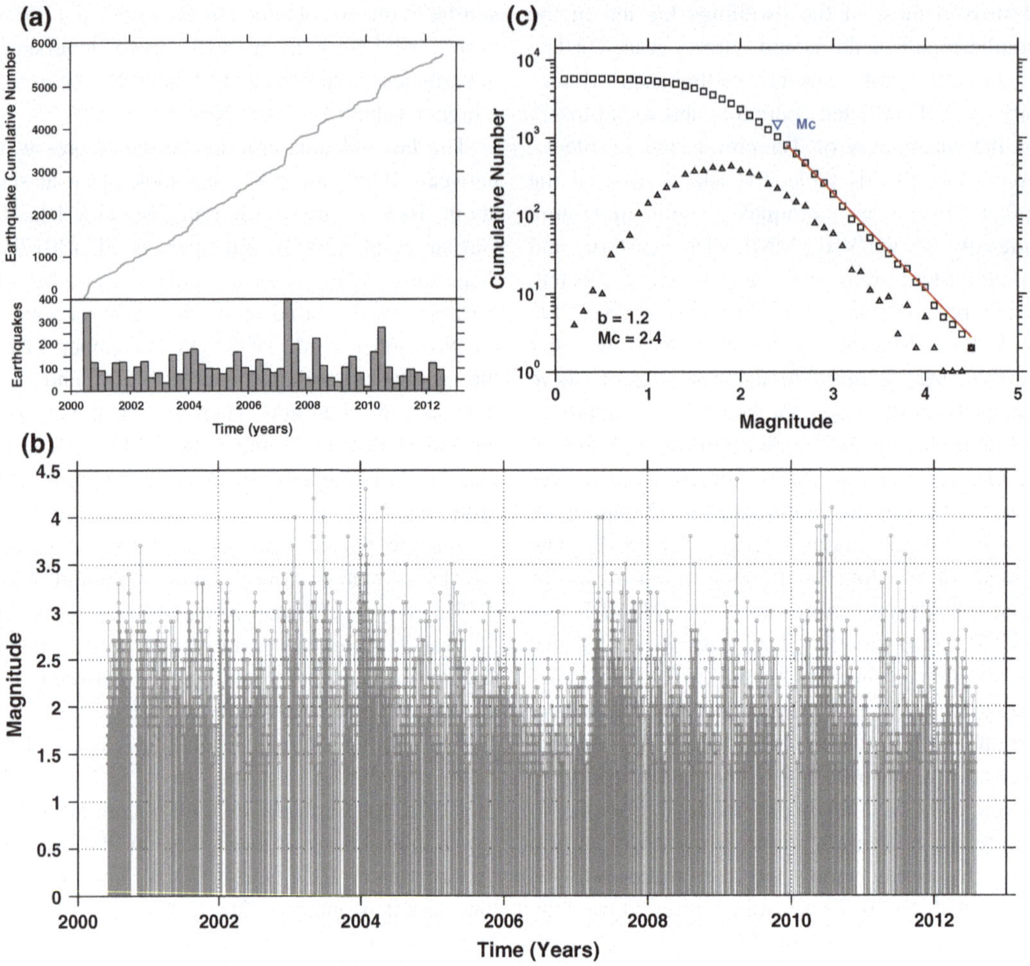

Fig. 11 Temporal distribution of earthquake in cell 6. **a** cumulative earthquake number (top) and annual distribution (bottom). **b** Annual number of event. The lack of seismicity for the first semester 2000 is due to the shift on reported magnitude. In December 2010, data are not available. **c** Frequency magnitude distribution. Triangles are noncumulative frequency magnitude distribution and empty squares are cumulative frequency magnitude distribution. The error in magnitude completeness is ±0.2 and b-value ±0.1. The red line is the fitted power law beginning at the minimum magnitude completeness (triangle)

corresponding values of the Gutenberg-Richter law are: Mc = 2.5 ± 0.25: a = 5.8 and b = 1.3 ± 0.24.

CELL 8 covers the North and South Hirondelle basins, Dom João de Castro Bank located between these two basins, São Miguel Island and Povoação Basin to the west of São Miguel Basin.

Dom João de Castro Bank is an active volcano located in a segment of the Terceira Rift. The single known eruption was in December 1720. The population of the islands of Terceira and São Miguel felt some earthquakes associated to the eruption. Since 1933, several moderate earthquakes have been reported around the volcano (electronic supplement S7, earthquakes marked with ✳). Some seismic swarms (e.g. in October 1988 and in June 1997) had more than 2,000 earthquakes in the first month, and 45 of them had a magnitude higher than 4 (Gaspar 1997).

Based on the focal mechanism of Azores earthquakes, Borges et al. (2007) established a seismotectonic model to the west of Terceira

Island. In this area, the seismicity is oriented NW-SE, with a normal type of faulting system with a horizontal tension axis trending NE-SW, normal to the Terceira Rift. Study of the focal mechanism of the 28 June 1997 earthquake shows a normal fault with a rupture process divided into three sub-events at focal depths ranging from 5 to 7 km (Borges et al. 2007). The total scalar seismic moment is 7.0×10^{17} Nm with source duration around 5 s.

São Miguel Island is seismically the most active. In the following, seismicity in different parts of the island is presented: On 22 October 1522, a massive landslide triggered by an earthquake buried the first capital of São Miguel Island—Vila Franca do Campo. The event caused heavy destruction on building stock and around 5,000 deaths, the deadliest event in the archipelago (Frutuoso 1998). Maximum intensity assigned was X on both intensity scales McS-17 and MM-31 (Dias 1945; Machado 1966) respectively, while Nunes et al. (2001) and Silveira et al. (2003) assign the same intensity value, applying MMI and EMS-98 scales, respectively. The epicentre was inland, to the north of Vila Franca do Campo, and the focal depth was at 12 km (Machado 1966). Table 2 lists the earthquakes that occurred in São Miguel after 1522. Some of them (1522, 1591 and 1852) could be related to the volcanic activity of the Fogo-Congro region.

Nine volcanic eruptions have been reported since the settlement of São Miguel Island in the 15th century, most of them preceded or followed by earthquakes (França et al. 2003). Around the Sete Cidades volcano in the west of the island, seismic swarms of more than 150 events are common, while on the Fogo-Congro volcano in the central region they are counted in the thousands.

During a seismic crisis near Sete Cidades volcano in September 1996, a total of 180 earthquakes were recorded in seven hours. All earthquakes were associated with the Mosteiros Graben, a tectonic element NW of the volcano. The epicentres were located at a depth of between 2 and 4 km (Forjaz et al. 1996). In August 1998, another seismic swarm manifested by 120 earthquakes recorded in three hours; five of them felt by the population with a maximum magnitude of 3.1 and MMI of V (Gaspar and Wallenstein 1998). One month later, a new swarm of about 120 earthquakes in four hours occurred; five of them were felt by the population and one had an MMI of V.

The Fogo volcano includes a complex system of faults that accommodates a tectonic structure in a graben known as the Congro Fracture Zone. The first recorded swarm in this region was in 1922, and lasted two months, with tens of earthquakes felt by the population (Nunes and Oliveira 1999). In 1967, a second swarm occurred and lasted for four days. For both of them, the maximum intensity was VII (Nunes and Oliveira 1999). From 1988 until July–September 1989, more than 8,000 events were recorded with magnitude (M_d) up to 3.8 (Nunes and Oliveira 1999). From May 2005 to late 2006, thousands of earthquakes were again recorded.

The Fogo volcano includes a geothermal reservoir where two geothermal power plants

Table 2 Historical earthquakes in São Miguel Island since 1522 (modified from Nunes et al. 2001; Silveira et al. 2003)

Date	Intensity	Human toll	Region most affected
22/10/1522	X (MMI and EMS)	~5,000	Vila Franca do Campo, São Miguel Island
26/07/1591	VIII/IX (MMI and EMS)	Unknown	Vila Franca do Campo and Água de Pau, São Miguel Island
16/04/1852	VIII (MMI and EMS)	9	Ribeira Grande, São Miguel Island
05/08/1932	VIII (EMS); VII (MMI)	3,000 homeless	Povoação, São Miguel Island
27/04/1935	IX (EMS); VII (MMI)	1	Povoação São Miguel Island
26/06/1952	VIII (MMI and EMS)	600 homeless	Ribeira Quente and Povoação, São Miguel Island

have been installed. Due to the economic importance of this renewable energy, several studies were developed in this region. Dawson et al. (1985) searched for anomalies in seismic velocity that may reflect the presence of a magma chamber or hydrothermal system beneath the volcano. They found a low-velocity area in 5 km depth, 10 km long in E-W direction and 5 km in N-S direction. This region might be a hydrothermal reservoir or a body of partial melt, or a combination of both.

Zandomeneghi et al. (2008) found anomalies in P-wave velocity model and P- and S- wave velocity ratio (Vp/Vs) in the central part of São Miguel known as Congro. In the region Fogo-Congro, they found low velocity (Vp) and Vp/Vs ratio in the NW and NNE of Fogo volcano. Areas of high velocity and Vp/Vs ratio are found in the south of Fogo and the central region. However, the authors did not find shallow bodies of partial melt in the Fogo area as described by Dawson et al. (1985).

Using data from 2003–2007, Fontiela and Nunes (2008) identified sub-volumes with high *b*-values (1.6–1.7) at depths around 8–9 km, compared to surrounding zones with lower *b*-values around 0.6. These high *b*-values could be explained by the presence of hydrothermal systems or dyke intrusions. A seasonal correlation between rainfall and velocity patterns is found by Martini et al. (2009). It is related to local pressure changes at depth in response to surface rainfall, when the system is critically stressed and/or interacts with the geothermal system. Fogo Volcano is dominated by a normal regional stress field, while the Congro region exhibits a highly heterogeneous stress field with permutations between σ_2 and σ_1 (Silva et al. 2012). For the period 2002–2012, the stress tensor is characterised by a maximum compressive sub-horizontal stress axis (σ_1) striking WNW-ESE and minimum compressive stress axis (σ_3) striking NNE-SSW.

Furnas Volcano has very low seismic activity. GPS data show low rates of deformation (Jónsson et al. 1999) which could be explained, firstly, by spreading in the region between the Eurasian and Nubian plates, and secondly, by a source of inflation located on the northwest of the caldera. Nunes and Forjaz (1998) describe two seismic swarms that occurred inside Furnas caldera in June 1989 and in January 1992. The first seismic swarm had 111 events with magnitude lower than 2.4, the strongest event occurring near the fumarole field in Furnas village. The second swarm had 30 events with magnitude lower than 2.4 but released more energy than the first. Zandomeneghi et al. (2008) found low Vp values and Vp/Vs ratio, related to an active geothermal system embedded in low density pyroclastic deposits.

The study of seismic hazard assessment in São Miguel shows that for the annual exceedance probabilities of 0.01 and 0.002, the central part of São Miguel is likely to experience an earthquake of VI and VIII MMI, respectively (Oliveira et al. 1990).

In terms of seismic activity, cell 8 has a signature completely different to CELLs 1–7. The cumulative number of earthquakes (Fig. 12a) highlights low seismic activity up to 2004, followed by a period of increasing seismicity becoming intense until 2007 and declining later. The period 2004–2006 is characterised by seismic swarms located in the Fogo-Congro region. From 2005 to 2007, more than 8,300 events with magnitude (M_L) lower than 4 were reported in the Seismological Bulletins. Comparing cumulative numbers of earthquakes (Fig. 12a) with those of the seismic swarm of 1989 (see Fig. 5b of Nunes and Oliveira 1999), both swarms have the same features; initially, there is a subtle increase in the number of earthquakes followed by a peak of seismic activity during a short period of time, then a return to normal background activity. The range of magnitude is between 0.5 and 1.3.

The temporal distribution of events in the graph of Fig. 12b shows periods of low activity before and after the seismic swarms of 2004 and 2005–2006, M_L magnitude varying in the range of 1.5–3.5. A low number of events is found between 2007 and 2012 with $0.5 < M_L < 4.2$. Magnitude for the seismic swarm varies between 0.1 and 4.0.

Figure 12c shows that the FMD fits the power law for magnitude range of between 2.2 and 3.4.

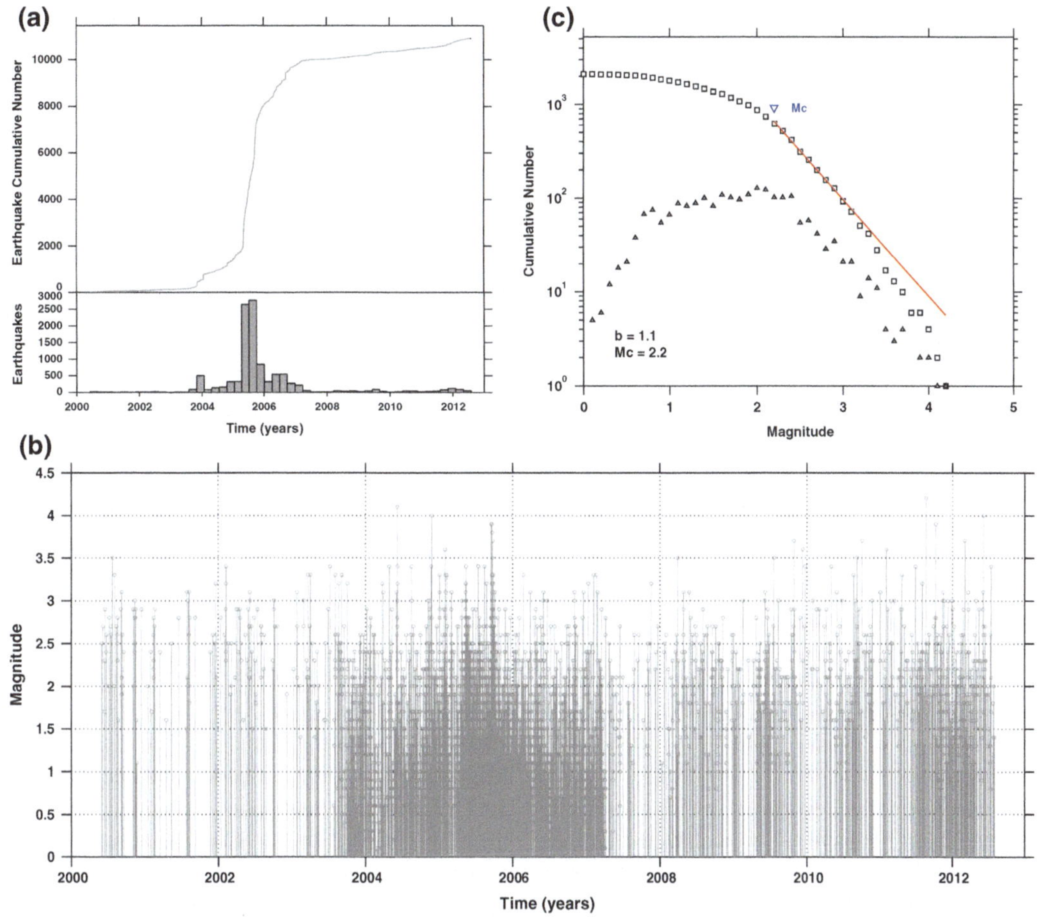

Fig. 12 Temporal distribution of earthquake in cell 8. **a** cumulative earthquake number (top) and annual distribution (bottom). **b** Annual number of event. The lack of seismicity for the first semester 2000 is due to the shift on reported magnitude. In December 2010, data are not available. **c** Frequency magnitude distribution. Triangles are noncumulative frequency magnitude distribution and empty squares are cumulative frequency magnitude distribution. The estimated error in magnitude completeness is ±0.3 and b-value ±0.2. The red line is the fitted power law beginning at the minimum magnitude completeness (triangle)

The values obtained in this cell were: $Mc = 2.2 \pm 0.3$; $a = 5.6$ and $b = 1.0 \pm 0.2$.

CELL 9 includes São Miguel and Santa Maria Islands, Povoação and São Miguel basin, and the Formigas Islets; all of them known to be seismically active areas. This cell is in the transition of the Terceira Rift to the Gloria Fault. Searle (1980) stated that the western end of the Gloria Fault stops abruptly at latitude 36°48′N and longitude 24°30′W.

Most of the significant earthquakes recorded in cell 9 are around M_L 5–5.9, with few earthquakes of higher magnitude. One located east of Santa Maria Island on 8 May 1939 of $M_b = 7$ caused heavy damage on Santa Maria, and another on 5 April 2007 of $M_w = 6.2$ (Global CMT) occurred in the Formigas Islets.

Seismicity in cell 9 (electronic supplement S8) is like that in cell 8 (Fig. 12a) up to the first quarter of 2007. Thenceforth, seismicity has been continuous above background values, up to now particularly, due to the earthquake of April 2007. Magnitude distribution is similar to that of cell 8 for events with magnitudes up to 4, but the number of earthquakes with magnitude greater than 3 is higher. FMD fits the power law for magnitude ranging from 2.2 to 3.5. The values obtained were: Mc = 2.2 ± 0.49; a = 4.8 and b = 0.8 ± 0.17. However, we must remember that some seismogenic sources are in areas where the seismic network has poor azimuthal coverage.

CELL 10 covers the transition of the Terceira Rift to the Gloria Fault (Fig. 10). The seismicity was very low until 2007, and increased after the earthquake of April 5th, 2007 (M_w = 6.2, Global CMT), remaining higher than before. The range of magnitude for most of the events is 1.8–2.6. Evolution of magnitude (M_L) with time exhibits a period with few earthquakes, with magnitude lower than 3, between mid-2000 and the first quarter of 2003. From then on, the seismicity slightly increases up to the end of 2006 with magnitudes ranging from 2 to 3.5, up to 4.5. Since the M_W 6.2 event of April 5th, 2007 the seismic rate has remained high. FMD is correlated with the power law for magnitudes up to 4.1. The *Mc*, *a*- and *b*-values are: $Mc = 2.4 \pm 0.5$, $a = 4.6$ and $b = 0.8 \pm 0.2$, respectively.

Table 3 summarizes the *a*- and *b*-values and *Mc* calculated for each cell, as well as the latitude and longitude coordinates of the upper left corner and lower right corner of each cell.

4 Discussion and Conclusions

The different seismicity parameters estimated for the Archipelago of Azores exhibit variability in different regions. The deployment of a new generation of broadband sensors as well as improvements in data treatment by the seismological observatories have increased the quality of data, as well as the level of detectability of the seismic network. It is noticeable a decrease of Mc (from 3.2 in cell two to 2.2 in cell nine) during the 12 years under analysis. This decrease is related to a better azimuthal coverage of some seismic clusters provided by the seismic network. Nevertheless, some significant seismic clusters cannot be adequately surveyed because of the geographic disposition of the islands, such as the ones located towards the edges of the strip of the Azores; to the west of Graciosa, São Jorge and Faial, and to the east of São Miguel and Santa Maria. Furthermore, even between the islands there are clusters with poor azimuthal coverage such as between the islands of Terceira and São Miguel. In fact, our results demonstrate that cells 1, 2, 3, 4 and 5 with poor azimuthal coverage have high Mc (2.8–3.2). Comparing cells 5 and 6, which overlap 0.5°, we notice a decrease from 2.8 to 2.4, mostly due to a better azimuthal

Table 3 *a*- and *b*-value and minimum magnitude of completeness (*Mc*) calculated for each cell

Cell	Northwest		Southeast		*b*-value	*a*-value	Magnitude of completeness (*Mc*)
	Lat.	Long.	Lat.	Long.			
1	41.9	−30.1	39.9	−28.1	1 ± 0.2	5.2	3.1 ± 0.2
2	40.9	−31.1	38.9	−29.1	1.4 ± 0.3	6.7	3.2 ± 0.3
3	39.9	−31.1	37.9	−29.1	1.2 ± 0.2	5.9	2.8 ± 0.2
4	38.9	−32.1	36.9	−30.1	0.9 ± 0.3	4.1	2.8 ± 0.2
5	39.9	−30.1	37.9	−28.1	1.3 ± 0.2	6.5	2.8 ± 0.2
6	39.4	−29.1	37.4	−27.1	1.2 ± 0.1	5.9	2.4 ± 0.2
7	38.9	−28.1	36.9	−26.1	1.3 ± 0.2	5.8	2.5 ± 0.2
8	38.4	−27.1	36.4	−25.1	1.1 ± 0.1	5.6	2.2 ± 0.1
9	37.9	−26.1	35.9	−24.1	0.8 ± 0.2	4.8	2.2 ± 0.5
10	36.4	−25.1	35.4	−23.1	0.8 ± 0.1	4.6	2.4 ± 0.4

Estimated errors for each parameter is given. The coordinates correspond to the NW and SE corners of the rectangular cells coverage between the triangle formed by São Jorge, Graciosa and Terceira. In cell 6 the Mc is 2.4 but in cell 5 it increases due to poor coverage between Terceira and São Miguel. The next two cells (8 and 9) have the lowest Mc in the Azores region. In cell 10 Mc increases to 2.4, mostly due to poor azimuthal coverage to the east of São Miguel and Santa Maria. In fact, Fontiela and Nunes (2008) studied a seismic cluster in the area of Fogo-Congro (in the central region of São Miguel), and obtained *Mc* values of between 0.9 and 1.5. The Mc values obtained herein highlight that in some regions it is possible to perform studies with microseismicity.

Concerning *b*-values, it is known that they vary in space even in short distances (Ogata et al. 1995; Wiemer and Benoit 1996; Wiemer and McNutt 1997; Wiemer and Wyss 1997; Wyss et al. 1997; Power et al. 1998; Murru et al. 1999; Wiemer and Katsumata 1999; El-Isa 2013; El-Isa and Eaton 2014). Another factor that should be taken into consideration is the tectonic complexity that coexists with volcanic activity in the Azores region. In this complexity of tectonic and magmatic processes, we discuss the results under the hypothesis that *b* is inversely proportional to the mean magnitude (Aki 1965). Despite this approach we consider that our FMD of each cell fits well to a power law.

Experiments in rock failure depict a prevalence of small earthquakes due to the increased heterogeneity of rock samples (Mogi 1962), or due to lower stress (Scholz 1968), which can be described as a state of high heterogeneity of stress increased *b*-value. Consequently, low *b*-values are due to an increase in applied shear stress (Scholz 1968), or an increase in effective stress (Wyss 1973). In the Azores region, we found cells with high as well as low *b*-values. In a coarse analysis, a decrease from west toward east is noticeable. Nevertheless, some cells have poor azimuthal coverage which may introduce some bias on the *b*-value. Thus, the physical meaning of the *b*-values of cells 1, 2, 3, 4, 10 will not be discussed. The significant seismicity in the Azores is to be found in cells 5–9. Fontiela et al. (2014) studied seismic clusters in the Azores, and found *b*-values varying from 0.7 to 1.6.

The *b*-values of the western part of the Azores obtained in this study and those of Fontiela et al. (2014) show that different stress regimes cohabit in the region. In fact, the volcanism emplacement is marked by two distinct buoyant mantle upwelling's (Adam et al. 2013, see also Vogt and Jung, Chapter "The "Azores Geosyndrome" and Plate Tectonics: Research History, Synthesis, and Unsolved Puzzles", O'Neill and Sigloch, Chapter "Crust and Mantle Structure Beneath the Azores Hotspot—Evidence from Geophysics"). One is located in a wide area that comprises the islands of Graciosa, São Jorge, Faial and Pico, and the another one between Terceira and São Miguel. In reality, these islands and the whole region have had more than one eruption since settlement, with exception of Graciosa. This buoyant region can justify the high *b*-values found. Moreover, the crust is likely to be cracked as illustrated by mean magnitude and the high number of small events. Another aspect that we cannot discard is the thermal gradient which increases the *b*-value (Warren and Latham 1970). Forjaz (1994) identified high-enthalpy geothermal fields (temperature > 250 °C) at shallow depths (<2.5 km) in some parts of São Miguel, Terceira, Graciosa, Faial and Pico islands. To the west of Terceira b-values are >1, but here *b* is very low (0.7—Fontiela et al. 2014). However, our analytical approach does not allow identification of a low b-value in this island. Anyway, this could be interpreted as an increase of shear stress, or effective stress due to the lateral compression detected in the island by Navarro et al. (2003) and Miranda et al. (2012).

The *b*-values to the east of Terceira Island exhibit behaviour different from that of the previous zone. In fact, Borges et al. (2007) identified from the total seismic moment tensor that the Azores region is divided into two zones. The western zone (30°W–27°W) is consistent with strike-slip faulting, with horizontal pressure in an E-W direction and extension N-S; and the eastern zone (27°W–23°W) corresponds to normal faulting, with a horizontal tension axis in a NE-SW direction. The region between Terceira and São Miguel (cell 8) has high *b*-values as shown in the study by Fontiela et al. (2014).

These high *b*-values suggest more strongly the presence of buoyant mantle upwelling, as referred to previously. Another feature is the presence of a submarine volcano with its last known eruption in 1720, and a shallow crater rim at a depth of 12 m (Nunes et al. 1998). The temperature in the fumarole field within the crater varies between 39 and 83 °C. According to these facts we can link the high b-values to high pore pressure and high geothermal gradient.

Moving further east (cell 9) the b-values change in relation to the previous cells, decreasing to values less than 1. The seismicity between 2000 and 2012 is located, essentially, in the central part of São Miguel, and in a wide region that comprises the southeast of São Miguel up to the transition of the TR with the Gloria Fault. In this region Fontiela et al. (2014) identified three clusters, all of them with b less than 1, which agrees with the value obtained by Carvalho et al. (2001). From the b-values we infer a change in the stress regime. The non-existence of a plume or plumbing system supports the low b-values in this part of the Azores region. In contrast, studies have identified intense geothermal activity in the central part of São Miguel, from Vp/Vs ratios and seismic tomography (Dawson et al. 1985; Zandomeneghi et al. 2008). This is the same area where Fontiela and Nunes (2008) identified volumes with b-values around 1.6 between 2003 and 2006. As already noted, spatial variation of b-values of 1 to even more are commonly found in volcanic areas at distances of only a few kilometres (Wyss et al. 2001; Wyss and Stefansson 2006; Cerdeña et al. 2011). Another aspect to take into consideration is the temporal variation of the b-value (see references listed in Wiemer and Wyss 2002). Yet another feature that may help explain the low b-values is the hypothesis that 75% of the relative displacement between the Eurasian and Nubian plates is found in a narrow, 10–15 km zone in the central part of São Miguel Island (Jónsson et al. 1999). The focal mechanisms in this region (Silva et al. 2012) show that the compression axis oriented WNW-ESE and the NNE-SSW tension axis are in agreement with the stress regime proposed by Madeira and Ribeiro (1990). Additionally, Silva et al. (2012) state that a stress regime at shallow depths (<5 km) is different from one at depths greater than 5 km. Surely the central part of São Miguel Island exhibits a complex tectonic fabric because of the boundary interactions between the Eurasian and Nubian plates. These interactions may explain the low b-values; however, they do not explain the observed geothermal gradient (Forjaz 1994), nor the presence of hydrothermal fluids at depth.

The low b-values persist to the southeast of São Miguel in a wide seismically active area. This could be related to the fact that this wide area is magmatically extinct (Hübscher et al. 2016). In this case, the low b-value may be explained by the tectonic regime. In fact, the focal mechanism in this region corresponds to normal faulting with a horizontal axis NE–SW normal the TR (Borges et al. 2007) which could accommodate the rifting activity at Povoação Basin (Weiß et al. 2015), increasing the shear or the effective stress and decreasing the b-value.

The *a*-values indicate the total or the annual seismicity rate of a region, and do not have such a clear physical meaning as the *b*-value. The *a*-values in Table 3 refer to the annual seismicity rate. Cells 1, 4, 6, 7 and 8 have an a-value between 4 and 4.9, while cells 2 and 5 have values around 5.5, and the remainder lie between 3 and 3.7.

The geodynamic context of the Azores plays a fundamental role in the geomorphology, the type of volcano activity, and also in the seismicity of the region. In terms of seismology, the western islands (Flores and Corvo) lie in a stable tectonic environment. The remaining seven islands lie on the plate boundary between the Eurasian and the Nubian, under the influence of regional deformation whose seismicity is constrained between those two plates. In fact, the seismicity in the Azores region could be divided into two main zones driven by different stress regimes as we move away from the MAR in the direction of the Gloria Fault.

Acknowledgements The authors would like to thank Max Wyss for his valuable comments. We also thank Mourad Bezzeghoud and Francisco Cota Rodrigues for their support throughout and comments. We acknowledge the contribution of an anonymous reviewer to improve the initial version of the manuscript. We also thank Ulrich Kueppers and Christoph Beier for their comments and inviting us to contribute on the book. João Fontiela was supported by project M3.1.2/F/060/2011 funded of the Regional Science Fund of the Azores Regional Government. This work is co-funded by the European Union through the European Fund of Regional Development, framed in COMPETE 2020 (Operational Competitiveness Programme and Internationalization) through the ICT project (UID/GEO/04683/2013) with the reference POCI-01-0145-FEDER-007690.

References

Adam, C., Madureira, P., Miranda, J. M., Lourenço, N., Yoshida, M., & Fitzenz, D. (2013). Mantle dynamics and characteristics of the Azores plateau. *Earth and Planetary Science Letters, 362,* 258–271.

Agostinho, J. (1927a). O terramoto de 31 de Agosto nas ilhas dos Açores. *Labor, 8,* 229–235.

Agostinho, J. (1927b). The earthquake in the Azores Islands, on the 31st August, 1926. *Labor, 8,* 229–235.

Agostinho, J. (1955a). Relato da sismicidade dos Açores e história sísmica do arquipélago com vista principalmente à delimitação das zonas onde são de aconselhar maiores precauções anti-sísmicas nas construções. Boletim da Ordem dos Engenheiros, 4 (21). *Memória, 108,* 1–4.

Agostinho, J. (1955b). Os abalos sísmicos na ilha Terceira em Dezembro de 1950 e em Janeiro de 1951. Boletim da Ordem dos Engenheiros, 4 (22). *Memória, 115,* 1–6.

Aki, K. (1965). Maximum likelihood estimate of b in the formula logn = a-bm and its confidence limits. *Bulletin of the Earthquake Research Institute, 43,* 237–239.

Álvarez-Gómez, J. A. (2014). FMC: A one-liner Python program to manage, classify and plot focal mechanisms. In *Geophysical Research Abstracts* (Vol. 16), EGU2014-10887.

Arquivo dos Açores. Editors: Ernesto do Canto and Francisco Afonso Chaves. Edition of 1981–1986. Instituto Universitário dos Açores e Universidade dos Açores. Ponta Delgada. Vol. I to XIV.

Arroyo, A. L., & Udías, A. (1972). Aftershock sequence and focal parameters of the February 28, 1969 earthquake of the Azores-Gibraltar fracture zone. *Bulletin of the Seismological Society of America, 62*(3), 699–720.

Bessone, P. (1932). *Dicionário Cronológico dos Açores* (p. 446p). Cambridge: Massachussets.

Bezzeghoud, M., Adam, C., Buforn, E., Borges, J. F., & Caldeira, B. (2014). Seismicity along the Azores-Gibraltar region and global plate kinematics. *Journal of Seismology, 18*(2), 205–220.

Boletim Sismológico Preliminar dos Açores (2001–2014), SIVISA, Ponta Delgada.

Borges, J. F., Bezzeghoud, M., Buforn, E., Pro, C., & Fitas, A. J. S. (2007). The 1980, 1997 and 1998 Azores earthquakes and some seismo-tectonic implications. *Tectonophysics, 435*(1–4), 37–54.

Borges, J. F., Fitas, A. J. S., Bezzeghoud, M., & Teves-Costa, P. (2001). Seismotectonics of Portugal and its adjacent Atlantic area. *Tectonophysics, 331*(4), 373–387.

Buforn, E., Udías, A., & Colombás, M. A. (1988). Seismicity, source mechanisms and tectonics of the Azores-Gibraltar plate boundary. *Tectonophysics, 152* (1–2), 89–118.

Carvalho, A., Sousa, M. L., Oliveira, C. S., Campos-Costa, A., Nunes, J. C., & Forjaz, V. H. (2001). Seismic hazard for central group of the Azores Islands. *Bollettino di Geofisica Teorica ed Applicata, 42,* 89–105.

Cerdeña, I. D., del Fresno, C., & Rivera, L. (2011). New insight on the increasing seismicity during Tenerife's 2004 volcanic reactivation. *Journal of Volcanology and Geothermal Research, 206*(1), 15–29.

Costa Nunes, J. (1986). Sismicidade Histórica e Instrumental do Arquipélago dos Açores. Catálogo Preliminar: 1444–1980. Relatório INMG/LNEC. Lisboa, 149p.

Dawson, P. B., da Silva, A. M. R., Iyer, H. M., Evans, J. R. (1985). Seismic study of the Água de Pau geothermal project, São Miguel, Azores. In *Transactions of the Geothermal Resources Council* (Vol. 9, Part II, pp. 401–406).

Dias, A. M. M. (1945). O sismo de 1522 em S. Miguel. *Insulana,* Ponta Delgada, I(4), 5–15.

Dias, N. (2005). Estudo da sequência sísmica gerada pelo sismo do Faial de 9 de Julho de 1998: Anisotropia crustal, inversão tomográfica e caracterização sismotectónica. Ph.D. Thesis. Universidade de Lisboa, 540p.

Dias, N. A., Matias, L., Lourenço, N., Madeira, J., Carrilho, F., & Gaspar, J. L. (2007). Crustal seismic velocity structure near Faial and Pico Islands (AZORES), from local earthquake tomography. *Tectonophysics, 445*(3–4), 301–317.

Drumond, F. F. (1859). Anais da Ilha Terceira. Editor: Região Autónoma dos Açores – Secretaria Regional de Educação e Cultura, Edition of 1981, Vol. I to III. Angra do Heroísmo

El-Isa, Z. H. (2013). Continuous-cyclic variations in the b-value of the earthquake frequency-magnitude distribution. *Earthquake Science, 26*(5), 301–320.

El-Isa, Z. H., & Eaton, D. W. (2014). Spatiotemporal variations in the b-value of earthquake magnitude–frequency distributions: Classification and causes. *Tectonophysics, 615–616,* 1–11.

Fontiela, J., Bezzeghoud, M., Rosset, P., Borges, J. F., & Cota Rodrigues, F. (2014). Azores seismogenic zones. *Comunicações Geológicas, 101*(I), 29–35.

Fontiela, J., & Nunes, J. C. (2008). The b values on the Congro Fracture Zone (S. Miguel Island) during the period 2003–2007: A first approach. In C. S. Oliveira, M. A. Ferreira, & J. C. Nunes (Eds.), *Azores 1998—International seminar on seismic risk and rehabilitation of stone masonry housing* (pp. 1–6). Faial: GRA.

Forjaz, V. H. (1994). Geologia económica e aplicada da ilha de S. Miguel (Açores): Recursos vulcano-geotérmicos. Tese de doutoramento, Universidade dos Açores, 573 pp.

Forjaz, V. H, Nunes, J. C., & França, Z. (1996). Swarm sísmico no vulcão das Sete Cidades (Set.96). *Açoreana, 8*(2), 27–277.

França, Z., Cruz, J. V., Nunes, J. C., & Forjaz, V. H. (2003). Geologia dos Açores: Uma perspectiva actual. *Açoreana, 10*(1), 11–140.

Frohlich, C., & Davis, S. (1993). Teleseismic b-values: Or, much ado about 1.0. *Journal of Geophysical Research, 98,* 631–644.

Frutuoso, G. (1998). Saudades da Terra. Editor: Instituto Cultural de Ponta Delgada. Ponta Delgada. Edição 1998. Livros I to VI.

Gaspar, J. L. (1997). Dão João de Castro Bank (Azores Islands) Magnitude 5.5 earthquake and associated seismic swarm. *Smithsonian Institution Bulletin of the Global Volcanism Network, 22*(7), 9.

Gaspar, J. L., Queiroz, G., Pacheco, J.-M., Ferreira, T., Wallenstein, N., Almeida, M. H., & Coutinho, R. (2003). Basaltic lava balloons produced during the 1998–2001 Serreta Submarine Ridge eruption (Azores). In J. D. L. White, J. L. Smellie, & D. A. Clague (Eds.), *Explosive Subaqueous Volcanism, Geophysical Monography Series* (Vol. 140, pp. 205–212). Washington D.C.: AGU.

Gaspar, J. L., & Wallenstein, N. (1998). Sete Cidades (Azores) Seismic swarm on submarine flank. *Smithsonian Institution Bulletin of the Global Volcanism Network, 23*(9), 8–9.

Grimison, N. L., & Chen, W.-P. (1986). The Azores-Gibraltar Plate Boundary: Focal mechanisms, depths of earthquakes, and their tectonic implications. *Journal Geophysical Research, 91* (B2), 2029.

Grimison, N. L., & Chen, W. P. (1988). Source mechanisms of four recent earthquakes along the Azores-Gibraltar plate boundary. *Geophysical Journal, 92,* 391–401.

Gutenberg, B., & Richter, C. F. (1944). Frequency of earthquakes in California. *Bulletin of the Seismological Society of America, 34,* 185–188.

Habermann, R. E. (1987). Man-made changes of seismicity rates. *Bulletin of the Seismological Society of America, 77*(1), 141–159.

Habermann, R. E. (1991). Seismicity rate variations and systematic changes in magnitudes in teleseismic catalogues. *Tectonophysics, 193,* 277–289.

Habermann, R. E., & Creamer, F. (1994). Catalogue errors and the M8 earthquake prediction algorithm. *Bulletin of the Seismological Society of America, 84,* 1551–1559.

Hirn, A., Haessler, H., Trong, P. H., Wittlinger, G., & Mendes-Victor, L. A. (1980). Aftershock sequence of the January 1st, 1980, earthquake and present-day tectonics in the Azores. *Geophysical Research Letters, 7*(7), 501–504.

Hübscher, C., et al. (2016). Azores Plateau—Cruise No. M113/1—December 29, 2014–January 22, 2015 —Ponta Delgada (Portugal)—Ponta Delgada (Portugal), METEOR-Berichte, M113/1, 31. https://doi.org/10.2312/cr_m113_1.

International Seismological Centre, On-line Bulletin, http://www.isc.ac.uk, International Seismological Centre, Thatcham, United Kingdom, 2013.

Ishimoto, M., & Iida, K. (1939). Observations of earthquakes registered with the microseismograph constructed recently. *Bulletin Earthquake Research Institute, 17,* 443–478.

Jónsson, S., Alves, M., & Sigmundsson, F. (1999). Low rates of deformation of the Furnas and Fogo Volcanoes, São Miguel, Azores, observed with the Global Positioning System, 1993–1997. *Journal of Volcanology and Geothermal Research, 92*(1–2), 83–94.

Junior, F. J. C., Monjardino, J. I. A., & Drummond, F. F. (1983). Memória histórica do horrível terramoto de 15.VI.1841 que assolou a Vila da Praia da Vitória. Editor Câmara Municipal da Praia da Vitória. Praia da Vitória, 281p.

Kearey, P., Klepeis, K. A., & Vine, F. J. (2009). *Global tectonics* (3rd ed., pp. xiii+482). Chichester: Wiley-Blackwell. ISBN 978-1-4051-0777-8.

Kueppers, U., Nichols, A. R. L., Zanon, V., Potuzak, M., & Pacheco, J. M. R. (2012). Lava balloons—Peculiar products of basaltic, submarine eruptions. *Bulletin of Volcanology, 74,* 1379–1393.

Lourenço, N. (2007). Tectono-Magmatic Processes at the Azores Triple Junction. Ph.D. Thesis. Faculdade de Ciências do Mar e Ambiente, Universidade do Algarve. 267 p.

Lourenço, N., Miranda, J. M., Luís, J. F., Ribeiro, A., Victor, L. A. M., Madeira, J., et al. (1998). Morphotectonic analysis of the Azores Volcanic Plateau from a new bathymetric compilation of the area. *Marine Geophysical Researches, 20*(1998), 141–156.

Luís, J. F., Miranda, J. M., Galdeano, A., Patriat, P., Rossignol, J., & Mendes Victor, L. A. (1994). The Azores triple junction evolution since 10 Ma from an aeromagnetic survey of the Mid-Atlantic Ridge. *Earth and Planetary Science Letters, 125*(1–4), 439–459.

Luís, J. F., Miranda, J. M., Galdeano, A., & Patriat, P. (1998). Constraints on the structure of the Azores spreading center from gravity data. *Marine Geophysical Researches, 20*(3), 157–170. https://doi.org/10.1023/A:1004698526004.

Luís, J. F., & Neves, M. C. (2006). The isostatic compensation of the Azores Plateau: A 3D admittance and coherence analysis. *Journal of Volcanology and Geothermal Research, 156*(1–2), 10–22.

Macedo, A. L. S. (1871). História das quatro ilhas que formam o distrito da Horta. Editor: Região Autónoma dos Açores – Secretaria Regional de Educação e Cultura – Direcção Regional dos Assuntos Culturais, Edition of 1981, Vol. I to III. Angra do Heroísmo.

Machado, F. (1948). Frequência dos sismos sentidos nas ilhas do Faial e do Pico. *Açoreana, 4,* 236–245.

Machado, F. (1949). O terramoto de S. Jorge de 1757. *Açoreana, 4*(4), 1–14.

Machado, F. (1966). Anomalias das intensidades do terramoto de S. Miguel (Açores) em 1522. Boletim da Museu e Laboratório Mineralógico e Geológico da Faculdade de Ciências de Lisboa *10*(2), 109–117.

Machado, F., Parsons, W. H., Richards, A. F., & Mulford, J. W. (1962). Capelinhos eruption of Fayal volcano, Azores, 1957–1958. *Journal of Geophysical Research, 67*(9), 3519–3529.

Madeira, J. (1998). Estudos de neotectónica nas ilhas do Faial, Pico e S. Jorge: Uma contribuição para o conhecimento geodinâmico da junção tripla dos Açores. Tese de doutoramento, Faculdade de Ciências da Universidade de Lisboa, 481 pp.

Madeira, J., & Ribeiro, A. (1990). Geodynamic models for the Azores triple junction: A contribution from tectonics. *Tectonophysics, 184*(3–4), 405–415.

Marques, R., Queiroz, G., Coutinho, R., & Zêzere, J. L. (2007). Actividade geomorfológica desencadeada pela crise sísmica de 2005 no Vulcão do Fogo (S. Miguel, Açores): avaliação da susceptibilidade com recurso a regressão logística, in 3rd Congresso de Geomorfologia - Dinâmicas geomorfológicas. *Metodologias e Aplicação, 5,* 47–61.

Martini, F., Bean, C. J., Saccorotti, G., Viveiros, F., & Wallenstein, N. (2009). Seasonal cycles of seismic velocity variations detected using coda wave interferometry at Fogo volcano, São Miguel, Azores, during 2003–2004. *Journal of Volcanology and Geothermal Research, 181*(3–4), 231–246.

Matias, L., Dias, N. A., Morais, I., Vales, D., Carrilho, F., Madeira, J., et al. (2007). The 9th of July 1998 Faial Island (Azores, North Atlantic) seismic sequence. *Journal of Seismology, 11*(3), 275–298.

McKenzie, D. (1972). Active tectonics of the Mediterranean region. *Geophysical Journal International, 30*(2), 109–185.

Miranda, J. M. L. A., Mendes Victor, J. Z., Simões, J. F., Luís, L., Matias, H., Shimamura, H., et al. (1998). Tectonic setting of the Azores Plateau deduced from a OBS survey. *Marine Geophysical Researches, 20,* 171–182.

Miranda, J. M., Navarro, A., Catalão, J., & Fernandes, R. M. S. (2012). Surface displacement field at Terceira Island deduced from repeated GPS measurements. *Journal of Volcanology and Geothermal Research, 217–218,* 1–7.

Mogi, K. (1962). Magnitude-frequency relation for elastic shocks accompanying fractures of various materials and some related problems in earthquakes. *Bulletin Earthquake Research Institute, 40,* 831–853.

Mogi, K. (1967). Earthquakes and fractures. *Tectonophysics, 5*(1), 35–55.

Murru, M., Montuori, C., Wyss, M., & Privitera, E. (1999). The locations of magma chambers at Mt. Etna, Italy, mapped by b-values. *Geophysical Research Letters, 26*(16), 2553–2556.

Navarro, A., Catalão, J., Miranda, J. M., & Fernandes, R. M. S. (2003). Estimation of the Terceira Island (Azores) main strain rates from GPS data. *Earth Planets Sp., 55*(10), 637–642.

Nunes, J. C. (1991). Microssismos e neotectónica – Contribuição para o seu estudo nos Açores. Provas de Aptidão Pedagógica e Capacidade Científica. Monografia. Departamento de Geociências. Universidade dos Açores. Ponta Delgada, 245p.

Nunes, J. C., & Forjaz, V. H. (1998). Study of two seismic swarms on Furnas Laboratory Volcano, Azores Islands. In: R. Casale, M. Fytikas, G. Sigvaldason, & G. Vougiokalakis (Eds.), *Proceedings of the 2nd European Laboratory Volcanoes Workshop. Santorini (Greece). European Commission-Volcanic Risk. DG XII-Environment and Climate Programme,* EUR 18161 EN; 549–557.

Nunes, J. C., Forjaz, V. H., Alves, J. L., & Bernardes, A. C. (1998). Caracterização vulcanológica do banco D. João de Castro (Açores): Novos Dados, Ciências da Terra, (5 - n.º especial), pp. D55–D58.

Nunes, J. C., Forjaz, V. H., & França, Z. (1997). Novos elementos sobre a crise sísmica de 1973, Ilha do Pico (Açores). In Instituto Superior Técnico (Ed.), 3º Encontro Sobre Sismologia e Engenharia Sísmica - Proceedings. Lisboa. Instituto de Engenharia de Estruturas, Território e Construção – DECivil, 39–48.

Nunes, J. C., Forjaz, V. H., & França, Z. (2001). Principais sismos destrutivos no arquipélago dos Açores – uma revisão. In M. R. Fragoso (Ed.), 5th Encontro Nacional de Sismologia e Engenharia Sísmica – SÍSMICA 2001, Ponta Delgada (Açores). Laboratório Regional de Engenharia Civil, 119–131.

Nunes, J. C., Forjaz, V. H., & Oliveira, C. S. (2004). Catálogo Sísmico da Região dos Açores. Versão 1.0 (1850–1998). Universidade dos Açores (Ed.). Ponta Delgada. Edição CD-ROM.

Nunes, J. C., & Oliveira, C. S. (1999). Actividade sísmica do vulcão do Fogo, Ilha de S. Miguel. In *Actas do I Simpósio de Meteorologia e Geofísica - Comunicações de Geofísica,* pp. 55–60.

Ogata, Y., Utsu, T., & Katsura, K. (1995). Statistical features of foreshocks in comparison with other earthquake clusters. *Geophysical Journal International, 121*(1), 233–254.

Oliveira, C. S., Costa, A. C., Forjaz, V. H., & Nunes, J. C. (1990). Seismic hazard analysis in zones of time and space interdependence: An application to São Miguel Island, Azores. *Natural Hazards, 3*(1), 15–29.

Power, J. A., Wyss, M., & Latchman, J. L. (1998). Spatial variations in the frequency-magnitude distribution of earthquakes at Soufriere Hills Volcano, Montserrat, West Indies. *Geophysical Research Letters, 25*(19), 3653.

Reasenberg, P. (1985). Second-order moment of central California seismicity, 1969–1982. *Journal of Geophysical Research, 90*(B7), 5479–5495.

Rydelek, P. A., & Sacks, I. S. (1989). Testing the completeness of earthquake catalogues and the hypothesis of self-similarity. *Nature, 337,* 251–253.

Schilling, J.-G. (1975). Rare-Earth variations across 'normal segments' of the Reykjanes Ridge, 60°–53° N, Mid-Atlantic Ridge, 29°S, and East Pacific Rise, 2°–19°S, and evidence on the composition of the underlying low-velocity layer. *Journal Geophysical Research, 80*(11), 1459–1473.

Scholz, C. H. (1968). The frequency-magnitude relation of micro-fracturing in rock and its relation to earthquakes. *Bulletin of the Seismological Society of America, 58,* 399–415.

Searle, R. (1976). Lithospheric structure of the Azores Plateau from Rayleigh-Wave dispersion. *Geophysical Journal Royal Astronomy Society, 44,* 537–546.

Searle, R. (1980). Tectonic pattern of the Azores spreading centre and triple junction. *Earth Planetary Science Letters, 51,* 415–434.

Senos, L., Alves, P., Vales, D., Cruz, J., Silva, M., Carrilho, F. (2008). O sismo de 9 Julho de 1998 nos Açores e a crise sísmica associada - dez anos depois. In Sismo 1998 - Açores, Uma década depois. Editors C. S. Oliveira, A. Costa, J. C. Nunes, SPRHI, S. A. Horta.

Silva, R., Havskov, J., Bean, C., & Wallenstein, N. (2012). Seismic swarms, fault plane solutions, and stress tensors for São Miguel Island central region (Azores). *Journal of Seismology, 16*(3), 389–407.

Silveira, D., Gaspar, J. L., Ferreira, T., & Queiroz, G. (2003). Reassessment of the historical seismic activity with major impact on S. Miguel Island (Azores). *Natural Hazards and Earth System Science, 3*(6), 615–623.

Silveira, G., Vinnik, L., Stutzmann, E., Farra, V., Kiselev, S., & Morais, I. (2010). Stratification of the Earth beneath the Azores from P and S receiver functions. *Earth and Planetary Science Letters, 299*(1–2), 91–103.

Steinmetz, L., Whitmarsh, R. B., & Moreira, V. S. (1977). Upper mantle structure beneath the Mid-Atlantic Ridge north of the Azores based on observations of compressional waves. *Geophysical Journal International, 50*(2), 353–380.

Taylor, D. W. A., Snoke, J. A., Sacks, I. S., & Takanami, T. (1990). Non-linear frequency-magnitude relationship for the Hokkaido corner, Japan. *Bulletin of the Seismological Society of America, 80,* 340–353.

Udías, A. (1980). Seismic stresses in the region Azores—Spain—Western Mediterranean. In A. E. Scheidegger (Ed.), *Tectonic stresses in the Alpine-Mediterranean region* (Vol. 9, pp. 75–84). Vienna: Springer.

Udías, A., Arroyo, A. L., & Mezcua, J. (1976). Seismotectonic of the Azores-Alboran region. *Tectonophysics, 31*(3–4), 259–289.

Vogt, P. R. R., & Jung, W. Y. Y. (2004). The Terceira Rift as hyper-slow, hotspot-dominated oblique spreading axis: A comparison with other slow-spreading plate boundaries. *Earth and Planetary Science Letters, 218*(1–2), 77–90.

Vogt, P. R. R., & Jung, W. Y. Y. (2017). The "Azores Geosyndrome" and Plate Tectonics: Research History, present understanding, and unsolved puzzles. In U. Kueppers & C. Beier (Eds.), *Volcanoes of the Azores*. Springer. ISBN 978-3-642-32226-6.

Warren, N. W., & Latham, G. V. (1970). An experimental study of thermally induced microfracturing and its relation to volcanic seismicity. *Journal of Geophysical Research, 75*(23), 4455–4464.

Weiß, B. J., Hübscher, C., & Lüdmann, T. (2015). The tectonic evolution of the southeastern Terceira Rift/São Miguel region (Azores). *Tectonophysics, 654,* 75–95.

Wiemer, S. (2001). A software package to analyze seismicity: ZMAP. *Seismological Research Letters, 72*(2), 374–383.

Wiemer, S., & Benoit, J. P. (1996). Mapping the B value anomaly at 100 km depth in the Alaska and New Zealand Subduction Zones. *Geophysical Research Letters, 23*(13), 1557.

Wiemer, S., & Katsumata, K. (1999). Spatial variability of seismicity parameters in aftershock zones. *Journal Geophysical Research, 104*(B6), 13135–13151.

Wiemer, S., & McNutt, S. R. (1997). Variations in the frequency-magnitude distribution with depth in two volcanic areas: Mount St Helens, Washington, and Mt. Spurr, Alaska. *Geophysical Research Letters, 24*(2), 189–192.

Wiemer, S., McNutt, S. R., & Wyss, M. (1998). Temporal and three-dimensional spatial analyses of the frequency-magnitude distribution near Long Valley Caldera, California. *Geophysical Journal International, 134*(2), 409–421.

Wiemer, S., & Wyss, M. (1997). Mapping the frequency-magnitude distribution in asperities: An improved technique to calculate recurrence times? *Journal of Geophysical Research, 102*(B7), 15115–15128.

Wiemer, S., & Wyss, M. (2000). Minimum magnitude of completeness in earthquake catalogs: Examples from Alaska, the Western United States, and Japan. *Bulletin of the Seismological Society of America, 90*(4), 859–869.

Wiemer, S., & Wyss, M. (2002). Mapping spatial variability of the frequency-magnitude distribution of earthquakes. *Advances in Geophysics, 45*(5), 259–302.

Wyss, M. (1973). Towards a physical understanding of the earthquake frequency distribution. *Geophysical Journal Royal Astronomy Society, 31,* 341–359.

Wyss, M. (1991). Reporting history of the central Aleutians Seismograph network and the quiescence preceding the 1986 Andreanof Island earthquake. *Bulletin of the Seismological Society of America, 81,* 1231–1254.

Wyss, M., Klein, F., Nagamine, K., & Wiemer, S. (2001). Anomalously high *b*-values in the South Flank of Kilauea volcano, Hawaii: Evidence for the distribution of magma below Kilauea's East rift zone. *Journal of Volcanology and Geothermal Research, 106*(1–2), 23–37.

Wyss, M., Shimazaki, K., & Wiemer, S. (1997). Mapping active magma chambers by b values beneath the off-Ito volcano, Japan. *Journal of Geophysical Research, 102*(B9), 20413–20422.

Wyss, M., & Stefansson, R. (2006). Nucleation points of recent mainshocks in southern Iceland, mapped by b-values. *Bulletin of the Seismological Society of America, 96*(2), 599–608.

Zandomeneghi, D., Almendros, J., Ibáñez, J. M., & Saccorotti, G. (2008). Seismic tomography of Central São Miguel, Azores. *Physics of the Earth and Planetary Interiors, 167*(1–2), 8–18.

Zbyszewski, G., Pacheco, T., & Rosa, D. (1977). Os sismos de 1964 na Ilha de S. Jorge (Açores): Registo diário. *Ciências da Terra, 3,* 33–59.

Zuniga, F. R., & Wiemer, S. (1999). Seismicity patterns: Are they always related to natural causes? *Pure and Applied Geophysics, 155,* 713–726.

The Marine Fossil Record at Santa Maria Island (Azores)

Sérgio P. Ávila, Ricardo S. Ramalho, Jörg M. Habermann and Jürgen Titschack

Abstract

Santa Maria is the oldest island of the Azores Archipelago and is remarkably rich in exposed marine fossiliferous sediments and submarine volcanic sequences. This chapter summarises the geological history of the island and reports on the most important palaeontological studies done on the outcrops of Santa Maria since the early studies, during the 19th century. The most important early Pliocene and late Pleistocene (Last Interglacial) fossiliferous deposits are described and palaeoecological reconstructions are presented for each sedimentary succession. The most abundant, diversified and well-studied fossil groups are also reviewed, namely the algae, vertebrates (cetaceans and selaceans), and invertebrates (molluscs, echinoderms, brachiopods, crustaceans

S. P. Ávila (✉)
InBIO Laboratório Associado, CIBIO, Centro de Investigação em Biodiversidade e Recursos Genéticos, Pólo dos Açores, Azores, Portugal
e-mail: avila@uac.pt

S. P. Ávila
Departamento de Biologia, Faculdade de Ciências e Tecnologias, Universidade dos Açores, 9501-801 Ponta Delgada, Azores, Portugal

S. P. Ávila
MPB - Marine PalaeoBiogeography Working Group of the University of the Azores, Rua da Mãe de Deus, 9501-801 Ponta Delgada, Azores, Portugal

R. S. Ramalho
Faculdade de Ciências, Instituto Dom Luiz, Universidade de Lisboa, 1749-016 Lisbon, Portugal

R. S. Ramalho
School of Earth Sciences, University of Bristol, Wills Memorial Building, Queen's Road, Bristol BS8 1RJ, UK

R. S. Ramalho
Lamont-Doherty Earth Observatory, Columbia University, Comer Geochemistry Building, PO Box 1000, Palisades, NY 10964-8000, USA

J. M. Habermann
Faculdade das Ciências Humanas e Sociais, Interdisciplinary Center for Archaeology and Evolution of Human Behaviour (ICArEHB), Universidade do Algarve, Campus Gambelas, 8005-139 Faro, Portugal

J. M. Habermann
Institute of Cognitive and Evolutionary Anthropology (ICEA), University of Oxford, 64 Banbury Road, Oxford OX2 6PN, UK

J. M. Habermann
GeoZentrum Nordbayern, Friedrich-Alexander-Universität Erlangen-Nürnberg, Schlossgarten 5, 91054 Erlangen, Germany

J. Titschack
MARUM – Center for Marine Environmental Sciences, University of Bremen, Leobener Strasse, 28359 Bremen, Germany

J. Titschack
Marine Research Department, Senckenberg am Meer, 26382 Wilhelmshaven, Germany

U. Kueppers and C. Beier (eds.), *Volcanoes of the Azores*, Active Volcanoes of the World, https://doi.org/10.1007/978-3-642-32226-6_9

and ostracods). We also discuss the palaeontological significance of carbonate sequences in reefless volcanic oceanic islands. Finally, we discuss the importance of applied palaeontology, with products specifically designed for tourism, such as the Museum "House of the Fossils", the "Route of the Fossils" project, and the "PalaeoPark Santa Maria", all of them aiming to protect and conserve fossil sites, allowing at the same time, its sustainable use by locals and tourists.

1 Introduction

Seamounts and oceanic islands are prominent volcanic features that rise abruptly from the deep sea to shallower waters, and often constitute true oases of biodiversity in the otherwise relatively poorly diverse open ocean. Seamounts, especially when they rise above the photic zone, provide substratum to a variety of benthic organisms—such as algae, corals and sponges—that in turn provide shelter and nutrients for a series of other benthic and pelagic organisms, allowing the establishment of more complex ecosystems as time passes. Oceanic islands, as they emerge above sea-level, allow the formation of coastlines, further increasing the diversity of available shallow marine environments. Likewise, the formation of landmasses in the vast open ocean is generally followed by a selective terrestrial biological colonization which is a wonder as many naturalists and scientists have recently noted. Thus, with time, mid-ocean volcanic edifices typically host unique ecosystems that reflect coeval environmental conditions and may be the result of different evolutionary pathways concerning patterns of colonization, adaptive and non-adaptive radiation, speciation, etc. The study of such ecosystems has been at the basis of many of the ground-breaking advances in the biological sciences in the last 250 years, and is at the forefront of several disciplines such as biogeography, ecology, evolution, etc.

Ocean island volcanoes typically grow through intermittent summit eruptions, gradually building up to form prominent edifices. Volcanic activity is, generally, discontinuous in space and time, allowing for the establishment of biological communities away from active vents (see Vogt and Jung, Chapter "The "Azores Geosyndrome" and Plate Tectonics: Research History, Synthesis, and Unsolved Puzzles" for an overview on the Azores). In a similar fashion, in these areas, erosion and sedimentation gradually contribute to the formation of soils and other sedimentary bodies that further enhance the possibilities for biological colonization. However, when volcanic activity shifts back to these areas, sedimentary sequences and biological remains may be incorporated into the volcanic edifice and thus preserved from subsequent erosion. The ongoing but intermittent volcanic activity—although highly disruptive for living communities—leads to a more generalised preservation of taphonomical assemblages into the fossil record at oceanic islands, allowing the study and reconstruction of past biological communities and their surrounding environmental conditions. This is especially true for marine environments, because the rapid cooling of lava flows imposed by the subaqueous environment minimizes the destruction associated with the extrusion of these flows, leading to the generation of submarine volcano-sedimentary sequences that are very informative of the environment in which they were formed (e.g. Johnson et al. 2012, 2017; Meireles et al. 2013; Ávila et al. 2015a; Rebelo et al. 2016a; Uchman et al. 2016).

The marine fossil record at oceanic islands is, however and at a first glance, relatively poor when compared to the marine fossil record on continents; the lack of fossiliferous exposures at oceanic islands sharply contrasts with the nearest continental coasts. Nevertheless, the marine fossil record at oceanic islands is of the upmost importance for the comprehension of life's dispersal in the oceans and adjacent coastlines, and may provide useful constraints to disciplines such as palaeo-climatology and palaeo-oceanography (e.g., Ávila et al. 2015b, 2016a). But why are marine fossiliferous exposures so rare in oceanic islands? The answer lays in the subsidence trend that most oceanic islands experience.

The growth and decay of volcanic island edifices is typically accompanied by vertical movements of diverse nature (e.g. Walcott 1970; McNutt and Menard 1978; Watts and ten Brink 1989; Ramalho et al. 2010a, b, c; Madeira et al. 2010). Commonly, island edifices subside markedly during their initial stages of growth, as the weight of the rapidly-growing edifices induce a downwards flexure of the "rigid" lithospheric plate underneath (Brotchie and Silvester 1969; Walcott 1970; Watts and ten Brink 1989; Huppert et al. 2015). Likewise, as the tectonic plate on which the island edifices rest gradually cools with age, it slowly subsides and so also the edifices with it, resulting in a longer-term subsidence response (Stein and Stein 1992; Morgan et al. 1995, see also Vogt and Jung, Chapter "The "Azores Geosyndrome" and Plate Tectonics: Research History, Synthesis, and Unsolved Puzzles"). In a similar fashion, the bathymetric swell associated with many hotspots (whose origins are still highly debated) and in which many islands rest, may experience a slow topographic decay, further contributing to long-term island subsidence (Dietz and Menard 1953; Crough 1978; Sleep 1990; Morgan et al. 1995). Thus, not surprisingly, oceanic islands are typically ephemeral features in the vastness of the geological time, subsidence and erosion being the main contributors to their relatively short-lived existence above sea-level.

The short- and long-term subsidence trends that most oceanic islands experience—together with ongoing vigorous volcanic activity during the initial stages—gradually bury and/or submerge coastlines as the edifices themselves age. As a consequence, the coastal sediments and the skeletal remains of existing marine living communities that eventually get incorporated in the subsiding volcanic edifice, become inaccessible, especially because eustatic sea level was rarely above the present level in the last 5–10 Ma, a reasonable life-time for a subsiding island (see Miller et al. 2005). Hence, the natural occurrence of exposed marine fossiliferous sequences at oceanic islands require particular conditions that are associated with the rare edifices that experienced a long-term vertical trend different than most islands—like many of the Cape Verdes and the Canaries, Porto Santo (Madeira archipelago), and Santa Maria (Azores), for example (Madeira et al. 2010; Ramalho et al. 2010a, b, c, 2015, 2017; Ramalho 2011; Menéndez et al. 2008; Klügel et al. 2005, 2015; Ávila et al. 2012). These islands all have undergone vertical movements that resulted in a net uplift, at least during part of their evolution. This uplift, coupled with marine and fluvial erosion, exposed sequences that otherwise would be deeply buried in the interior of the volcanic edifice and below present sea level.

2 The Marine Fossil Record of Santa Maria as a Result of Exposure by Uplift and Erosion

Santa Maria Island—the oldest and southeasternmost island in the Azores (Fig. 1a)—is remarkably rich in exposed marine fossiliferous sediments and submarine volcanic sequences (Fig. 1b), even within the North Atlantic context (Mitchell-Thomé 1976; Serralheiro et al. 1990; Madeira et al. 2007). This is mostly due to several factors: (1) Santa Maria's edifice experienced a relatively long but intermittent volcanic life intercalated by periods of quiescence, erosion and sedimentation, leading to the deposition and preservation of thick marine volcano-sedimentary sequences; (2) the island experienced pronounced uplift during the Pliocene-Quaternary that, coupled with coastal and fluvial erosion, exposed numerous submarine sequences.

The emergence of Santa Maria's volcanic edifice above sea level occurred during the Late Miocene (Mitchell-Thomé 1976; Serralheiro et al. 1987; Serralheiro and Madeira 1990; Serralheiro 2003; Sibrant et al. 2015; Ramalho et al. 2017). The oldest volcanostratigraphic units in the island attest to this transition between submarine and subaerial volcanic activity and correspond to remains of surtseyan and strombolian monogenetic cones (Cabrestantes and Porto Formations; Serralheiro et al. 1987), dated to ~6.0 Ma (Ramalho et al. 2017).

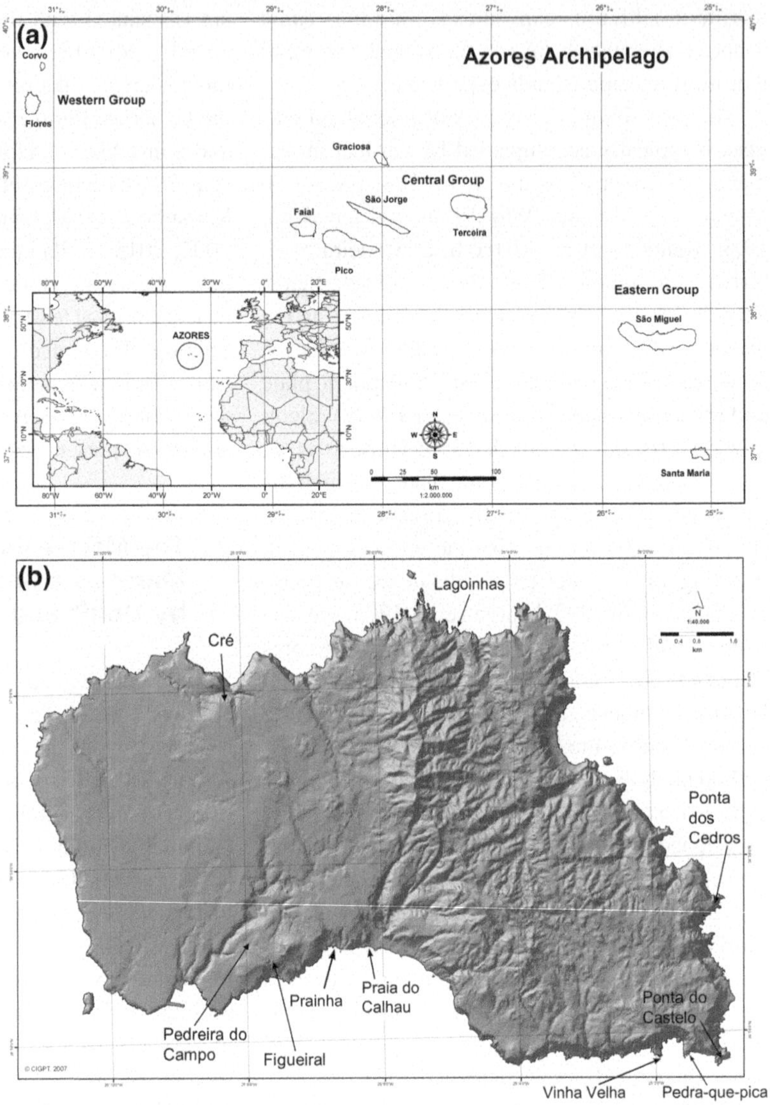

Fig. 1 **a** Map of the Azores (insert) and geographical location of Santa Maria Island. © Secção de Geografia, Departamento de Biologia, Universidade dos Açores. **b** Map of Santa Maria with the location of the most important and well-studied early Pliocene and Pleistocene (MIS 5e) outcrops

The next stage in the development of Santa Maria was the extrusion of a basaltic subaerial shield volcano, through a number of fissural vents. The remains of this largely effusive edifice correspond to the Anjos Volcanic Complex (Serralheiro et al. 1987) extruded 5.8–5.3 Ma (Ramalho et al. 2017). The former dimensions of this edifice are difficult to constraint but it partially corresponded to the western half of present-day Santa Maria, as its eroded products do not extend to its eastern portion (Serralheiro 2003; Sibrant et al. 2015; Ramalho et al. 2017).

Subsequently to the extrusion of a basaltic shield volcano, the edifice experienced a period of general quiescence and erosion during which it was truncated by marine erosion and probably totally submerged again (Ávila et al. 2012; Meireles et al. 2013). This stage is attested to by the Touril Complex that comprises a thick (up to 130 m) sedimentary sequence of terrestrial and marine sediments with increasingly marine characteristics towards the top. On the eastern side of the island, submarine lavas and surtseyan tuffs occur intercalated within the sediments indicating that low-volume submarine volcanic activity occurred alongside erosion and sedimentation on this side of the edifice (Serralheiro et al. 1987; Serralheiro 2003; Ávila et al. 2012; Meireles et al. 2013; Ramalho et al. 2017). The Touril sequence was deposited during the Early Pliocene (5.3–4.1 Ma), as the overall volcanostratigraphy, fossil content and existing isotopic ages suggest (see Feraud et al. 1980; Serralheiro et al. 1987; Serralheiro and Madeira 1990; Serralheiro 2003; Kirby et al. 2007; Sibrant et al. 2015; Ramalho et al. 2017). The marine sediments of the Touril Complex constitute the most important and richest source of marine fossil assemblages on Santa Maria, reflecting a period during which the island represented a wide shallow and sandy submarine bank where marine life thrived (Ávila et al. 2012, 2015a, c, 2016a; Ramalho et al. 2017).

The following volcanostratigraphic unit—the Pico Alto Volcanic Complex (Serralheiro et al. 1987)—suggests that towards the beginning of the Pliocene, volcanic activity resumed with increased intensity. This new volcanic edifice started entirely submarine but as it grew upwards it breached sea-level forming a new island, elongated along a NNW-SSE direction, on the eastern portion of the old edifice (Serralheiro and Madeira 1990; Ávila et al. 2012; Ramalho et al. 2017). This renewed period of volcanic activity rapidly covered a large portion of the pre-existing shallow sandy bank with extensive submarine sheet flows, lava deltas and surtseyan deposits, effectively protecting the Touril sedimentary sequence from marine erosion and contributing to the preservation of such a rich coastal and shelf fossiliferous record. Likewise, the later onset of effusive subaerial activity and the formation of lava deltas along the existing coastlines further contributed to the preservation of more recent and discrete shelf sedimentary bodies that developed in depocentres away from active volcanic vents.

With the demise of Pico Alto volcanic activity, Santa Maria's edifice eroded briefly during which shore platforms formed and subaerial reliefs decayed. Alluvial/fluvial deposits and remains of marine terraces that were later covered by the products of the last eruptive stage of Santa Maria's volcanic life—corresponding to the Feteiras Formation (Serralheiro et al. 1987)—attest to this period. This brief period of volcanic activity took place during the Late Pliocene and involved low volumes of erupted material; it was restricted to a set of monogenic magmatic and hydromagmatic cones and their proximal and distal deposits, centred along the western portion of the then existing edifice, with volcanic activity seemingly ceasing at ~2.8 Ma (Serralheiro 2003; Sibrant et al. 2015; Ramalho et al. 2017).

Since the extrusion of the Feteiras volcanics, Santa Maria entered a long period of volcanic quiescence, uplift and erosion, as indicated by the presence of well-developed staircase shore platforms up the western portion of the present-day edifice. These marine terraces, found at elevations from 210–230 m down to 7–11 m above the present sea level, confirms that the edifice underwent a slow uplift from ~3.5 Ma to the Holocene, at a rate of ~60 m/Ma (Ramalho et al. 2017). This recent uplift, coupled with marine and fluvial erosion, was responsible for the exposure of the rich and diverse submarine volcanic and sedimentary sequences that otherwise would be inaccessible. These circumstances make Santa Maria one of the best places to study the Neogene marine fossil record in the North Atlantic region. Furthermore, the fossiliferous outcrops on Santa Maria are of utmost importance to gain a better understanding of how coeval living communities relate to the broader evolutionary and biogeographic history of the Atlantic basin during the late Neogene (Kirby et al. 2007; Ávila et al. 2015b, 2016a).

3 A Brief Overview About the Palaeontological Studies at Santa Maria Island

The marine fossils of Santa Maria Island have been known since the 16th century. Gaspar Frutuoso (1522–1591), a Portuguese priest, first reported on the fossils of Santa Maria in his third volume of "Saudades da Terra" (Frutuoso 1983). He described a quarry at Figueiral, where the extracted calcareous sandstones had "seafood shells glued on it" (Madeira et al. 2007). A long lapse of time passed until the late 19th century studies by Bronn (1860), Hartung (1860) and Morelet (1860). The first two authors extensively described the geomorphology of the island and, together with the work of Reiss (1862), Hartung (1864) and Mayer (1864), established the background for the palaeontological research in Santa Maria during the first half of the 20th century (Madeira et al. 2007). During the 1950s, a series of studies published by Berthois (1950, 1951, 1953a, b, c), Ferreira (1952, 1955) and Krejci-Graf et al. (1958) was followed by the important palaeontological studies of Zbyszewski et al. (1961), da Ferreira (1961) and Zbyszewski and Ferreira (1961, 1962a, b), which predominantly concentrated on the Pliocene fossils. In 1990, a paper by García-Talavera raised attention to the Pleistocene Prainha outcrop, located on the southern coast of the island. Callapez and Soares (2000) described another outcrop of the same age from Lagoinhas on the northern shores.

From 1998 on, a series of expeditions to Santa Maria undertaken by the first author of this chapter were followed by the birth of a palaeontological research group—the Marine PalaeoBiogeography Working Group (MPB)—based at the Department of Biology of the University of the Azores. Its main objectives were to achieve an understanding of the palaeoecology and palaeobiogeography of the Pleistocene and Pliocene outcrops, and to explore the legal protection of the geological legacy of Santa Maria Island. As a result, a checklist of the Pleistocene marine molluscs of Lagoinhas and Prainha (Ávila et al. 2002), and a technical report for the protection of the outcrops of Pedreira do Campo and Figueiral (Cachão et al. 2003), the first "Regional Natural Monument" of the Azores, were produced. Since 2002, the outcrops of Santa Maria became the target of the international workshops "*Palaeontology in Atlantic Islands*", which have taken place on a yearly basis. During these workshops, researchers studied the geology and geomorphology of the island, the petrology, geochemistry and chronology of magmatic rocks, the volcanostratigraphy, the sedimentology and the systematic palaeontology of several phyla, the palaeoecology of the fossil assemblages and the palaeobiogeography of diverse invertebrate groups embedded within these strata. This research effort resulted in several publications covering a wide range of topics, such as the evolution of marine organisms in oceanic islands (Ávila 2013), systematics (Ávila et al. 2002, 2007, 2012; Estevens and Ávila 2007; Janssen et al. 2008; Kroh et al. 2008; Madeira et al. 2011; Meireles et al. 2012), palaeoecology and geochemistry (Kirby et al. 2007; Ávila et al. 2008b, 2009a, 2015a, c; Habermann 2011; Stöckhert 2011a, b), palaeobiogeography (Ávila et al. 2008a, 2009b, 2016a), geology (Habermann 2010; Ramalho et al. 2017), sedimentology (Meireles et al. 2013; Johnson et al. 2017), ichnology (Uchman et al. 2016), conservation and sustainable touristic management (Cachão et al. 2003; Madeira and Ávila 2006; Calado et al. 2007; Nunes et al. 2007), educational purposes (Ávila 2009; Ávila and Monteiro 2009; Ávila et al. 2010; Ávila and Rodrigues 2013), geoconservation (Ávila et al. 2016b) and historical reviews (Madeira et al. 2007). The majority of these papers dealt with the fossils of marine molluscs, but other groups of fossils also were studied, e.g.: calcareous algae (Amen et al. 2005; Rebelo et al. 2014, 2016a, b), cetaceans (Estevens and Ávila 2007; Ávila et al. 2015c), brachiopods (Kroh et al. 2008), (…) crustaceans (barnacles: Winkelmann et al. 2010), ostracods: Meireles et al. 2012), echinoderms (Madeira et al. 2011; Santos et al. 2015), and fish (sharks; Ávila et al. 2012). Two congresses entitled "*Atlantic Islands Neogene*" were also organised by MPB members in June 2006 and in September 2008 aiming at: (1) reviewing the accumulated scientific knowledge on the palaeontology of the Atlantic oceanic islands and (2) outlining future strategies of research in the Azores (Ávila and Martins 2007).

4 The Early Pliocene Outcrops

4.1 Pedreira do Campo and Figueiral

Pedreira do Campo (Fig. 3a) and Figueiral (Fig. 4a) are two historic quarries located in the area southwest of Pico Facho, a 254 m high tuff-cone ~2 km east of Villa do Porto close to the southern coast of the island, in which up to 20 m thick successions of the sedimentary Touril Complex and the overlying volcanic Pico Alto Complex (sensu Serralheiro et al. 1987) crop out. The outcrops are situated at altitudes between 95–115 m and comprise major deposits of fossiliferous calcarenites and limestones (Krejci-Graf et al. 1958; Zbyszewski and Ferreira 1962b; Abdel-Monem et al. 1975; Serralheiro et al. 1987, 1990; Serralheiro and Madeira 1990; Kirby et al. 2007; Madeira et al. 2007).

4.1.1 Pedreira do Campo

At Pedreira do Campo, located ~1.4 km southwest of Pico Facho, a ~20 m thick succession of basanitic pillow lavas of the Pico Alto Complex is exposed in a cross-section with a lateral extent of ~150 m (Fig. 3a). Underlying strata only crop out locally, comprising deposits of the Touril and Pico Alto Complexes (Serralheiro et al. 1987, 1990; Cachão et al. 2003; Kirby et al. 2007; Madeira et al. 2007). The base of the Touril deposits is not exposed.

Three lithological facies (1–3) are distinguished at Pedreira do Campo (Fig. 2a). The lowermost unit (facies 1) consists of a white, impure, volcaniclastic limestone, which forms the top of the Touril Complex at this locality (Serralheiro et al. 1987; Figs. 2a, 5a–c). Well-rounded volcaniclastic pebbles and cobbles are scattered within the limestone. Volcaniclastic components make up ~5% of these compact rocks and are commonly concentrically encrusted by coralline algae (rhodoliths). The basaltic cores of these rhodoliths are small, ranging from 10 to 40 mm (average = 20.8 mm; 56 rhodoliths); 32% of the 82 rhodoliths sampled had no nucleus. This rhodolith floatstone exhibits a packstone matrix, rich in mollusc fragments (mainly bivalves and a few gastropods), spines of echinoderms (Fig. 3b), bryozoans and benthic foraminifers (mostly amphestigenids that are less than 2 mm in length). The rhodolith floatstone is unconformably overlain by a ~4.5 m thick unit (facies 2; cf. Fig. 2a) of light brown, well-stratified surtseyan tuffs (previously classified as lithic arenites by Kirby et al. 2007) that were assigned to the Pico Alto Complex according to Serralheiro et al. (1987, 1990). The tuffs are moderately sorted and show planar to low-angle cross-bedding (Fig. 3c). Influx of epiclastic materials, particularly moulds of bivalves and gastropods, are restricted to the basal 10 cm (Kirby et al. 2007). The tuffs are succeeded by about 20 m of basanitic pillow lavas (facies 3; Fig. 3d), which are assigned to the Pico Alto Complex (Serralheiro et al. 1987, 1990). The basal contact is abrupt or made up of a thin discontinuous horizon of pillow breccias and hyaloclastites. The pillows are approximately 1–2 m wide and abundantly display typical concentric structures with disjunctive radial solidification fractures and altered glassy quenching rims at their surfaces.

4.1.2 Figueiral

Known since the 16th century (Frutuoso 1983), the Figueiral outcrop (Fig. 4a)—located on the south-western flank of Pico do Facho, ~500 m east of Pedreira do Campo—was repeatedly mentioned in literature for its fossiliferous marine strata (Hartung 1860; Reiss 1862; Mayer 1864; Ferreira 1952, 1955, 1961; Zbyszewski et al. 1961; Zbyszewski and Ferreira 1962a; Cachão et al. 2003; Madeira et al. 2007).

Eight lithological facies (1–8) are distinguished through the exposed deposits (Fig. 2b). The outcrop starts with ~5 m thick, unstratified, poorly sorted, volcaniclastic cobble breccia (facies 1). The base of this unit is not exposed. Towards the top, increasing numbers of rounded cobbles and pebbles occur, which are frequently encrusted by coralline red-algae (rhodoliths). The volcaniclastic matrix of the breccia grades into a strongly cemented bioclastic pack- to floatstone (facies 2) rich in fragments of bivalves, gastropods, bryozoans, coralline algae, echinoids,

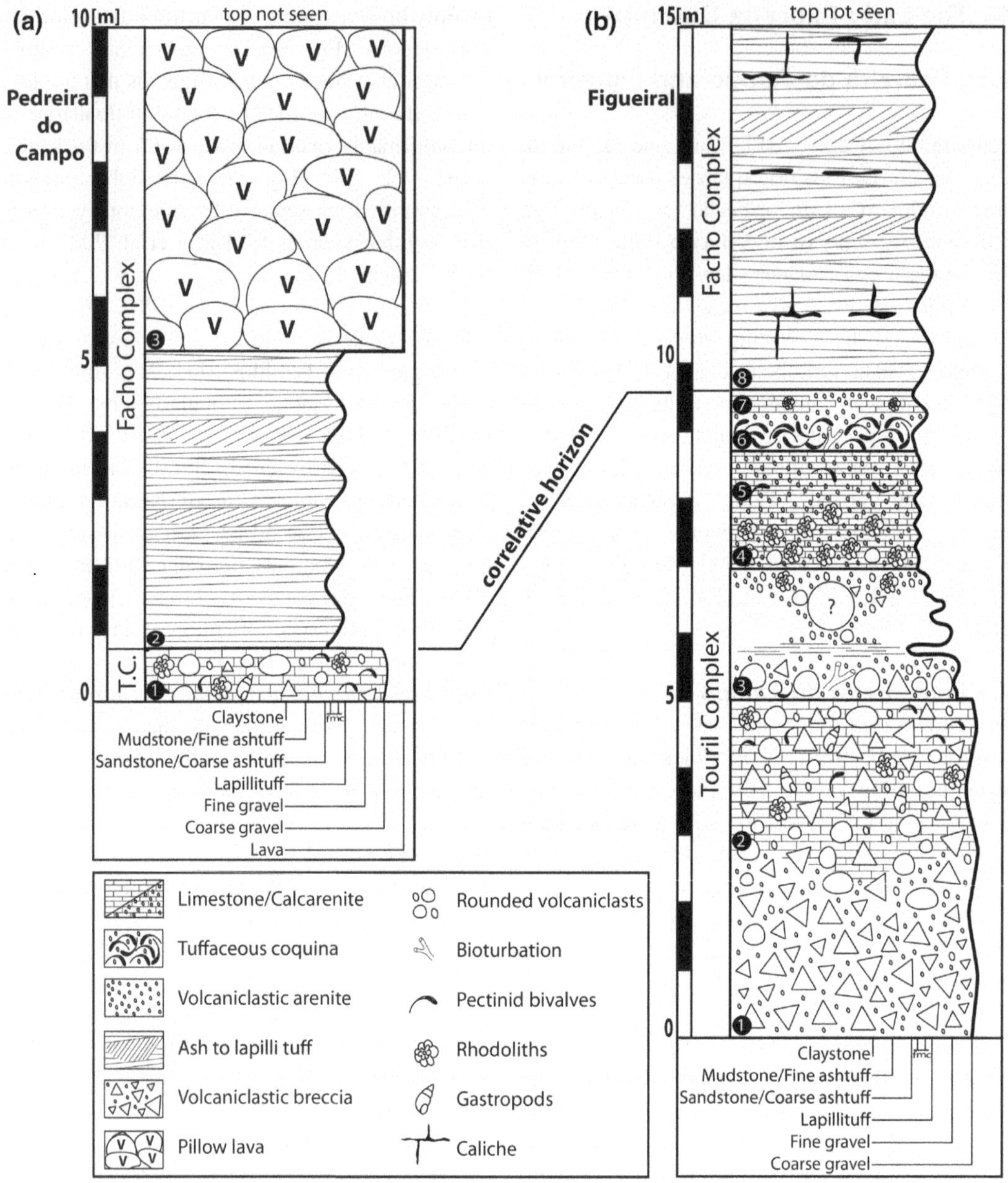

Fig. 2 Stratigraphic sections at Pedreira do Campo (**a**; modified after Kirby et al. 2007) and Figueiral (**b**; after Habermann 2010), displaying the Pliocene sedimentary and volcanic successions of Touril and Pico Alto Volcanic Complexes (sensu Serralheiro et al. 1987, 1990)

benthic and planktonic foraminifers (amphestigenids, globigerinids) and with a micritic lime-mud matrix (Figs. 2b, 5d–f). The overlying 2 m thick beds are poorly exposed on the surface. Small outcrops in this interval suggest a heterogeneous composition of volcaniclastic gravel and boulders and increased amounts of arenitic and argillaceous matrix, generally lacking lime-mud. Bioturbated sandstone lenticules, thin horizons of brown clays, as well as scattered rhodoliths and pectinid shells occur within this 'transition zone' (facies 3). The basal bed exposed in the Figueiral

Fig. 3 Pedreira do Campo outcrop. **a** The former quarry for basalts. **b** Detail of the fossil assemblage rich in mollusc fragments, spines of echinoderms, bryozoans and benthic foraminifers. **c** Ash and lapilli tuffs of the Pico Alto Complex with planar to low-angle cross-bedding. **d** basanitic pillow lavas of the Pico Alto Complex

quarry consists of a rhodolith-floatstone (facies 4). Most volcaniclastic pebbles that are scattered in a white grainstone matrix are well rounded and encrusted by red algae. The abundance of volcaniclastic pebbles and rhodoliths decreases upwards, while matrix amounts increase and the floatstone grades into a ~1.5–2 m thick, unstratified, well-sorted grainstone (facies 5), mainly consisting of benthic and planktonic foraminifers (amphestigenids up to 1 mm; globigerinids), fragments of bryozoans and red algae (Fig. 5g–i). Larger fossil-fragments and volcaniclastic detritus are accessories, lime mud is absent. Towards the top, the amount of pyroclastic materials considerably increases. The overlying ~50 cm thick tuffaceous coquina (facies 6) is rich in disarticulated, well-preserved pectinid shells <5 cm in length (dominantly *Aequipecten macrotis* (Sowerby) and *Aequipecten opercularis* (Linnaeus); cf. Fig. 4b). Poorly-sorted palagonitic lapilli form the major matrix constituents besides variable accessory amounts of skeletal fragments of bivalves, gastropods, bryozoans and foraminifers. Barnacle fragments of *Zullobalanus santamariaensis* Buckeridge and Winkelmann also are observed

Fig. 4 Figueiral outcrop. **a** The historical quarry seen from the outside. **b** Shells of the dominant bivalves *Aequipecten macrotis* (Sowerby) and *Aequipecten opercularis* (Linnaeus) (facies 6). **c** Note the abundant benthic and planktonic macroforaminifers (amphestigenids and globigerinids) (facies 5). **d** The interior of one of the galleries exploited for limestone. **e** ash and lapilli tuffs of the Pico Alto Complex (facies 8)

(Winkelmann et al. 2010). A sharp boundary characterizes the transition to the overlying 5–10 cm thick, laterally discontinuous amphestigenid-grain- to rudstone horizon (facies 7) with rhodoliths, benthic foraminifers, bryozoans and accessory echinoid fragments (Fig. 5j–l). This unit represents the top of the Touril Complex at Figueiral, and is abruptly overlain by ash and lapilli tuffs of the Pico Alto Complex with a thickness of >10 m at this outcrop (facies 8). Fossil content and the amount of epiclastic materials rapidly decrease at the base of this unit. The pyroclastic succession abundantly shows primary sedimentary structures. Horizontally stratified, subparallel, thin-bedded subsets (up to 1–2 m thick) are alternatingly interbedded with tabular- to tangential low-angle cross-stratified channels. Diagenetic nodular limestone crusts are commonly intercalated within the tuffs.

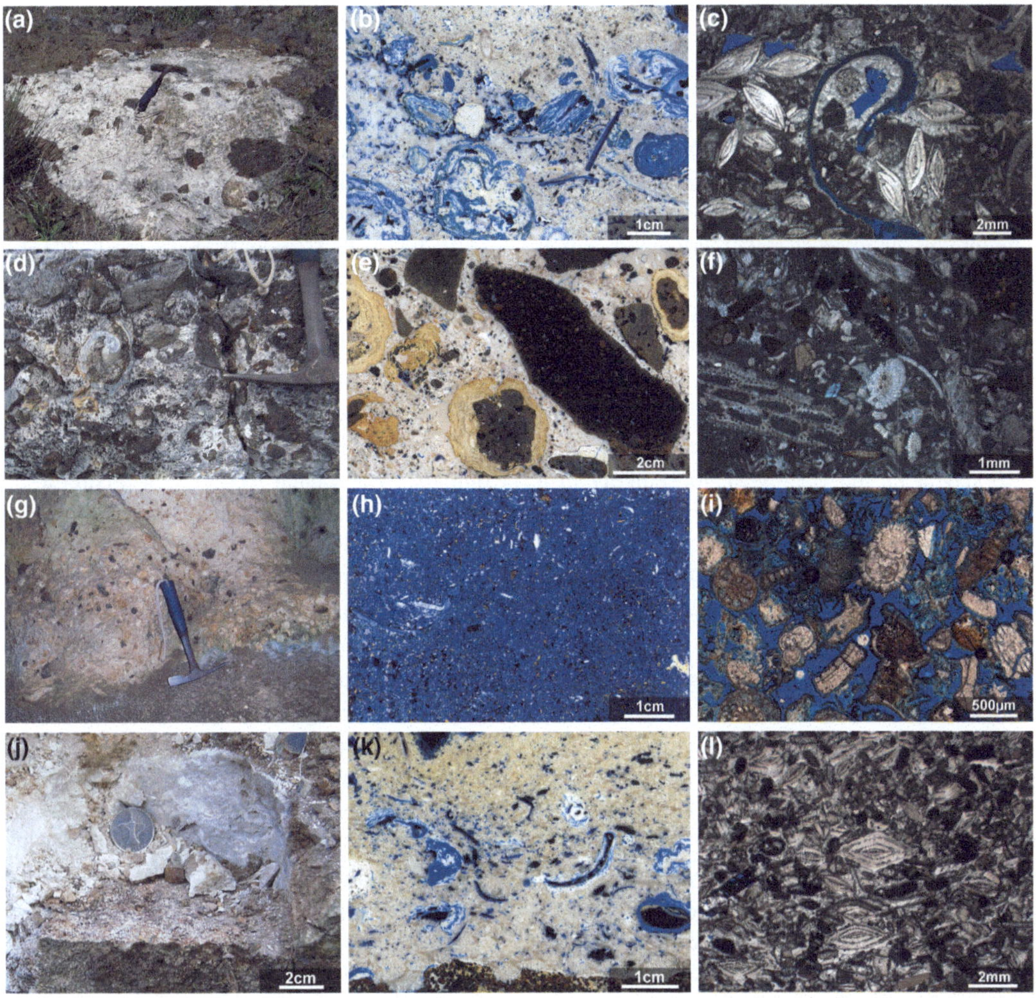

Fig. 5 Outcrop photographs and photomicrographs of Pedreira do Campo and Figueiral limestones. **a**–**c** Pedreira do Campo (facies 1). **d**–**f** Figueiral (facies 2). **g**–**i** Figueiral (facies 5). **j**–**l**: Figueiral (facies 7). The blue stain is due to resin impregnation. For detailed facies descriptions and stratigraphic sample localities see text and Fig. 4

4.1.3 Interpretation and Discussion

On Santa Maria, sediment supply to the shelf mostly comes from marine erosion during sea-level highstands, while sediment supply by river systems is negligible (Ávila et al. 2008b). This allowed the development of carbonate factories during phases of minor volcanic activities, which were preserved in small depocentres, such as at Figueiral and Pedreira do Campo during the Lower Pliocene.

At both localities, carbonate factories developed in a shallow-marine photic environment. Stratigraphic relationships and the faunal composition suggest that the Pedreira do Campo limestone (facies 1) stratigraphically corresponds to the uppermost limestone horizon (facies 7) at Figueiral (Fig. 2a, b). Facies differences between both limestones likely reflect the high facies variability due to the variability in palaeo-shelf morphology between two small depocentres or within such a small depocentre.

The sedimentary succession within these outcrops is interpreted as a retrogradational-progradational cycle that reached open-ocean conditions, but remained within the photic zone, during maximal retrogradation. While the retrogradational phase is represented by the Touril deposits, the progradational phase is formed by the tuffs of the Pico Alto Complex. The basal breccia at Figueiral (facies 1 and 2) is interpreted as a debris flow with short transport distance (no rounding of angular clasts). Clast composition indicates a provenance from the Anjos Complex basement lavas. Because the lower part of this unit lacks indicators for submarine deposition, a subaerial depositional environment is suggested. In the course of a subsequent rise in (relative) sea-level (retrogradation), the debris flow was reworked at its top (facies 2), most likely in a shallow-intertidal, high-energy marine environment, as indicated by the occurrence of rounded pebbles and rhodoliths. A carbonate factory developed and the bioclastic pack- to floatstones, which form the matrix of the volcaniclastic breccia, were deposited. During the ongoing retrogradation, a second debris flow likely formed the 'transition zone' sediments (facies 3). Marine sedimentation continued afterwards and the graded float- to grainstone units (facies 4 and 5), exposed in the Figueiral quarry, were deposited. These sediments represent the continuing transgression and a deeper, probably open-ocean environment, during which the input of siliciclastic sediment from emerged land was limited. The overlying tuffaceous coquina horizon was probably deposited as a debris flow, based on the lack of sedimentary structures and shell orientation. The incorporated tuffs indicate a first, short period of submarine eruptive activity. During the subsequent eruptive recreation phase, the overlying amphestigenid-grain- to rudstone horizon (facies 7) formed. The horizon represents the top of the Touril Complex and was likely deposited during the maximal retrogradation of the sequence exposed at Figueiral. Abundant phototrophic organisms indicate that deposition took place within the photic zone. According to Kirby et al. (2007), the limestone unit at Pedreira do Campo probably formed on a shallow bank or shoal far from any subaerial sources of terrigenous sediment, due to the dominance of photosynthetic organisms (rhodoliths and symbiont-bearing benthic foraminifera) and the absence of fine-grained terrigenous material. The small size of the basaltic cores of the rhodoliths, also points to some distance to the shore (Santos et al. 2012). These conclusions were supported by cathodoluminescence studies done on rhodoliths from Figueiral and Pedreira do Campo (Rebelo et al. 2016b), with both sites interpreted as being located far from the islands's palaeo-shore (at about 2.9 and 2.8 km, respectively). At Pedreira do Campo, at least, a period of consolidation and erosion of the marine sequence took place before burial by the extrusion of the surtseyan tuffs of the Pico Alto Complex. The planar bedding at Pedreira do Campo might suggest their formation in a shoreface environment (Kirby et al. 2007). With the onset of tuff deposition, which cut off the previous carbonate factory on the shelf, sediment supply to the shelf was tremendously increased and the sedimentary system switched from retrogradation to progradation, most likely without a significant change in relative sea level. The topmost pillow lavas at Pedreira do Campo (Fig. 3d) are absent at Figueiral, which suggests that their distribution was likely morphologically controlled and/or restricted to the vicinity of the local feeder vent.

4.2 Pedra-que-pica

4.2.1 Geological Setting

The Pedra-que-pica outcrop, situated in a coastal area at the southeastern corner of Santa Maria Island (Serralheiro 2003; Kirby et al. 2007; Ávila et al. 2015a; Fig. 7), contains a 10–11 m thick succession of marine and very fossiliferous sediments, of which only the uppermost 3–4 m are presently exposed above sea level, that lies in between basalt pillow lavas and a volcano-sedimentary sequence on top. The lower part of the outcrop is presently submerged and its area was estimated to be 23,436 m^2 (Ávila et al. 2015a), whereas the intermediate part of the section is exposed on a wave cut platform, extending for $\sim$3,179 m^2 in the intertidal zone, the upper part extending below the slope deposits

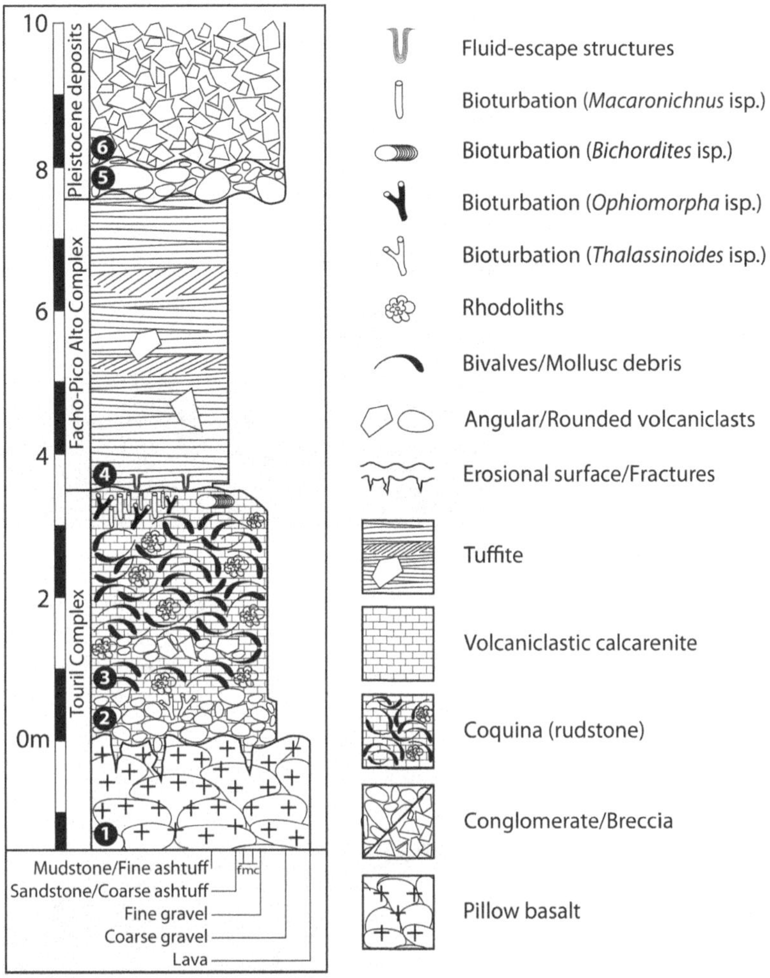

Fig. 6 Stratigraphic section at Pedra-que-pica (after Ávila et al. 2015a), displaying the Pliocene sedimentary and volcanic successions of Touril and Pico Alto Volcanic Complexes (sensu Serralheiro et al. 1987, 1990)

at the base of the present steep cliff. The basal volcanic sequence and the overlying shell accumulation (coquina) are assigned to the Touril Complex, whereas the overlying tuffites and the succeeding basalt flow are assigned to the volcanic Pico Alto Complex (sensu Serralheiro et al. 1987, 1990; Ávila et al. 2015a; Fig. 6). $^{87}Sr/^{86}Sr$ isotope age estimates of three mollusc shells collected from the coquina yielded an average age of 5.51 ± 0.21 Ma (Kirby et al. 2007); the age of the deposit, however, is slightly younger given that the sediments intercalate basaltic pillow lavas yielding $^{40}Ar/^{39}Ar$ ages of 4.78 ± 0.13 Ma and 4.13 ± 0.19 Ma respectively, for the base and the top of the sequence (Ramalho et al. 2017). The base of the section is formed by limited remains of strata that predate the coquina, which can be found in spaces in between the pillow lavas (facies 1), forming sandstone pockets composed of fine-grained, light grey calcarenites, where trace fossils belonging to *Macaronichnus segregatis* Clifton and Thompson 1978 are abundant (Ávila et al. 2015a; Uchman et al. 2016). The pillow basalts show an erosional top. A laterally discontinuous 0.5–1 m thick conglomerate (facies 2; Fig. 6), composed of volcaniclastic pebbles and cobbles in a volcaniclastic calcarenite matrix, unconformably overlies the lavas. The conglomerate grades upwards into a 3–4 m thick unstructured coquina (facies 3; Fig. 6). The very poorly-sorted coquina-rudstone is rich in large,

Fig. 7 Pedra-que-pica outcrop. **a** Aerial view of the southeastern shores of Santa Maria Island (photo by Paulo Henrique Silva/SRAM). **b** View from the sea off Pedra-que-pica outcrop area, just before loading areas

mainly disarticulated valves of bivalves (dominated by pectinids, ostreids and spondylids), and also contains echinoids, barnacles, brachiopods, bryozoans, calcareous algae (rhodoliths), bryoliths (nodules entirely composed by bryozoans), corals, and rare teleost fish teeth and shark teeth, rare gastropods (e.g., moulds and stomatofossils of *Persististrombus coronatus*; Ávila et al. 2016a), and whale bones (Kirby et al. 2007; Madeira et al. 2011; for a detailed list of species, see Ávila et al. 2015a). Volcaniclastic calcarenites, composed of moderately sorted, predominantly sand-grade skeletal bioclasts (fragments of molluscs, echinoids, coralline algae, bryozoans and planktonic foraminifers) and well-rounded volcaniclasts (coarse palagonitic ash and tachylitic to lithic clasts) form the matrix of the coquina. The majority of larger pectinid shells (e.g., *Gigantopecten* with shell length over 10 cm) is oriented in concave-down position, whereas shells of other smaller bivalve species showed a chaotic disposition of the disarticulated valves (Ávila et al. 2015a). Most shells are encrusted by bryozoans, balanids, oyster serpulids and coralline red algae, and bioeroded (*Gastrochaenolithes* isp. and *Entobia* isp.) at external shell surfaces. The preservation, however, can still be considered as good. The top of the coquina exhibits a fining-upward trend and is succeeded by coarse-grained, faintly bedded volcaniclastic calcarenites (Fig. 6). The arenites are ~10–60 cm thick and show a subhorizontal, very regular, bioturbated top. The most common trace fossil is *Asterosoma* isp. *Bichordites* isp. burrows occur about 1–2 cm below the top of this unit in the eastern part of the outcrop. The topmost 10 cm contain abundant spines of the echinoderm *Eucidaris tribuloides* (Lamarck) and *Porites* sp. coral fragments. The arenites are abruptly overlain by a 36 m thick unit of well-stratified, fine- to coarse-grained vitric ash to lapilli tuff, generated by a nearby surtseyan eruption (facies 4; Ávila et al. 2015a; Fig. 6). The tuffs show water-escape structures in the basal 20 cm. The sedimentary structures within the tuffs include thin planar lamination to medium-thick bedding, low-angle cross-bedding and internal erosive surfaces where the planar bedding is discordant. Multiple reverse and normal graded beds are also present. Angular lava blocks and basalt pebbles are rarely scattered in this unit. Pebble- to boulder-sized lithic clasts of sedimentary origin (bioclastic tuffaceous sandstone to almost pure limestone) are subordinately present within the lowermost layers of the tuffs. Locally, the tuffs are truncated by a Pleistocene shore platform and beach deposit, and show a thickness of ~4 m only. The erosional unconformity and the overlying 0.5–1 m thick boulder conglomerate (facies 5; Fig. 6), however, pinch out ~50 m to the east, where a wave-cut notch is developed. This conglomerate is topped by a >20 m thick, wedge-shaped talus deposit, consisting predominantly of boulder breccias (facies 6; Fig. 6). Beyond the erosional unconformity, the tuffs reach a thickness of about 32 m and show less distinct bedding and more volcaniclastic boulders and blocks (Kirby et al. 2007). The tuffs are overlain by a ~0.5–1 m thick conglomerate horizon. A faint erosive relief occurs along the contact between this conglomerate and the overlying basalt flows. The passage zone between the foreset unit of pillow lavas and hyaloclastites, and the overlying topset of flat-lying subaerial lavas occurs at an altitude of ~55 m within these basaltic flows (Ramalho et al. 2017).

4.2.2 Interpretation and Discussion

The Pedra-que-pica sedimentary succession is interpreted as a retrogradational-progradational cycle, of which the retrogradational phase is represented by carbonates that were deposited during a transgressional period of rising relative sea-level, while the progradational phase is documented by tuffs. During the late retrogradation, a coquina was formed, which is interpreted as the result of a succession of several debris-fall deposits (sensu Titschack et al. 2005) whose redeposition was triggered by major storm events that removed the sediments from its original nearshore setting to a local depocenter below fair-weather wave base. The retrogradation phase was prematurely ended by the rapid progradational deposition of the tuffs, which filled up the available accommodation space on the shelf.

The basal pillow basalt (facies 1) indicates submarine volcanism. The overlying pockets of bioturbated calcarenite that include the trace fossil *Macaronichnus segregatis* indicate intertidal to shallow subtidal conditions (i.e., upper foreshore). With rising sea level, a carbonate factory development and a coquina with a volcaniclastic calcarenite matrix was deposited in a local depocenter. The coquina exclusively comprises allochthonous, poorly-sorted components, which exhibit a low degree of fragmentation/reworking. Thus, the fossil assemblage within this deposit indicates a proximal shallow-marine provenance and deposition of the coquina below fair-weather wave base, where low-energy conditions prevented reworking and shell fragmentation (cf. Ávila et al. 2015a). The occurrence of few articulated *Gigantopecten* shells, together with the low degree of shell fragmentation and sorting, indicate a proximal provenance of the coquina constituents. The volcaniclastic calcarenite matrix might suggest that Pedra-que-pica was close to emerged land (Kirby et al. 2007). Increased bioturbation towards the top of the overlying calcarenite corroborates a retrogradational (deepening-upward) trend in that the calcarenites represent the deepest environment of the Touril sequence. The sharply developed transition to the overlying tuffs (facies 4) indicates an abrupt change in sediment supply, which is related to the onset of a new volcanic phase in this area (Ávila et al. 2015a). The limestone clasts within the tuffs are classified as "accidental clasts", which were produced by the disruption of pre-existing and already consolidated subvolcanic rocks on the seafloor during the explosive hydromagmatic volcanic eruption. A continuous and voluminous supply of pyroclastic material in a relatively short period of time likely caused a switch in the sedimentary system from retrogradational to progradational, which filled up the available accommodation space. Most likely, no significant changes in relative sea level occurred during that period. The absence of ballistic emplacement structures of volcanic bombs (e.g., impact pits, bomb sags) suggests a submarine origin of these pyroclastics, classifying them as surtseyan tuffs or water-settled tuffs, formed by hydromagmatic volcanic activity in shallow waters (Fisher and Schmincke 1984).

The subaquatic deposition of this unit is further corroborated by the common occurrence of cross-bedding and small cross-cutting channels, as these sedimentary structures are interpreted as secondary structures related to the marine environment and the resedimentation of the tuffs. The unconformity at the top of the tuffs at an altitude of ~35 m and the overlying conglomerate indicate erosion (although minor) and reworking in shallow waters and thus formation in a high-energy environment. The tuffs, therefore, describes a progradational trend, caused by an extreme enhancement in sediment supply and the rapid fill up of the local depocentres. Following the formation of the conglomerate on top of the tuffs, another period of volcanism emplaced the overlying basaltic lava flows that occur at altitudes between ~32–60 m. Low relief, which indicates a hiatal surface due to sediment reworking, characterizes the transition between the latter units. Pillows and palagonitised hyaloclastites are developed in the basal part of these lavas, indicating submarine volcanism. The submarine lavas are immediately topped by flat-lying massive subaerial flows. The transition between submarine and subaerial morphologies occurs at an altitude of ~55 m above present sea level and represents a palaeo-sea-level marker (Meireles et al. 2013; Ramalho et al. 2017). The conglomerate (facies 5) overlying the erosional unconformity (wave cut bench) carved on the lowermost tuffs (facies 4) is interpreted as a Pleistocene boulder-beach conglomerate, most likely formed during the Marine Isotopic Stage 5e (MIS 5e) interglacial period (Ávila et al. 2008a, 2015b). Breccias above the conglomerate are interpreted as multi-phase slope deposits (facies 6). Modern (Holocene) counterparts to both the latter Quaternary deposits (facies 5 and 6) occur at the footslope of the cliff east of Pedra-que-pica.

4.3 Ponta do Castelo

4.3.1 Geological Setting

Ponta do Castelo constitutes the southeastern-most headland on Santa Maria Island and comprises a typical lava delta sequence, whose bottom-set unit comprises marine fossiliferous sediments. The overall sequence is thought to be Early Pliocene in age and is attributed to the Pico Alto Volcanic Complex (Serralheiro et al. 1987). It is comprised from the base to the top (see Meireles et al. 2013) by: (1) fossiliferous marine sediments with a high volcaniclastic content that partially correspond to the remobilization of tuffs and water-settled tuffs (the same that cover the Pedra-que-pica outcrop) from the surtseyan cone whose remains can still be seen farther west at Rocha Alta; (2) a steeply dipping foreset unit of pillow lavas and hyaloclastites, prograding to the E; (3) a topset unit of flat-lying subaerial lava flows; and (4) pillow lavas of another lava delta stacked on top of the previous, but whose topset unit was locally removed by erosion. The passage zone between the forests of submarine lavas and the overlying subaerial lava flows of the main lava delta is presently located at ~55 m a.s.l and marks the contemporaneous sea level. It is thus possible to infer very accurately the palaeo-water depth for the deposition of the bottomset sediments. The exposed sedimentary body at the base of the sequence exhibits a sigmoidal or wedge-shaped geometry, and comprises a set of 4 or 5 sandstone units amounting to a total thickness of 9 m. A complete and detailed description for this sequence can be found in Meireles et al. (2013), from where the following summary was extracted. The first 1 m thick unit is composed of coarse- to medium-grained sandstones that show large-scale cross-stratification with centimetre- to decimetre-thick, slightly wavy sets, forming low but wide swales and possibly hummocks. This unit is eroded at the top and exhibits a centimetre- to decimetre-high palaeo-relief. These sediments are unconformably overlain by a wedge-shaped sedimentary unit that gradually thins towards the ENE, from about 1.2 to about 0.5 m. The sediments grade from very coarse-grained sandstones at the base, into coarse- to medium-grained, chaotic-bedded sandstones in the middle (both very rich in shell debris), and medium- to fine-grained sandstones with plane-parallel bedding and small-scale bioturbation at the top. A shallow but wide erosional channel unconformably cuts through the topmost plane-parallel beds and marks the top of this unit.

The subsequent up to 1.5 m thick sedimentary unit comprises medium-grained, poorly-sorted fossiliferous sandstones that exhibit diffuse bedding. These sediments fill the above-mentioned erosional channel. Rip-up clasts that originate from the previous unit occur within the channel-fill sandstone. Fossils and shell debris are generally dispersed in random orientation and remains of solitary corals and rock-encrusting bryozoan colonies that were ripped from their hard-substrate occur "floating" within the sandstones (Björn Berning, pers. comm.).

A pronounced erosive unconformity deeply cuts through all previous units, forming an irregular palaeo-relief characterised by side-by-side channels up to 3–4 m deep and about as wide. This palaeo-relief was filled and overlain by a 4–5 m thick unit of fossiliferous, medium-grained sandstones exhibiting 1–3 m wide, 0.3–1.5 m high slightly asymmetric hummocks and swales that exhibit onlapping contacts along the steep channel sides; beds typically fan out from conformable almost-parallel fine beds, internal truncation surfaces are rare, and hummocks are almost ubiquitously preserved. Towards the top bedding becomes almost plane-parallel, with occasional smaller-scale ripples. Fossil content is composed of bivalves, gastropods, echinoids, fragments of endemic barnacles, and remains of rock-encrusting bryozoan colonies that were ripped from their hard-substrate; shells and shell fragments are generally dispersed within the sediment and show no preferred orientation.

Above the previous sediments and bounded by two faint erosive surfaces, a 0.5 m thick unit of unstratified/massive, fossil-rich, medium- to coarse-grained sandstones occurs. This unit is very rich in microgastropods, bivalves, fragments of barnacles, bryozoans, echinoids and rhodoliths. Shell debris is randomly/chaotically distributed within these sediments, and exhibits no preferential orientation.

The topmost sedimentary unit is about 3.7 m thick and shows vertical fining upward from medium- to fine-grained fossiliferous sandstones. These exhibit a faint cross-stratification at the base that gradually passes to an almost imperceptible plane-parallel sparse stratification. The uppermost ~1.5 m show abundant bioturbation by burrowing organisms (*Thalassinoides* isp., *Diplocraterion* isp., *Crossopodia* isp., *Ophiomorpha* isp. and *Rhizocorallium* isp.; Meireles et al. 2013). Basaltic pillow lavas of the overlying lava delta cap the sequence, imprinting load casts in the once soft sediments. These were dated to 4.13 ± 0.19 Ma (Ramalho et al. 2017).

4.3.2 Interpretation and Discussion

The sediments at Ponta do Castelo are interpreted as tempestite deposits that represent a rapid succession of 4 or 5 individual energetic events. Thereby, each event caused massive sediment transport from shallow waters to greater depths with intense bottom erosion and rapid redeposition along the steep shelf of Santa Maria by unidirectional and/or combined flows. Final deposition took place above or close to storm wave base (Meireles et al. 2013). The first unit is interpreted as a tempestite formed under wave oscillations (or combined flows) above or close to storm wave base, as suggested by the presence of swaley and hummocky stratification. The second unit is thought to represent a density-induced flow that brought sediments from shallower levels (as suggested by the presence of allochthonous littoral faunas), initially eroding the sea-bed and eventually settling as it gradually lost energy. After a short period of fair weather—attested to by the presence of bioturbation—another tempestite was deposited with strong basal erosion, creating a channel, which was subsequently filled. The existence of well-preserved bryozoan colonies in the sediment, which were ripped from their shallow-water rocky substrate, attests the vigorous energy of the event and the short transport distance from the shallow-water/littoral environment. Another even more massive event followed the previous one, cutting deeply into the existing sequence and resulting in the very rapid deposition of the fourth unit. The resulting deeply irregular palaeo-relief (with 3–4 m deep channels) attests the vigorous erosion necessarily associated to very energetic bottom currents. In contrast, the distinct mega-hummocky stratification with preserved metric, slightly asymmetrical hummocks with rare internal truncations, attests to the subsequent deposition under extremely high aggradation rates—perhaps corresponding to a deposition time of just a few hours—as a result of combined flow conditions above or at storm wave base (see Dumas and Arnott 2006; Meireles et al. 2013). The generation of such a deep basal erosional unconformity and the deposition of sediments under extremely high aggradation rates is interpreted as the result of strong downwelling currents (combining wave oscillation with unidirectional flow dynamics) associated with a very violent storm event that remobilised, transported and deposited large amounts of littoral sediments onto a deeper part of the shelf (Meireles et al. 2013). Finally, the last sedimentary unit represents the transition to the fair-weather suite. The intense bioturbation by burrowing organisms in the uppermost part of the sequence is related to the biological colonization that typically occurs at such depths (~50 m). Sedimentation was subsequently interrupted by the rapid progradation of a costal lava-fed delta sustained by an eruption on land. The prograding lavas thus preserved the underlying strata from subsequent erosion (Meireles et al. 2013). As the basaltic pillow lavas of the overlying lava delta imprinted load casts in the once soft sediments, the volcanic sequence is penecontemporaneous of the underlying sediments.

The sequence at Ponta do Castelo thus suggests that storm wave base in steep and narrow insular shelves may reach down to depths of 50–60 m and that deposition in this setting is mainly controlled by storm-events, during which sediments are being remobilised across the shelf, from the nearshore to the offshore, by strong ebb-return currents (Meireles et al. 2013).

4.4 The Palaeontological Significance of Carbonate Sequences in Reefless Volcanic Oceanic Islands

Sedimentary systems on reefless volcanic oceanic islands are subject to multiple controls, which differ considerably from other open shelf systems (e.g. Reuter and Piller 2011 and references therein), even when they are also influenced by volcanic activity (Wilson and Lokier 2002). The general steep relief, the limited sediment supply by river systems (Ávila et al. 2008b), the windward/leeward asymmetry in energy conditions, the influence of volcano-tectonic processes (uplift/subsidence; cf. Ramalho et al. 2013, 2017), frequent eruptive events and their related processes with localised ash-tuff-lava flow input result in small depocentres with a high facies variability and potentially even with an individual depositional evolution controlled by the input of volcanic material. Furthermore, volcanic events might influence carbonate systems by: (1) covering partly or completely the carbonate producing organisms, which results in physical stress, tissue necrosis and/or death of these organisms (Fortes 1991; Heikoop et al. 1996a, b;

(2) decreasing light penetration when suspended in the water, thus affecting photosynthetic communities (Short and Wyllie-Echeverria 1996; Wilson and Lokier 2002); and (3) changing the water chemistry (Frogner et al. 2001; Ralph et al. 2006; Duggen et al. 2010). Additionally, shelves at reefless oceanic islands are frequently exposed to the full force of recurrent storms that inevitably disrupt the sedimentary cycle and frequently force the remobilization, transport and redeposition of existing sediments (see Meireles et al. 2013 and references therein). All these processes should be taken into account when interpreting carbonate sequences at volcanic islands.

In conclusion, the early Pliocene carbonate deposits of Santa Maria highlight the importance of the interaction of different processes for the development and evolution of a carbonate system on reefless volcanic oceanic islands (the present-day dominant calcitic mineralogy of the carbonates at Pedra-que-pica, Pedreira do Campo and Figueiral is regarded as a taphonomical artefact and is attributed to the diagenesis of the aragonitic shells). Besides volcanic quiescence, as already suggested by Wilson and Lokier (2002), a rise in sea-level that provides accommodation space on the steep island shelf is generally necessary for the development and preservation of carbonate factories. Hence, marine deposits on reefless volcanic oceanic islands should be predominantly formed and preserved during transgressions and sea-level highstands, and their structure and distribution frequently reflect the highly energetic environment in which they were formed. Comparable to other structured island shelves, the high variability in volcanic island shelf morphology results in a complex pattern of small local depocentres (patchy outcrop patterns in sedimentary prisms), enhances the number of microhabitats and biocoenoses, and provokes a highly discontinuous facies variability (see also Titschack and Freiwald 2005; Titschack et al. 2005, 2013). In reefless insular shelves, sedimentary bodies typically have small residence times as these almost invariably get to be remobilised by storm-induced processes such ebb-return currents, onto deeper waters (Quartau et al. 2012, 2014, 2015; Meireles et al. 2013) or across the shelf (Johnson et al. 2017). Combined with the island's narrow and steep shelf, which provides little space for the deposition of sediments, the remobilisation of the transgressive/highstand deposits during subsequent regressive trends over the shelf edge into adjacent deep-sea basins, seems to represent their most likely fate (Ávila et al. 2008b; Ávila 2013). However, eruptive events, which generally cut off marine carbonate factories and interrupt marine sedimentary processes, might locally prevent the erosion of these marine carbonate deposits by their rapid and abrupt burial, as in the cases of Ponta do Castelo, Malbusca (see Rebelo et al. 2016a), and of the other above-mentioned outcrops. Consequently, volcanic events might limit carbonate production but might on the other hand be responsible for the effective fossilization and preservation of these deposits on reefless volcanic oceanic island shelves.

5 The Pleistocene MIS 5e Outcrops

Scattered around the shores of the island of Santa Maria, wave-cut platforms, wave-cut notches, and the remains of former boulder and sandy beaches (Fig. 8) are preserved at altitudes ranging from 3–4 up to 9–10 m above present mean sea level (Ávila et al. 2015b). These outcrops are interpreted as a series of sea-level stands and their formation during the last interglacial maximum sea level, hence the MIS 5e (Marine Isotopic Stage 5e, ~ 120–130 ka in age) is indicated (Ávila et al. 2008a, 2009a, 2010). Some of these outcrops are very rich in marine fossils. The best studied outcrops are Prainha and Praia do Calhau (both on the southern shores; Zbyszewski and Ferreira 1961; García-Talavera 1990; Ávila et al. 2002, 2007, 2008a, 2009a, 2015b; Amen et al. 2005; Ávila 2005), Vinha Velha (also located on the south; Ávila et al. 2015b) and Lagoinhas (northern shores of Santa Maria; Callapez and Soares 2000; Ávila et al. 2002, 2007, 2009a, 2015b; Ávila 2005; cf. Fig. 1b). Other known Pleistocene outcrops, some of them not yet

Fig. 8 Indicators of the last interglacial maximum sea-level (MIS 5e). **a** Wave-cut platform at Prainha (+4 m). **b** Boulder conglomerates lined up at Lagoinhas (+6 to +7 m). **c** The MIS 5e sea-level nearby Vinha Velha (note the boulder conglomerates, and the wave-cut notch at the base of the palaeocliff (bottom right). **d** The MIS 5e sea-level at Ponta do Castelo (+4 to +5 m). **e** Wave-cut notch at Baía da Rocha Alta (+8 m). **f** Wave-cut notches at Ponta do Pesqueiro Alto (+7 to +10 m). **g** The MIS 5e deposit at Ponta do Cedro (+6 to +7 m). **h** The MIS 5e deposit at Vinhas do Sul (+3 to +4 m)

studied nor dated are: a small outcrop west of Pedra-que-pica (see Ávila et al. 2015b), Ponta do Castelo, and Ponta do Cedro.

The Prainha outcrop was first reported by Zbyszewski and Ferreira (1961) and later by García-Talavera (1990), with all authors paying special attention to the marine molluscs. Callapez and Soares (2000) discovered the Lagoinhas deposit and published a palaeoecological interpretation of this outcrop. Ávila et al. (2002) redescribed the stratigraphic sequence of both Lagoinhas and Prainha and revised the marine molluscs' checklist, whereas Amen et al. (2005) wrote a detailed account of the fossil coralline–algal framework of Prainha. Ávila (2005) and Ávila et al. (2007, 2009a, b, 2015b) reviewed the Pleistocene mollusc fossils of Prainha and Lagoinhas. Based on U/Th ages on shells of *Patella* spp., and on faunal, facies and geomorphological considerations, Ávila et al. (2008a) assigned the basal conglomerate of the Prainha outcrop to Marine Isotopic Substage 5e (MIS 5e). Due to taphonomic problems and the wet climatic conditions prevailing on the island, the U-series measurements from their fossilised remains can only help constrain the age of the uranium taken up diagenetically (Ávila et al. 2008a, 2009a). U-series data from Santa Maria Island indicate an open system, providing mean ^{230}Th ages that are much younger than the true age of the unit. They suggest relatively steady, more or less continuous, diagenetic U-uptake by the fossils since their deposition in the Pleistocene (Ávila et al. 2009a). Therefore, accepting an age of 120–130 ka for the deposits, Ávila et al. (2008a) estimated that within the embedded biogenic carbonates U-uptake rates varied from c. 2.5 to 5 μg $gCaCO_3^{-1}$ kyr^{-1}. This was based on ^{230}Th-ages in the 60–70 ka range, for shells that would have started accumulating diagenetic U some 130–120 ka ago (i.e. during the MIS 5e; Ávila et al. 2008a).

5.1 Prainha and Praia do Calhau

These sections consist of unconsolidated shallow-water deposits overlying a basement formed by basaltic lavas (facies 1; Fig. 9). The bases of the outcrops are formed by irregular shore platforms on top of the ankaramitic basalts of the Anjos Complex, which occur at an altitude of 3–4 m (cf. Figs. 9 and 10; Serralheiro et al. 1987; Serralheiro 2003), and exhibit a lateral extent of ~800 m. Roughly circular depressions, 2.7–7.0 cm in diameter (mean = 4.26 cm, SD = 1.01, n = 48), were observed at this surface (Fig. 10e). Ávila et al. (2009a) attributed these structures to bioerosion traces of epilithic sea urchins and Ávila et al. (2015b) assigned them to the ichnofossil *Circolites kotocensis* Mikulás. The traces were recently destroyed, after Hurricane Gordon passed the Azores, in August 2012. The unconformity is overlain by a 40 cm thick, strongly cemented conglomerate (facies 2; Figs. 9 and 10a). A calcareous coralline algal biostrome (facies 3; Fig. 9) with a maximum thickness of ~50 cm covers the conglomerate or, locally, directly on the unconformity on top of the basalts. This biostrome is built by encrusting and warty to fruticose non-geniculate Corallinales (Rhodophyta) growing one over the other. Fragments of mollusc shells, bryozoans, and echinoderms, are accessory components. Amen et al. (2005) reported four species of Corallinaceae (Rhodophyta): *Spongites fruticulosus* Kützing (the main builder of the algal framework), *Lithophyllum incrustans* Philippi, *Neogoniolithon brassica-florida* (Harvey) Setchell and Mason, and *Titanoderma pustulatum* (Lamouroux) Nägeli. Of these, only *T. pustulatum* is recorded for the recent algal flora of the Azores (Neto 1994). The algal crust abundantly shows macrobioerosion structures, mostly clavate borings, assigned to the ichnogenus *Gastrochaenolites* Leymarie. Remains of the producer of these borings, the endolithic bivalve *Myoforceps aristatus* (Dillwyn), may still be found in situ inside most of the borings (Fig. 11a). At least two species of Corallinaceae described in this facies appear with a superimposed growth in thin sections (Fig. 11b). The inter-skeletal spaces in the algal framework and bioerosion structures were later filled by two internal sediments (Fig. 11c). The basal one consists of angular, poorly sorted rock fragments, marphic mineral grains', and

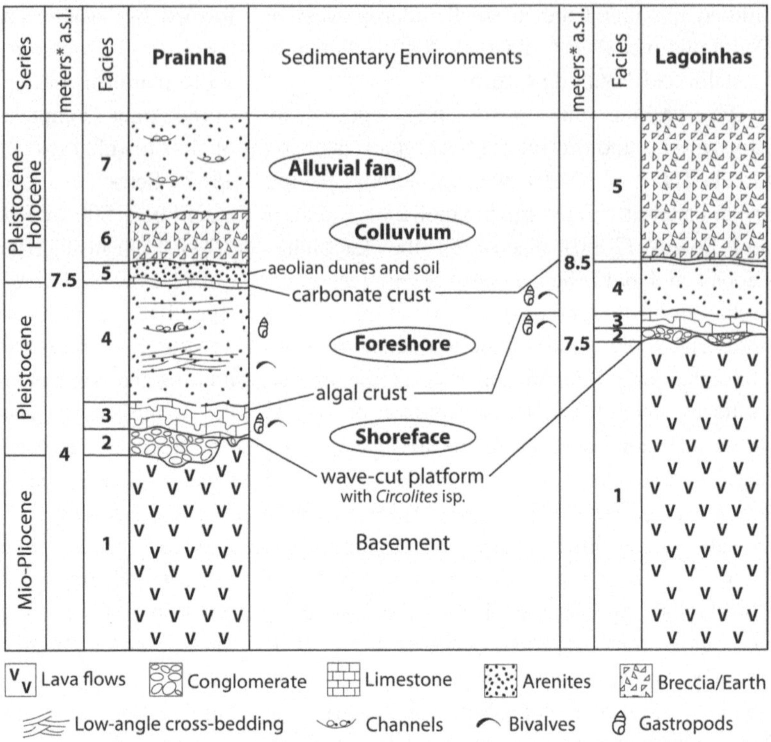

Fig. 9 Stratigraphic sections at Prainha (modified after Ávila et al. 2002, 2009a, 2010, 2015b) and Lagoinhas (after Ávila et al. 2009a, 2009a, 2015b), displaying the Pleistocene (MIS 5e) sedimentary successions

bioclasts (foraminifers, echinoids, coralline algae and scarce molluscs) of medium- to coarse-grained size in a micritic matrix. The succeeding internal sediment is richer in micrite. The conceptacles of the red algae are filled with spherulitic aragonite cement. These cements are slightly weathered, with limited superficial dissolution of aragonite. Locally, towards the top of the biostrome, accumulations of iron oxide micro-nodules occur between coralline thalli. Traces of iron oxides and interruptions in algal growth are observed in thin sections. In some places, where the thickness of the biostrome is enhanced, 3–4 discontinuity surfaces are macroscopically observable by the increased presence of bioerosive structures at these surfaces (Ávila et al. 2009a). Ávila et al. (2002) reported fractures in the algal crust (facies 3) that were initially attributed to "subaerial exposure and desiccation". Ávila et al. (2009a) reinterpreted these fractures on the upper surface of the algal framework and along the entire vertical section of the fossil algae as the result of an extensional process fracturing the algal framework after its formation. Fracturing by local extension can take place underwater with no subaerial exposure. In fact, formation of neptunian dykes (open fractures filled by marine deposits; cf. Fig. 10c) by extension is a common process in submarine settings (Winterer and Sarti 1994).

The biostrome (facies 3) is overlain by 1.3 m thick, yellowish, partly cross-laminated, volcaniclastic to bioclastic, uncemented sands (facies 4; Figs. 9 and 10b) that also fill most of the fractures penetrating facies 3 (cf. Fig. 10c). Carbonate contents range from 56.2 to 67.3% (in the samples investigated) (Fig. 10b). The grain-size distribution is dominated by the 125–250 μm fraction. The bioclasts consists almost exclusively of small mollusc fragments. Lenses with ripple marks, trace-fossils and/or root casts are preserved locally (Ávila et al. 2002) and are better lithified. The sediments correspond to a beach foreshore (intertidal) facies and show cross-lamination (cf. Fig. 10b). A thin carbonate crust of pedogenic origin occurs at the top of the

Fig. 10 The Pleistocene deposits at Praia do Calhau and Prainha. **a** One of the most representative sections studied at Praia do Calhau: over an irregular shore platform carved on top of the ankaramitic basalts of the Anjos Complex, a strongly cemented conglomerate was stabilised by a calcareous coralline algal biostrome. **b** Partly cross-lamination sands that overlain the calcareous coralline algal biostrome, covered by colluvial-alluvial deposits. **c** Fossil coralline algal crust with abundant signs of bioerosion by endolithic bivalves (clavate borings), and neptunian dykes (large fractures affecting the algal framework), filled with sand; both structures are present along the entire vertical section of the fossil algae. **d** The algal framework at Prainha. **e** Bioerosion traces of epilithic sea urchins, probably made by *Paracentrotus lividus* (Lamarck) and assigned to the ichnofossil *Circolites kotocensis* Mikulás

bioclastic sandy sediments (facies 5, Fig. 9). The crust mostly consists of micrite that precipitated together with clay and other silt-sized impurities and exhibits a clotted texture (Fig. 11d). Sediment below the carbonate crust (Fig. 11e) includes bioclasts (molluscs, echinoid spines, branching coralline algae), poorly-sorted volcanic grains and very poorly-sorted rock fragments. Aeolian dunes consisting of a mixture of biogenic sand particles derived from molluscs, echinoderms, bryozoans and coralline algae with black, basaltic sand (facies 5; Fig. 9) and colluvial–alluvial deposits (facies 6 and 7; Fig. 9) cover the carbonate crust (Ávila et al. 2009a).

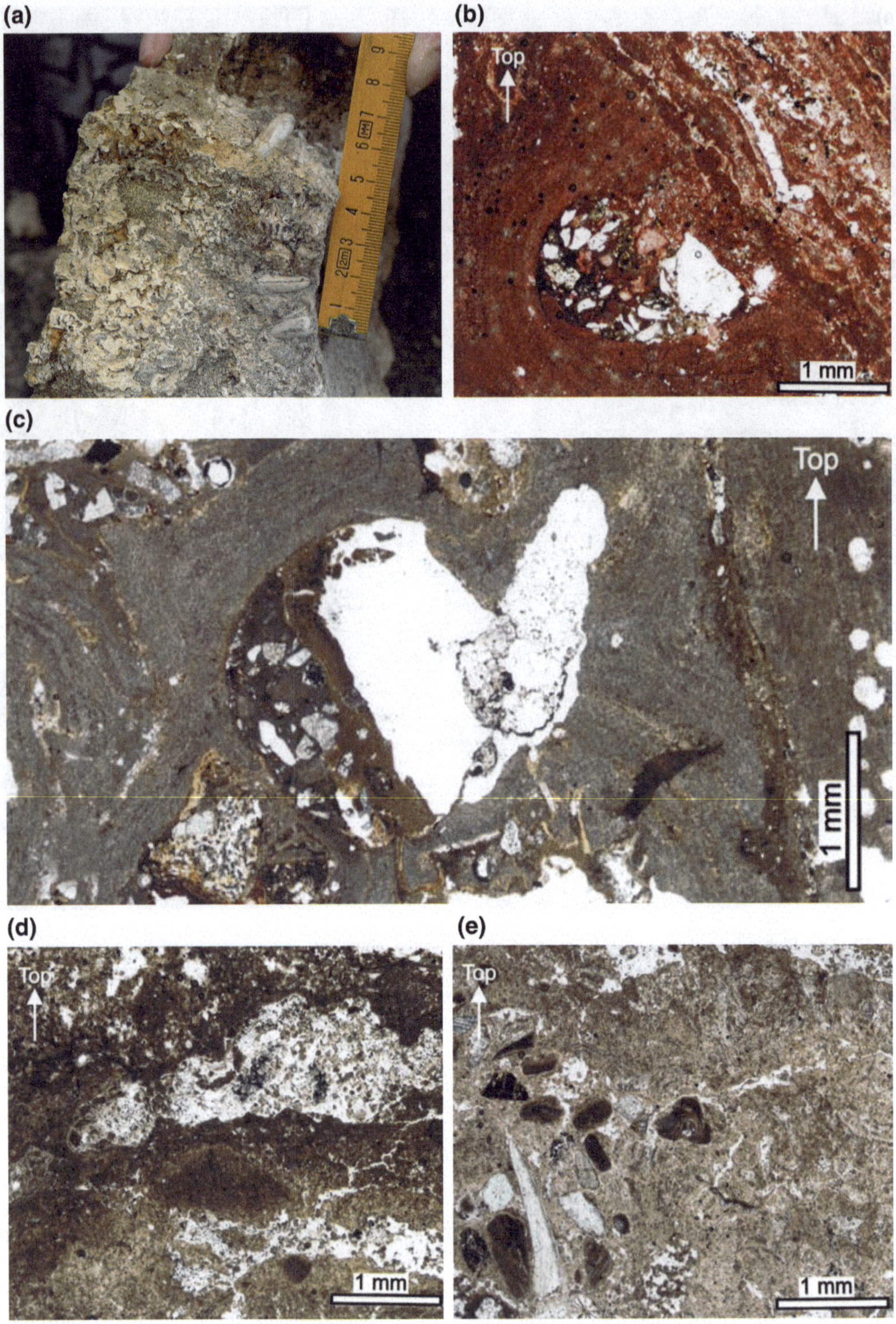

Fig. 11 **a** At Prainha, the endolithic bivalve *Myoforceps aristatus* (Dillwyn) is sometimes found in situ inside some of its borings. **b**, **c** Thin section microphotographs of the algal framework collected at Prainha. **d**, **e** Thin section microphotographs of the bioclastic sand sediments collected at Prainha

5.2 Lagoinhas

The Lagoinhas outcrop is about 100 m wide and its base lies ~7.4 m above present sea-level (Ávila et al. 2002, 2009a). The base is formed by an irregular shore platform on top of the ankaramitic basalts of the Touril Complex (facies 1; Fig. 9; Serralheiro et al. 1987; Serralheiro 2003). A ~10 cm thick, strongly cemented basal-conglomerate (facies 2) passes laterally and upward into a ~30 cm thick calcareous red algal biostrome (facies 3). The coralline algal framework is similar to the one found at Prainha, but was not yet studied in detail. The faunal assemblage is dominated by molluscs, with bryozoans and echinoderms as accessory components. As at Prainha, macrobioerosion structures in the biostrome are assigned to the ichnogenus *Gastrochaenolites* Leymarie and were produced by the endolithic bivalve *Myoforceps aristatus* (Dillwyn) (Ávila et al. 2009a). The overlying white-yellowish unconsolidated sands (facies 4) exhibit thicknesses of up to ~0.7 m and are rich in very-well preserved fossil assemblages, which are dominated by molluscs. The grain-size distribution of these sands is dominated by the 250–500 μm fractions. A thin carbonate crust of pedogenic origin occurs at the top of these bioclastic sands (cf. Fig. 9). This crust is covered by colluvial deposits (facies 5; Fig. 9).

5.3 Interpretation and Discussion

Although structures similar to the Prainha and Lagoinhas coralline algal biostromes are not known from other islands of the Azores, their occurrence is well documented from shallow-water rocky substrates in the Mediterranean Sea (e.g., Pérès and Picard 1964; Hofrichter 2001) and Western Atlantic, where they form flat calcareous patches, stabilizing shore conglomerates (Thorton et al. 1978; Bosence 1983, 1984). These formations develop primarily on shores with a narrow tidal amplitude (Adey 1986), as is the case for all the Azores. No known direct analogues of the Prainha and Lagoinhas shallow-water fossil algal frameworks exist in other regions of the world (Ávila et al. 2009a). According to Amen et al. (2005), the algal biostromes were formed in shallow water depths, probably less than 10 m deep. Today, *Lithophyllum incrustans*, *L. pustulatum* (= *Titanoderma pustulatum*) and *Neogoniolithon brassica-florida* are most common in the shallowest infralittoral zone within the Mediterranean, while *Spongites fructiculosus* shows a wider depth range (Bressan and Babbini 2003). The mastophoroid genera *Spongites* and *Neogoniolithon* predominate in the Neogene algal assemblages, when the western Mediterranean experienced tropical/subtropical climatic conditions (Braga and Aguirre 2001). The Mediterranean and western Atlantic "trottoirs" are more similar as they are structures a few decimetres thick and locally developed on abrasion terraces (Thorton et al. 1978; Adey 1986). *Neogoniolithon brassica-florida* and *Lithophyllum incrustans*, abundant in the Prainha framework (Amen et al. 2005), can be main components of the Mediterranean "trottoirs", together with *Lithophyllum byssoides* (frequently mentioned as *Lithophyllum tortuosum*) (Adey 1986).

Based on stratigraphic, petrologic, and faunal analysis of the Lagoinhas and Prainha outcrops, Ávila et al. (2009a, 2015b) suggested that during a marine transgression, an irregular shore platform on volcanic rocks developed. Subsequently, during the ongoing transgression, a conglomerate (facies 2) and the overlying red algal biostrome (facies 3) formed in a shallow-marine, high-energy (shoreface) environment (Fig. 9). The hydrodynamic setting is further supported by the presence of sea urchin burrows likely being attributable to the activity of the epilithic grazer *Paracentrotus lividus* (Lamarck), of which one test fragment and numerous primary spine fragments were found at Lagoinhas (Madeira et al. 2011). In the present-day, this sea urchin is extremely common on the shores of the islands of the Azores, at depths of 1–2 m, always on rocky substrate, where it produces similar shallow excavations (Ávila et al. 2009a). The bioerosion structures—*Gastrochaenolites* isp.—that characterize the algal reef in both outcrops are also indicative of shallow water depths (less than

5 m). The endolithic bivalve *Myoforceps aristatus*, found in life-position, is the producer of these structures. Today, the congeneric *M. aristatus* lives in shallow-marine surf benches (called "trottoirs" in the Mediterranean). Maximal transgression with minimal sediment input was reached on top of the coralline algal biostrome (top of facies 3). At the start of the subsequent regression, the poorly consolidated yellow fossiliferous sands (facies 4) were deposited. Sedimentary structures such as wave-ripple cross-bedding and cross-lamination indicate a shoreface to foreshore depositional setting. The palaeoecological evidence derived from the analysis of the bioerosion structures, the characteristics of the coralline algal assemblages, the taxonomic composition of the fossil assemblage and the bathymetrical zonation of the species represented in it, indicate a change from a transgressive marine depositional setting in shallow infralittoral (facies 2 and 3) to a regressive trend (foreshore/beach; facies 4). The regressive, shallowing upwards trend proceeds with dune sediments, palaeosols (micritic crusts) and colluvial–alluvial deposits (facies 5, 6 and 7; Fig. 9). The position of the Pleistocene sediments, not taking the apparent uplift trend of Santa Maria Island and glacio-isostatic adjustments into account, suggests a palaeo-sea-level highstand about 4–6 m above the modern sea-level. This suggests the formation of these deposits during the interglacial sea-level highstand of MIS 5e (Ávila et al. 2009a, 2015b).

6 The Fossils

Zbyszewski and Ferreira (1962b) reported 188 taxa (179 invertebrates and 9 vertebrates) from the early Pliocene deposits of Santa Maria Island. Some groups were recently revised, e.g., the Brachiopoda, the Echinodermata, the Chordata (sharks and cetaceans), and the Ostracoda (Kroh et al. 2008; Madeira et al. 2011; Ávila et al. 2012, 2015c; Meireles et al. 2012). Other groups are presently under review, such as the Bryozoa, the Cnidaria (Anthozoa) and also the Mollusca, the latter representing the most abundant fossil group. The MPB group has reviewed the Zbyszewski and Ferreira checklist and over 40 taxa of fossil molluscs need revision. Table 1 provides a revised list of species with a conservative number of 191 taxa (13 of which are considered as endemic species) from the early Pliocene outcrops of Santa Maria. With 99 taxa accepted as valid, the Mollusca exhibit the highest number of species. Of these 99 taxa, 7 are considered as endemic to Santa Maria.

Regarding the Pleistocene (MIS 5e) outcrops, Ávila et al. (2015b) reported 136 species of marine molluscs. Twenty-two of the gastropods are endemic, and all of them still live in the waters of the Azores. Considering all groups, this chapter updates the latest checklists and increases the total number of Pleistocene species to 146 (Table 1).

6.1 Fossil Algae

The Pleistocene layer of calcareous algae described in detail in Sect. 5.1 was first reported by Berthois (1953b, c) and by Zbyszewski and Ferreira (1961) who named this specific layer as "*Lithothamnium*". The first phycological study of the algal components of the fossil framework of Prainha was made by Amen et al. (2005) who, based on morphological and anatomical characters, as well as reproductive structures, reported 4 species: *Spongites fruticulosus* Kützing (the most abundant algal species), *Lithophyllum incrustans* Philippi, *Neogoniolithon brassica-florida* (Harvey) Setchell and Mason, and the rare *Titanoderma pustulatum* (J.V. Lamouroux) Nägeli. No *Lithothamnion* species were found at Prainha and structures similar to these algal frameworks are not known in the present times on the Azores (Amen et al. 2005).

The calcareous fossil algae present on rhodoliths were recently studied by Rebelo et al. (2014, 2016a, b), who reported 6 taxa. Her fieldwork reports fossil rhodolith forming coralline algae from all outcrops but Cré (Ana Rebelo, pers. comm).

Table 1 Number of specific taxa and number of endemic species collected and identified at the early Pliocene and Pleistocene (MIS 5e) fossiliferous deposits from Santa Maria Island (Azores)

	Early Pliocene		Pleistocene (MIS 5e)	
	Total number spp.	Endemic spp.	Total number spp.	Endemic spp.
Algae	6		4	
Annelida (Polychaeta)	2			
Brachiopoda	3			
Bryozoa	5			
Chordata (Mysticeti)	1		1	
Chordata (Odontoceti)	1			
Cnidaria (Anthozoa)	7	1		
Coelenterata	2			
Crustacea (Cirripedia)	1	1		
Crustacea (Decapoda)	1			
Crustacea (Ostracoda)	13	1		
Echinodermata	8		3	
Fish (Other)	3			
Fish (Sharks)	7			
Foraminifera	32			
Mollusca (Bivalvia)	51	2	24	
Mollusca (Gastropoda)	31	3	114	22
Mollusca (Gastropoda: Heteropoda and Pteropoda)	17	2		
Total	**191**	**10**	**146**	**22**

6.2 Fossil Vertebrates

The Azores are well-known as one of the best spots for whale-watching nowadays, with 28 species of cetaceans reported around the archipelago (Ávila et al. 2011). In Santa Maria, the fossils of whales have been known for a long time, being called "ossos de gigante" (bones of giants) by the inhabitants of the island. However, the first published information on this matter was done only in the 19th century by Boid (1835) who wrote "In a part (…) (of the) N.W. side scarcely accessible, is to be seen an immense fossil thighbone". This bone, located at Ponta do Pesqueiro, was later identified by Bedemar (1837) as a whale bone (Estevens and Ávila 2007). Other fossil fragments of cetacean vertebrae and ribs, reported by subsequent authors (Cotter 1888–1892; Teixeira 1950; da Ferreira 1955; Zbyszewski and Ferreira 1962b) were never the subject of a proper study in order to identify these remains.

6.2.1 Cetaceans

The first thoroughly systematic study on the fossil whales of Santa Maria Island was done by Estevens and Ávila (2007). These authors reported 10 occurrences of fossil cetaceans belonging to two suborders: Odontoceti (Ziphiidae), with one *Mesoplodon* species collected at Assumada outcrop; and several bone fragments of ribs, vertebrae and other skeletal parts, assigned to undetermined Mysticeti (?Balaenopteridae). All of these samples were collected in sediments from the Touril Complex (Estevens and Ávila 2007). Albeit meager, the fossil Pliocene whales of Santa Maria were compared with contemporaneous faunas (Messinian-Zanclean) from the western

Atlantic—Eastover (7.2–6.1 Ma) and Yorktown (4.8–3.0 Ma) formations from the Middle Atlantic Coastal Plain, the Palmetto Fauna (5.2–4.5 Ma) from Florida—and from the eastern Atlantic—the Kattendijk (5.0–4.4 Ma) and Lillo (4.2–2.6 Ma) formations from Belgium (see references in Estevens and Ávila 2007: 157). Remarkably, all of these faunistic associations share a noticeable modern character, being dominated by living groups such as the beaked whales (Ziphiidae) and true dolphins (Delphinidae) among the odontocetes, and the rorquals (Balaenopteridae) and right whales (Balaenidae) among the mysticetes, similar to the Azorean fossil fauna described by Estevens and Ávila (2007) (Fig. 12a–d).

Fig. 12 Neogene fossil cetaceans from Santa Maria (Estevens and Ávila 2007; Ávila et al. 2015c). **a** Small rib fragment assigned to Cetacea indet (Ponta Negra). **b** Small rib fragment assigned to Cetacea indet (Cré). **c**, **d** Fragment of a large left rib assigned to ? Balaenopteridae indet. (DBUA-F 401) in lateral view (**c**) and cross-section at intermediate break (**d**) (Cré). **e**, **f** Large fragment ($\sim$1 m) of the mandible of a medium- to large-sized baleen-bearing mysticete (Chaeomysticeti) (MIS 5e of Prainha)

Recently, in September 2012, after the passage of the Hurricane Gordon, a large fragment ($\sim$1 m in length) from the mandible of a medium- to large-sized baleen-bearing mysticete (Chaeomysticeti) was collected at Praia do Calhau, in the Pleistocene (MIS 5e) sandy layer (Fig. 12e, f). This finding was reported by Ávila et al. (2015c) as the first world record of fossil whales for the last interglacial (MIS 5e), in outcrops located in oceanic islands, thus increasing the relevance of the fossil record of Santa Maria Island for the palaeobiogeography of the North Atlantic.

6.2.2 Selaceans

Reports of fossil shark teeth from oceanic islands are few to date. In the Atlantic, shark teeth are reported from fossiliferous deposits of Jamaica (Donovan and Gunter 2001), Puerto Rico (Nieves-Rivera et al. 2003), Cape Verde (Serralheiro 1976), and the Azores (Zbyszewski and Ferreira 1962b; Ávila et al. 2012). In the Caribbean, Cuba has 15 taxa of fossil sharks from the Miocene-Pliocene (Iturralde-Vinent et al. 1996) and 5 species are known from the Miocene of Carriacou Island (the Grenadines, Lesser Antilles; Portell et al. 2008).

Zbyszewski and d'Almeida (1950) and Zbyszewski and Ferreira (1962b) were the first to report fossil sharks from Santa Maria. Ávila et al. (2012) added three new records, increasing the number of fossil sharks to seven species: *Notorynchus primigenius* (Agassiz), *Carcharias acutissima* (Agassiz), *Cosmopolitodus hastalis* (Agassiz), *Paratodus benedenii* (Le Hon), *Isurus oxyrinchus* Rafinesque, *Megaselachus megalodon* (Agassiz in Charlesworth), and *Carcharhinus* cf. *leucas* (Valenciennes in Müller and Henle) (Fig. 13a–q). The Azorean Pliocene selachian fauna clearly differs from those described from sediments deposited on continental shelves, in which batoids (i.e., electric rays, sawfishes, guitarfishes, skates and stingrays) and small benthic sharks (e.g., scyliorhinids) are usually well represented. Several attempts were made to recover small selachian teeth by screen-washing (sieve sizes: 0.5 and 1 mm) from bulk sediment samples from most of the studied localities, but that effort yielded no teeth of batoids or small sharks (Ávila et al. 2012). The absence of batoids in the Pliocene coastal environments of Santa Maria Island is not in accordance with the diversity of the recent batoid fauna from the Azores, including about 15 species of Torpediniformes, Rajiformes and Myliobatiformes (Santos et al. 1997; Barreiros and Gadig 2011).

During the early Pliocene, subtropical to warm-temperate seas were prevalent in the area of the Azores, as deduced from palaeontological, geological and isotopic data, all indicating a warmer climate than today (Kirby et al. 2007; Ávila et al. 2015b). Five species are extinct nowadays: *C. acutissima*, *C. hastalis*, *N. primigenius*, *M. megalodon* and *P. benedenii*. All these species disappeared by the end of the Pliocene, including *N. primigenius*, formerly thought having went extinct during the Miocene (Ávila et al. 2012).

6.3 Fossil Invertebrates

6.3.1 Mollusca

Molluscs are by far the most diverse and abundant fossil group from the outcrops of Santa Maria (see Table 1). The particularly good fossil record of molluscs can be explained by a combination of initial abundance, taphonomic biases and selective investigations (most palaeontological studies on Santa Maria during the 19th and 20th centuries focused on molluscs).

Notwithstanding the many studies on the early Pliocene deposits (Bronn 1860; Hartung 1860; Reiss 1862; Mayer 1864; Cotter 1888–1892; Ferreira 1952, 1955; Zbyszewski and Ferreira 1962b), the fossil molluscs have not been recently studied systematically/taxonomically, with the exceptions of Janssen et al.'s (2008) paper on the early Pliocene heteropods and pteropods, and Habermann's (2011) study on the bivalve *Gigantopecten latissimus* (Brocchi). Much material has been collected since the beginning of the international workshops "*Palaeontology in Atlantic Islands*", in 2002, with a special emphasis on the micromolluscs.

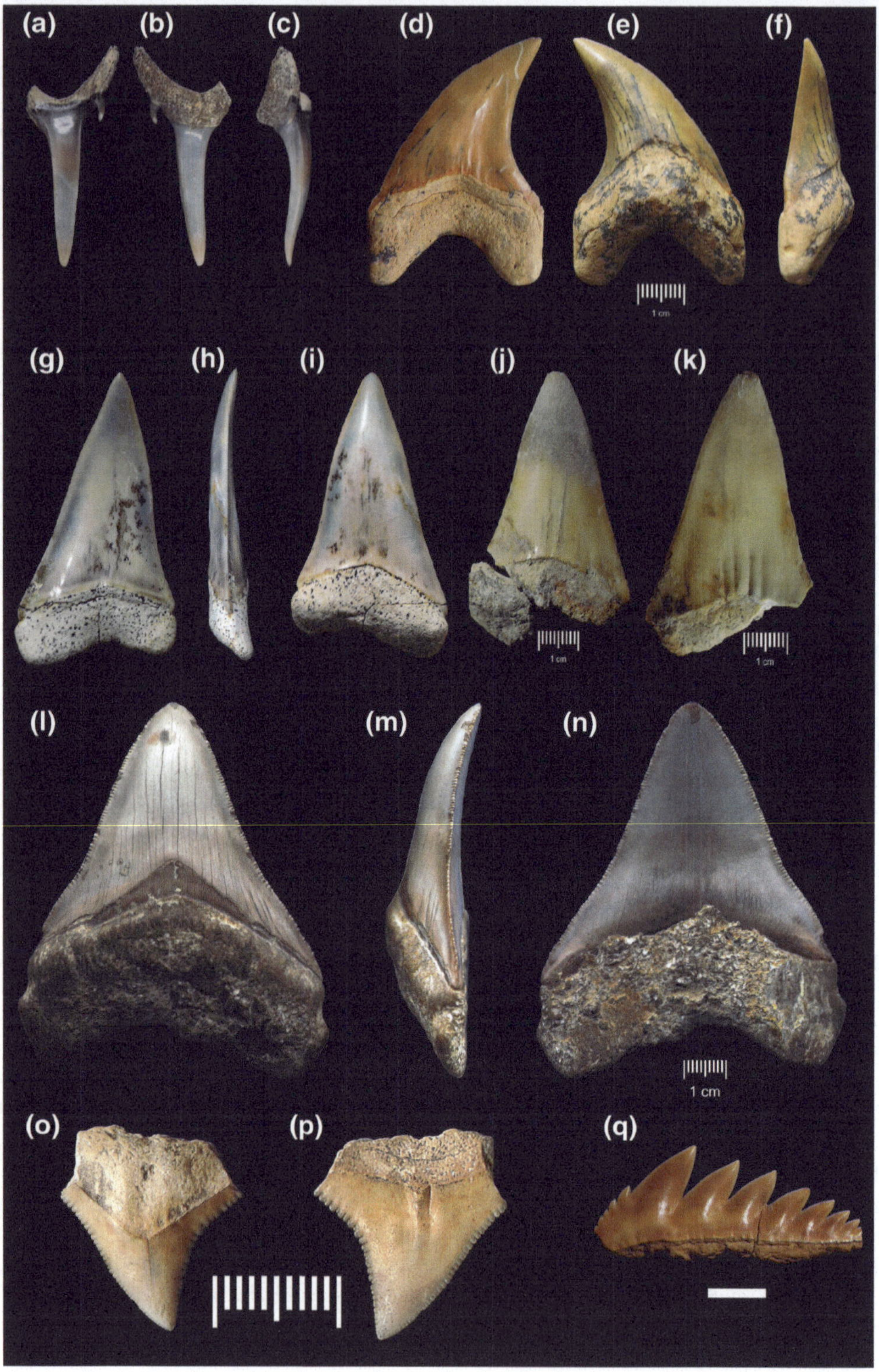
(a)
(b)
(c)
(d)
(e)
(f)
1 cm
(g)
(h)
(i)
(j)
(k)
1 cm
1 cm
(l)
(m)
(n)
1 cm
(o)
(p)
(q)

◀ **Fig. 13** Neogene fossil sharks from Santa Maria (Ávila et al. 2012). **a–c** *Carcharias acutissima* (Agassiz). Upper-anterior tooth in labial (**a**), lingual (**b**) and lateral view (**c**) (H = 27.0 mm, W = 9.3 mm). **d–f** *Paratodus benedenii* (Le Hon) in labial (**d**), lingual (**e**), and distal view (**f**) (H = 52.0 mm, W = 37.0 mm). **g–k** *Cosmopolitodus hastalis* (Agassiz). **g–i**: labial (**g**), mesial (**h**), and lingual (**i**) view (H = 55.0 mm, W = 35.0 mm). **j**, **k** Lingual (J) and labial view (**k**) (H = 50.0 mm, W = 36.0 mm). **l–n** *Megaselachus megalodon* (Agassiz) in lingual (**l**), lateral (**m**), and labial view (**n**) (H = 88.3 mm, W = 74.5 mm). **o**, **p** *Carcharhinus* cf. *leucas* (Valenciennes) in lingual (**o**) and labial view (**p**). **q** *Notorynchus primigenius* (Agassiz). Lower antero-lateral tooth in labial view (H = 9.8 mm, W = 23.5 mm)

The revision of the 105 historically reported taxa and the formal description of many new species (Ávila, unpublished data) are of utmost importance and present the major research task of the MPB for the next years.

For the Pleistocene (MIS 5e) malacofauna, Ávila et al. (2002) reviewed the old literature (Berthois 1951, 1953b; Zbyszewski and Ferreira 1961; García-Talavera 1990; Callapez and Soares 2000) and added new records, reporting 89 taxa, including 75 gastropods and 14 bivalves. Further work by Ávila (2005) and Ávila et al. (2007, 2009b, 2010, 2015b) increased the number of Pleistocene mollusc species to 138 taxa (114 gastropods and 24 bivalves). During the course of the last glaciation, 23 of these species locally disappeared from the Azores, of which 14 were thermophilic species that were not able to cope with the drop of the sea temperatures, and 7 species were associated with shallow sandy substrates that disappeared from the island when sea-level dropped below the shelf-edge (cf. Ávila et al. 2008b, 2015b for further information). Of the 138 taxa of molluscs, 22 are endemic gastropods, all of them still living along the coastlines of the Azores (Table 1).

6.3.2 Echinodermata

The most recent review on the fossil echinoid material collected from the fossiliferous deposits of Santa Maria (both the early Pliocene deposits as well as the Pleistocene deposits) was published by Madeira et al. (2011). These fossils are predominantly disarticulated skeletal material (primary spines and coronal fragments). Only subordinate complete tests are preserved. At least 7 taxa are reported by Madeira et al. (2011) from the Upper Miocene to lower Pliocene beds: *Eucidaris tribuloides* (Lamarck), *Echinoneus* cf. *cyclostomus* Leske, *Clypeaster altus* (Leske), *Echinocyamus pusillus* (Müller), *Echinocardium* sp. 1, *Echinocardium* sp. 2, and *Schizobrissus* sp (cf. Fig. 14a–m). Undetermined spatangoids were also found, but the poor preservation of the fossils did not allow their identification. Irregular echinoids such as spatangoids, clypeasteroids and *Echinoneus* usually live in soft-bottom environments, whereas the regular echinoid *Eucidaris tribuloides* preferentially lives epifaunally on hard substrates.

The Upper Pleistocene outcrops (MIS 5e) yielded three regular echinoid species: *Sphaerechinus granularis* (Lamarck), *Arbacia lixula* Linnaeus and *Paracentrotus lividus* (Lamarck) (Fig. 14n–v). The presence of taxa typical of tropical seas in the Mio-Pliocene sediments contrasts with the Pleistocene and modern echinoid fauna, which is warm-temperate in composition (Madeira et al. 2011). The diversity and specific composition of echinoids preserved in the Pliocene deposits of Santa Maria differs markedly from that of the island's Pleistocene fossil record. Worldwide, Pleistocene faunas tend to be closely related to the living biota of any area (Madeira et al. 2011), and the Azorean Pleistocene is no exception.

6.3.3 Brachiopoda

Although brachiopods are common members of shallow-water communities in Miocene and Pliocene sediments of the Mediterranean or Central Paratethys (Ruggiero 1994; Bitner and Dulai 2004), only two species are known from the Pliocene sediments of Santa Maria: *Terebratulina retusa* (Linnaeus), reported by Bronn (1860) and by Ferreira (1955) as *T. caput-serpentis*; and *Novocrania turbinata* (Poli), reported by Kroh et al. (2008; Fig. 15a–c). Today, none of

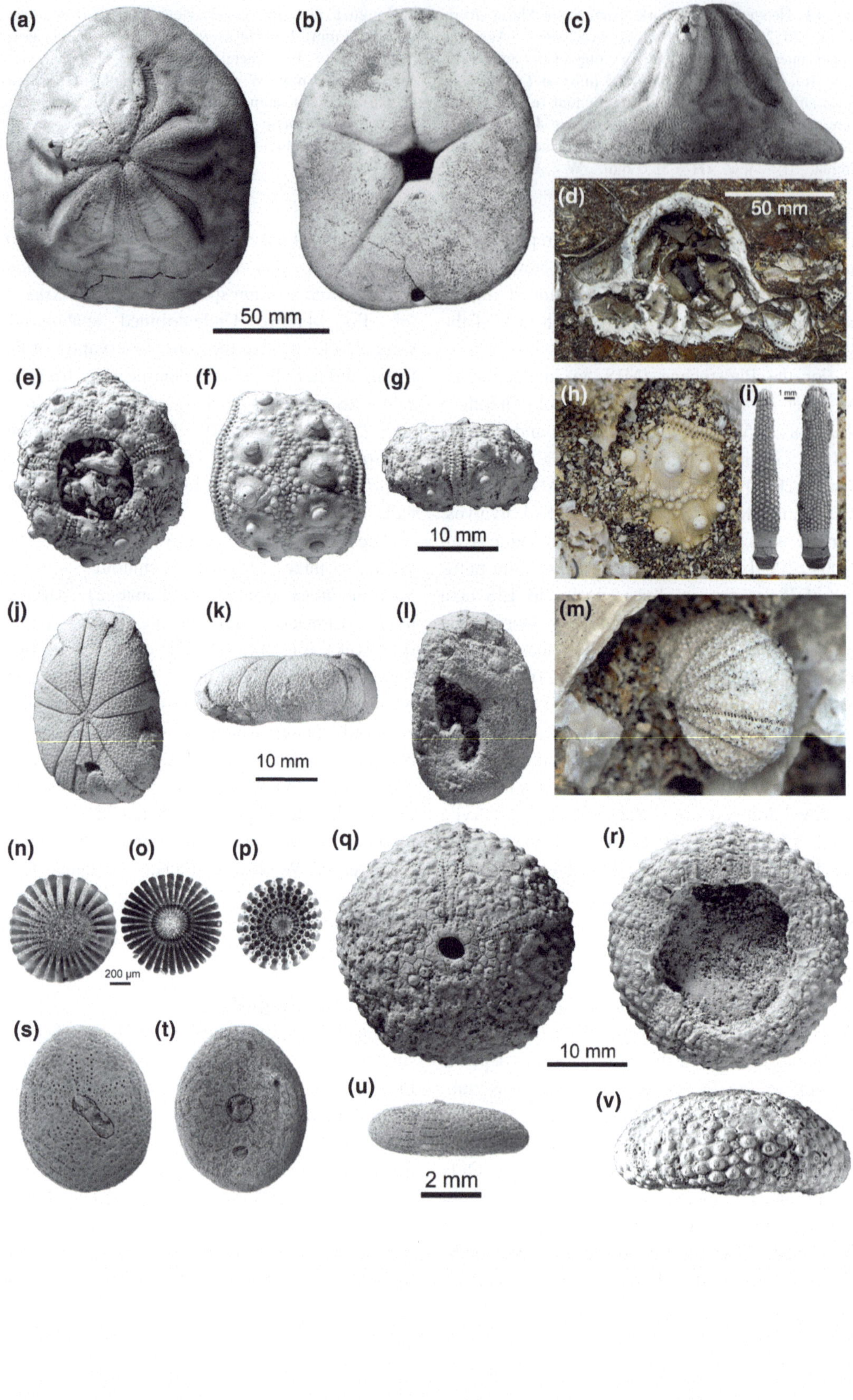
(a)
(b)
(c)
50 mm
(d)
50 mm
(e)
(f)
(g)
10 mm
(h)
(i)
1 mm
(j)
(k)
(l)
10 mm
(m)
(n)
(o)
(p)
200 µm
(q)
(r)
10 mm
(s)
(t)
(u)
2 mm
(v)

◄ **Fig. 14** Neogene fossil echinoderms from Santa Maria (Madeira et al. 2011). **a–d** *Clypeaster altus* (Leske) in aboral (**a**), oral (**b**), and left lateral view (**c**). **d** Cross view of a specimen photographed at Pedra-que-pica. **e–i** *Eucidaris tribuloides* (Lamarck) in oral (**e**) and lateral view (**f**, **g**). **h** Fragment of a specimen photographed at Pedra-que-pica. **i** Spines. **j–m** *Echinoneus* cf. *cyclostomus* Leske in aboral (**j**), lateral (**k**) and oral view (**l**). **m** specimen photographed at Pedra-que-pica. **n–p** Cross section of spines from the Pleistocene of Lagoinhas. *Arbacia lixula* Linnaeus (**n**), *Sphaerechinus granularis* (Lamarck) (**o**), and *Paracentrotus lividus* (Lamarck) (**p**). **q**, **r**, **v** *Arbacia lixula* in aboral (**q**), oral (**r**), and lateral view (**v**). **s–u** *Echinocyamus pusillus* (Müller) in aboral (**s**), oral (**t**) and left lateral view (**u**) (Madeira et al. 2011)

these species occurs in the Azores. *N. turbinata* presently lives in warmer waters (the Cape Verde Islands, off the north-western African coast and the Mediterranean), and is a non-molluscan example of the local disappearance of thermophilic taxa from the Azores since the Pliocene (Kroh et al. 2008).

6.3.4 Crustacea

Fossil barnacles were reported by Zbyszewski and Ferreira (1962b) from the Pliocene of Santa Maria (as *Balanus laevis* Bruguière) and by Callapez and Soares (2000) of the Pleistocene deposits of Lagoinhas (as *Balanus* sp.). However, these fossil cirripedes were not described in detail. Winkelmann et al. (2010) described a new species, *Zullobalanus santamariaensis* Buckeridge and Winkelmann (Fig. 15d–f), endemic from the early Pliocene outcrops of the island. This endemic barnacle lived in shallow-water, and probably, the *Balanus laevis* record was misidentified with *Zullobalanus santamariaensis* by Zbyszewski and Ferreira (1962b). The ancestral *Z. santamariaensis* might have reached the Azores passively attached to whale barnacles that were subsequently dislodged near Santa Maria Island, ultimately establishing a viable population (Winkelmann et al. 2010). Zbyszewski and Ferreira (1962b) also reported *Neptunus granulatus* Edwards, based on a chela collected at Lombo Gordo outcrop (Fig. 15g).

6.3.5 Ostracoda

The first study on the fossil ostracods from the Azores archipelago was recently published by Meireles et al. (2012). These authors reported 13 taxa from the Pliocene of the Malbusca section, representing seven families and 12 genera: *Xestoleberis* cf. *paisi* Nascimento, *Loxoconcha stellifera* Müeller, *Loxoconcha rhomboidea* (Fischer), *Callistocythere oertlii* Nascimento, *Leptocythere azorica* Meireles and Faranda, *Pachycaudites* cf. *armilla* Ciampo, *Dameriacella* cf. *dameriacensis* (Keij), *Aurila* sp., ?*Quadracythere* sp., *Heliocythere magnei* (Keij), *Neonesidea rochae* Nascimento, *Paracypris* sp., and *Cyamocytheridea* sp. All these species are typical of warm waters and epi-neritic habitats ($\sim$ 10–50 m of depth) (Meireles et al. 2012). *Xestoleberis* cf. *paisi*, *L. rhomboidea*, *L. stellifera*, *C. oertlii* and *N. rochae* still occur nowadays in the shallow-water environments (down to 40 m depth) of the Azores. All other ostracod taxa recorded from Malbusca (including the new species *L. azorica*) disappeared from the Azores and the Atlantic domain probably at the end of the Zanclean age, around 3.6 Ma, as suggested by Ávila et al. (2016a) for marine molluscs. Meireles et al. (2012) considered these ostracods as warm-water taxa that did not survive the Plio-Pleistocene climate deterioration.

7 The Future: The PalaeoPark Santa Maria

In 1996, Richard Lane and David L. Bruton argued during the 30th International Geological Congress (IGC) in Beijing (China) that protection and conservation of fossil sites around the world should constitute a major objective of the International Palaeontological Association (IPA). This initiative was supported by four workshops that took place during the 32nd International Geological Congress (IGC—2004) in Florence (Italy), at the 2nd International Palaeontological Congress (IPC—2006) in Beijing (China), the 33rd IGC (2008) in Oslo (Norway) and the 3rd IPC (2012) in London (UK). From 2004 on, Jere Lipps promoted the concept of the "PalaeoParks"

Fig. 15 **a–c** The brachiopod *Novocrania turbinata* (Poli). **d–f** The crustacean *Zullobalanus santamariaensis* Buckeridge and Winkelmann. Lateral (**d**) and apical view (**f**) of a specimen from Malbusca outcrop. 1: carina; 2: carinolatus; 3: latus; 4: rostrum Winkelmann et al. (2010). **e** *Zullobalanus santamariaensis* photographed in situ, on a rhodolith, at Malbusca outcrop. **g** Chela of the crustacean *Neptunus granulatus* Edwards from the Pliocene of Lombo Gordo outcrop

internationally, with the aim of fast-protecting and conserving fossil sites around the world. According to Lipps (2009), the preservation of fossil specimens in museums, as a rule being performed correctly, is insufficient. Instead, fossiliferous deposits themselves must be preserved because they provide data that cannot be preserved in museums, such as: the characteristics and structures of their hosting sedimentary rocks, the existence of trace fossils, data on adjacent sediment layers, and possibility of collecting new data by using techniques and methodologies not currently available or even developed. Thus, this PalaeoPark Initiative not only protects fossil-bearing sites but, perhaps more importantly, aims to "preserve the history of life on Earth to provide materials for scientific research now and in the future" (Lipps 2009: 1–2).

The Azores Geopark was created in 2010 as a non-profit association charged with the conservation and promotion of the natural heritage of the Azores with particular emphasis on the islands' geological heritage. The challenging concept of the Azores Geopark—"9 Islands, 1 Geopark"—represents the main goal of the association. By definition, a Geopark "is a territory with well-defined boundaries, which has an important Geological Heritage, coupled with a sustainable development strategy. Thus, Geoparks include a significant number of sites of geological interest which, due to their rarity or peculiarities, have scientific, educational, cultural, economic (e.g. tourism), scenic or aesthetic (e.g. landscape) value (or importance), or can be considered as geosites. These sites may also contain other points of interest and value (e.g. ecological, historical and cultural), theme parks and other related infrastructures, which should be interconnected in a network by paths and trails" (Geoparque Açores 2013). In the Azores, 121 geosites were considered important and 57 were selected as representative of the geodiversity of the archipelago. These geosites are distributed throughout the islands, as well as on seamounts and seafloor that surround them.

On Santa Maria Island, 15 geosites were classified and 5 were selected as deserving priority status for the development of geoconservation strategies and implementation of socioeconomical development. Of these five geosites, only two (Pedreira do Campo and Ponta do Castelo) contain fossils. Moreover, no Pleistocene outcrops were selected at this stage, notwithstanding the recent inclusion of the Prainha outcrop into the first Azorean Regional Natural Monument (besides Pedreira do Campo and Figueiral; DR 47 2008). In 2012, new legislation, produced by the Azores Regional Government with the scientific support of the MPB group, has protected all fossiliferous sites on Santa Maria (DR 39 2012). Today, in the entire island, no fossils may be collected and only professionals may get licenses to sample for scientific purposes.

In order to further promote the exceptional palaeontological heritage of Santa Maria Island, the MPB group proposed to the IPA the classification of Santa Maria Island as a PalaeoPark, which was accepted in late 2012. The aim of PalaeoPark Santa Maria is to emphasize in a synergetic way the outstanding palaeontological record of such a small island within the broader "umbrella" of the Azores Geopark. Thus, in addition to the thematic routes proposed by the Azores Geopark (Volcanic Caves Route, Viewpoints Route and Hiking Trails Route), the creation of the Fossils Route was presented by the MPB group to the Azores Regional Government in 2010. This project was funded and started in July 2012. The main objective of the Fossils Route is the sustainable touristic use of the palaeontological heritage of Santa Maria Island and to showcase the relatively recent concept of biogeodiversity. For this purpose, five paths were outlined on Santa Maria, which were mapped and characterised in relation to biodiversity, geodiversity and various palaeontological aspects, such as: which fossil species are the most abundant, a detailed description of the palaeoecology of the deposits crossed by the paths and perhaps the most interesting point, the individual history of each of the deposits. One of these paths is distinguished by being entirely marine. This "pedestrian" marine path, unique in the Azores Autonomous Region, consists of a tour around the island by boat, disembarking (sea conditions permitting) at exposed deposits previously selected for this purpose. Along the way, tourists have access to scientific information provided by fully accredited guides, such as the description of the various geological units that comprise the island of Santa Maria, a description of some of the geological features visible on the coast, as well as the pinpointing of spectacular pillow lava and prismatically jointed lava sections. All this is complemented by beautiful scenery and, of course, marine fossils.

The scientific certification of the guides, who have received specialised biological, palaeontological and geological training adapted to the volcanic and sedimentary sequences of Santa Maria Island, will be mandatory for the success of this PalaeoPark. Today, with the approval of the "Parque de Ilha de Santa Maria", more than ten guides work with school children, local inhabitants and tourists. The raising of public awareness to the value of the fossiliferous outcrops is being achieved by: (1) addressing the island's inhabitants with a special emphasis on local schools, whose students could attend excursions to the most significant outcrops led by MPB researchers; (2) spreading the scientific knowledge and the importance of science with scientific lectures done during the annual workshops "*Palaeontology in Atlantic Islands*" in cafes ("Clube Naval de Santa Maria" and "Central-Pub") and Science Centers ("Centro de Interpretação Ambiental Dalberto Pombo"); (3) writing popular science books related to the fossils of Santa Maria Island (Ávila 2009; Ávila and Monteiro 2009; Ávila et al. 2010; Ávila and Rodrigues 2013); (4) reports published in newspapers, national and regional magazines, as well as with television reports; (5) the recent "House of Fossils" Museum at Vila do Porto, a multimedia museum that also portraits the most important fossils discovered at Santa Maria; and (6) reaching tourists through the "Fossils Route" trails. These fossil-related trails have caught the interest of the maritime-touristic companies located on the island, as they provide an interesting alternative for tourists, when sea conditions do not allow diving or whale-watching activities.

On Santa Maria, fossils have always been part of the local people's traditional life. However, with the exception of a few more educated people, this heritage was overlooked until recently. Thanks to the efforts of the scientific community, this paradigm has changed from 2000 onward. At the present time, the fossiliferous deposits have an intrinsic value that goes far beyond the scientific one, as tourist attractions, for environmental, recreational and even artistic education (see Ávila et al. 2016b). As a consequence, a genuine development strategy for tourism based on the local palaeontology and biogeodiversity is now underway on Santa Maria Island. Perhaps the most visible result of all these scientific research/publishing efforts is evident when, while landing at Santa Maria on the local aircraft company SATA, the air hostess can be heard saying: "Dear passengers: welcome to Santa Maria, where the fossil heritage is a natural story-teller". The final surprise is reserved for the airport's baggage claim hall, where our attention is drawn to photographs of fossils decorating the floor. Ten years ago, no one spoke about the fossils of Santa Maria… today, they are a distinguishing feature of the "Island of the Sun"… or should that be the "Island of the fossils"?

Acknowledgements We thank the Direcção Regional da Ciência, Tecnologia e Comunicações (Regional Government of the Azores), Direcção Regional do Ambiente e do Mar (Regional Government of the Azores), FCT (Fundação para a Ciência e a Tecnologia) of the Portuguese government for financial support, and we also thank the Clube Naval de Santa Maria and Câmara Municipal de Vila do Porto for field assistance. We are grateful to the organizers and participants of several editions of the International Workshop "Palaeontology in Atlantic Islands" who assisted with fieldwork (2005–2012). Sérgio P. Ávila acknowledges his IF/00465/2015 research contract funded by Fundação para a Ciência e a Tecnologia (Portugal). This research received support from the SYNTHESYS Application DE-TAF-1071 "Systematics and taxonomy of the Miocene malacofauna of Santa Maria Island (Azores)" (http://www.synthesys.info/) which is financed by European Community Research Infrastructure Action under the FP7 "Capacities" Program. Ricardo Ramalho's funding was provided by

DFG-The German Research Foundation, through the project TH1530/6-1 "Investigation of ocean island uplift in the Azores Island Region", and later though his FP7-PEOPLE-2011-IOF Marie Curie Fellowship. The contribution of Jörg Habermann is based on his master thesis and Prof. Dr. André Freiwald and Prof. Dr. Karsten Haase are gratefully acknowledged for their supervision. Furthermore, Jörg Habermann and Jürgen Titschack benefitted from the financial support by the DFG project BE4459/1-1 "Magmatic and sedimentological evolution of Oceanic Plateaus: constraints from the Azores". We are grateful to the referee work of Jere Lipps and Markes Johnson, whose comments helped to clarify relevant parts of the manuscript. This work was also supported by FEDER funds through the Operational Programme for Competitiveness Factors—COMPETE and by National Funds through FCT—Foundation for Science and Technology under the UID/BIA/50027/2013 and POCI-01-0145-FEDER-006821.

References

Abdel-Monem, A. A., Fernandez, L. A., & Boone, G. M. (1975). K-Ar ages from the eastern Azores group (Santa Maria, São Miguel and the Formigas Islands). *Lithos, 8,* 247–254.

Adey, W. H. (1986). Coralline algae as indicators of sea-level. In O. van de Plasche (Ed.), *Sea-level research, a manual for the collection and evaluation of data* (pp. 229–279). Amsterdam: Free University of Amsterdam.

Amen, R. G., Neto, A. I., & Azevedo, J. M. N. (2005). Coralline-algal framework in the Quaternary of Prainha (Santa Maria island, Azores). *Revista Española de Micropaleontologia, 37,* 63–70.

Ávila, S. P. (2005). *Processos e Padrões de Dispersão e Colonização nos Rissoidae (Mollusca: Gastropoda) dos Açores* (x + 329 pp.) (Unpublished Ph.D. Thesis). Universidade dos Açores, Ponta Delgada.

Ávila, S. P. (2009). Geologia e os fósseis. In Publiçor (Eds.), *Açores. Santa Maria, a Ilha Amarela* (pp. 16–23). Ponta Delgada: Publiçor Editores.

Ávila, S. P. (2013). Unravelling the patterns and processes of evolution of marine life in oceanic islands: a global framework. In J. M. Férnández-Palacios, L. Nascimento, J. C. Hernández, S. Clemente, A. González, & J. P. Díaz-González (Eds.), *Climate change: past, present and future perspectives. A global synthesis from the Atlantic* (pp. 85–115). La Laguna: Universidad de La Laguna, Tenerife.

Ávila, S. P., Amen, R. G., Azevedo, J. M. N., Cachão, M., & García-Talavera, F. (2002). Checklist of the Pleistocene marine molluscs of Prainha and Lagoinhas (Santa Maria Island, Azores). *Açoreana, 9*(4), 343–370.

Ávila, S. P., Ávila, S. P., Silva, L., Ramalho, R., Quartau, R., Hipólito, A., et al. (2015a). A review of the MIS 5e highstand deposits from Santa Maria Island (Azores, NE Atlantic): Palaeobiodiversity, palaeoecology and palaeobiogeography. *Quaternary Science Reviews, 114,* 126–148.

Ávila, S. P., Cachão, M., Ramalho, R. S., Botelho, A. Z., Madeira, P., Rebelo, A. C., et al. (2016a). The palaeontological heritage of Santa Maria Island (Azores: NE Atlantic): a re-evaluation of geosites in GeoPark Azores and their use in geotourism. *Geoheritage, 8,* 155–171.

Ávila, S. P., Cordeiro, R., Rodrigues, A. R., Rebelo, A. C., Melo, C., Madeira, P., et al. (2015c). Fossil Mysticeti from the Pleistocene of Santa Maria Island, Azores (NE Atlantic Ocean), and the prevalence of fossil cetaceans on oceanic islands. *Palaeontologia Electronica, 18.2.27A,* 1–12. palaeo-electronica.org/content/2015/1225-oceanic-island-fossil-cetacean.

Ávila, S. P., Madeira, P., García-Talavera, F., Silva, C. M., Cachão, M., & Martins, A. M. F. (2007). *Luria lurida* (mollusca: gastropoda), a new record for the Pleistocene of Santa Maria (Azores, Portugal). *Arquipélago, 24,* 53–56.

Ávila, S. P., Madeira, P., Mendes, N., Rebelo, A. C., Medeiros, A., Gomes, C., et al. (2008a). Mass extinctions in the Azores during the last glaciation: Fact or myth? *Journal of Biogeography, 35,* 1123–1129.

Ávila, S. P., Madeira, P., Silva, C. M., Cachão, M., Landau, B., Quartau, R., et al. (2008b). Local disappearance of bivalves in the Azores during the last glaciation. *Journal of Quaternary Science, 23,* 777–785.

Ávila, S. P., Madeira, P., Zazo, C., Kroh, A., Kirby, M., Silva, C. M., et al. (2009a). Palaeocology of the Pleistocene (MIS 5.5) outcrops of Santa Maria Island (Azores) in a complex oceanic tectonic setting. *Palaeogeography, Palaeoclimatology, Palaeoecology, 274,* 18–31.

Ávila, S. P., & Martins, A. M. F. (Eds.) (2007). Palaeontology in Atlantic Islands. In *Proceedings of the First Atlantic Islands Neogene, International Congress, Açoreana* (Suppl. 5, pp. 1–172).

Ávila, S. P., Medeiros, A., Martins, A. M. F., Silva, A., Melo, C., Gomes, C., et al. (2011). *Lajes do Pico "À ban-baxe-muro"* (pp. 1–345). Ponta Delgada: Editora Publiçor.

Ávila, S. P., Melo, C., Berning, B., Cordeiro, R., Landau, B., & da Silva, C. M. (2016b). *Persististrombus coronatus* (Mollusca: Strombidae) in the early Pliocene of Santa Maria Island (Azores: NE Atlantic): Palaeoecology, palaeoclimatology and palaeobiogeographic implications on the NE Atlantic Molluscan Biogeographical Provinces. *Palaeogeography, Palaeoclimatology, Palaeoecology, 441,* 912–923.

Ávila, S. P., & Monteiro, P. (2009). *Açores, o Império dos Fósseis* (pp. 1–272). Porto: Edições Caixotim, Lda.

Ávila, S. P., Ramalho, R., Habermann, J., Quartau, R., Kroh, A., Berning, B., et al. (2015b). Palaeoecology, taphonomy, and preservation of a lower Pliocene shell bed (coquina) froma volcanic oceanic island (Santa

Maria Island, Azores). *Palaeogeography, Palaeoclimatology, Palaeoecology, 430,* 57–73.

Ávila, S. P., Ramalho, R., & Vullo, R. (2012). Systematics, palaeoecology and palaeobiogeography of the Neogene fossil sharks from the Azores (Northeast Atlantic). *Annales de Paléontologie, 98,* 167–189.

Ávila, S. P., Rebelo, A. C., Medeiros, A., Melo, C., Gomes, C., Bagaço, L., et al. (2010). *Os fósseis de Santa Maria (Açores). 1. A jazida da Prainha* (pp. 1–103). Lagoa: OVGA – Observatório Vulcanológico e Geotérmico dos Açores.

Ávila, S. P., & Rodrigues, J. A. (2013). *PaleoPark Santa Maria* (pp. 1–120). Ponta Delgada: Letras Lavadas edições.

Ávila, S. P., Silva, C. M., Schiebel, R., Cecca, F., Backeljau, T., & Martins, A. M. F. (2009b). How did they get here? Palaeobiogeography of the Pleistocene marine molluscs of the Azores. *Bulletin of the Geological Society of France, 180,* 295–307.

Barreiros, J. P., & Gadig, O. B. F. (2011). *Catálogo ilustrado dos tubarões e raias dos Açores* (pp. 1–195). Angra do Heroísmo: Instituto Açoriano de Cultura.

Berthois, L. (1950). Sur la présence d'une microfaune dans le calcaire de Santa Maria (Açores). *Açoreana, 4* (4), 277–287.

Berthois, L. (1951). Sur la présence de basses terrasses marines dans l'Archipel des Açores. *Comptes Rendus du 76éme Congrès des Sociétés Savantes à Rennes,* 101–106.

Berthois, L. (1953a). Sur la repartition granulométrique des éléments calcaires dans les sables des Açores (Portugal). *Açoreana, 5*(1), 54–63.

Berthois, L. (1953b). Terrasses marines d'altitude +5 à +8 m dans l'Archipel des Açores. *Açoreana, 5*(1), 64–70.

Berthois, L. (1953c). Contribution à l'étude lithologique de l'Archipel des Açores. *Comunicações dos Serviços Geológicos de Portugal, 34,* 1–198.

Bitner, M. A., & Dulai, A. (2004). Revision of Miocene brachiopods of the Hungarian Natural History Museum, with special regard to the Meznerics collection. *Fragmenta Palaeontologica Hungarica, 22,* 69–82.

Boid, E. (1835). *A description of the Azores or Western Islands* (pp. 1–373). London: Edward Churton.

Bosence, D. W. J. (1983). Coralline algal reef frameworks. *Journal of the Geological Society of London, 140,* 365–376.

Bosence, D. W. J. (1984). Construction and preservation of two Recent coralline algal reefs, St. Croix, Caribbean. *Palaeontology, 27,* 549–574.

Braga, J. C., & Aguirre, J. (2001). Coralline algal assemblages in upper Neogene reef and temperate carbonates in southern Spain. *Palaeogeography, Palaeoclimatology, Palaeoecology, 175,* 27–41.

Bressan, G., & Babbini, L. (2003). Biodiversità marina delle coste Italiane: Corallinales del Mar Mediterraneo: guida all deteminazione. *Biologia Marina Mediterrânea, 10*(Suppl. 2), 1–237.

Bronn, H. G. (1860). Die fossilen Reste von Santa Maria, der südlichsten der Azorischen Inseln. In G. Hartung (Ed.), *Die Azoren in ihrer äusseren Erscheinung und nach ihrer geognostischen Natur* (pp. 1–350). Leipzig: Verlag von Wilhelm Engelmann.

Brotchie, J. F., & Silvester, R. (1969). On Crustal Flexure. *Journal of Geophysical Research (Solid Earth), 74* (22), 5240–5252.

Cachão, M., Madeira, J., Silva, C. M., Azevedo, J. M. N., Cruz, A. P., Garcia, C., et al. (2003). Pedreira do Campo (Santa Maria, Açores): monumento natural. Actas do VI Congresso Nacional de Geologia. *Ciências da Terra (UNL), 5,* 120–123.

Calado, H., Ávila, S. P., & Madeira, P. (2007). The Coastal Zone Management Plan of Santa Maria as a chance for fossiliferous outcrops management. In S. P. Ávila & A. M. F. Martins (Eds.), *Proceedings of the "1st Atlantic Islands Neogene", International Congress.* Ponta Delgada, 12–14 June 2006. *Açoreana, Suplemento 5,* 162–172.

Callapez, P., & Soares, A. F. (2000). Late Quaternary marine mollusks from Santa Maria (Azores); paleoecologic and paleobiogeographic considerations. *Ciências da Terra (UNL), 14,* 313–322.

Cotter, J. C. B. (1888–1892). Noticia de alguns fosseis terciarios da Ilha de Santa Maria no archipelago dos Açores. *Communicações da Commissão dos Trabalhos Geológicos de Portugal, 2,* 255–287.

Crough, S. (1978). Thermal origin of mid-plate hot-spot swells. *Geophysical Journal International, 55*(2), 451–469.

Bedemar, V. (1837). *Resumo de observações geologicas feitas em uma viagem ás ilhas da Madeira Porto Santo e Açores nos annos de 1835 e 1836* (pp. 1–25). Lisboa: Impressão de Galhardo Irmãos.

Dietz, R., & Menard, H. (1953). Hawaiian swell, deep, and arch, and subsidence of the Hawaiian Islands. *The Journal of Geology, 61*(2), 99–113.

Donovan, S. K., & Gunter, G. C. (2001). Fossil sharks from Jamaica. *Bulletin of the Mizunami Fossil Museum, 28,* 211–215.

DR 47. (2008). Decreto Legislativo Regional 47/2008/A de 7 de Novembro. Parque Natural da Ilha de Santa Maria. *Diário da República,* 1.ª série, nº 217 – 7 de Novembro de 2008, 7787–7801.

DR 39. (2012). Decreto Legislativo Regional n.º 39/2012/A. Primeira alteração ao Decreto Legislativo Regional n.º 47/2008/A, de 7 de novembro (Parque Natural da Ilha de Santa Maria). *Diário da República,* 1.ª série, nº 182 – 19 de setembro de 2012, 5282–5310.

Duggen, S., Olgun, N., Croot, P., Hoffman, L., Dietze, H., Delmelle, P., et al. (2010). The role of airborne volcanic ash for the surface ocean biogeochemical iron-cycle: A review. *Biogeosciences, 7,* 827–844.

Dumas, S., & Arnott, R. W. C. (2006). Origin of hummocky and swaley cross-stratification—The controlling influence of unidirectional current strength and aggradation rate. *Geology, 34,* 1073–1076.

Estevens, M., & Ávila, S. P. (2007). Fossil whales from the Azores. In S. P. Ávila & A. M. F. Martins (Eds.), *Proceedings of the "1st Atlantic Islands Neogene", International Congress,* Ponta Delgada, 12–14 June 2006. *Açoreana, Suplemento, 5,* 140–161.

Feraud, G., Kaneoka, I., & Allegre, C. J. (1980). K-Ar ages and stress pattern in the Azores: Geodynamic implications. *Earth and Planetary Science Letters, 46,* 275–286.
Ferreira, O. V. (1952). Os pectinídeos do Miocénico da ilha de Santa Maria (Açores). *Revista da Faculdade de Ciências de Lisboa, 2ª Série, C, 2*(2), 243–258.
Ferreira, O. V. (1955). A fauna Miocénica da ilha de Santa Maria. *Comunicações dos Serviços Geológicos de Portugal, 36,* 9–44.
Ferreira, O. V. (1961). Afloramentos de calcário miocénico da Ilha de Santa Maria (Açores). *Comunicações dos Serviços Geológicos de Portugal, 45,* 467–478.
Fisher, R. V., & Schmincke, H.-U. (1984). *Pyroclastic rocks* (pp. 1–427). Berlin: Springer.
Fortes, M. D. (1991). Seagrass-mangrove ecosystems managment: A key to marine coastal conservation in the ASEAN region. *Marine Pollution Bulletin, 22,* 113–116.
Frogner, P., Gislason, S. R., & O'Skarsson, N. (2001). Fertilizing potential of volcanic ash in ocean surface water. *Geology, 29,* 486–490.
Frutuoso, G. (1983). *Livro Terceiro das Saudades da Terra* (pp. 1–300). Ponta Delgada: Instituto Cultural de Ponta Delgada.
García-Talavera, F. (1990). Fauna tropical en el Neotirreniense de Santa Maria (I. Azores). *Lavori della Società Italiana di Malacologia, 23,* 439–443.
Geoparque Açores. (2013). http://www.azoresgeopark.com/. Accessed 26 February 2013.
Habermann, J. (2010). *Thesis Mapping—Santa Maria, Azores. Area: Vila do Porto – Almagreira – Praia* (Unpublished Diplom mapping, 84 pp). GeoZentrum Nordbayern, University of Erlangen-Nürnberg.
Habermann, J. (2011). *Gigantopecten latissimus (Brocchi, 1814). Microstructure and stable isotope sclerochronology of an Upper Miocene Gigantopecten shell (Bivalvia: Pectinidae) from Santa Maria Island (Azores, Portugal)* (Unpublished Diplom thesis, xi + 107 pp). GeoZentrum Nordbayern, University of Erlangen-Nürnberg.
Hartung, G. (1860). *Die Azoren in ihrer äusseren Erscheinung und nach ihrer geognostischen Natur* (pp. 1–350). Leipzig: Verlag von Wilhelm Engelmann.
Hartung, G. (1864). *Geologische Beschreibung der Inseln Madeira und Porto Santo* (pp. 1–299). Leipzig: Verlag von Wilhelm Engelmann.
Heikoop, J. M., Tsujita, C. J., Heikoop, C. E., Risk, M., & Dickin, A. P. (1996a). Effects of volcanic ashfall recorded in ancient marine benthic communities: Comparison of a nearshore and an offshore environment. *Lethaia, 29,* 125–139.
Heikoop, J. M., Tsujita, C. J., Risk, M. J., & Tomascik, T. (1996b). Corals as proxy recorders of volcanic activity: Evidence from Banda Api, Indonesia. *Palaios, 11,* 286–292.
Hofrichter, R. (2001). *Das Mittelmeer: Fauna, Flora, Ökologie – Allgemeiner Teil, 1* (p. 607). Heidelberg: Spektrum.
Huppert, K. L., Royden, L. H., & Perron, J. T. (2015). Dominant influence of volcanic loading on vertical motions of the Hawaiian Islands. *Earth and Planetary Science Letters, 418,* 149–171.
Iturralde-Vinent, M., Hubbell, G., & Rojas, R. (1996). Catalogue of Cuban fossil Elasmobranchii (Paleocene to Pliocene) and paleogeographic implications of their Lower to Middle Miocene occurrence. *Journal of the Geological Society of Jamaica, 31,* 7–21.
Janssen, A. W., Kroh, A., & Ávila, S. P. (2008). Early Pliocene heteropods and pteropods (Mollusca, Gastropoda) from Santa Maria Island (Azores, Portugal): Systematics and biostratigraphic implications. *Acta Geologica Polonica, 58,* 355–369.
Johnson, M. E., Baarli, B. G., Santos, A., & Mayoral, E. (2012). Ichnofacies and microbial build-ups on Late Miocene rocky shores from Menorca (Balearic Islands), Spain. *Facies, 57,* 255–265.
Johnson, M. E., Uchman, A., Costa, P. J. M., Ramalho, R. S., & Ávila, S. P. (2017). Intense hurricane transport sand onshore: example from the Pliocene Malbusca section on Santa Maria Island (Azores, Portugal). *Marine Geology, 385,* 244–249.
Kirby, M. X., Jones D. S., & Ávila, S. P. (2007). Neogene shallow-marine paleoenvironments and preliminary Strontium isotope chronostratigraphy of Santa Maria Island, Azores. In S. P. Ávila & A. M. F. Martins (Eds.), *Proceedings of the "1st Atlantic Islands Neogene", International Congress,* Ponta Delgada, 12–14 June 2006. *Açoreana, Suplemento 5,* 112–125.
Klügel, A., Hansteen, T., & Galipp, K. (2005). Magma storage and underplating beneath Cumbre Vieja volcano, La Palma (Canary Islands). *Earth and Planetary Science Letters, 236*(1–2), 211–226.
Klügel, A., Longpré, M.-A., García-Cañada, L., & Stix, J. (2015). Deep intrusions, lateral magma transport and related uplift at ocean island volcanoes. *Earth and Planetary Science Letters, 431,* 140–149.
Krejci-Graf, K., Frechen, J., Wetzel, W., & Colom, G. (1958). Gesteine und Fossilien von den Azoren. *Senckenbergiana Lethaea, 39*(5/6), 303–351.
Kroh, A., Bitner, M. A., & Ávila, S. P. (2008). *Novocrania turbinata* (Brachiopoda) from the Early Pliocene of the Azores (Portugal). *Acta Geologica Polonica, 58,* 473–478.
Lipps, J. H. (2009). PaleoParks: Our paleontological heritage protected and conserved in the field worldwide. In: J. H. Lipps, & B. R. C. Granier (Eds.), *PaleoParks—The protection and conservation of fossil sites worldwide.* Brest: Carnets de Géologie/Notebooks on Geology. http://paleopolis.rediris.es/cg/CG2009_BOOK_03/CG2009_BOOK_03_Chapter01.pdf.
Madeira, P., & Ávila, S. P. (2006). Jazidas Fósseis. In H. Calado, J. Porteiro, M. Pereira, & A. Z. Botelho (Coordenação Científica). *LITOSOST: Gestão Sustentável do Desenvolvimento Social, Económico e ecológico das Áreas Litorais da Macaronésia* (pp. 74–76). Plano de ordenamento da Orla Costeira – Santa Maria. Fase I – Caracterização e Diagnóstico

Madeira, P., Kroh, A., Cordeiro, R., Meireles, R., & Ávila, S. P. (2011). The fossil echinoids of Santa Maria Island, Azores (Northern Atlantic Ocean). *Acta Geologica Polonica, 61,* 243–264.

Madeira, P., Kroh, A., Martins, A. M. F., & Ávila, S. P. (2007). The marine fossils from Santa Maria Island (Azores, Portugal): an historical overview. In S. P. Ávila & A. M. F. Martins (Eds.), *Proceedings of the "1st Atlantic Islands Neogene", International Congress*, Ponta Delgada, 12–14 June 2006. *Açoreana, Suplemento 5,* 59–73.

Madeira, J., Mata, J., Mourão, C., Silveira, A. B., Martins, S., Ramalho, R., et al. (2010). Volcano-stratigraphic and structural evolution of Brava Island (Cape Verde) from 40Ar/39Ar, U/Th and field constraints. *Journal of Volcanology and Geothermal Research, 196,* 219–235.

Mayer, K. (1864). *Die Tertiär-Fauna der Azoren und Madeiren. Systematisches Verzeichniss der fossilen Reste von Madeira, Porto Santo und Santa Maria nebst Beschreibung der neuen Arten,* VI + 107 pp. Zürich.

McNutt, M., & Menard, H. W. (1978). Lithospheric exure and uplifted Atolls. *Journal of Geophysical Research (Solid Earth), 83*(B3), 1206–1212.

Meireles, R. P., Costanza, F., Gliozzi, E., Pimentel, A., Zanon, V., & Ávila, S. P. (2012). Late Miocene marine ostracods from Santa Maria Island, Azores (NE Atlantic): palaeoecology and palaeobiogeography. *Révue de Micropaléontologie, 55,* 133–148.

Meireles, R. P., Quartau, R., Ramalho, R., Rebelo, A. C., Madeira, J., Zanon, V., et al. (2013). Depositional processes on oceanic island shelves—Evidence from storm-generated Neogene deposits from the Mid North Atlantic. *Sedimentology (accepted article).* https://doi.org/10.1111/sed.12055.

Menéndez, I., Silva, P., Martin-Betancor, M., Perez-Torrado, F., Guillou, H., & Scaillet, S. (2008). Fluvial dissection, isostatic uplift, and geomorphological evolution of volcanic islands (Gran Canaria, Canary Islands, Spain). *Geomorphology, 102*(1), 189–203.

Miller, K. G., Kominz, M. A., Browning, J. V., Wright, J. D., Mountain, G. S., Katz, M. E., et al. (2005). The phanerozoic record of global sea-level change. *Science, 310,* 1293–1298.

Mitchell-Thomé, R. C. (1976). *Geology of the middle Atlantic Islands* (351 pp.). Berlin: Gebrüder Borntraeger.

Morelet, A. (1860). *Notice sur l'Histoire Naturelle des Açores suivie d'une description des Mollusques terrestres de cet Archipel* (pp. 1–216). Paris: J.-B. Baillière et Fils.

Morgan, J., Morgan, W., & Price, E. (1995). Hotspot melting generates both hotspot volcanism and a hotspot swell. *Journal of Geophysical Research (Solid Earth), 100*(B5), 8045–8062.

Neto, A. I. (1994). Checklist of the benthic marine macroalgae of the Azores. *Arquipélago, 12A,* 15–34.

Nieves-Rivera, A. M., Ruizyantín, M., & Gottfried, M. D. (2003). New record of the lamnid shark Carcharodon megalodon from the Middle Miocene of Puerto Rico. *Caribbean Journal of Science, 39,* 223–227.

Nunes, J. C., Lima E. A. & Medeiros, S. (2007). Os Açores, ilhas de geodiversidade: o contributo da ilha de Santa Maria. In S. P. Ávila & A. M. F. Martins (Eds.), *Proceedings of the "1st Atlantic Islands Neogene", International Congress,* Ponta Delgada, 12–14 June 2006. *Açoreana, Suplemento 5,* 74–111.

Pérès, J., & Picard, J. M. (1964). Nouveau manuel de bionomie benthique de la mer Méditerranée. *Recueil Travaux Station Marine Endoume, 31*(47), 1–131.

Portell, R. W., Hubbel, G., Donovan, S. K., Green, J. L., Harper, D. A. T., & Pickerill, R. (2008). Miocene sharks in the Kendeace and Grand Bay formations of Carriacou, The Grenadines, Lesser Antilles. *Caribbean Journal of Science, 44,* 279–286.

Quartau, R., Hipólito, A., Romagnoli, C., Casalbore, D., Madeira, J., Tempera, F., et al. (2014). The morphology of insular shelves as a key for understanding the geological evolution of volcanic islands: Insights from Terceira Island (Azores). *Geochemistry, Geophysics, Geosystems, 15,* 1801–1826.

Quartau, R., Madeira, J., Mitchell, N. C., Tempera, F., Silva, P. F., & Brandão, F. (2015). The insular shelves of the Faial-Pico Ridge (Azores archipelago): A morphological record of its evolution. *Geochemistry, Geophysics, Geosystems, 16,* 1401–1420.

Quartau, R., Tempera, F., Mitchell, N. C., Pinheiro, L. M., Duarte, H., Brito, P. O., et al. (2012). Morphology of the Faial Island shelf (Azores): The interplay between volcanic, erosional, depositional, tectonic and mass-wasting processes. *Geochemistry, Geophysics, Geosystems, 13,* Q04012. https://doi.org/10.1029/2011GC003987.

Ralph, P. J., Tomasko, D., Moore, K., Seddon, S., & Macinnis-Ng, C. M. O. (2006). Human impacts on seagrasses: eutrophication, sedimentation and contamination. In A. W. D. Larkum, R. J. Orth, & C. M. Duarte (Eds.), *Seagrasses: Biology, ecology and conservation* (pp. 567–593). Dordrecht: Springer.

Ramalho, R. (2011). *Building the Cape Verde Islands.* Berlin/Heidelberg: Springer.

Ramalho, R. S., Brum da Silveira, A., Fonseca, P., Madeira, J., Cosca, M., Cachão, M., et al. (2015). The emergence of volcanic oceanic islands on a slow-moving plate: the example of Madeira Island, NE Atlantic. *Geochemistry, Geophysics, Geosystems, 16,* 522–537.

Ramalho, R. S., Helffrich, G., Cosca, M., Vance, D., Hoffmann, D., & Schmidt, D. N. (2010a). Vertical movements of ocean island volcanoes: Insights from a stationary plate environment. *Marine Geology, 275,* 84–95.

Ramalho, R. S., Helffrich, G., Cosca, M., Vance, D., Hoffmann, D., & Schmidt, D. N. (2010b). Episodic swell growth inferred from variable uplift of the Cape Verde hot spot islands. *Nature Geoscience, 3,* 774–777.

Ramalho, R. S., Helffrich, G., Madeira, J., Cosca, M., Thomas, C., Quartau, R., et al. (2017). Emergence and evolution of Santa Maria Island (Azores)—The conundrum of uplifted islands revisited. *Geological Society of America Bulletin, 129*(3–4), 372–390.

Ramalho, R. S., Helffrich, G., Schmidt, D., & Vance, D. (2010c). Tracers of uplift and subsidence in the Cape Verde Archipelago. *Journal of the Geological Society of London, 167,* 519–538.

Ramalho, R., Quartau, R., Trenhaile, A.S., Mitchell, N.C., WOODROFFE, C.D., & Ávila, S.P. (2013). Coastal evolution in oceanic islands: A complex interplay between volcanism, erosion, sedimentation and biogenic production. *Earth Science Reviews, 127,* 140–170. http://doi.org/10.1016/j.earscirev.2013.10.007.

Rebelo, A. C., Meireles, R. P., Barbin, V., Neto, A. I., Melo, C., & Ávila, S. P. (2016a). Diagenetic history of lower Pliocene rhodoliths of the Azores archipelago (NE Atlantic): application of cathodoluminescence techniques. *Micron, 80,* 112–121.

Rebelo, A. C., Rasser, M. W., Kroh, A., Johnson, M. E., Ramalho, R. S., Melo, C., et al. (2016b). Rocking around a volcanic island shelf: Pliocene rhodolith beds from Malbusca, Santa Maria Island (Azores, NE Atlantic). *Facies, 62*(3), 1–31.

Rebelo, A. C., Rasser, M. W., Riosmena-Rodríguez, R., Neto, A. I., & Ávila, S. P. (2014). Rhodolith forming coralline algae in the Upper Miocene of Santa Maria Island (Azores, NE Atlantic): a critical evaluation. *Phytotaxa, 190*(1), 370–382.

Reiss, W. (1862). *Mittheilungen über die tertiären Schichten von Santa Maria, der südlichsten der Azoren, und ihre organischen Einschlüsse* (pp. 1–22). Geognosie, Geologie und Petrefactenkunde: Neues Jarhrbuch für Mineralogie.

Reuter, M., & Piller, W. E. (2011). Volcaniclastic events in coral reef and seagrass environments: Evidence for disturbance and recovery (Middle Miocene, Styrian Basin, Austria). *Coral Reefs, 30,* 889–899.

Ruggiero, E. T. (1994). Neogene Salento brachiopod palaeocommunities. *Bollettino della Società Paleontologica Italiana, 33,* 197–213.

Santos, A. G., Mayoral, E., Dumont, C. P., da Silva, C. M., Ávila, S. P., Baarli, B. G., et al. (2015). Role of environmental change in rock-boring echinoid trace fossils. *Palaeogeography, Palaeoclimatology, Palaeoecology, 432,* 1–14.

Santos, A. G., Mayoral, E., Johnson, M. E., Baarli, B. G., Silva, C. M., Cachão, M., et al. (2012). Basalt mounds and adjacent depressions attract contrasting biofacies on a volcanically active Middle Miocene shoreline (Porto Santo, Madeira Archipelago, Portugal). *Facies, 58,* 573–585.

Santos, R. S., Porteiro, F. M., & Barreiros, J. P. (1997). Marine fishes of the Azores: An annotated checklist and bibliography. *Arquipélago, 1,* 1–242.

Serralheiro, A. (1976). *A Geologia da Ilha de Santiago (Cabo Verde)* (Unpublished Ph.D. thesis). Faculdade de Ciências da Universidade de Lisboa.

Serralheiro, A. (2003). A geologia da ilha de Santa Maria, Açores. *Açoreana, 10,* 141–192.

Serralheiro, A., Alves, C. A. M., Forjaz, V. H., & Rodrigues, E. B. (1987). Carta Vulcanológica dos Açores - Ilha de Santa Maria. Centro de Vulcanologia do I.N.I.C.

Serralheiro, A., & Madeira, J. (1990). Stratigraphy and geochronology of Santa Maria Island (Azores). *Açoreana, 7,* 575–592.

Serralheiro, A., Madeira, J., Alves, C. A. M., & Fernandez-Rubio, R. (1990). *Stratigraphy and geochronology of Santa Maria Island* (pp. 357–376). Azores: Livro de Homenagem ao Prof. Carlos Romariz. Departamento de Geologia da Faculdade de Ciências da Universidade de Lisboa.

Short, F. T., & Wyllie-Echeverria, S. (1996). Natural and human-induced disturbance of seagrasses. *Environmental Conservation, 23,* 17–27.

Sibrant, A.L.R., Hildenbrand, A., Marques, F.O., & Costa, A.C.G. (2015). Volcano-tectonic evolution of the Santa Maria Island (Azores): Implications for paleostressevolution at the western Eurasia-Nubia plate boundary: *Journal of Volcanology and Geothermal Research, 291,* 49–62.

Sleep, N. (1990). Hotspots and mantle plumes: Some phenomenology. *Journal of Geophysical Research (Solid Earth), 95*(B5), 6715–6736.

Stein, C., & Stein, S. (1992). A model for the global variation in oceanic depth and heat flow with lithospheric age. *Nature, 359*(6391), 123–129.

Stöckhert, F. (2011a) *Kartierung des Gebietes zwischen Praia und Malbusca auf Santa Maria, Azoren* (Unpublished master mapping thesis). GeoZentrum Nordbayern, University of Erlangen-Nürnberg.

Stöckhert, F. (2011b) *Geochemical evolution of an intraplate volcanic island using the example of Santa Maria – Azores* (Unpublished master thesis). GeoZentrum Nordbayern, University of Erlangen-Nürnberg.

Teixeira, C. (1950). Notas sobre a geologia das Ilhas Atlântidas. *Anais da Faculdade de Ciências do Porto, 23*(3–4), 193–233.

Thorton, S. E., Pilkey, O. H., & Lynts, G. W. (1978). A lagoonal crustose coralline algal microridge: Bahiret el Bibane, Tunisia. *Journal of Sedimentary Petrology, 48*(3), 743–750.

Titschack, J., Bromley, R. G., & Freiwald, A. (2005). Plio-Pleistocene cliff-bound, wedge-shaped, warm-temperate carbonate deposits from Rhodes (Greece): sedimentology and facies. *Sedimentary Geology, 180,* 29–56.

Titschack, J., & Freiwald, A. (2005). Growth, deposition, and facies of Pleistocene bathyal coral communities from Rhodes, Greece. In A. Freiwald & J. M. Roberts (Eds.), *Cold-Water Corals and Ecosystems* (pp. 41–59). Heidelberg: Springer.

Titschack, J., Joseph, N., Fietzke, J., Freiwald, A., & Bromley, R. G. (2013). Record of a tectonically-controlled regression captured by changes in carbonate skeletal associations on a structured island shelf (mid-Pleistocene, Rhodes, Greece). *Sedimentary Geology, 283,* 15–33.

Uchman, A., Johnson, M., Rebelo, A. C., Melo, C., Cordeiro, R., Ramalho, R. S., et al. (2016). Vertically-oriented trace fossil *Macaronichnus segregatis* from Neogene of Santa Maria Island (Azores; NE Atlantic) records vertical fluctuations of the coastal groundwater mixing zone on a small oceanic island. *Geobios, 49,* 229–241.

Walcott, R. (1970). Flexure of the lithosphere at Hawaii. *Tectonophysics, 9*(5), 435–446.

Watts, A., & ten Brink, U. (1989). Crustal structure, exure, and subsidence history of the Hawaiian Islands. *Journal of Geophysical Research (Solid Earth), 94* (B8), 10473–10500.

Wilson, M. E. J., & Lokier, S. W. (2002). Siliciclastic and volcaniclastic influences on equatorial carbonates: insights from the Neogene of Indonesia. *Sedimentology, 49,* 583–601.

Winkelmann, K., Buckeridge, J. S., Costa, A. C., Dionísio, M. A. M., Medeiros, A., Cachão, M., et al. (2010). *Zullobalanus santamariaensis* sp. nov. a new late Miocene barnacle species of the family Archeobalanidae (Cirripedia: Thoracica), from the Azores. *Zootaxa, 2680,* 33–44.

Winterer, E. L., & Sarti, M. (1994). Neptunian dykes and associated features in southern Spain: Mechanics of formation and tectonic implications. *Sedimentology, 41,* 1109–1132.

Zbyszewski, G., & d'Almeida, F. M. (1950). Os peixes miocénicos portugueses. *Comunicações dos Serviços Geológicos de Portugal, 31,* 309–423.

Zbyszewski, G., & Ferreira, O. D. V. (1961). La faune marine des basses plages quaternaires de Praia et Prainha dans l'ile de Santa Maria (Açores). *Comunicações dos Serviços Geológicos de Portugal, 45,* 467–478.

Zbyszewski, G., & Ferreira, O. D. V. (1962a). Étude géologique de l'île de Santa Maria (Açores). *Comunicações dos Serviços Geológicos de Portugal, 46,* 209–245.

Zbyszewski, G., & Ferreira, O. D. V. (1962b). La faune Miocène de l'île de Santa Maria (Açores). *Comunicações dos Serviços Geológicos de Portugal, 46,* 247–289.

Zbyszewski, G., Ferreira, O. D. V., & de Assunção, C. T. (1961). *Carta Geológica de Portugal na escala 1/50 000. Notícia explicativa da folha de Ilha de Santa Maria (Açores).* Lisboa: Serviços Geológicos de Portugal.

Petrology of the Azores Islands

Patricia Larrea, Zilda França, Elisabeth Widom
and Marceliano Lago

Abstract

This chapter presents an overview of the petrology of the Azores Archipelago, based on a review of the published literature on the mineralogy, petrology and major element geochemistry of the Azores islands and Formigas islets. In this chapter, we describe the mineralogy, petrology and major element chemistry of xenoliths/enclaves including peridotites, gabbros and syenites, and their roles in the magmatic evolution of the islands in which they occur. Where sufficient temporal data are available, we further describe the petrogenesis of the islands within a geochronological context. This synthesis and comparison of the petrology of each island is further combined with new modeling results for depth of melting and melt evolution paths to better understand the origin of the distinct petrologic characteristics of individual islands and the archipelago as a whole.

Keywords

Azores · Volcanic rocks · Petrology
Xenoliths · Fractional crystallization

1 Introduction

The Azores Archipelago is located in the Atlantic Ocean between ~37°–40°N latitude and 25°–31° W longitude. The Archipelago spans a width of ~600 km with a WNW-ESE orientation, and comprises nine volcanic ocean islands, the Formigas islet reef and several seamounts built on the Azores submarine plateau (Fig. 1 in Beier et al., Chapter "Melting and Mantle Sources in the Azores"). It is geographically divided into three groups: (i) the Eastern Group, formed by Santa Maria and São Miguel islands, (ii) the Central Group, which includes Terceira, Graciosa, Faial, Pico and São Jorge islands, and (iii) the Western Group, comprising two islands, Flores and Corvo.

Starting as early as the mid-fifteenth century, there has been interest in the volcanism of the

P. Larrea (✉)
School of Geological Sciences and Engineering, Yachay Tech, Urcuquí, Ecuador
e-mail: plarreamarquez@gmail.com

P. Larrea · E. Widom
Department of Geology and Environmental Earth Science, Miami University, Oxford, OH, USA

Z. França
Geosciences Department, Azores University, São Miguel, Portugal

M. Lago
Department of Earth Sciences, Faculty of Sciences, Universidad de Zaragoza, Zaragoza, Spain

U. Kueppers and C. Beier (eds.), *Volcanoes of the Azores*, Active Volcanoes of the World, https://doi.org/10.1007/978-3-642-32226-6_10

Azores Archipelago. Gaspar Fructuoso (XVI century), a columnist, historian, humanist and researcher of international renown, was the first to bring to the forefront various aspects of the geology of the Azores Archipelago. His work “Saudades da Terra” is considered to be the first publication describing the rocks of the archipelago, and including observations on the mineralogy and petrology (Fructuoso 1583). Subsequently, many other important naturalists, including Webster (1821), Buch (1825), Bedemar (1837), Hartung (1860), Fouqué (1873), Darwin (1844), Mügge (1883), Castro (1888) and Lacroix (1893) contributed to the literary heritage and scientific understanding of the origin and evolution of the archipelago. In the twentieth century, Esenwein (1930) carried out the first geochemical studies on rocks of the archipelago. Several studies focusing on the petrography, mineralogy, and major element chemistry of the islands followed thereafter, including those of Berthois (1953), Jeremine (1957), de Assunção (1961), de Assunção and Canilho (1970), Girod and Lefèvre (1972), Schmincke and Weibel (1972) and Schmincke (1973). Taken together, these studies documented characteristics of the Azores islands that are unusual compared to many other ocean islands, including their strongly alkaline and in some cases highly potassic character, the significant volume fraction of highly evolved rocks, and the diversity of the evolved rocks, which include both silica saturated and silica undersaturated rock types (Ridley et al. 1974). White et al. (1979) subsequently published a key archipelago-wide study of the petrology and geochemistry of the Azores islands, which demonstrated for the first time, based on petrography, whole rock geochemistry and isotopic analyses, the complexity and variation among the islands in their mantle sources, primitive magma compositions and magmatic evolution trends.

Since these early seminal studies, and continuing to the present day, modern analytical tools have been applied extensively towards documenting the mineralogy, petrology and geochemistry of the volcanic rocks of the archipelago. Many of these studies have been detailed investigations focusing on single eruptions, individual eruptive centres, or single islands. Some islands have been studied significantly more thoroughly than others, and few of the recent studies have investigated the archipelago as a whole. The most extensive documentation is available for the islands of São Jorge, Graciosa, Faial, Pico and Corvo. São Miguel and Terceira have also been studied extensively, but investigations have mainly focused on individual volcanic centres, and the petrology of the older volcanic centres is less well documented. The petrologic development of Santa Maria and Flores, as well as the Formigas Islets, are the least well documented.

In this chapter, we present an overview of the petrology of the Azores Archipelago, based on a review of the published literature on the mineralogy, petrology and major element geochemistry of the Azores islands and Formigas islets. We also describe the mineralogy, petrology and major element chemistry of xenoliths/enclaves including peridotites, gabbros and syenites, and their roles in the magmatic evolution of the islands in which they occur. Where sufficient temporal data are available, we further describe the petrogenesis of the islands within a geochronological context. This synthesis and comparison of the petrology of each island is further combined with new modelling results for depth of melting and melt evolution paths to better understand the origin of the distinct petrologic characteristics of individual islands and the archipelago as a whole. However, there is a need for future studies that focus comprehensively on the Archipelago petrology and petrogenesis through time, to fully evaluate several outstanding and fundamental archipelago-wide questions that will be touched on in the discussion.

2 Petrology of the Azores Islands

2.1 Santa Maria Island

Santa Maria is the southern- and eastern-most island of the Azores Archipelago, forming part of the Eastern Group and located at ~36° 58′N latitude and 25° 5′W longitude, ~100 km south of São Miguel (Fig. 1 in Beier et al., Chapter "Melting and Mantle Sources in the Azores").

Santa Maria is the oldest island of the Azores Archipelago with the first subaerial volcanic deposits considered Late Miocene to Pliocene in age (Abdel-Monem et al. 1968, 1975; Feraud et al. 1980, 1981; Salgueiro 1991). The first references to the volcanostratigraphy (Friedlander 1929; Agostinho 1937) mentioned the presence of a sedimentary formation separating two volcanic sequences, which led to the publication of the first geological map of the island (scale 1:50,000) based exclusively on lithological criteria (Zbyszewski et al. 1961b; Zbyszewski and Ferreira 1962). The subsequent geological map (1:15,000) was produced based on stratigraphic criteria (Serralheiro et al. 1987; Serralheiro and Madeira 1990, 1993; Serralheiro 2003). Serralheiro (2003) defines two main sequences comprising seven stratigraphic units separated by a major stratigraphic unconformity.

The Lower sequence corresponds to a major basaltic shield volcano that forms the western part of Santa Maria and comprises the Cabrestantes, Porto and Anjos Units:

(1) The Cabrestantes Formation is composed of basaltic hydromagmatic tuffs from the remains of a surtseyan cone cropping out near the NW littoral, and probably related to the emerging phase of the shield volcano.
(2) The Porto Formation was formed by two hawaiian-strombolian cinder cones exposed in the northern and southeastern coast.
(3) The Anjos Complex includes a thick sequence of subaerial basaltic lava flows covering the oldest units, usually strongly weathered and separated by thin and scarce pyroclastic layers and paleosols.
    Following the formation of this early Lower sequence, there was a period of reduced volcanic activity and intense erosion, during which abundant terrestrial and marine clastic sediments were produced.
    The Upper sequence comprises the Touril, Facho—Pico Alto, Feteiras and Pleistocene—Holocene units:
(4) The Touril Complex comprises subaerial conglomerates (lahars), and fossiliferous marine conglomerates and calcarenites. Rare subaerial lavas are intercalated with the sediments, and towards the top of the unit and to the east, submarine lava flows and pyroclastic deposits become more abundant.
(5) The Facho—Pico Alto unit includes a NNW-SSE elongated shield volcano, currently represented by the Pico Alto ridge on the eastern half of the island. The volcanic products are submarine and composed of basaltic sheet flows, pillow lavas, pillow breccias and hyaloclastites. Fossiliferous marine sediment lenses and conglomerates of a likely lahar origin are frequently intercalated in the volcanic sequences.
(6) The Feteira Formation represents the latest of the volcanic manifestations on the island. It comprises a set of highly degraded scoria cones, which produced abundant pyroclast deposits that cover the entire island.
(7) The Pleistocene—Holocene Unit is exclusively sedimentary and is composed of conglomerates, sandstones, clays, limestones, calcarenites, sands and aluviall deposits.

Santa Maria has not been extensively studied from a petrological point of view. Anderson (1983) and Rodrigues et al. (1985), together with the petrologic studies by Zbyszewski and Ferreira (1962) and Zbyszewski et al. (1961b) are the only works dealing exclusively with the petrology and geochemistry of the island, although the high alkalinity of Santa Maria lavas was first noted by White et al. (1979). The lavas belong to the sodic series and, based on the TAS classification of Le Bas et al. (1986) (Fig. 1), range from picrobasalt to mugearites with basanites being the most abundant rocks on the island.

According to Anderson (1983), the rocks from Santa Maria are moderately to coarsely phyric, with most phenocrysts larger than 0.2 mm.

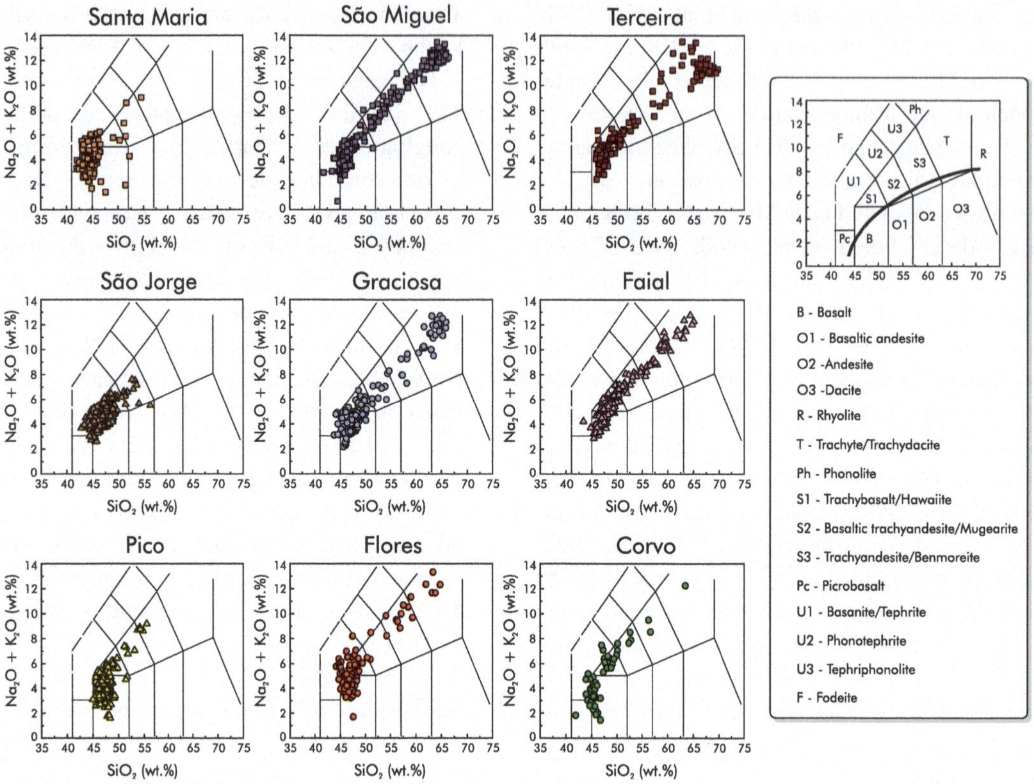

Fig. 1 Total Alkalis versus Silica (TAS) diagram after Le Bas et al. (1986) for Azores island lava flows. Published data from: **Santa Maria**: Esenwein (1930), Berthois (1953), Schmincke and Staudigel (1976), Furnes and Fridleifsson (1978), White et al. (1979), Feraud et al. (1981), Madeira (1986), Beier et al. (2012a, b) **São Miguel**: Esenwein (1930), Berthois (1953), Girod and Lefèvre (1972), Schmincke and Weibel (1972), White et al. (1979), Fetter (1981), Storey (1982), Fidczuk (1984), Wolff and Storey (1984), Davies et al. (1989), Widom (1991), Widom et al. (1992), Turner et al. (1997), Widom et al. (1997), Renzulli and Santi (2000), Claude-Ivanaj et al. (2001), Snyder (2005), Beier (2006), Beier et al. (2006), (2007), (2008), Elliott et al. (2007), Watanabe (2010), Zanon (2015). **Terceira**: Esenwein (1930), Berthois (1953), Schmincke and Weibel (1972), Self (1973), Rosenbaum (1974), Baker (1975), Self and Gunn (1976), Glitsch and Allegre (1979), White et al. (1979), Fidczuk (1984), Davies et al. (1989), Mungall and Martin (1995), Turner et al. (1997), França (2000), Claude-Ivanaj et al. (2001), Madureira (2006), Pimentel (2006), Beier et al. (2008), Madureira et al. (2011), Yu (2011), Hildenbrand et al. (2014), Zanon and Pimentel (2015). **São Jorge**: Machado (1977), White et al. (1979), Turner et al. (1997), França (2000), Hildenbrand et al. (2008), Millet et al. (2009), Ribeiro (2011), Yu (2011), Beier et al. (2012a), Hildenbrand et al. (2014). **Graciosa**: White et al. (1979), Maund (1985), Gaspar (1996), Almeida (2001), Beier et al. (2008), Yu (2011), Larrea (2014), Larrea et al. (2014a). **Faial**: Esenwein (1930), Canilho (1970), Girod and Lefèvre (1972), White et al. (1979), Chovelon (1982), Wolff and Storey (1984), Lemarchand (1987), Metrich et al. (1981), Turner et al. (1997), França (2000), Claude-Ivanaj et al. (2001), Machado et al. (2008), Lima et al. (2011), Yu (2011), Beier et al. (2012a), Zanon and Frezzotti (2013), Zanon et al. (2013), Hildenbrand et al. (2014). **Pico**: Esenwein (1930), Berthois (1953), White et al. (1979), Chovelon (1982), Hart and Ravizza (1996), Turner et al. (1997), França (2000), Claude-Ivanaj et al. (2001), França et al. (2006c), Elliott et al. (2007), Prytulak and Elliott (2009), Yu (2011), Beier et al. (2012a), Zanon et al. (2013), Metrich et al. (2014). **Flores**: Girod and Lefèvre (1972), Glitsch and Allegre (1979), White et al. (1979), Azevedo (1998), Genske (2012), Genske et al. (2012b), Larrea (unpublished data). **Corvo**: Esenwein (1930), White et al. (1979), Dias (2001), França et al. (2002), Genske (2012), Genske et al. (2012b), Larrea et al. (2013), Larrea (2014)

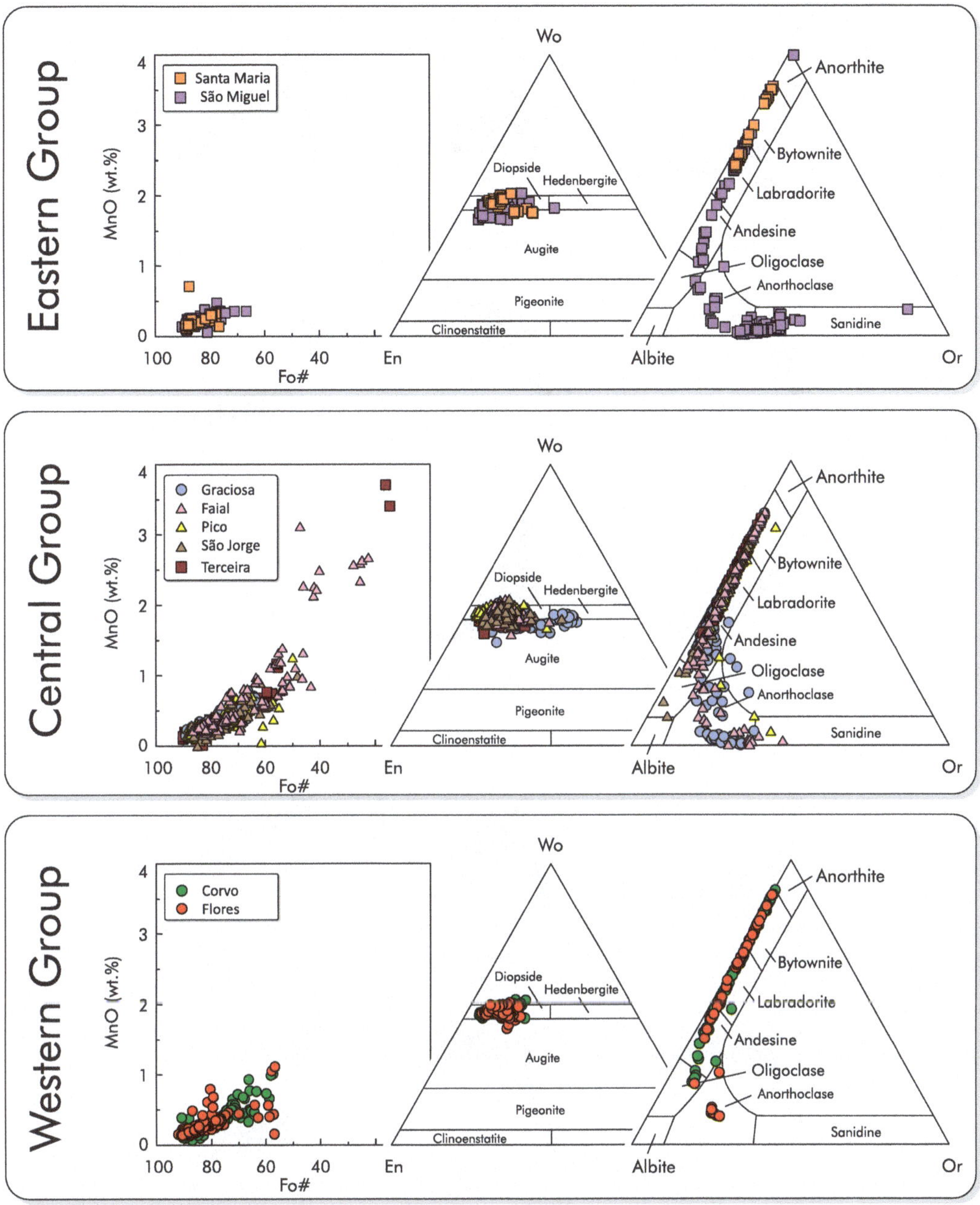

Fig. 2 Composition of minerals from Azorean lava flows including **a** MnO (wt%) versus Fo# for olivines. **b** Ternary classification diagram (wollastonite—enstatite—ferrosilite; Morimoto et al. 1988) for clinopyroxenes and **c** Ternary classification diagram (Ab–An–Or) for feldspars showing the significant chemical variability. Previous published data from: **Santa Maria**: Anderson (1983). **São Miguel**: Esenwein (1930), Storey (1982), Wolff and Storey (1983), Widom et al. (1992), Renzulli and Santi (2000), Beier et al. (2006), Watanabe (2010), Zanon (2015), Widom (unpublished data). **Terceira**: Schmincke and Weibel (1972), Shimizu (1978), Madureira et al. (2011), Zanon and Pimentel (accepted). **São Jorge**: Ribeiro (2011). **Graciosa**: Larrea et al. (2014a) **Faial**: Canilho (1970), Brousse et al. (1981), Metrich et al. (1981), Beier et al. (2012a), Zanon and Frezzotti (2013), Zanon et al. (2013). **Pico**: Hart and Ravizza (1996), França (2000), Beier et al. (2012a), Zanon and Frezzotti (2013). **Flores**: Girod and Lefèvre (1972), Genske et al. (2012b). **Corvo**: Genske et al. (2012b), Larrea et al. (2013)

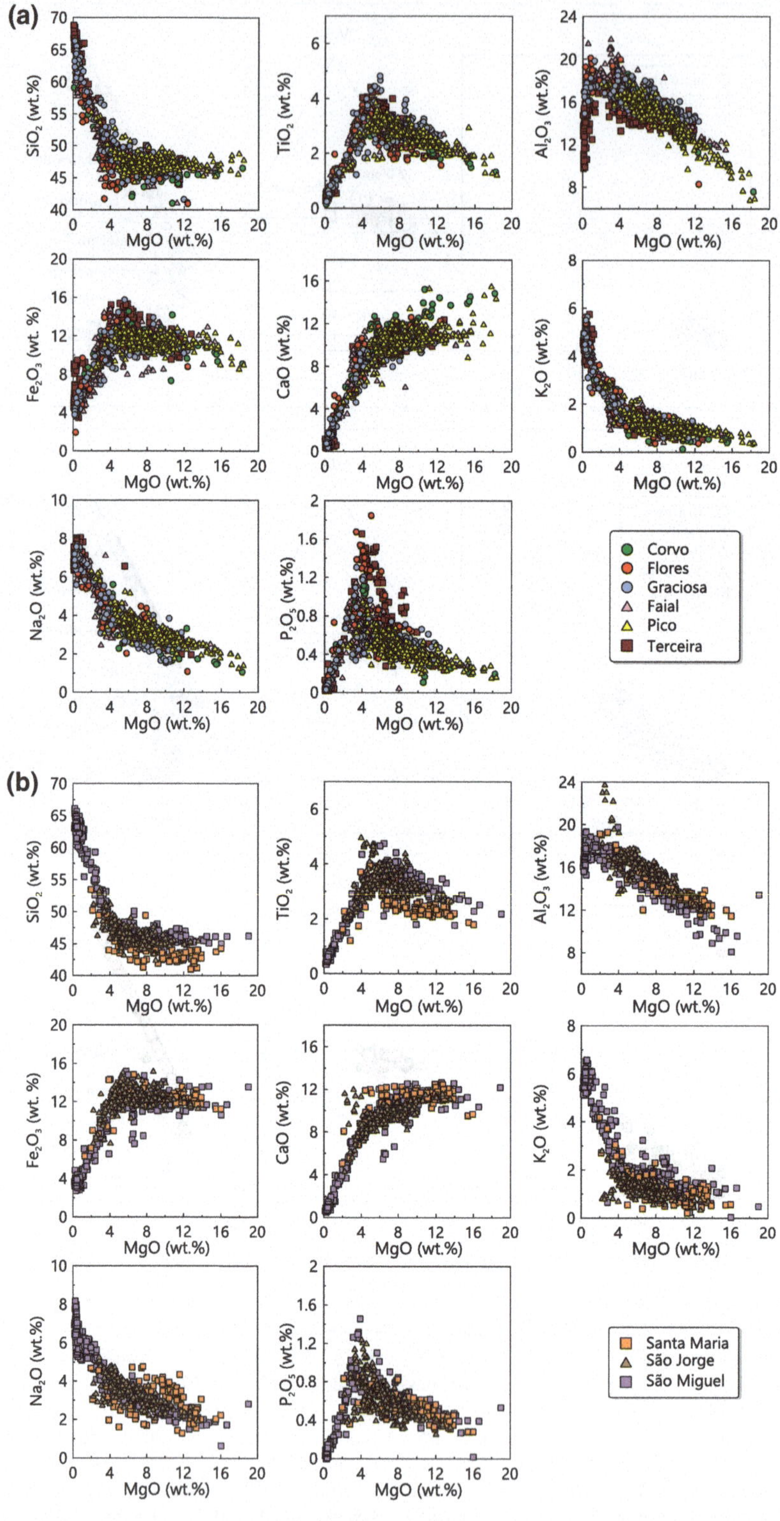

Fig. 3 **a**, **b** Major element concentrations versus MgO for Azores lava flows. Published data from Azores as in Fig. 1

Vesicles are common and most samples are relatively unweathered. The most common phenocrysts are olivine and clinopyroxene, followed by plagioclase, oxides and amphibole (see Fig. 2). Olivine crystals range from Fo_{89} to Fo_{78} and are often fully or partially replaced by iddingsite or clay minerals. Clinopyroxene phenocrysts are classified as augite—diopside, and have irregularly shaped inner cores with overgrowth rims. Smaller pyroxenes are often glomerophyric. Plagioclase phenocrysts range from An_{88} to An_{61} (bytownite—labradorite). Plagioclase crystals often have rounded, optically discontinuous cores, and tend to occur in glomerophyric clots. Amphibole phenocrysts are classified as kaersutite and are usually rimmed with fine-grained oxides. Megacrysts (up to 2 cm long) of olivine, clinopyroxene, plagioclase and amphibole are present occasionally and are characterised by frequently corroded and embayed edges, suggesting disequilibrium with their host melts. In contrast, the groundmass is composed of clinopyroxene, plagioclase, oxides, glass, apatite and nepheline. Olivine rarely appears as microcrysts, but rather has been completely replaced by iddingsite.

Compositional ranges for Santa Maria whole rocks are shown in Fig. 3b as concentrations of major element oxides versus MgO (wt%). Magnesium contents range from 16.51 to 1.27 wt%. Silica, Al_2O_3, K_2O and P_2O_5 concentrations increase with decreasing MgO, whereas CaO and Na_2O decrease with decreasing MgO, although the latter are somewhat scattered. In contrast, TiO_2 and Fe_2O_3 initially increase with decreasing MgO, until ~4 wt% MgO, when they exhibit abrupt changes in slope and concentrations then decrease with decreasing MgO.

Based on the above geochemical and petrographic observations, Anderson (1983) proposed that polybaric fractionation was the dominant process controlling the magmatic evolution of the island. This hypothesis was tested with a mass balance calculation, which showed that initial phases of magma generation underwent high-pressure fractionation (not quantified) with a mineral assemblage mainly composed of olivine and clinopyroxene, whereas the generation of the most evolved melts was dominated by plagioclase fractionation at lower pressures (not quantified). Recently, in a manuscript dealing primarily with the mantle source beneath Santa Maria, Beier et al. (2012b) discuss the magmatic processes controlling the evolution of the island. They proposed that the primitive magmas on Santa Maria have MgO contents of ~12 wt%, and that those lavas with MgO contents >12 wt% likely accumulated olivine. In contrast, more evolved samples, with MgO contents <12 wt%, are related to the primary magmas by fractional crystallization. As a group, the Santa Maria magmas follow similar compositional trends in most major elements to those observed on the other Azores islands, indicating fractionation of olivine and clinopyroxene until ~5 wt% MgO, when plagioclase and Fe–Ti oxides join the fractionating mineral assemblage.

Rodrigues et al. (1985) further linked the petrogenetic evolution of the island to the active tectonics operating during the formation and evolution of Santa Maria. They suggested based on the geochemical composition of the samples that the early formation of the Anjos Complex was related to the East Azores Fracture Zone with the melts forming this unit emanating directly from the mantle, with rapid ascent to the surface and minor fractionation. In contrast, the extensive range in degree of evolution and eruptive style of magmas from the later Facho—Pico Alto unit were controlled by the Terceira Rift system. These melts are thought to have experienced low to high rates of fractional crystallization in several magma chambers located at different depths, resulting in a wide range of rock compositions.

2.2 São Miguel Island

São Miguel, the largest island of the archipelago, is situated near the eastern extent of the Azores platform at ~37° 48′N latitude and 25° 12′W longitude, and, together with Santa Maria, forms the Eastern Group (Fig. 1 in see Beier et al., Chapter "Melting and Mantle Sources in the Azores").

A lithological study of the island lead to the preparation of a 1:50,000 geological map by Zbyszewski et al. (1958, 1959b). A preliminary volcanological map of São Miguel was made by Forjaz (1976), who also defined the main volcanostratigraphic units for this island (Forjaz 1984). Moore (1990, 1991) presented a new geologic map (scale 1:50,000) and summarised the geology of the island based on radiometric dating (Shotton 1969; Shotton and Williams 1971; Moore 1991).

São Miguel consists of four large stratovolcanoes (from west to east: Sete Cidades, Água de Pau (also referred to as Fogo), Furnas and Nordeste; Moore 1990), in which magmatic compositions range from basalts to trachytes, interspersed by two waist zones that dominantly erupt mafic material. These waist zones are mainly formed by quaternary monogenetic cinder cones and lava flows (Booth et al. 1978; Moore 1990). All four stratovolcanoes have been active in the last 100,000 years, although the Nordeste volcano in the eastern sector is extinct and significantly eroded (Abdel-Monem et al. 1975; Feraud et al. 1980; Johnson et al. 1998). Moore (1990) divided the island into six geologically distinct zones, described below:

(1) The Sete Cidades volcano is an ~210 ky stratovolcano that occupies about 110 km^2 on the western end of the island. A circular caldera about 5 km in diameter with walls up to 400 m high truncates its summit. Sete Cidades was formed by three major phases of volcanic activity (Moore 1990, 1991; Beier et al. 2006) including: a dominantly alkali basaltic, pre-caldera shield-building phase (~210 ka); a trachytic caldera-forming phase (~22 ka); and a mainly effusive trachytic post-caldera phase associated with small basaltic flank eruptions (<20 ka).
(2) The western waist zone encompasses an area of ~180 km^2 between the upper eastern flank of Sete Cidades volcano and the western flank of Água de Pau volcano. It is dominated by two chains of cinder cones, mainly of Holocene age (from >30 ka to 1652 AD). It is primarily composed of mafic lavas, basanitoids and ankaramites, although four trachyte eruptions have been also recognised (Moore 1990; Haase and Beier 2003).
(3) The Água de Pau volcano occupies about 150 km^2 in the center of São Miguel, and is largely (~90%) composed of trachytic pumices. A caldera of ~3 km diameter with 300 m high walls truncates its summit. The topographic margin of an older and larger caldera is preserved locally on the western, northern and eastern sides of the volcano. The oldest dated rock from the unit is a 181 ka trachytic dome (Gandino et al. 1985). The formation of the outer caldera took place between 46 and 26.5 ka, whereas the inner caldera forming eruption took place at about 15.2 ka (Moore 1990).
(4) The eastern waist zone comprises about 80 km^2 between Água de Pau and Furnas volcanoes; lava flows from the Nordeste Complex define its northeastern boundary. This waist zone is latest Pleistocene to Holocene in age, and consists of ankaramites, basanitoids, hawaiites, tristanites and trachytes (Moore 1990; Haase and Beier 2003).
(5) Furnas volcano is the smallest stratovolcano, occupying ~75 km^2 in the east-central part of São Miguel. Furnas volcano is truncated by a large caldera, about 6 km in diameter with walls up to 500 m high, situated on the western outer flank of the older Povoação caldera. The Furnas caldera was formed as the result of at least two distinct collapses, estimated to have occurred from 30 to 10 ka (Guest et al. 1999). It is essentially trachytic in composition, and most of the activity has been explosive accompanied by dome formation, being mainly built up from 100 to 5 ka (Guest et al. 1999).
(6) The Nordeste Complex, in the eastern part of the island, comprises an ~180 km^2 eroded shield volcano as well as the Povoação caldera, with ~6 km diameter and 700 m high walls, which truncates part of the shield volcano on its southern side. The Nordeste

Complex is mainly composed of mafic volcanic products (~90%). Nevertheless, five distinct stratigraphic sequences have been defined, including (from oldest to youngest): lower basalts, Nordeste ankaramites, upper basalts, tristanites and trachytes (Fernandez 1980, 1982). K–Ar dating of the lower basalts (4.01 Ma) and the upper basalts (1.86 Ma) suggested a time span of 2.15 Ma during which most of the exposed Nordeste Complex was emplaced (Abdel-Monem et al. 1975), although more recent Ar–Ar dates of the Nordeste Complex indicates that it may be substantially younger, ranging from 0.88 to 0.78 Ma (Johnson et al. 1998).

Subaerial and submarine samples from São Miguel range in composition from alkaline picrobasalt to trachyte based on the TAS classification of Le Maitre et al. (2002) (Fig. 1), and is the most potassic island of the archipelago (Schmincke and Weibel 1972; Schmincke 1973). No major compositional gap is observed in the TAS diagram; instead, these rocks describe a well-defined evolution trend from mafic to evolved compositions.

The mafic rocks include picrobasalt, basanites and basalts with MgO contents >5 wt%. They are characterised by variable volume percent of phenocrysts with olivine > clinopyroxene > plagioclase, included in an intergranular to intersertal groundmass composed of the same mineral phases and opaque microcrysts (e.g., Fernandez 1982; Moore 1991; Renzulli and Santi 2000; Beier et al. 2006). São Miguel samples with MgO contents >12 wt% show large olivine and clinopyroxene (and minor plagioclase) crystals with corroded cores and overgrowth rims in optical and chemical disequilibrium with their host rocks, suggesting that they are incorporated by accumulation; these rocks are commonly called ankaramites (Moore 1991). Intermediate rocks from São Miguel include trachybasalt to trachyandesite, although the latter is relatively less frequent. They show aphyric intersertal to pilotaxitic or hyalophitic textures. They contain variable proportions of olivine, clinopyroxene, plagioclase, titanomagnetite, ilmenite, amphibole, biotite and apatite. In contrast to the basic rocks, these intermediate rocks contain up to 30% feldspars and 3% magnetite in the groundmass. Evolved rocks (trachytes) occur as lavas as well as pyroclastic material (pumices and ash). The lavas show trachytic textures and contain phenocrysts of alkali feldspar (generally sanidine or anorthoclase), plagioclase, biotite, amphibole, opaque minerals (mainly ilmenite and titanomagnetite), aegirine—augite pyroxene and minor zircon and apatite. Trachytic pumices are glassy and essentially aphyric, and comprise pyroclastic fall and flow deposits that characterize the dominant volcanism from the stratovolcanoes (e.g., Moore 1991; Beier et al. 2006).

The available information on mineral chemistry (Fig. 2) is somewhat sporadic among the different São Miguel volcanic systems. The best studied volcano is Sete Cidades, for which Renzulli and Santi (2000) and Beier et al. (2006) conducted detailed studies of the various mineral phases. Some mineral chemistry has been published for the Nordeste Complex (Fernandez 1980) and Água de Pau volcano (e.g., Marriner et al. 1982; Storey 1982; Wolff and Storey 1983; Widom et al. 1992; Watanabe 2010), but these studies do not cover the whole range of rock compositions present in these volcanoes. Relatively little mineral chemistry is available for Furnas volcano or the waist zones, apart from some isolated microprobe data reported by Rowland-Smith (2007), Zanon (2015) and Widom (unpublished data). Based on the available data for mafic and intermediate rocks, olivine crystals range from Fo_{90-59}, with the higher Fo contents generally characteristic of phenocrysts and large, non-equilibrium crystals in the most mafic rocks. Clinopyroxenes are classified as augite and diopside and they are chiefly normally zoned with decreasing MgO contents from the core to rim. The feldspars present a compositional range from anorthite to andesine and the Fe–Ti oxides are mainly titanomagnetites. Some mafic and intermediate samples also contain Cr-spinel microcrysts and kaersutite phenocrysts. The São Miguel evolved rocks present augite crystals with the lowest concentrations of Al_2O_3 and TiO_2 and abundant Ti-rich kaersutite, ilmenite and titanomagnetite. The feldspars in these trachytic rocks show two compositional groups, one from labradorite to andesine and a second from anorthoclase to sanidine. Biotite crystals together with F-apatite microcrysts are also observed in these evolved rocks.

Bivariate plots of selected whole rock major element compositions versus MgO wt% of São Miguel rocks are shown in Fig. 3B. In general, as MgO decreases, SiO_2, Al_2O_3, K_2O and Na_2O increase, and CaO decreases. In contrast, TiO_2, Fe_2O_3 and P_2O_5 increase until ~5 wt% MgO, and then decrease towards lower MgO contents. These samples show MgO concentrations from 0.15 to >18 wt%. Samples with MgO greater than 12 wt% appear to have accumulated mafic crystals; accordingly, samples with about 12 wt % MgO are interpreted to most closely represent the primary magma compositions of São Miguel (e.g., Fernandez 1982; Moore 1991; Beier et al. 2006).

Each São Miguel volcanic system lies on a single liquid line of descent indicating progressive fractionation from primary magma compositions. Fractional crystallization is proposed as the key magmatic process controlling the evolution of the different São Miguel volcanic systems (e.g., Fernandez 1982; Marriner et al. 1982; Moore 1991; Renzulli and Santi 2000; Beier et al. 2006). Based on the observed compositional trends for major element oxides (Fig. 3b), olivine and clinopyroxene fractionation are inferred to be the primary fractionating phases in the mafic and intermediate rocks from São Miguel. However, at approximately 5 wt% MgO, plagioclase, amphibole and Fe–Ti oxides join the fractionating assemblage, followed by a minor apatite fractionation in the very late stages. Several studies have focused in detail on the magmatic evolution of three of the four stratovolcanoes form São Miguel, the results of which are summarised below.

Sete Cidades volcano was studied by Beier et al. (2006). Major element modelling using MELTS (Ghiorso and Sack 1995; Asimow and Ghiorso 1998) and trace element modeling assuming Rayleigh fractionation (Rayleigh 1896) shows that Sete Cidades volcano has evolved primarily by olivine, clinopyroxene, magnetite, ilmentite, feldspars, apatite, amphibole and biotite fractional crystallization, with limited variability in mantle source composition or degree of partial melting throughout its eruptive history. Basaltic magmas appear to start fractionating close to the crust-mantle boundary (~16–5 km depth), whereas the most evolved magmas fractionated in a shallower magma chamber at ~3–5 km depth beneath the volcano. Renzulli and Santi (2000) reached similar conclusions using mass balance calculations on major elements to evaluate the fractional crystallization process.

The petrogenesis of the Nordeste Complex was studied in detail by Fernandez (1980, 1982). The proposed fractional crystallization model was tested using the Wright and Doherty (1970) least squares method, supporting a fractional crystallization process controlled by high and variable contents of olivine and pyroxene, and minor plagioclase, titanomagnetite and ilmenite as the dominant mechanism for the origin of the Nordeste Complex.

The magmatic evolution of the entire suite of rocks from Água de Pau volcano was studied by Marriner et al. (1982) and Storey et al. (1989). They propose that the most evolved trachytes could be related to primitive basalt by 80–90% fractional crystallization using the Rayleigh (1896) fractional crystallization equation with olivine, clinopyroxene, plagioclase, alkali feldspars and opaque minerals as fractionating mineral assemblage. Nevertheless, they notice the existence of hybrid lavas characterised by intermediate silica content and the presence of large xenocrysts, which cannot be accounted for solely by crystal fractionation of basaltic to trachytic melts, but rather indicate that mixing between trachytic and less evolved melts may have occurred beneath Água de Pau volcano (Storey 1981, 1982). Later studies by Widom et al. (1992) investigated the formation of the chemically zoned Fogo A trachytes, and demonstrated by least squares major element modelling that the least evolved trachytic melts could be derived directly by fractional crystallization of an alkali basaltic parent magma. The chemical variation within the Fogo A trachyte deposit is attributed to further fractional crystallization (~70%) of the least evolved trachytic composition, with the fractionating assemblage dominated by sanidine with minor biotite, clinopyroxene, Fe–Ti oxides, apatite and zircon.

Plutonic rocks

Plutonic rocks from São Miguel have been described by Cann (1967), França (1993), Widom et al. (1993), França and Rodrigues (1994), Johansen (1996), Mattioli et al. (1997) and Almeida (2001). They can be divided into three groups according to their mineral assemblage: ultramafic, mafic and syenite xenoliths.

Abundant mafic and ultramafic xenoliths are found in several vent deposits and lava flows from Furnas, Água de Pau and Sete Cidades volcanoes. These xenoliths have been studied in detail by Johansen (1996), Mattioli et al. (1997) and Almeida (2001) (Fig. 4). They are divided

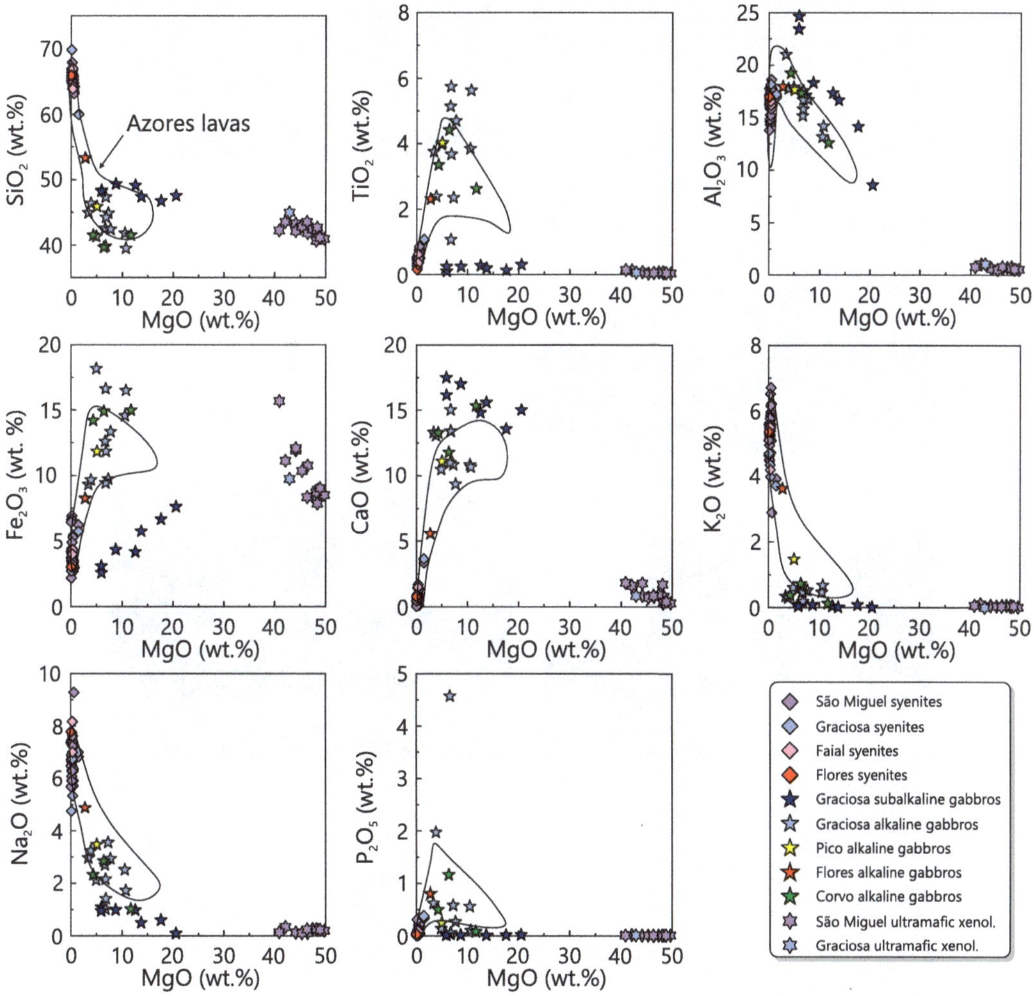

Fig. 4 Major element concentrations versus MgO for ultramafic, mafic (alkaline and subalkaline gabbros) and syenite xenoliths from the Azores archipelago. Azores lava composition fields as in Fig. 3. Previous published data from: **São Miguel**: Esenwein (1930), Berthois (1953), Cann (1967), França (1993), Widom et al. (1993), Johansen (1996), Almeida (2001). **Graciosa**: França (1993), Larrea et al. (2014a). **Faial**: França (1993). **Pico**: Esenwein (1930). **Flores**: Esenwein (1930), França (1993). **Corvo**: Larrea et al. (2013)

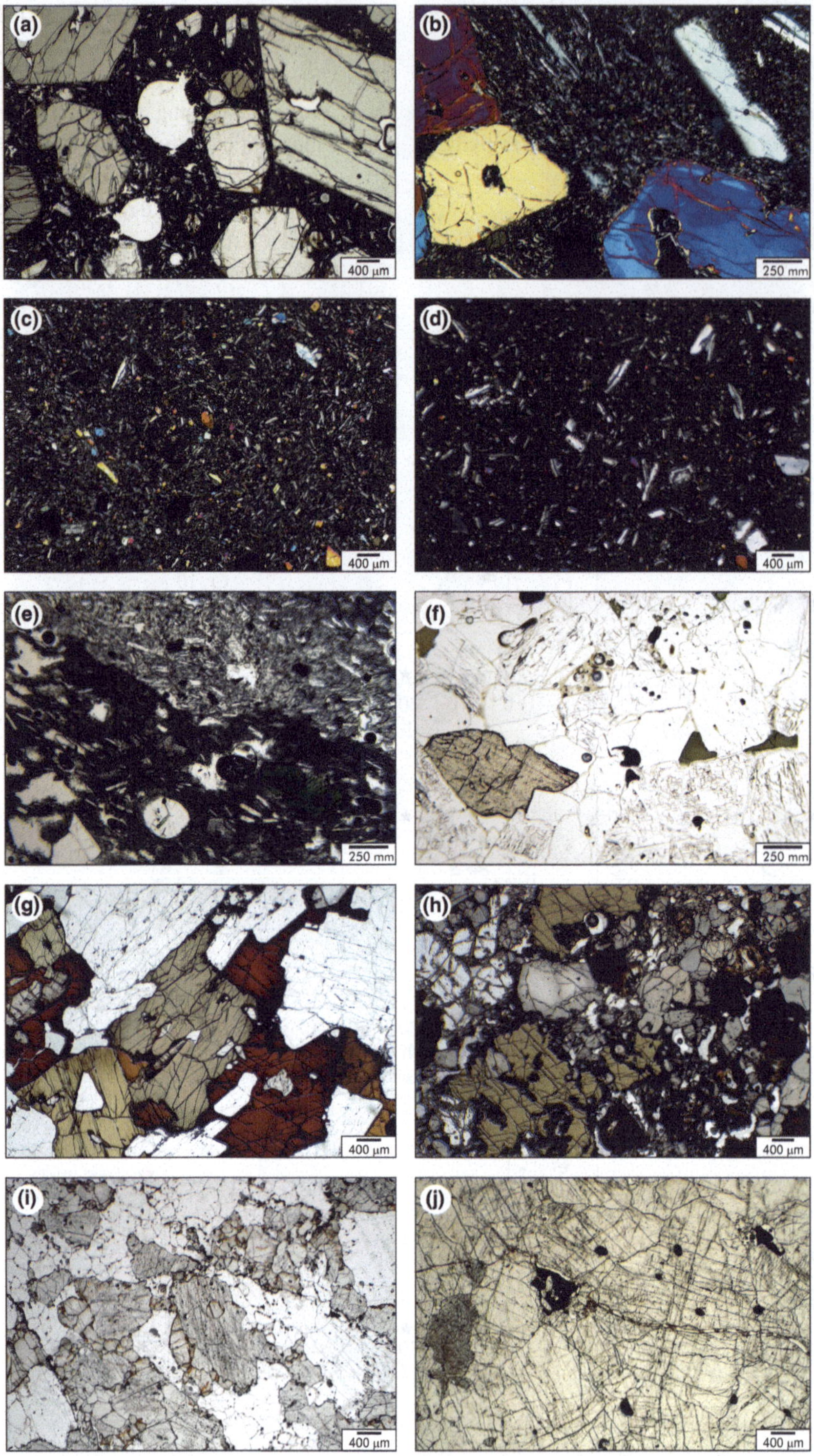
(a)
400 μm
(b)
250 mm
(c)
400 μm
(d)
400 μm
(e)
250 mm
(f)
250 mm
(g)
400 μm
(h)
400 μm
(i)
400 μm
(j)
400 μm

◀ **Fig. 5** Representative photomicrographs of volcanic and plutonic rocks from the Azores Archipelago in plane- and cross-polarised transmitted light. **a** Porphyritic basalt from Corvo, lava sample COR-12. **b** Porphyritic basalt from Terceira, sample TERZF-2. **c** Microporphyritic hawaiite from Flores, sample WAFL-4. **d** Microporphyritic mugearite from Graciosa, sample GRZF-16. **e** Trachyte from Terceira, sample WAT-29. **f** Syenite from São Miguel, Sample EC-SMC-26Syenite8. **g** Alkaline gabbro from Graciosa, sample GRZF1 × 1. **h** Alkaline gabbro from Flores, sample FL-ENC-3. **i** Subalkaline gabbro from Graciosa, sample GRQUx4. **j** Dunite xnolith from Graciosa, sample GRQUx1. Photos courtesy of George Daly, Elise Conte and Patricia Larrea

into two groups based on their texture and mineral chemistry. The first group comprises spinel harzburgites, lherzolites, dunites and wherlites with porphyroclastic textures (Johansen 1996). They are interpreted as fragments of the upper mantle that have been exposed to both partial melting and metasomatic enrichment. The second group presents granular textures with a great modal variability: dunite, wherlite, clinopyroxenite, gabbros and diorites, but all of the studied samples are orthopyroxene-free. Most of these samples are considered to be cumulates derived from alkali basaltic melts of the same composition as the Azorean lavas (Mattioli et al. 1997), although some of them are likely to be cumulates derived from melts with a very high Ti/Al ratio (Johansen 1996). In accordance, Renzulli and Santi (2000) suggest that the presence of these plutonic nodules in Sete Cidades volcano is broadly compatible with the fractional crystallization process proposed to explain the magmatic evolution of Sete Cidades.

Syenite samples were studied from the Furnas, Água de Pau and Sete Cidades volcanoes. Some whole rock compositions were reported by Esenwein (1930), Berthois (1953), Cann (1967), França (1993) and Widom et al. (1993) (Fig. 4). They are mainly found within trachyte pumice deposits, ignimbrites, scoria and lava flows, and in alluvial deposits within the calderas or on the flanks of the volcanoes. Most of them are 5–50 cm in diameter and generally medium-grained with hypidiomorphic granular textures (Fig. 5). They contain variable modes of sanidine, arfvedsonite, quartz, aegirine, augite and opaque minerals, as well as minor biotite, zircon, apatite, pyrochlore, sphene, dalyite, astrophyllite, sodalite and lavenite (Cann 1967; França 1993; Widom et al. 1993). Cann (1967) noted that the syenite compositions matched the compositions of the extrusive rocks, and therefore pointed to their probable cognate origin. This hypothesis was later verified by Widom et al. (1993), who proposed a bulk trachyte liquid origin for these syenites. Similar results have been found for syenites studied at Santa Bárbara volcano (Terceira island; Daly et al. 2012) and Vulcão Central Unit (Graciosa island; Larrea et al. 2014a).

2.3 Terceira Island

Terceira is the easternmost island of the Central Group, located on the Terceira Rift (Vogt and Jung, Chapter ""Azores Geosyndrome" and Plate Tectonics: Research History, Synthesis, and Unsolved Puzzles") between ~38° 37′N—38° 48′N latitude and 27° 02′W—27° 23′W longitude (Fig. 1 in Beier et al., Chapter "Melting and Mantle Sources in the Azores").

The first geological map on scale 1:25,000 of Terceira was produced by Zbyszewski et al. (1971), mainly based on petrographic observations. Self (1973, 1976) and Self and Gunn (1976) established four volcanostratigraphic units, which were recently redefined by Nunes (2000) and Madeira (2005). Based on Nunes (2000) stratigraphy, five main volcanic systems are recognised in order of decreasing age: Cinco Picos, Guilherme Moniz, Santa Bárbara, Pico Alto and the Fissural zone, with the three latter considered volcanically active.

(1) Cinco Picos comprises the oldest volcanic center (>370 ka; Calvert et al. 2006) forming the eastern part of the island, which is truncated by the largest caldera of the Azores Archipelago (~7 km diameter). Mugearitic lavas are most common with some mildly undersaturated peralkaline lavas exposed on

the flanks. Caldera formation was accompanied by explosive eruptions forming pumice-fall deposits and ignimbrite sheets. The caldera is partially filled with young lavas erupted from the fissure zone (Nunes 2000).

(2) Guilherme Moniz volcano is located in the southern-central portion of the island, and began erupting about 270 ka (Calvert et al. 2006). It comprises mainly trachyte domes, flows and minor pyroclastic deposits. Comenditic trachyte lavas are exposed in the caldera walls and on its southern flanks. The northern flank is covered by the more recent lava domes of Pico Alto Volcano.

(3) Santa Bárbara is a stratovolcano occupying the western end of the island that began erupting at ~100 ka (Calvert et al. 2006). A circular caldera formed approximately at 15 ka truncates its summit. Santa Bárbara Volcano is mainly composed of hawaiite and mugearite lavas, and after the caldera formed, rhyolite and trachyte lava have erupted both from within the caldera and from radial fissures in the flanks, and basaltic eruptions have been limited to the lower slopes.

(4) Pico Alto volcano is built on the northern flanks of Guilherme Moniz. It is composed only of highly evolved lavas and pyroclastic deposits, forming at least three ignimbrites of comenditic trachyte composition. No mafic rocks are exposed. The oldest rocks on Pico Alto may be ~18 ka and the youngest eruptions were 1000 years old (Calvert et al. 2006).

(5) The Fissural zone is marked by a line of monogenetic basaltic scoria and spatter cones, and associated lava flows across the central part and southeastern sector of the island (Cinco Picos caldera). The NW part has been most active during the past 50 ky and previous activity was mainly from the central and SE portion of the zone (Self and Gunn 1976). The last subaerial eruption was a hawaiite lava in 1761 BP (França 2000).

Terceira subaerial and submarine lavas range in composition from alkali basalt to trachyte following the TAS classification of Le Bas et al. (1986) (Fig. 1). It presents two main compositional trends: one from basalt to peralkaline trachyte (comendite and pantellerite), and a second from basalt to rhyolite (comendites) (e.g., Self 1973; Self and Gunn 1976; Allègre et al. 1977). Terceira is distinct from all of the other islands of the archipelago in two respects: (1) the noteworthy abundance of lava domes, and (2) the large compositional gap with a scarcity of rocks in the benmoreite—trachyte range in Santa Bárbara and Pico alto volcanoes. This gap is partially filled only by a small volume of eruptions during the last 23,000 years (Self 1973), and by plutonic rocks that are incorporated as xenoliths or cognate inclusions in some trachytic eruptions (Daly et al. 2012; Daly et al. in preparation).

Mafic and intermediate lavas from Terceira (Fig. 5) are predominantly porphyritic to seriate with major olivine, clinopyroxene and plagioclase phenocrysts, embedded in a microcrystalline groundmass mainly composed of plagioclase and opaque microcrysts (e.g., Self 1973; Rosenbaum 1974; Mungall and Martin 1995; Madureira et al. 2011). The proportion of phenocrysts can reach up to 30% by volume. The evolved lavas are aphyric to porphyritic (up to 20% crystals by volume), with trachytic or glomeroporphyritic textures (e.g., Self 1973; Pimentel 2006) (Fig. 5). Alkaline feldspars are the most abundant phenocrysts, followed by plagioclase, clinopyroxene, olivine (commonly iddingsitized) and oxides. The same mineral phases also occur as microcrysts. Biotite and kaersutite amphibole crystals also occur both as phenocrysts and microcrysts.

The chemistry of the mineral phases present in Terceira rocks (Fig. 2) have been studied only in basaltic lavas from Santa Bárbara and the Fissural System (Madureira et al. 2011; Zanon and Pimentel 2015), apart from some isolated microprobe data reported by Schmincke and Weibel (1972) and Shimizu (1978). Olivine phenocrysts range from Fo_{90} to Fo_{72} in the Fissural system and from Fo_{85} to Fo_{56} in Santa Bárbara volcano. Clinopyroxene is a major phase in Terceira lavas, occurring as phenocrysts and microcrysts, classified as diopsides and augites

(Morimoto et al. 1988). Feldspar phenocrysts display a wide range in composition from An_{85} to An_{40} (bytownite—andesine). Groundmass feldspar microcrysts extend the compositional range up to An_{30}. Multiple and complex oscillatory zoning surrounding resorbed cores are commonly found in the feldspar phenocrysts. Opaque minerals such as chromite and titanomagnetite can be found as inclusions in phenocryst phases, and as phenocrysts and microcrysts in the groundmass. Ilmenite is the dominant microcryst but can also occur as phenocrysts or as exsolution lamellae in titanomagnetite phenocrysts.

Bivariate plots of selected major-element oxides versus MgO wt% from Terceira rocks are shown in Fig. 3a. These lavas show a general increase in SiO_2, K_2O and Na_2O as MgO decreases. Calcium decreases over the entire MgO range, and TiO_2, Fe_2O_3 and P_2O_5 first increase to a maximum at 4–5 wt% MgO, after which they start decreasing sharply towards lower MgO contents. Aluminium shows a similar trend but Al_2O_3 contents start to decrease at ~1–2 wt% MgO.

Incompatible element ratios and isotopic signatures suggest that the Terceira volcanic systems are not all comagmatic (e.g., Madureira et al. 2011; Beier et al., Chapter "Melting and Mantle Sources in the Azores"). Thus, each volcanic system represents the evolution of similar non-cogenetic magma batches that undergo similar magmatic processes. Fractional crystallization is the main magmatic process controlling the evolution of the different Terceira volcanic systems (e.g., Self 1973; Mungall and Martin 1995; Pimentel 2006; Madureira et al. 2011). In general, clinopyroxene, olivine and plagioclase fractionation has controlled the formation of the mafic and intermediate compositions, whereas the evolution of the most evolved melts is mostly controlled by fractionation of feldspars, Fe–Ti oxides and amphibole, with apatite fractionation in the last stages of Terceira magmatic evolution. Santa Bárbara, Pico Alto and the Fissural System are the best studied volcanic systems in Terceira, whereas scarce information is reported about the genesis of the oldest volcanoes of Cinco Picos and Guilherme Moniz.

Madureira et al. (2011) focused on mafic and intermediate products from Santa Bárbara and the Fissural System. They modelled the fractional crystallization process using least-squares mass balance calculations for major elements and the Rayleigh (1896) fractional crystallization equation for trace elements. The results of these calculations suggest that a more mature magmatic plumbing system is feeding the Santa Bárbara volcano, and indicate that between 50 and 65% fractional crystallization is needed to generate the range of compositions observed in Santa Bárbara and the Fissural System volcanic systems. Moreover, Zanon and Pimentel (2015) studied $CO_2 + H_2O$ fluid inclusions hosted in mafic minerals from the Fissural system, revealing different storage and ascent conditions among the different fissure zone segments, with a maximum pressure of fluid trapping at the Moho Transition Zone (498–575 MPa or 20.3–21 km depth). Mungall and Martin (1995) developed a fractional crystallization model based on trace elements to test the fractional crystallization process controlling the evolution of Santa Bárbara and Pico Alto volcanic systems. Their results strongly support the hypothesis that the felsic magmas of Santa Bárbara and Pico Alto are derived by fractional crystallization of associated basaltic magmas. However, they indicate that these two distinct basaltic suites originate from distinct mantle sources and evolve along distinct P–T paths during ascent; Santa Bárbara magmas undergo an initial period of high-pressure (~800 MPa) fractionation before moving to shallower crustal levels, whereas Pico Alto magmas appears to evolve entirely at lower pressures. Moreover, Pimentel (2006) focused on the evolved lava domes from these two volcanic systems. He proposed that the different behaviour of some major oxides (i.e., Fe_2O_3, MnO and K_2O) in the magmatic series of Santa Bárbara and Pico Alto volcanoes is related to distinctive fO_2 conditions during the final stages of the magmatic evolution, resulting from differences in the degassing process; the pantelleritic Pico Alto suite probably evolved by fractional crystallization at low fO_2 conditions (<FMQ), whereas the comenditic Santa Bárbara suite seems to reflect somewhat higher fO_2 conditions.

Plutonic rocks

Gabbro and syenite xenoliths from Terceira have been studied by Self (1973), França (1993), França and Rodrigues (1994), Mungal and Martin (1995) and Daly et al. (2012), although whole rock chemistry of these xenoliths have not been published to date.

Self (1973) mentions the existence of cognate gabbroic xenoliths composed of clinopyroxene and plagioclase with minor olivine in a volcanic deposit from Santa Bárbara, but no further studies are reported. Syenites are commonly found in felsic pyroclastic deposits of Santa Bárbara and Pico Alto ignimbrites. These xenoliths show medium-grained hypidiomorphic inequigranular textures with subhedral shapes and micropertitic intergrowth textures, and display variable degrees of hydrothermal alteration (Fig. 4). They present variable proportions of alkali feldspar, augite, biotite, hornblende, plagioclase (andesine), magnetite and interstitial quartz. Accessory minerals such as apatite, sphene and zircon are also observed (e.g., França 1993; Daly et al. 2012). Daly et al. (2012) suggested that Santa Bárbara syenites are frozen melts, based on their negative Eu/Eu* anomalies and their similar trace element compositions to Santa Bárbara trachytes, thus interpreted in the same way as the syenite nodules studied in São Miguel and Graciosa islands (Widom et al. 1993; Larrea et al. 2014a). This idea was previously suggested by Self (1973) and Mungall and Martin (1995), who proposed that the compositional gap found in the intermediate compositions (between 54 and 64 wt% SiO_2) in Terceira is greatly reduced by the intermediate compositions of these syenites. Therefore, this compositional gap does not result from an absence of intermediate composition magmas, but rather from an inability of these magmas to erupt.

2.4 São Jorge Island

São Jorge is situated in the middle of the five islands forming the Central island Group of the Azores Archipelago, located between ~38° 33′ and 38° 45′N latitude and 27° 44′ and 28° 20′W longitude (Fig. 1 in Beier et al., Chapter "Melting and Mantle Sources in the Azores"). The island has an elongated shape, inherited from fissural volcanism, sub-parallel to the Terceira rift and sub-perpendicular to the MAR (Vogt and Jung, Chapter ""Azores Geosyndrome" and Plate Tectonics: Research History, Synthesis, and Unsolved Puzzles" and Miranda et al., Chapter "The Tectonic Evolution of the Azores Based on Magnetic Data").

Machado and Forjaz (1965) carried out the first geomorphologic, tectonic and geologic study on São Jorge. Further works by Forjaz et al. (1970, 1990), Forjaz and Fernandes (1970) and Madeira (1998) defined the volcanostratigraphy of São Jorge. Based on Forjaz et al. (1990), three main volcanic complexes are distinguished: Serra do Topo, Rosais and Manadas.

(1) Topo Volcanic Complex was the first to emerge and comprises the whole eastern half of the island. It is composed of thick volcanic piles mainly formed as a result of effusive volcanism due to the predominance of lava flows over pyroclastic materials. The volcanic sequences are well exposed in the cliffs and frequently cut by dikes at the base of slopes and on the exposed axial zone (Ribeiro 2011). In the summit areas, erosion smoothed the topography and weathering altered the rocks. Soils were formed during volcanic quiescence, and as a consequence, reddish baked-soils interbedded with lava flows are commonly found (França et al. 2005). Recent studies by Hildenbrand et al. (2008) and Ribeiro (2011) proposed that the Fajã de São João, with an age of ~1.3 Ma, is the oldest subaerial part of the proto-island comprising the remnants of an old volcanic system separated from the rest of the Topo volcanic sequence by a significant temporal gap.

(2) the Rosais Volcanic Complex formed to the west of Topo Volcanic Complex; the stratigraphic contact between Topo and Rosais volcanic complexes is not visible because it was covered by younger Manadas lavas. The volcanic activity was predominantly effusive, edifying thick volcanic piles. The lava sequences are partially exposed in the strongly eroded coastal cliffs and on better preserved summit areas from the western half of the island.
(3) the Manadas Volcanic Complex comprises the central part of the island. It is composed of numerous recent strombolian volcanic cones roughly constrained along the main WNW-ESE axis of the ridge. Lava flows partly buried the older cliff successions resulting in an unconformity and producing lava deltas at the shore level. This volcanic complex is considered volcanically active; the last two sub-aerial historic eruptions took place in 1580 (Fajã da Queimada) and 1808 (Urzelina).

Extensive sampling of São Jorge carried out by Hildenbrand et al. (2008), Millet et al. (2009) and Ribeiro (2011) has characterised the three main volcanic complexes, including subaerial lava flows and dikes, together with submarine samples from the southeast flank of the island [Ribeiro (2011) and Beier et al. (2012a)]. The TAS classification diagram (Fig. 1) reveals that São Jorge subaerial lavas are all alkaline ranging from picrobasalt to mugearite. However, the lavas sampled in the southeast submarine flank are exclusively basalts (Ribeiro 2011). The Fajã de São João sequence (the oldest part of the island), although alkaline in composition has lower alkali contents for a given SiO_2 content than the rest of the lavas from the island. The Topo Volcanic Complex is formed by basalts, hawaiites and basanites—tephrites. The Rosais Volcanic Complex is composed of basalts, trachybasalts, hawaiites and basanites. The youngest volcanic complex, Manadas, is composed of basalts, basanites—tephrites and more evolved lavas represented by hawaiites and mugearites. Sedimentary rocks, described as consolidated volcaniclastic breccias and bioclastic limestones, have also been found in the southeastern flank of São Jorge (Ribeiro 2011).

In terms of texture, the lavas from the three main volcanic complexes are quite similar (Ribeiro 2011), with the exception of the lavas from the Fajã de São João sequence (Hildenbrand et al. 2008; Ribeiro 2011) and the pillow lavas from the submarine flank. The São João lavas are mostly porphyritic with zoned plagioclase as the major phenocryst phase (often grouped in glomerocrysts), minor olivine and scarce clinopyroxene (augite). Submarine lavas are vesicular, porphyritic pillow lavas with a cryptocrystalline matrix with plagioclase, olivine and oxides as the main phenocryst phases (Ribeiro 2011). In contrast, the lavas from the three main volcanic complexes present textures ranging from microcrystalline to porphyritic with variable amounts of large crystals. The porphyritic lavas are formed by well-developed crystals embedded in a microcrystalline matrix where locally intergranular or glomeroporphyric textures are found (Ribeiro 2011). These lavas are vesicular with variable vesicle shapes and sizes. The phenocryst and microcryst assemblage is composed of olivine, pyroxene, feldspars and Fe–Ti oxide crystals with minor presence of kaersutite and biotite. Trachytic textures are also present, characterised by orientated plagioclase microlites.

Mineral compositions from São Jorge are plotted in Fig. 2. Olivine is ubiquitous in almost all lavas, either as phenocrysts and microcrysts. Olivine phenocrysts exhibit euhedral to subeuhedral shapes and are compositionally zoned (Fo_{87-60}) with distinct rims, with the exception of a few crystals that reach values of Fo_{29}. Most olivine crystals are fresh, but iddingsite alteration is observed in some crystals. Pyroxene phenocrysts and microcrysts are classified as diopside—augite (Morimoto et al. 1988). They are euhedral to anhedral phenocrysts showing oxide inclusions and corroded rims. In the matrix, pyroxene usually is not the dominant mineral but appears as prismatic microcrysts. Zoning in the pyroxenes is commonly observed (Ribeiro 2011). Pyroxene crystals with green cores surrounded by lighter colored rims have been found in a few lavas from Rosais Volcanic Complex, probably resulting from magma mixing as inferred for Terceira pyroxenes (Madureira 2006). Feldspar crystals are very abundant and range in size up to 1 cm; their shape is often subeuhedral to euhedral in prismatic or basal sections. The feldspar phenocrysts are commonly oscillatory zoned with multiple resorption surfaces and truncation of growth surfaces. In the matrix, prismatic plagioclase microlites are dominant in the majority of the samples, along with Fe–Ti oxides. Compositionally, the feldspars range from An_{81} to An_7, although most of the crystals are classified as bytwonite and labradorite (An_{81-55}). Fe–Ti oxides mainly occur as euhedral phenocrysts, microcrysts or small inclusions in phenocrysts. They are classified as titanomagnetite, magnetite, ilmenite and chromite (Ribeiro 2011). Minor amphibole crystals occur as phenocrysts and xenocrysts with reaction coronas, and are classified as kaersutite (Leake et al. 1997). Biotite crystals are only present in samples from the Topo Volcanic Complex, and appear as anhedral microcrysts showing evidence of disequilibrium, such as the typical speckled effect.

Major element variations among São Jorge volcanic products are shown in Fig. 3b as a function of MgO content. Topo Volcanic Complex samples are the most primitive and have the most extensive range in MgO, the lowest P_2O_5 and highest CaO, Fe_2O_3 and TiO_2 concentrations for a given MgO; in contrast, the oldest part of the island (the Fajã de São João sequence) has the lowest MgO contents. The samples from the Rosais and Manadas volcanic complexes range from primitive to somewhat evolved (MgO contents of ~4–11 wt%). The lavas from Rosais Volcanic Complex show the highest K_2O concentrations, while lavas from Manadas Volcanic Complex present an intermediate composition between Topo and Rosais samples. The submarine pillow lavas from the southeast flank are fairly similar to those from Topo Volcanic Complex, but show higher SiO_2 and lower TiO_2, CaO and P_2O_5 contents. These variations in major element concentrations between the different volcanic products of the island suggest that the lavas are derived from non-cogenetic magmatic liquids (Hildenbrand et al. 2008; Millet et al. 2009; Ribeiro 2011). In accordance, Ribeiro (2011) evaluated fractional crystallization as a potential mechanism for magmatic differentiation using Pearce diagrams for assessing major element variations and Rayleigh fractionation modelling of trace element variations (Rayleigh 1896). Olivine and clinopyroxene were found to be the main fractionating mineral phases controlling magma evolution in the Topo, Rosais and Manadas volcanic complexes, whereas olivine and plagioclase fractionation governs the evolution of the submarine pillow lavas. Plagioclase accumulation is also required to explain the evolution of the Fajã de São João sequence. Therefore, the geochemical evolution of the island is mainly controlled by fractional crystallization, with periodic replenishment of primitive magmatic melts. This geochemical evolution, together with geochronological constraints (Hildenbrand et al. 2008; Ribeiro 2011), allowed Ribeiro (2011) to define the magmatic evolution of São Jorge in two discrete volcanic phases.

(1) The first volcanic phase corresponds to the proto-island subaerial activity represented by the Fajã de São João sequence, which was active from approximately 1.32–1.21 Ma. The low degree of evolution of the melts allowed the formation of basalts, trachybasalts and basaltic trachyandesites with the presence of plagioclase as the dominant mineral phase (up to 45% of the volume). These lavas were formed as a result of fractional crystallization, gravitational segregation and crystal accumulation in a shallow magma chamber (1000–1100 °C and a depth of 17 km or 500 MPa based on thermobarometric estimates). After this period, volcanism stopped for approximately 450 ky.
(2) The second volcanic phase, which is still active, started ~757 ka and formed the three main volcanic complexes of the island. The Topo Volcanic Complex was formed

between 757 and 543 ka, and the migration of the volcanism to the west allowed the formation of the Rosais Volcanic Complex from 368–117 ka. The beginning of the Manadas volcanic activity, in the center of the island, is not well-constrained temporally, but it is currently active. The pillow lavas from the southern flank are also not temporally constrained, but the lack of seawater alteration might indicate a relative young age (Ribeiro 2011). The volcanic products of this second volcanic episode show compositions that range from predominately basaltic on Topo, to slightly more evolved and K-rich on Rosais, and finally more alkaline lavas on Manadas Volcanic Complex. These varied magmatic compositions suggest a non-comagmatic origin, and indicate that the melts were generated as distinct magma batches. However, fractional crystallization is the main magmatic process controlling the evolution of these lavas. Temperature and pressure estimates suggest that olivine and clinopyroxene began to fractionate at ~ 1190 °C and 1 GPa, and imply that fractional crystallization occurred at deeper levels than in the first volcanic phase.

2.5 Graciosa Island

Graciosa is the northernmost island of the Central Group, located on the Terceira Rift (Vogt and Jung, Chapter "The "Azores Geosyndrome" and Plate Tectonics: Research History, Synthesis, and Unsolved Puzzles") between $\sim 39°$ and 39° 06′N latitude and 27° 56′ and 28° 05′W longitude (Fig. 1 in Beier et al., Chapter "Melting and Mantle Sources in the Azores"). Its WNW-ESE elongated shape has been linked to the tectonic influence of the Terceira Rift (e.g., Hildenbrand et al. 2008).

The first regional geological survey on the island was undertaken by Zbyszewski (1970), who published a geological map of Graciosa on a scale of 1:25,000 (Zbyszewski et al. 1972), mainly based on petrographic observations. Forjaz and Pereira (1976) completed this study with stratigraphic data and established the preliminary 1:25,000 volcanic map. Gaspar (1996) established the final volcanostratigraphy of Graciosa based on the spatio-temporal development of the main eruptive centres and their associated volcaniclastic deposits. Three major volcanic complexes were defined, in order of decreasing age:

(1) The Serra das Fontes Volcanic Complex (1056.9 ± 28.0 ky, Larrea et al. 2014b) is the oldest subaerial part of the island. It comprises a shield volcano with its main eruptive center located in the current Serra Branca mountain range (Gaspar 1996).
(2) The Serra Branca Volcanic Complex (433.5 ± 12.9 ky, Larrea et al. 2014a) is formed by the evolution of the Serra das Fontes magmatic system, which eventually produced a central composite volcano (Gaspar 1996). Most of these products have been eroded and covered by younger volcanic rocks.
(3) The Vitória-Vulcão Central Volcanic Complex is the most recent and consists of two units: Vitória and Vulcão Central. The formation of Vitória Unit may have started after a long period of volcanic inactivity on the island (ca. 110 ky, Larrea et al. 2014b), during which the older volcanic edifices were eroded (Gaspar 1996). This unit is formed by monogenetic eruptive centres and their products, which make up the present NW platform. The Vulcão Central Unit currently comprises the main stratovolcano (prior to 50 ka, Larrea et al. 2014b) and its caldera, to the south of the oldest volcanic structures.

The various magmatic products across the volcanostratigraphic sequence, including subaerial and submarine lava flows, gabbroic xenoliths, syenites and dunites, have been the subjects of several studies including Almeida (2001), Beier et al. (2008) and Larrea et al. (2014a).

The composition of the lavas ranges from basalt to trachyte following the TAS classification by Le Bas et al. (1986) (Fig. 1). Basalts and hawaiites are the most abundant rock compositions in Serra das Fontes and the Vitória—Vulcão Central volcanic complexes, apart from some more evolved rocks (mugearite—trachyte) in the Vulcão Central Unit. In contrast, Serra Branca is entirely evolved in composition (benmoreite to trachyte) (Larrea et al. 2014a). These rocks are characterised by porphyritic, microporphyritic and trachytic textures (Fig. 5). The porphyritic rocks are characterised by the presence of 15–50% macrocrysts (2–10 mm) embedded in a holocrystalline to hypocrystalline groundmass. The macrocryst assemblage comprises subhedral to anhedral olivine, clinopyroxene, feldspars and less abundant amphibole and Fe–Ti oxides, which have been defined as phenocrysts in equilibrium with their host magma (Larrea et al. 2014a); The presence of antecrysts has also been recognised. Antecrysts are large crystals that, based on their chemical composition, likely did not share a common history or crystallize from the magma in which they are now hosted, but rather may have grown from a different magma batch within the same magmatic system; (Charlier et al. 2005; Gill et al. 2006; Davidson et al. 2007; Jerram and Martin 2008). In the Graciosa lavas, however, the antecrysts are relatively scarce, especially in comparison to Corvo (see Sect. 2.9). The groundmass is composed of microcrysts (<2 mm) of olivine, feldspars, clinopyroxene and Fe–Ti oxides. The microporphyritic rocks are holocrystalline to hypocrystalline and are composed of microcrysts of feldspars, olivine, clinopyroxene and Fe–Ti oxides with a bimodal size distribution; most microcrysts are smaller than 0.5 mm, although crystals of 0.5–2 mm are also recognised. Amphibole microcrysts are also present in the evolved rocks (benmoreites and trachytes). In some microporphyritic samples larger isolated antecrysts (>1 mm) are observed, including olivine, clinopyroxene and feldspar that are highly corroded and have rounded rims (Larrea et al. 2014a). Samples from Serra Branca exhibit a trachytic texture. They consist of isolated <1 cm tabular feldspar crystals embedded in a groundmass predominantly composed of sub parallel, minute (<0.5 mm), acicular to tabular feldspar microcrysts, with scarce clinopyroxene and opaque microcrysts.

All mineral phases present in lava flows have been analysed by Larrea et al. (2014a) (Fig. 2). Olivine phenocrysts and microcrysts show normal zoning, with decreasing forsterite (Fo_{89-60} and Fo_{85-55}, respectively) and Ni contents, and increasing MnO and CaO concentrations from core to rim. Phenocrysts and microcrysts of clinopyroxene are classified as augites, diopsides and hedenbergite (Morimoto et al. 1988). Feldspar phenocrysts and microcrysts are normally zoned or unzoned and range from bytownite to anorthoclase (An_{82-0} Ab_{17-66} Or_{0-34}). Amphibole shows quite varied compositions, classified as kaersutite, magnesiotaramite and riebeckite depending on their Ca and Na contents (Leake et al. 1997); they appear as scarce phenocrysts in basaltic lavas or microcrysts in the evolved rocks. Opaque minerals occur mainly as small crystals (<0.5 mm) although some isolated phenocrysts are found as well. They are primarily normally zoned Al-bearing titanomagnetite, but ilmenite and scarcer Cr-Spinel crystals are also recognised. Fluorapatite is found as microcrysts in the evolved rocks.

Bivariate plots of selected whole rock major element compositions versus MgO (wt%) from Graciosa rocks are plotted in Fig. 3a. Major elements including TiO_2, Fe_2O_3, and CaO decrease with decreasing MgO, whereas Na_2O and K_2O increase with decreasing MgO. In contrast, Al_2O_3 increases slightly over the range of MgO values. The major element compositions of the lavas fall along a single liquid line of descent. Although slight differences in incompatible trace element and radiogenic isotope ratios imply that the parental magmas might not be all comagmatic, they are likely similar in composition.

Larrea et al. (2014a) evaluated fractional crystallization as a potential mechanism for magmatic differentiation using major and trace elements. The major element compositions of Graciosa lavas were best modelled by a MELTS (Ghiorso and Sack 1995; Asimow and Ghiorso 1998) polybaric (500–10 MPa) fractional

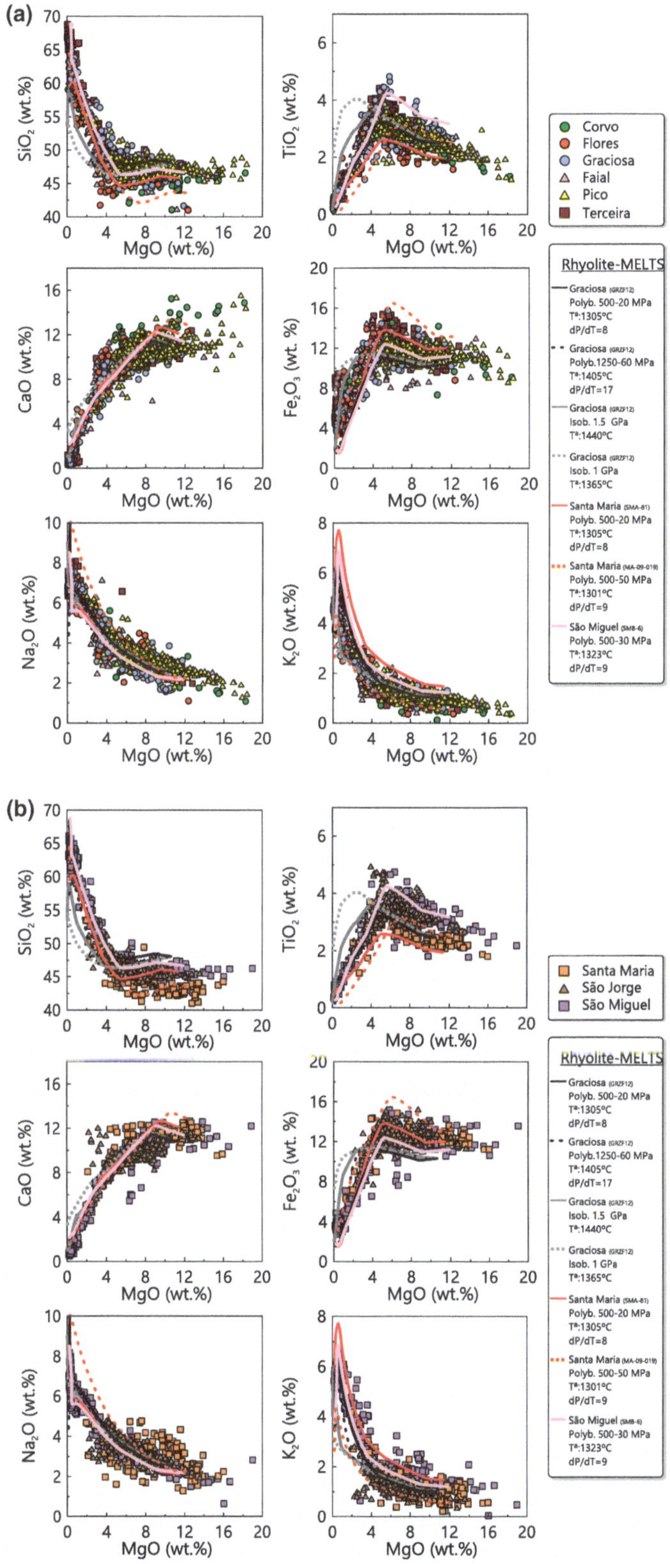
(a)
SiO2 (wt.%)
TiO2 (wt.%)
CaO (wt.%)
Fe2O3 (wt. %)
Na2O (wt.%)
K2O (wt.%)
MgO (wt.%)
Corvo
Flores
Graciosa
Faial
Pico
Terceira
Rhyolite-MELTS
Graciosa (GRZF12) Polyb. 500-20 MPa T°:1305°C dP/dT=8
Graciosa (GRZF12) Polyb.1250-60 MPa T°:1405°C dP/dT=17
Graciosa (GRZF12) Isob. 1.5 GPa T°:1440°C
Graciosa (GRZF12) Isob. 1 GPa T°:1365°C
Santa Maria (SMA-81) Polyb. 500-20 MPa T°:1305°C dP/dT=8
Santa Maria (MA-09-019) Polyb. 500-50 MPa T°:1301°C dP/dT=9
São Miguel (SMB-6) Polyb. 500-30 MPa T°:1323°C dP/dT=9
(b)
Santa Maria
São Jorge
São Miguel

◀ **Fig. 6** a, b Major element concentrations versus MgO for the Azores Archipelago. Published data from Azores as in Figs. 1 and 3. Curves represent the evolution paths of residual melts modeled using Rhyolite-MELTS (Gualda et al. 2012). Polybaric (Polyb.) and Isobaric (Isob.) fractional crystallization models were carried out at different initial pressures (as indicated for each model), with cooling steps of 5 °C and an initial water content of 0.5%, using GRZF12 (Larrea et al. 2014a) as the initial composition for Graciosa models, SMA-81 (Anderson 1983) and MA09-019 (Beier et al. 2012a) for Santa Maria, and SMB-6 (Widom et al. 1997) for São Miguel

crystallization process, starting at a pressure of 500 MPa with 0.5% H_2O (Fig. 6a), and with olivine, clinopyroxene, plagioclase, Fe–Ti oxides, kaersutite and minor apatite as fractionated solids. A two-stage Rayleigh (1896) fractional crystallization model was needed to reproduce the whole range of Graciosa lava trace element compositions. In agreement with the MELTS model, the modes of the alkaline xenoliths hosted by Graciosa lavas (see Graciosa plutonic rocks section) were used to govern the fractionating mineral phases. Degrees of fractional crystallization of <60% reproduce the compositions of mafic and intermediate lavas from Serra das Fontes and the Vitória-Vulcão Central, and >70% reproduce the composition of the evolved samples from the Serra Branca. Accordingly, the evolution of Graciosa magmatic products took place almost entirely at oceanic crustal depths from ~15 to 1 km (Escartín et al. 2001; Luis and Neves 2006; Matias et al. 2007). The most primitive magmas started to fractionate at ~15 km and polybaric fractionation continued up to shallower depths (<1 km) where the most evolved volcanic products (trachytic lavas and syenites; see Graciosa plutonic rocks section) were formed. Phenocrysts grew in equilibrium with the host melts and microcrysts grew almost entirely at surface conditions. Therefore, the existence of porphyritic or microporphyritic rocks in Graciosa depends exclusively on the initial composition of the melt, and its fractionation rate during magma chamber residence and ascent to surface prior to eruption (Larrea 2014).

This geochemical information together with the volcanostratigraphic sequence (Gaspar 1996) and the geochronology (Larrea et al. 2014b) of the island were used to investigate the magmatic evolution and ocean island construction stages (shield, erosional and rejuvenated) in Graciosa island (Larrea 2014). The shield stage started at least 1.05 Ma ago with the formation of the Serra das Fontes Volcanic Complex as a result of low to intermediate degrees of fractionation (<50%) of primitive melts at ~10–15 km depth. At some point around 450 ka, in the central part of the current island, the Serra Branca composite volcano was formed in response to evolution of the magmatic plumbing system. Trachytic products were produced by >70% fractional crystallization of primitive melts in a ~1 km magma chamber. This shield stage remained active for ca. 850 ky, followed by an erosional stage of ca. 110 ky that was characterised by volcanic inactivity and erosion. The rejuvenated stage generated two main volcanic units: Vitória and Vulcão Central. The volcanic activity restarted on the NW of the island associated with monogenetic eruptive centres, producing the NW Vitória Unit platform and covering most of the remaining volcanic units. This younger volcanism, mainly mafic in composition, indicates primitive magma recharge together with a low to intermediate degree of fractionation (<50%) in the plumbing system (~10–15 km). Simultaneously, the formation of a new island started to the south of the oldest volcanic structures associated with monogenetic cones of unknown age. Afterwards, a main eruptive center produced successive lava flows and volcaniclastic eruptions generated the Vulcão Central stratovolcano; this was followed by the formation of a caldera and a basaltic lava lake over a period of ~60 ky. Accordingly, the formation of the Vulcão Central Unit implies low to high degrees of fractionation (<90%) of primitive melts within the related magma plumbing system (~15–1 km depth).

Plutonic rocks

Plutonic rocks from Graciosa have been studied by Larrea (2014). They have been found in four outcrops: Baía da Folga and Enxudreiro within the Vulcão Central Unit, and Ponta da Pesqueira and Quitadouro within the Vitória Unit. The xenoliths are 3–20 cm long and show sub-rounded to angular shapes and sharp contacts with the host rocks without any apparent preferred orientation. These xenoliths are classified according to Le Maitre (2002) as syenite, dunite and gabbro and their whole rock compositions are plotted in Fig. 4; the latter comprises two types, alkaline and subalkaline gabbros, according to their mineral chemistry and whole rock compositions (Larrea et al. 2014a).

Dunites have a phaneritic texture (Fig. 5) constituted by an equigranular mosaic of 90% olivine, 5% spinel and 5% clinopyroxene, with olivine (Fo_{92-90}, CaO <0.05 wt% and 1493–3143 ppm Ni) and spinel (Cr# 0.43–0.54, Mg# 0.61–0.75 and $TiO_2 < 0.41$ wt%) compositions in the range of abyssal peridotites (e.g., Dick and Bullen 1984; Hekinian et al. 1993; Allan and Dick 1996). Together, the mineralogy, mineral chemistry, and whole rock chemistry of these dunites argue against either a cumulate or residual origin (Larrea et al. 2014a). Rather, they are consistent with an origin as replacive dunites (e.g., Batanova and Savelieva 2009), although the scarcity of dunites within the volcanic sequence and their small size precluded any more in-depth evaluation of their origin.

Alkaline gabbros were previously described by Larrea (2010) and Larrea et al. (2010). They display a medium- to coarse-grained orthocumulate texture (Fig. 5) and are composed of variable proportions of feldspars, clinopyroxene, amphibole, olivine, Fe–Ti oxides and apatite. Euhedral to subhedral feldspars, olivine and clinopyroxene are the main cumulus phases. Amphibole is the principal intercumulus phase in all the cases; when it develops large poikilitic crystals (up to 2 cm), the xenolith texture is described as heteroadcumulate (Larrea et al. 2014a). Their cumulate textures were the first indicator of their likely origin as magma chamber cumulates and therefore, their genetic relationship with the lavas by fractional crystallization. The fractionated mineral assemblage obtained in the MELTS model (see above) and the composition of these minerals was broadly consistent with that of the alkaline xenoliths. Therefore, Larrea et al. (2014a) concluded that these xenoliths were likely formed as magma chamber cumulates under a range of pressures from ~450 to 100 MPa, possibly in different magma chambers or pockets located at depths ranging from 15 to 3 km. Moreover, one gabbroic xenolith hosted in a lower lava flow of the Vulcão Central volcanic sequence yielded an age of 865.6 ± 61.1 ka (Larrea et al. 2014b), which suggested that its formation might be related to the Serra das Fontes—Serra Branca magmatic system.

Subalkaline gabbros consist of olivine, clinopyroxene, feldspars and accessory opaque minerals. These xenoliths present inequigranular textures characterised by a clear bimodal distribution of sizes (Fig. 5). Anhedral to subhedral crystals (1–2 mm in size) are surrounded by a framework of polygonal grains smaller than 0.5 mm. These mineral phases and their whole-rock compositions are distinguished by being much more primitive than those of the alkaline gabbros. Larrea et al. (2014a) interpreted their origin as cumulates, resulting from isobaric fractional crystallization of highly refractory melts at shallow depths (~3 km) with a very restricted range of temperature and limited degree of fractionation, although these melt compositions have not been identified within the Graciosa eruptive sequence.

The syenites were also studied petrographically by França (1993) and França and Rodrigues (1994). They occur as an inequigranular (0.1–2 mm), allotriomorphic mosaic composed primarily of alkaline feldspar (~90%), as well as amphibole, Fe–Ti oxides, and pyroxene, quartz, biotite, sphene and zircon as accessory minerals. Their whole rock major and trace element compositions are quite similar to the Graciosa trachyte lavas. This, together with their negative Eu/Eu* anomalies (<0.84), argue against their origin as an evolved cumulate, favoring a bulk trachyte liquid origin for these syenites, similar to the nodules studied by Widom et al. (1993) in

Água de Pau Volcano on São Miguel, and Daly et al. (2012) at Santa Bárbara volcano on Terceira (Larrea et al. 2014a).

2.6 Faial Island

Faial is the westernmost island of the Central Group, located between ~38° 30′ 56″ and 38° 38′ 40″ N latitude and 28° 35′ 55″ and 28° 50′ 06″ W longitude (Fig. 1 in Beier et al., Chapter "Melting and Mantle Sources in the Azores"), forming the seismically active WNW-ESE Faial-Pico fracture zone (Fernandes et al., Chapter "The Contribution of Space-Geodetic Techniques to the Understanding of the Present-Day Geodynamics of the Azores Triple Junction", Mitchell et al., Chapter "Volcanism in the Azores: A Marine Geophysical Perspective").

The first geological 1:25,000 map of Faial was made by Zbyszewski et al. (1959a). Afterwards, several volcanostratigraphic maps were proposed by Machado and Forjaz (1968), Forjaz (1977a, 1980a, 1980b), Chovelon (1982), Lemarchand (1987), Serralheiro et al. (1989), Madeira (1998), Coutinho (2000) and Pacheco (2001). However, the 1:15,000 map by Serralheiro et al. (1989), modified somewhat by several subsequent studies (e.g., Madeira 1998; Pacheco 2001; Forjaz 2008), is the basis for most of the more recent volcanostratigraphy. Based on the Pacheco (2001) stratigraphy, four stages are proposed for Faial formation, in order of decreasing age:

(1) The Ribeirinha Volcanic Complex (Ribeirinha Volcano) is located in the NE of the island. The formation of this unit began ~850,000 years ago (Hildenbrand et al. 2012) with the emplacement of a central shield volcano. It consists of a series of lava flows ranging from basaltic to benmoreitic in composition, as well as a pumiceous deposit, welded scoria, ignimbrites and basaltic pyroclasts. Now this volcano is highly eroded, and due to the tectonic displacements along the WNW-ESE trending Pedro Miguel Graben, there are currently only traces of the old Ribeirinha Volcano remaining.

(2) The Caldeira Volcano (also known as Cedros Volcanic Complex) is a polygenetic volcano formed ~410,000 years ago with a 2 km-diameter caldera (~1 ka), located to the SW of Riberinha Volcano. This complex was divided into Lower and Upper Groups (Pacheco 2001). The Lower Group is formed by hawaiian/strombolian eruptive styles with emission of basaltic to benmoreitic products (~16 ka). The Upper Group shows predominantly explosive eruptions (16–1.2 ka), which produced pyroclastic flows, fall and surge deposits of trachytic to benmoreitic compositions.

(3) The Horta-Caldeira Fissural System (also known as Horta Platform or Almoxarife Formation) constitutes a low altitude and smooth relief platform located to the SE of the island and formed by an 11,000 year old fissural system. It is composed of basalt to benmoreite lava flows interbedded with pumice layers and subaerial-submarine basaltic pyroclasts, which form some scoria and tuff cones.

(4) The Caldeira-Capelo Fissural System (Capelo Formation) includes all products from the western fissural volcanism and the historical eruptions of Cabeço do Fogo and Capelinhos (~1.2 ka to the present). It is composed of several WNW-ESE scoria cones that give rise to mafic volcanism (basalts to hawaiites) of low explosivity. The last eruption took place in 1957–1958 (Capelinhos surtseyan eruption), adding 1.5 km^2 to the island's area (Forjaz 2008).

Extensive petrographic and geochemical studies were reported by Chovelon (1982), Lemarchand (1984) and most recently updated by Machado et al. (2008), Lima et al. (2011), Beier et al. (2012a) and Zanon et al. (2013). Faial subaerial and submarine lavas are alkaline and belong to the sodic series without any Daly Gap. According to the TAS classification of Le Bas et al. (1986), Faial volcanic products range from basalts to trachytes (Fig. 1). The basalts are mainly porphyritic and less frequently aphyric. They contain variable amounts of olivine, clinopyroxene and plagioclase. Some of the large crystals found in these samples show resorbed rims and embayments, indicating disequilibrium with the surrounding melt and providing evidence for an antecrystic origin (Zanon et al. 2013). Olivine phenocrysts are occasionally cracked, corroded and iddingsitized. Microcumulate aggregates of clinopyroxene and plagioclase are common. The groundmass is usually holocrystalline, with the same mineralogy as the phenocrysts but with higher occurrence of Fe–Ti oxides. The hawaiites exhibit similar petrographic features; nevertheless, they are less porphyritic, more plagioclase-rich and contain dark-brown amphibole phenocrysts and microphenocrysts. The mugearites are characterised by aphyric to subaphyric textures with a small percentage of phenocrysts to microphenocrysts of mainly plagioclase and minor amounts of clinopyroxene, olivine and oxides. Amphibole and apatite crystals appear as accessory minerals. The benmoreites show aphyric and hyalopilitic textures where plagioclase is the dominant phase. Olivine, clinopyroxene, amphibole, apatite, biotite and oxides are present in minor proportions. The trachytes display trachytic textures mostly composed of plagioclase and alkali feldspar microphenocrysts, with minor occurrence of amphibole, biotite, apatite and opaque crystals.

All of these mineral phases have been analysed by Beier et al. (2012a), Zanon and Frezzotti (2013) and Zanon et al. (2013) (Fig. 2). Olivine crystals are euhedral to subhedral, with phenocryst compositions ranging from Fo_{87} to Fo_{72}, while the microcrysts from the groundmass have lower Fo contents ($Fo_{76–63}$). Clinopyroxene appears as anhedral phenocrysts and microcrysts, normally zoned and partly resorbed with embayments and reaction rims. Their core compositions range from augite to diopside ($Wo_{42–49}$, $En_{36–45}$, $Fs_{10–19}$), sometimes with aegirine—augite rims. Plagioclase phenocrysts are euhedral to subhedral, tabular, twinned and show oscillatory to normal zoning and occasionally embayments. Phenocryst and microphenocryst compositions from various samples with different degrees of evolution show a continuous trend from bytownite to andesine ($An_{83–30}$ $Ab_{17–66}$ $Or_{0–4}$). In some pumices, plagioclase phenocrysts show resorption, indicating disequilibrium with the surrounding melt. Alkali feldspars (anorthoclase—sanidine) are present as euhedral, tabular and zoned phenocrysts or microphenocrysts in pumices and evolved lavas. Opaque minerals are ubiquitous in all samples as euhedral to subhedral microphenocrysts in the matrix and as inclusions in olivines. They are classified as ilmenite, magnetite and titanomagnetite. Euhedral Cr-spinel microcrysts occur embedded in many olivine antecrysts and phenocrysts. Fluorapatite, kaersutite and biotite are present in many evolved samples as both phenocrysts and microcrysts in the groundmass. Quartz is also present in some trachytic pumices.

Variations in major elements versus MgO (Fig. 3a) indicate an overall continuous range from mafic to felsic rocks, with an abrupt slope change of some major oxides at ~4 wt% MgO. The highest MgO contents (>10 wt%) observed in some samples reflect accumulation of clinopyroxene and olivine antecrysts, as observed petrographically (Machado et al. 2008; Lima et al. 2011; Zanon et al. 2013). Major element variation with increasing degree of differentiation is characterised by enrichment in SiO_2, Na_2O, and K_2O and decreasing CaO contents. Titanium, Fe_2O_3, Al_2O_3 and P_2O_5 contents increase in the initial stages of evolution and decrease as the rocks become more evolved. Therefore, the evolutionary trend formed by Faial rocks suggests that fractional crystallization involving clinopyroxene, olivine, calcic plagioclase and titanomagnetite was the major process controlling the magmatic differentiation of the different evolutionary stages of Faial (Machado et al. 2008; França et al. 2009; Lima et al. 2011).

Zanon et al. (2013) tested this hypothesis using both trace element modelling and major element mass balance calculations, aiming to determine the amount of fractionation needed to reproduce the compositional ranges of the Caldeira volcano and the Fissure Systems, and the different conditions of magma storage and fractionation. According to Zanon et al. (2013), Faial magmas were formed by low degrees of partial melting of a variably metasomatised upper mantle in the garnet stability field (>85–90 km deep). Based on geothermobarometry data, the basaltic rocks from the Caldeira Volcano were formed at 760 MPa (~ 26 km) beneath the present edifice, evolving to generate the mugearite—benmoreite sequence at 159 MPa (~ 9 km). The benmoreite-trachyte series from the Caldeira Volcano was formed by ~ 70% fractional crystallization, in a shallower magma chamber located at ~ 3.5–5 km (100–132 MPa) beneath the Caldeira Volcano, via fractionation of olivine, clinopyroxene, feldspars, biotite and apatite. In contrast, the Horta-Caldeira and the Caldeira-Capelo Fissural Systems are composed of basalts—hawaiites formed by moderate degrees of fractional crystallization at high to intermediate pressures of crystallization, corresponding to the uppermost mantle (~ 15 to 28 km). In the case of the Horta-Caldeira Fissure system, fractional crystallization occurred at a shallower depth to generate magmas with a similar degree of evolution as the Caldeira-Capelo Fissural System. Therefore, geochemical and petrological data indicate polybaric fractionation of Faial magmas, and suggest the existence of reservoirs at different depths within the crust.

Plutonic rocks

Ultramafic, mafic and evolved plutonic rocks from Faial have been studied by Baker (1966), França (1993), França and Rodrigues (1994) and Zanon and Frezzotti (2013).

Ultramafic xenoliths were found in a flow from Capelo Fissure Zone, cropping out on the northern coastal cliff. The xenoliths range in size up to 20 cm, and are coarse-grained and composed of variable amounts of olivine, clinopyroxene, orthopyroxene and spinel. They are classified as websterite and wherlite, and have porphyroclastic to equigranular textures, where olivine and orthopyroxene grains are present as large porphyroclasts (up to 4 mm) and polygonal grains typically <1 mm, together with polygonal clinopyroxene grains in the wherlite. Spinel is interstitial (up to 0.5 mm in diameter) or present as inclusions in olivine. Based on mineral chemistry, Zanon and Frezzotti (2013) interpreted these websterite and wherlite xenoliths from Faial as fertile fragments of the upper mantle, representing crystallised melts or cumulates formed at mantle depth.

Baker (1966) reports the existence of mafic plutonic nodules at several localities on Faial, including Caldeira, Feteira, the 1957–1958 historic Capelinhos lavas and the shore to the north of Fajã da Praia do Norte. The plutonic nodules present sharp contacts with their host lava flows, and their mineral assemblage is composed of variable amounts of olivine, clinopyroxene and plagioclase. In accordance, these mafic xenoliths were divided into five groups: (1) Olivine nodules. The olivine is forsterite with some chromite inclusions. Chilled glass is frequently present around the crystals and along fractures. (2) Plagioclase, clinopyroxene and olivine nodules. The percentage of the different minerals and their degree of alteration is highly variable from sample to sample. Some olivine crystals have iddingsite bands and hematite filling fractures. The pyroxene is Cr-rich diopside, with orthopyroxene exsolution lamellae. The plagioclase shows patchy extinction. (3) Clinopyroxene and olivine nodules. Chromium-rich diopside is the dominant phase with rare discontinuous orthopyroxene exsolution lamellae. Olivine and clinopyroxene show irregular patchy extinction. (4) Clinopyroxene and plagioclase nodules. They are Cr-rich diopside-plagioclase clusters with common orthopyroxene exsolution lamellae. (5) Clinopyroxene nodules. They are composed of non-pleochroic clinopyroxene containing minor hematite, chromite and glass inclusions. Baker (1966) proposed a cumulate origin for these mafic nodules based on their mineralogy and textures, ruling out their mantle origin due to the absence of orthopyroxene.

Syenites found in fall deposits from the Caldeira Formation (Rinquim deposits) and Capelo Formation (Caldeira and Fajã deposits) were studied by França (1993) and França and Rodrigues (1994). They are <2 cm in size and generally medium-grained with subhedral shapes and consertal and micropertitic intergrowth textures. Minor radiated and corona textures are also found. They are composed of variable amounts of feldspars, biotite, clinopyroxene, fayalite-olivine, amphibole, zircon, apatite, sphene and oxides. No conclusions about their origin have been proposed.

2.7 Pico Island

Pico is the southernmost island of the Central Group, located between ~38° 22′ and 38° 36′N latitude and 28° and 28° 34′W longitude (Fig. 1 in Beier et al., Chapter "Melting and Mantle Sources in the Azores"). Pico island, together with the nearby island of Faial, forms a 160 km long volcanic ridge oriented WNW-ESE (Fernandes et al., Chapter "The Contribution of Space-Geodetic Techniques to the Understanding of the Present-Day Geodynamics of the Azores Triple Junction" and Mitchell et al., Chapter "Volcanism in the Azores: A Marine Geophysical Perspective").

Zbyszewski et al. (1962) published the first geological map of Pico on scale 1:50,000. The first volcanostratigraphical column was presented by Forjaz (1966), which was subsequently expanded upon by Forjaz (1977b, 1980a, b), Forjaz et al. (1990) and Madeira (1998). The first K/Ar determinations by Feraud et al. (1980) enabled a new stratigraphy to be established (Chovelon 1982) and a new volcanostratigraphic sequence for Pico (Cruz et al. 1995; França et al. 1995; Nunes et al. 1995), detailed in Nunes et al. (1999). Three main volcanic complexes were proposed for Pico (Nunes et al. 1999): Topo-Lajes, São Roque-Piedade and Montanha.

(1) The Topo-Lajes Volcanic Complex is the oldest volcanic system of the island, located on the central-south coast. Its formation began ~300,000 years ago in response to fissural volcanism, controlled by the WNW-ESE trending regional fracture. The last eruption of Topo volcano probably took place 5000 years ago. Currently, it represents the eroded remains of an older shield basaltic volcano.
(2) The São Roque-Piedade Volcanic Complex (also known as Planalto da Achada Fissure Zone) is a fissure system characterised by a 29 km long ridge located in the eastern half of the island, which started to form about 230,000 years ago. Small basaltic flows originate from a series of 170 identified scoria cones along the ridge (e.g., França et al. 2006b; Nunes et al. 2006). According to França (2000), eruptions in this volcanic complex were contemporaneous with those of the Topo and Montanha centres and were active until AD 1562.
(3) The Montanha Volcanic Complex is a basaltic composite volcano located in the western area of Pico. It comprises lavas that are generally <1500 years old (Nunes et al. 1999), including the most recent eruption in AD 1720. Three stages are recognised in the development of the composite volcano: (1) volcano growth up to 2050 m high and local collapse/formation of an 800 m diameter crater, (2) volcano growth up to 2250 m high filling the previous crater and (3) formation of a pit crater of ~590 m diameter with 25 m high walls (França et al. 2009). The eruptive center then drifted to the NE, and vigorous lava fountaining and extrusion built a 100 m-high lava cone (Piquinho) and a lava lake filled the pit crater. The most recent activity at the volcano's summit produced volcaniclastic lavas.

Detailed studies of representative samples from all three volcanic complexes were carried out by Flower et al. (1976), Chovelon (1982), França (2000), Claude-Ivanaj et al. (2001), França et al. (2006c), Prytulak and Elliott (2009), and Zanon and Frezzotti (2013). Pico lavas mostly belong to the alkaline series; according to the TAS classification of Le Bas et al. (1986), Pico volcanic products are mainly basalts (~78%), although hawaiites (~21%), mugearites and benmoreites are also present (Fig. 1). Only two samples have been classified as potassic trachybasalt: one from the Montanha and the other from the São Roque-Piedade volcanic complexes (França 2000; França et al. 2006c).

Three different types of textures have been described in Pico lava samples: porphyritic, aphyric and trachytic textures. Most basalts and hawaiites are porphyritic, although some of them are aphyric with small phenocrysts. In these rocks, the most common phenocryst phases are plagioclase, clinopyroxene and olivine, with the presence of minor oxides. The rock groundmass is composed of the same mineral assemblage with a higher fraction of oxides, and rare apatite and volcanic glass. Some porphyritic samples with abundant pyroxene and olivine megacrysts (ankaramites) were described by Zanon and Frezzoti (2013) in the São Roque-Piedade (<19% of megacryst) and Montanha (from 25 to 70% of megacrysts) volcanic complexes. These ankaramites have been interpreted by França et al. (2009) as products of crystal accumulation/settling, producing alternating crystal-rich and aphyric lava flow layers. In contrast, trachytic textures are found exclusively in mugearite and benmoreite samples from the 1718 AD eruption (Santa Luzia eruption). These samples are, both macroscopically and microscopically, characterised by oriented microlites and euhedral plagioclase crystals with a transparent to milky appearance. In addition to plagioclase, minor pyroxene, olivine, amphibole, opaque minerals and apatite phenocrysts are also found. The mineral assemblage forming the groundmass of these rocks is similar to the phenocryst assemblage.

All of these mineral phases have been analysed by França (2000), França et al. (2006c), Beier et al. (2012a) and Zanon and Frezzotti (2013) (Fig. 2). Olivine crystals are subhedral to euhedral and/or skeletal. Their core compositions range from Fo_{87} to Fo_{50}, with rims enriched in fayalite contents. The olivine megacrysts are commonly iddingsitized corroded, and have overgrowth rims, thus they are considered to be antecrysts (Zanon and Frezzotti 2013). Clinopyroxene is the most abundant mineral in Pico rocks, occurring as microcrysts and phenocrysts classified as augites and diopsides according to Morimoto et al. (1988). They are mostly zoned prismatic subhedral crystals, sometimes showing pristine basal sections, although embayment and corrosion features are sometimes observed in large crystals. As in the case of the olivine, the clinopyroxene megacrysts are classified as antecrysts (Zanon and Frezzoti 2013). Feldspar crystals are mainly euhedral and tabular plagioclases, showing oscillatory or reverse zoning with compositions ranging from bytownite to andesine (An_{82}–An_{31}) in the basaltic lavas (França 2000). Alkali feldspars (anorthoclase and sanidine) have also been identified in the most evolved lavas from the Santa Luzia eruption. Amphibole crystals are rare in Pico lavas, occurring as kaersutite crystals only in the most evolved lavas from Santa Luzia. They appear as isolated crystals or associated with other amphibole or clinopyroxene crystals. They are commonly subhedral to anhedral, with corroded rims that indicate instability relative to the surrounding liquid. Iron-titanium oxides are mainly present as microcrysts in the groundmass of the most evolved lavas (phenocrysts are rare), occurring as isolated crystals or included in olivine and clinopyroxene crystals. They are titanomagnetites and ilmenites, although some oxides occur as weathering products. Apatite crystals occur as microcrysts, usually as inclusions or otherwise associated with other minerals in the most evolved samples. Temperature and pressure calculations based on the olivine-augite geothermometer by Loucks (1996) and the experimental geothermobarometer by Albarède (1992) indicate equilibration at 1200 °C and 1.1 GPa for the basaltic samples, and 1100 °C and 600 MPa for the most evolved rocks from Pico (França 2000).

Regarding their whole-rock compositions, Pico lavas show an overall continuous range from primitive to intermediate-evolved rocks, with an abrupt change in slope on bivariate diagrams of some major oxides at ~5 wt% MgO. There is a negative correlation between MgO and SiO_2, TiO_2, Al_2O_3, Na_2O, K_2O and P_2O_5 contents, and a positive correlation with CaO contents, whereas the total Fe_2O_3 contents are relatively constant over a wide range of MgO values (from 16 to 3%) (Fig. 3a). França (2000) pointed to fractional crystallization as the dominant process generating Pico lavas, although the geochemical behaviour of some incompatible element ratios in these lavas was also interpreted to result from variable degrees of partial melting of the mantle source beneath Pico (França et al. 2006c). The basaltic lavas evolved primarily via fractionation of clinopyroxene, Mg-rich olivine, and Ca-rich plagioclase. The TiO_2, P_2O_5 and Fe_2O_3 contents of the most evolved lavas suggest that ilmenite and apatite fractionation played an important role in their formation. However, Prytulak and Elliott (2009) indicate that mixing between a mafic and more evolved magma is also required to explain the geochemical behaviour of these evolved lavas from the Santa Luzia flow.

This geochemical information, together with geothermobarometric estimates, was further used to delineate the complex plumbing system beneath Pico (França 2000). The most primitive melts started to fractionate at depths of ~ 40 km, and evolved by fractional crystallization forming the hawaiites during ascent within volcanic conduits. However, the mugearite—bemoreite rocks were formed by fractional crystallization at shallower depths (~ 20 km). The fact that the volume of mugearitic/benmoreitic lava erupted is small (21×10^6 m^3; Cruz et al. 1995), and the absence of a significant negative gravimetric anomaly under the Montanha composite volcano (Camacho et al. 1999), seem to be consistent with the absence of a well-developed magma chamber beneath Pico; however, França (2000) and França et al. (2006c) propose the existence of a magma chamber at an early stage formation.

Plutonic rocks

Plutonic rocks from Pico have been studied by Esenwein (1930), França et al. (1995), França (2000) and Zanon and Frezzotti (2013). Esenwein (1930) reports the geochemical composition of only one gabbro xenolith from Pico (Fig. 4); however, the ultramafic xenoliths have been thoroughly studied from a petrologic point of view. They comprise ultramafic xenoliths found in two different outcrops, including the Santa Luzia flow from Montanha Volcanic Complex and a lava that crops out at the base of the northern cliff (>50 ka), covered by the AD 1562 lava flow from the São Roque-Piedade Volcanic Complex. These ultramafic xenoliths are <4 cm in size and present polygonal shapes with planar boundaries, surrounded by a thin lava layer. They are coarse-grained, equigranular harzburgites composed of variable amounts of olivine, clinopyroxene, orthopyroxene and spinel. Olivine and orthopyroxene grains are present as large porphyroclasts (up to 4 mm) and polygonal grains typically <1 mm. Both olivine and orthopyroxene exhibit deformation features along with secondary fluid inclusions. Clinopyroxene is present as interstitial grains and exsolution lamellae on orthopyroxenes. Spinel is interstitial (up to 0.5 mm in diameter) or included in olivine. Based on mineral chemistry, Zanon and Frezzotti (2013) interpreted these ultramafic xenoliths as mantle restites; i.e., fragments of upper mantle that have undergone partial melting and melt extraction.

2.8 Flores Island

Flores, along with the island of Corvo, forms the Western Group of the Azores Archipelago (Miranda et al., Chapter "The Tectonic Evolution of the Azores Based on Magnetic Data"). It is located within the American plate at $\sim 39°$ 28′N latitude and 31° 13′W longitude (Fig. 1 in Beier et al., Chapter "Melting and Mantle Sources in the Azores").

The first work carried out on Flores was the 1:25,000 geological map of the island by Zbyszewsky et al. (1968). Initial petrographic studies on the island were carried out by Azevedo et al. (1986), Morisseau (1987), Azevedo (1990) and Azevedo et al. (1991). Later mapping was extended to the entire island (1:15,000 map) and the volcanostratigraphic sequence was defined based on new geochronologic, petrographic and geochemical data, as well as the tectonic and geomorphological evolution (Azevedo 1998; Azevedo and Ferreira 2006). Based on Azevedo and Ferreira (2006), two major volcanic complexes were defined on Flores:

(1) The Basal Volcanic Complex includes products and structures from both submarine and subaerial volcanism of Plio-Pleistocene age ($\sim$2.2–0.7 Ma), which are largely observed in the lower units along the north and south coast. This complex is mainly formed by palagonitised volcaniclastic deposits (a mixture of hyaloclastite and hydroclastite), breccias and tuffs interbedded with basaltic flows.
(2) The Upper Volcanic Complex represents the main subaerial activity of the island. It includes three main subunits: Lower, Middle and Upper. The Lower subunit was formed between 0.67 and 0.55 Ma and consists of extensive lava flows (basaltic to trachytic) alternating with pyroclastic deposits. The Middle subunit is composed of basaltic and hawaiitic lava flows and associated pyroclastic deposits produced from 0.4 to 0.2 Ma. The Upper subunit was emplaced between 4–3 ka (Morrisseau and Traineau 1985) and is exclusively pyroclastic, including strombolian pyroclastic cones and a layer of widespread ash enriched in lithic clasts.

Flores is one of the least well studied islands in the Azores Archipelago from a petrological point of view. Zbyszewski et al. (1968) and Morisseau (1987) were the first to describe the petrography of Flores volcanic rocks, but the first in depth petrographic study of the whole lava sequence was carried out by Azevedo and Ferreira (2006). Recently, Genske et al. (2012b) published a study on the petrology and geochemistry of the Western Group islands, which included lavas from two major volcanic complexes of Flores.

As mentioned above, the lavas from Flores range in composition from alkali basalt to trachyte following the TAS classification of Le Bas et al. (1986), although there is a compositional gap in the intermediate compositions (Fig. 1). Genske et al. (2012b) indicate that the most primitive lavas are alkali-basalts characterised by megacrysts and phenocrysts of normally zoned clinopyroxene, olivine and plagioclase embedded in a plagioclase-dominated cryptocrystalline to finely crystalline groundmass. The most evolved rocks (hawaiite—trachyte) are higly vesicular and belong to the youngest volcanic units. They are characterised by abundant olivine, clinopyroxene, magnetite and plagioclase phenocrysts and microcrysts (Fig. 5). As the rocks become more evolved, plagioclase becomes the dominant phase. Moreover, some accessory minerals are recognised in the groundmass of the trachytes: hypidiomorphic amphiboles (kaersutite) replacing original clinopyroxenes, biotite as small patches around the feldspars and disseminated single crystals, and apatite needles (<1 mm).

All of these mineral phases have been analysed by Genske et al. (2012b) (Fig. 2). Olivine cores range from Fo_{91} to Fo_{69} and olivine rim compositions extend to Fo_{64}. Some of these olivine crystals are in chemical disequilibrium with their host lavas, and are interpreted by Larrea et al. (2012) as antecrysts. The clinopyroxenes are slightly zoned Ti-augites (Morimoto et al. 1988) with quite restricted compositions and decreasing MgO and increasing TiO_2 and FeO contents from core to rim; some of these clinopyroxene megacrysts have also been classified as antecrysts (Larrea et al. 2012). The feldspars are mostly bytownite—andesine crystals ($An_{88-41}Ab_{12-66}Or_{0-3}$), although anorthoclase crystals ($Ab_{67-70}An_{10-13}Or_{19-23}$) are found in the trachytic lavas. Iron-titanium oxides are classified as ilmenite, titanomagnetite and chrome-spinel and are commonly found as microcrysts in the groundmass, and around the rims or as inclusions in the mafic mineral phases.

Bivariate plots of selected major-element oxides versus MgO wt% (Fig. 3a) were used by Genske et al. (2012b) to investigate the evolutionary trends of the lavas from Flores. As MgO decreases from 12 to 4 wt%, SiO_2 increases slightly from 40 to 50 wt%, but then abruptly increases (up to 63 wt%) in lavas with 4–1 wt% MgO. Calcium decreases over the entire MgO range, and TiO_2, Fe_2O_3 and P_2O_5 first increase to a maximum at 4 wt% MgO, after which they decrease significantly as the compositions evolve to lower MgO contents. Aluminium, K_2O and Na_2O contents increase towards lower MgO.

The mineralogy, textures and bulk-rock compositions of the lavas are consistent with fractional crystallization being the dominant differentiation process. The decrease in size of the olivine and clinopyroxene phenocrysts and the accompanying increase in the size of plagioclase phenocrysts support a polybaric fractionation process; however, Genske et al. (2012b) suggest that not all the samples are cogenetic and that other processes, such as magma mixing or assimilation combined with fractional crystallization, may be involved in the formation of these lavas.

Genske et al. (2012b) tested this hypothesis with a single fractional crystallization model using the MELTS algorithm (Ghiorso and Sack 1995; Asimow and Ghiorso 1998) and the most primitive sample from Corvo as a near-primary magma composition. The polybaric MELTS model, starting at 1375 °C and 1.2 GPa with fractionating solids based on petrographic observations, provided the best fit to Corvo whole rock lava compositions. Flores whole rock lava data are significantly more scattered than Corvo data, however, Genske et al. (2012b) suggest that this model may also be applied to Flores. The fractional crystallization process was further appraised using a Rayleigh trace-element fractionation model. Mineral modes for the fractionating assemblage were extracted from the MELTS outputs. Unfortunately, this trace-element model did not provide a particularly robust test, because the trace-element trends display substantial scatter in the Flores lava suite. According to this model, crystallization commenced close to the lithosphere-asthenosphere boundary and a total of 70% crystallization of parental magma is needed to explain the most evolved compositions in Flores (Genske et al. 2012b).

The existence of crystals with disequilibrium textures and reverse zoning suggests that, in addition to simple fractional crystallization, there are likely more complex processes operating in the Flores magmatic system. One possibility that has been proposed is mixing of mafic melts with crystallizing ascending magmas as well as conduit convection and small-scale stagnation (Genske et al. 2012b). Alternatively, Larrea et al. (2012) have demonstrated a similar textural distribution in Flores rocks as that observed in Corvo (see Sect. 2.9 below; Larrea et al. 2013), including the existence of microporphyritic and trachytic lavas related by fractional crystallization and antecryst-rich porphyritic lavas formed by crystal accumulation, suggesting that both magmatic processes control the evolution of Flores.

Plutonic rocks

Gabbroic rocks from Flores were sampled from the base of the Middle subunit of the Upper Volcanic Complex (França et al. 2008). They have rounded shapes and sharp contacts with their host lava. Only one whole rock composition is reported by França (1993) (Fig. 4). Based on their petrologic and mineralogic compositions, two groups were defined: olivine bearing gabbros and olivine bearing gabbronorites.

The olivine bearing gabbros are fine to medium grained with hypidiomorphic granular and seriate textures (Fig. 5). Their mineral assemblage is composed of variable proportions of olivine, clinopyroxene, plagioclase, Fe–Ti oxides, kaersutite and accessory apatite. The mineral assemblage and mineral compositions indicate an alkaline affinity for these xenoliths, suggesting a genetic relationship with Flores lavas, as has been proposed for the alkaline xenoliths studied in the Graciosa and Corvo islands.

The olivine bearing gabbro norites have textural similarities to the olivine bearing gabbros, but the presence of orthopyroxene indicates a different mineralogic and chemical composition. These consist of olivine, clinopyroxene, orthopyroxene, plagioclase, amphibole and Fe–Ti oxides. The occurrence of both orthopyroxene and clinopyroxene is indicative of a subalkaline affinity, and the occurrence of olivine suggests that they crystallised from a silica undersaturated tholeiitic melt. This tholeiitic affinity, in contrast to the alkaline affinity of the archipelago, suggests that the olivine bearing gabbro norites probably crystallised at deeper volcanic levels or they are fragments of the oceanic crust beneath the island (França et al. 2008).

In addition, a preliminary study by França (1993) and França and Rodrigues (1994) described syenite samples found in two maar deposits from the Upper Volcanic Complex of Flores (Caldeira Funda das Lajes and Lagoa Comprida). They are <2 cm in size and present a medium granular xenomorphic texture composed of variable amounts of feldspar, biotite, pyroxene, zircon, opaque minerals, quartz and apatite. No further conclusions about their origin have been discussed.

2.9 Corvo Island

Corvo, the smallest island of the Azores archipelago, is the north-westernmost island of the Azores Archipelago. It is located in the Atlantic Ocean between ~39° 40′ and 39° 44′N latitude and 31° 05′ and 31° 08′W longitude (Fig. 1 in Beier et al., Chapter "Melting and Mantle Sources in the Azores"). Corvo, together with the island of Flores, makes up the Western Group islands (Miranda et al., Chapter "The Tectonic Evolution of the Azores Based on Magnetic Data").

The first geological work published on Corvo was the geological map (Zbyszewski et al. 1967) in which five volcanic units were proposed for the evolution of the island. At the beginning of the 21st century, a new field campaign was carried out on Corvo in order to publish a 1:10,000 geological map, together with a neotectonic and geomorphologic characterization of the island. These studies led to the establishment of a volcanostratigraphic sequence (Dias 2001; Azevedo et al. 2003). Concurrently, França et al. (2002) proposed a new straightforward volcanostratigraphic succession based on the caldera-formation event and relative chronostratigraphy. This led to the establishment of three main units in Corvo, in order of decreasing age:

(1) The pre-caldera unit is formed by submarine volcanism (proto-island) followed by two subaerial phases. The Lower subaerial unit includes a primitive shield volcano, strombolian deposits, the upper lava flows and the associated dikes. The Upper unit comprises secondary cones, buried cones and their related lava flows.
(2) The syn-caldera unit includes plinian to sub-plinian pumiceous deposits, lahars, surges and other pyroclastic flows associated with the stratovolcano collapse and the caldera formation.
(3) The post-caldera unit comprises the most recent lava flow (Pão de Açúcar), the intra-caldera pyroclastic and spatter cones and the Cortinhas scoria cone and associated lava flows.

An in-depth investigation of magmatic processes based on petrology, mineralogy and major and trace element data was carried out by Larrea et al. (2013) in accordance with the preliminary work by França et al. (2006b). In addition, Genske et al. (2012b) published a contribution on the petrology and geochemistry of the Western Group islands. Together, these studies (Genske et al. 2012b; Larrea et al. 2013) have characterised all of the diverse igneous rocks on Corvo, including lava flows, dikes and xenoliths, in the context of the volcanostratigraphy proposed by França et al. (2002).

The lava flows and dikes from the three volcanostratigraphic units range in composition from alkali picrobasalts to trachytes following the TAS classification of Le Bas et al. (1986) (Fig. 1). They appear to show a relatively well defined alkaline trend in the TAS diagram, although some basaltic samples plot in the tholeiitic basalt field and were identified in the field as clinopyroxene-dominated cumulate rocks (Genske et al. 2012b). Two types of lava flows and dikes from the pre-, syn- and post-caldera stages were clearly distinguished based on their texture: porphyritic rocks and microporphyritic rocks (Larrea 2014). The former group includes samples belonging to the pre-caldera and post-caldera units; the latter includes samples from all of the units (pre-, syn- and post-caldera). In accordance with the petrographic observations, compositional differences are observed between the porphyritic and microporphyritic rocks (Larrea et al. 2013). The porphyritic rocks range from alkaline picrobasalts to alkaline basalts showing the most primitive compositions, whereas the microporphyritic rocks range from hawaiites to trachytes with lower MgO contents. The porphyritic rocks are holocrystalline and characterised by the presence of a variable proportion of large crystals (up to 15 mm in size) that are set in a fine-grained groundmass (Fig. 5). The macrocryst assemblage mainly comprises subhedral to anhedral olivine, clinopyroxene, and feldspars; some macrocrysts are corroded and present overgrowth rims. The groundmass is composed of feldspars, olivine, clinopyroxene and opaque microcrysts (<2 mm). The volume fraction of macrocrysts in these samples ranges from 80 to 8%, varying significantly even within different areas of the same sample. The recognition of disequilibrium conditions between these macrocrysts and their host rocks led to the interpretation of their origin as non-cogenetic antecrysts (Larrea et al. 2013). In contrast, the microporphyritic rocks include lava flows and dikes without antecrysts, resembling the groundmass of the porphyritic rocks (Larrea et al. 2013). Those belonging to the pre-caldera and post-caldera units present a holocrystalline to hypocrystalline microporphyritic texture, sometimes indicating flow. They are mainly composed of feldspars, minor olivine, clinopyroxene, opaque minerals, scarce amphibole and occasional glass. Most groundmass microcrysts are smaller than 0.5 mm, although scarce crystals of 0.5–2 mm are also recognised. The syn-caldera samples are highly vesicular pumices mainly composed of glass with scarce microcrysts of feldspars, olivine, clinopyroxene and opaque minerals.

All of these mineral phases were analysed from lava flow and dike samples (Larrea et al. 2013 and Fig. 2). Olivine microcrysts show normal zoning, with decreasing forsterite contents (Fo_{78-57}) and increasing MnO concentrations from core to rim. The olivine antecrysts are weakly zoned from core (Fo_{90}) to rim (Fo_{81}). They have the lowest MnO concentrations, and are more primitive in composition than the microcrysts. Clinopyroxene antecrysts and microcrysts are classified as augite and diopside according to Morimoto et al. (1988), but they are clearly distinct in Ti, Cr, Al and mg* (Mg/(Mg + Fe^{2+} + Fe^{3+} + Mn)) contents. Olivine and clinopyroxene antecrysts are characterised by an overgrowth rim similar in composition to the groundmass microcrysts. Feldspars microcrysts range from labradorite to anorthoclase (An_{69-22} Ab_{30-70} Or_{1-8}). The feldspar antecrysts have more primitive compositions that range from anorthite to bytownite (An_{88-70} Ab_{12-29} Or_{0-1}). Amphibole crystals are very scarce in Corvo, occurring as kaersutite, magnesiokatophorite and gedrite microcrysts (Leake et al. 1997). Isolated amphibole antecrysts have been found in the evolved syn-caldera pumices and in the oldest lava flow of the pre-caldera unit, and are classified as kaersutite and hastingsite. Iron-titanium oxides occur mainly as ilmenite and Al-bearing titanomagnetite microcrysts, with hematite-maghemite microcrysts present as weathering products.

The porphyritic rocks display an increase in TiO_2, Al_2O_3, Fe_2O_3, Na_2O and K_2O and a decrease in CaO with decreasing MgO, whereas in the microporphyritic rocks TiO_2, Al_2O_3, Fe_2O_3 and CaO decrease with decreasing MgO (Fig. 3a). The porphyritic rocks from Corvo contain variable volume fractions of antecrysts, and have MgO contents >4.93 wt%. In contrast, the microporphyritic (antecryst-free) rocks have MgO contents <5.55 wt%. The highest MgO contents correlate with the highest proportions of mafic antecrysts in the rocks, which implies that the Mg-enrichment of the porphyritic rocks is controlled by the presence and proportion of mafic antecrysts (Larrea et al. 2013). This fact was also noticed by Genske et al. (2012b), who reported that samples with MgO concentrations higher than 12 wt% showed evidence of olivine and clinopyroxene accumulation. In addition, Larrea et al. (2013) developed a least squares model using the MINSQ software (Herrmann and Berry 2002) to quantify the proportions of each antecryst phase and groundmass needed to reproduce the whole rock composition of the porphyritic rocks, the results of which were consistent with the petrographic estimates (Larrea et al. 2013). Therefore, Corvo porphyritic rocks are composed of a groundmass equivalent to the microporphyritic rocks plus accumulation of variable proportions of olivine, clinopyroxene and plagioclase antecrysts (from 80 to 8% of the total volume fraction). This leads to the conclusion that the whole rock compositions of the magmas that accumulated a large fraction of antecrysts appear to be much more primitive than they would be without this accumulation process.

In contrast, the microporphyritic rocks are composed only of microcrysts, thus they represent liquid compositions, and are suitable for identifying primary magmatic processes affecting their evolution. The compositions of the microporphyritic rocks lie within a single liquid line of descent, pointing to a comagmatic origin and a common differentiation process by fractional crystallization. This process has been tested both for major and trace elements (Genske et al. 2012b; Larrea et al. 2013). Major element modelling (Genske et al. 2012b; Larrea et al.) was carried out using the MELTS algorithm (Ghiorso and Sack 1995; Asimow and Ghiorso 1998). The data wer2013e best modeled by polybaric fractionation starting at a pressure of 500 MPa (ca. 15 km depth) with a H_2O content of 1%, obtaining as fractionated phases olivine, plagioclase, clinopyroxene, opaque minerals, apatite and kaersutite. This fractionated mineral assemblage and the composition of these minerals is broadly consistent with that of Corvo xenoliths (see plutonic rocks section below). Furthermore, trace element modelling was carried out by applying the Rayleigh (1896) fractional crystallization equation, using as fractionating mineral phases the modes of the xenoliths. Good agreement was observed between the model and the microporphyritic lava flows and dikes from the pre-caldera stage (30–50% fractional crystallization), post-caldera stage (30–40% fractional crystallization) and syn-caldera stage (50–90% fractional crystallization). These results are in general accord with those of Genske et al. (2012b), and together both studies demonstrate that polybaric fractional crystallization is the dominant process controlling variations in major and trace elements of Corvo microporphyritic samples.

This geochemical information, together with the relative geochronological sequence (França et al. 2002), was used to investigate the magmatic evolution beneath the island (Larrea et al. 2013). Porphyritic rocks are formed by antecryst accumulation. These antecrysts are likely crystallised from primitive magmas in a deep magma chamber frequently replenished by primitive liquids and active during the entire evolution of the island. Injection of these antecryst-rich magmas into a shallower magma chamber in which fractionation predominates would provide antecrysts to ascending pre- and post-caldera melts. Depending on the availability of antecrysts within the magmatic system and the eruptive power of the magma, antecryst-bearing or antecryst-free magmas erupted and generated porphyritic or microporphyritic lava flows on the surface, respectively. In contrast, the pre- and post-caldera microporphyritic lava flows and dikes are petrographically and compositionally equivalent, and related to the same ~15 km depth magma chamber (Larrea 2014). Pre-caldera products were primarily erupted from the main caldera vent, constructing ~90% of the volume of the volcanic island, whereas most post-caldera products were erupted from secondary cones on the caldera flanks. The syn-caldera stage, more evolved compositionally, was related to the pre-caldera melts by fractional crystallization in a different and shallower magma chamber. The existence of the caldera provides evidence of this shallow magma chamber beneath the center of the volcano before the collapse. Residual evolved magmas of the syn-caldera stage erupted within the caldera after its collapse, generating a few, isolated intra-caldera pyroclastic and spatter cones, which constitute the post-caldera stage.

Plutonic rocks

Mafic plutonic rocks from Corvo were studied by França et al. (2006a), Lago et al. (2007) and Larrea et al. (2013). They were found in a pre-caldera lava flow in the southwest area of the island and occur without any apparent preferred orientation in the field. They are 5–10 cm long and show subrounded shapes and sharp contacts with the host lava flow. They display a coarse-grained heteroadcumulate texture and are composed of variable proportions of olivine, clinopyroxene, feldspars, amphibole, opaque minerals and accessory apatite. Clinopyroxene is present as 200 μm to 1 cm subhedral-anhedral crystals, feldspar develops single or intergrown 200 μm to 9 mm subidiomorphic crystals and olivine, opaque minerals and apatite display 100–500 μm subrounded crystals. Amphibole is present as poikilitic crystals up to 2 cm in size surrounding the aforementioned mineral phases. The xenoliths fall into three groups based on the classification of Le Maitre (2002): olivine-bearing gabbro, pyroxene kaersutite gabbro and kaersutite gabbro.

Their cumulate textures provided distinct petrographic evidence that they are magma chamber cumulates. The fractionating mineral assemblage and the composition of these minerals obtained by the MELTS model (see discussion above) are consistent with the mineralogy and mineral compositions of the Corvo xenoliths. All of these lines of evidence suggest that these gabbros represent cumulates resulting from fractional crystallization of the evolving Corvo magmas (Larrea et al. 2013).

2.10 Formigas Islets

The Formigas islets, with an age of 4.65 Ma (Abdel-Monem et al. 1975), are located approximately 32 km to the NE of Santa Maria, between ~39° 16′ 13 N latitude and 24° 46′ 51 W longitude (Fig. 1 in Beier et al., Chapter "Melting and Mantle Sources in the Azores"). These islets are a group of multiple reefs with a total length of approximately 1 km, aligned along a N–S direction.

Zbyszewski et al. (1961a) carried out the first and only petrologic study of these islets. They studied two samples collected on the lighthouse islet; one corresponding to the main lava flow forming the islet and a dike that cuts it. At the outcrop scale, all open spaces within the lava flow are filled by a fossiliferous Miocene limestone. The lava flow was described as basanite with a porphyritic texture formed by numerous phenocrysts of Ti-rich augite, iddingsitized olivine and labradorite—bytownite crystals, usually small and sometimes forming clusters. The groundmass is mainly composed of andesine—labradorite and oligoclase—andesine microcrysts, although minor augite, magnetite, ilmenite and serpentinised olivine crystals are also present. The dike is a black porphyritic basalt composed of minor small labradorite crystals penetrated by the matrix. The groundmass is composed of augite, magnetite, ilmenite, serpentinised olivine and andesite—labradorite and oligoclase—andesine feldspar microcrysts. No other petrologic information is currently available for the Formigas Islets.

3 Origin of Magma in the Azores Archipelago

In this section, we synthesize and compare the petrogenesis of each island with new modeling results for melt evolution paths and depth of melting to further understand the origin of the distinct petrologic characteristics of individual islands and the archipelago as a whole. Moreover, we emphasize the need for future studies that focus broadly on the Archipelago petrology and petrogenesis through time, to wholly evaluate several unresolved and fundamental archipelago-wide questions.

3.1 Plutonic Rocks from the Azores

Plutonic rocks have been documented in all of the Azores islands except for São Jorge and Santa Maria. Three types of xenoliths are recognised: ultramafic rocks, gabbros and syenites. A summary of the available information for these xenoliths was compiled for each of the islands, including their mineralogy, geochemistry and petrology, with special emphasis on their formation and origin. In this section, we compare all of the available information in order to provide a comprehensive overview of the Azorean xenoliths.

The study of plutonic rocks from the Azores has been relatively limited, with the exception of dissertations by França (1993), Johansen (1996), Almeida (2001) and Larrea (2010). Most of the available information is descriptive, with inferences about the likely origin of these xenoliths based on their textural characteristics, mineral chemistry and/or magmatic affinity; e.g., ultramafic xenoliths from Pico and Faial (Zanon and Frezzotti 2013), Flores gabbros (França et al. 2008), and syenites from São Miguel, Terceira, Graciosa, Faial and Flores (França 1993). However, in recent work on Graciosa and Corvo islands (Larrea et al. 2013, 2014a), the origin of these xenoliths has been modelled based on whole-rock major, trace and isotope geochemistry, to understand the role that these xenoliths play in the magmatic evolution of the islands.

Ultramafic xenoliths have been found in São Miguel, Graciosa, Faial and Pico islands. Several publications have documented their mineralogy and textures, with limited geochemical data for those of São Miguel (Johansen 1996; Almeida 2001) and Graciosa (Larrea et al. 2014a) (Fig. 4). The ultramafic xenoliths comprise variable amounts of olivine, orthopyroxene, clinopyroxene and spinel (Fig. 5). A non-cumulate origin has been inferred in the majority of cases, with the exception of some samples from Sete Cidades in São Miguel, for which an origin related to fractional crystallization of Sete Cidades lavas has been proposed based on the mineral chemistry of the samples (Johansen 1996; Mattioli et al. 1997; Renzulli and Santi 2000). For the other ultramafic xenoliths, a mantle origin has been proposed, in which the xenoliths represent either fertile fragments of the upper mantle (Faial; Zanon and Frezzotti 2013), mantle restites (São Miguel and Pico; Johansen 1996; Zanon and Frezzotti 2013) or, in the absence of orthopyroxene, replacive dunites (Graciosa; Larrea et al. 2014a). However, comprehensive mineralogical and geochemical studies of these ultramafic xenoliths would be required in order to better understand the depths of origin, melt extraction ages, metasomatic reactions and timing, and/or cumulate processes that are operating beneath the Azores.

Gabbroic xenoliths are found in São Miguel, Terceira, Graciosa, Faial, Pico, Flores and Corvo; however, those from Terceira and Pico have not been studied in detail. The gabbroic xenoliths from the other islands have been described petrographically, and are primarily of alkaline affinity with cumulate textures composed of variable amounts of olivine, clinopyroxene, feldspars, amphibole, Fe–Ti oxides and apatite (Fig. 5). Subalkaline gabbros with inequigranular plutonic textures have been also recognised in Graciosa and Flores; they consist of olivine, clinopyroxene, feldspars and opaque minerals, with those from Flores also containing orthopyroxene crystals (Fig. 5). Few whole-rock compositions have been reported for Azores gabbros, but data are available for alkaline gabbros from Graciosa, Pico, Corvo and Flores, and Graciosa subalkaline gabbros (Fig. 4). These gabbroic compositions are plotted against the composition of the Azores lavas in Fig. 4. The alkaline gabbro compositions are similar to those of the lavas for most of the major elements, although some xenolith samples have slightly higher TiO_2, Fe_2O_3, P_2O_5 and slightly lower K_2O and Na_2O contents. These minor compositional differences are most likely due to non-representative samples (i.e., small xenoliths) or variable crystal accumulation, in which the mineral phases are in slightly different proportions. Mineral and whole rock chemistry suggest that the alkaline xenoliths were formed as cumulates by fractional crystallization of Azorean melts. This process has been quantified for Graciosa (Larrea et al. 2014a) and Corvo xenoliths (Larrea et al. 2013) using the MELTS software. Accordingly, these gabbros were likely formed as magma chamber cumulates under a range of pressures from $\sim$450 to 100 MPa, possibly in different magma chambers or pockets located at depths ranging from 15 to 3 km. Similar studies of alkaline gabbros from other islands are necessary to better understand the range of magma chamber processes controlling the evolution of the different islands. In contrast, the subalkaline gabbro compositions from Graciosa are clearly distinct from the alkaline gabbros and the Azores lavas (Fig. 4). Their subalkaline affinity, in contrast to the alkaline affinity of the Azorean lavas, suggests that their genesis cannot be linked to the differentiation of primitive Azores magmas. Larrea et al. (2014a) suggest that they may originate as cumulates resulting from fractional crystallization of highly refractory melts, even though such melt compositions have not been identified within the archipelago. The absence of subalkaline gabbros from other islands and their scarcity and small size in Flores, preclude a more in-depth evaluation of their origin at this time.

The syenites have been studied more thoroughly than other plutonic xenoliths from the Azores. Syenites have been found in São Miguel, Terceira, Graciosa, Faial, Flores and Corvo, being commonly associated with highly explosive volcanism (i.e., plinian, sub-plinian and hydromagmatic eruptions) and less frequently found within lava flows. In contrast, syenites do not exist in those islands characterised exclusively by mafic and intermediate volcanism (i.e., Santa Maria, São Jorge and Pico islands). Syenites have been well described petrographically (e.g., França 1993), although whole-rock analyses are sparse and available only for syenites from São Miguel, Terceira (unpublished data), Graciosa, Faial and Flores (Fig. 4). The syenites are generally medium-grained with hypidiomorphic granular textures (Fig. 5), and display variable degrees of high temperature hydrothermal alteration (Morisseau 1987). The mineralogy of the syenites varies from island to island, although they all contain variable modes of feldspars, pyroxene, amphibole, biotite, sphene, Fe–Ti oxides, zircon and apatite. Quartz is abundant in Graciosa and Terceira islands, while is rare or totally absent in São Miguel, Faial and Flores islands. São Miguel syenites display distinct accessory mineral phases like dalyite, astrophyllite, pyrochlore, sodalite and lavenite. Glass is only present in Flores island syenites. In terms of geochemical composition, the syenites are the most evolved plutonic xenoliths, and are consistent with the overall compositional trends exhibited by the most evolved trachyte lavas from the Azores (Fig. 4). In particular, Graciosa syenites display the highest SiO_2, TiO_2 and P_2O_5 as do the trachytes from Graciosa, consistent with the presence of the highest amounts of quartz, sphene and opaque minerals, respectively. São Miguel syenites show the highest contents in K_2O, Al_2O_3 and Na_2O, which is also a characteristic of the evolved rocks from São Miguel. Accordingly, the syenites from each island reflect the distinct geochemical characteristics of the magmas from the respective island. There has been considerable discussion about their origin, in order to determine whether the syenites could be evolved cumulates or if they represent frozen liquid compositions. Based on their textural characteristics, major and trace element compositions and their strong negative Eu/Eu* anomalies, syenites from São Miguel (Widom et al. 1993), Graciosa (Larrea et al. 2014a) and Terceira islands (Daly et al. 2012) have been identified as bulk trachyte liquid in origin. In addition, some syenites from Santa Bárbara (Terceira) seem to partially fill the Daly Gap that characterizes Terceira volcanism, indicating that this compositional gap does not result from an absence of intermediate compositions; this may reflect an inability of such intermediate magmas to erupt, although the cause of the eruption barrier for these intermediate-evolved magmas remains unresolved.

3.2 Compositional Range of the Azores Magmatic Rocks

The Azores islands are alkaline in affinity, exhibiting the full range of compositions from mafic to felsic rocks, except for Santa Maria, São Jorge and Pico, which are formed exclusively by mafic and intermediate volcanic rocks (Figs. 1 and 3a, b). Although some isolated samples from the islands of Pico, Flores and Corvo fall within the subalkaline basalt field (Fig. 3a), they are apparently altered and/or cumulate rocks.

The mafic to intermediate volcanic rocks are principally nepheline-normative alkali basalts, hawaiites and mugearites, most belonging to the sodic series. Santa Maria rocks are the most strongly nepheline-normative and sodic within the archipelago, whereas São Miguel is the most potassic. Mafic rocks from Terceira are also distinct in that some samples are transitional hypersthene-normative. Felsic rocks, which range from nepheline-normative to quartz-normative, are particularly abundant in the Azores Archipelago. Most of the felsic rocks are classified as trachytes, with lesser comenditic trachytes, comendites and pantellerites; the latter two occur only on Terceira, where the rocks tend to be more silica-saturated and peralkaline than those on other islands.

An outstanding question is why there is such a large volume of highly evolved rocks in the Azores in comparison to most other ocean islands (e.g., Hawaii). Large volumes of silicic volcanism can potentially be explained by fractional crystallization of parental basaltic magmas or by partial melting of pre-existing oceanic crust. In Iceland, for example, the origin of silicic rocks is associated with hotspot activity in conjunction with significant reprocessing and recycling of the heterogeneous ocean crust beneath Iceland (e.g., Jónasson 2007 and references therein). In contrast, the silicic rocks in the Azores Archipelago are best explained by fractional crystallization of parental magmas (see fractional crystallization models below; White et al. 1979; Mungall and Martin 1995; Pimentel 2006). The strongest evidence in favour of the fractional crystallization process is that the chemical characteristics that distinguish basaltic compositions from different islands are also present in the felsic lavas of the same island, implying a genetic relationship between them. However, the acceptance of this mechanism for the genesis of the evolved compositions in the Azores leaves the question of the origin of the Daly compositional gap partially unresolved (see plutonic rocks from the Azores section above; Daly et al. 2012).

Bohrson and Reid (1997) have suggested based on a compilation of studies of silicic peralkaline volcanism in ocean islands globally, that evolved peralkaline magmas are probably generated by a combination of three key conditions: a mildly extensional tectonic setting, a shallow magma reservoir, and the availability of parental basaltic magmas. Although these tectono-magmatic conditions have not been proven to control the formation of non-peralkaline silicic magmas, in the particular case of the Azores Archipelago, the association of all of the islands with either local or regional rifts (Fig. 1 in Beier et al., Chapter "Melting and Mantle Sources in the Azores"), and the presence of calderas indicating the existence of shallow magma chambers beneath some islands (e.g., Sete Cidades, Fogo, Furnas and Povoação in São Miguel; Vulcão da Caldeira in Faial; The Caldeirão in Corvo), provide evidence of the existence of similar tectono-magmatic conditions that may be conducive to the generation of the silicic volcanism in the Azores. The lack of silicic volcanism in Santa Maria, São Jorge and Pico islands may indicate that they are in an early evolutionary stage (i.e., pre-caldera stage), mainly characterised by mafic-intermediate volcanism. If these three islands follow the magmatic evolution of the rest of the archipelago, one might expect that shallow magma chambers would allow the formation of evolved melts, and consequently calderas, to form on these islands in the future. However, for Santa Maria, the lack of eruptions over the past ~ 1.5 Ma may indicate that magmatism on this island has ceased, and that this island will not complete the same evolutionary path.

3.3 Textural and Compositional Characteristics—The Antecryst Accumulation Effect

The lavas from the archipelago are characterised by three different types of textures including porphyritic, microporphyritic and trachytic, which appear to be linked respectively to mafic, mafic-intermediate and evolved rock compositions (Fig. 5). Abundant porpyritic rocks from the archipelago contain large crystals, which are frequently corroded and present overgrowth rims. These large crystals are interpreted as antecrysts and have been studied in detail in Graciosa (Larrea et al. 2014a) and Corvo islands (Larrea et al. 2013). Antecrysts have been observed previously in other Azorean islands, although different terminologies have been used: xenocrysts in São Miguel (Beier et al. 2006), São Jorge (Ribeiro 2011) and Pico islands (França 2000; França et al. 2006c), or phenocrysts with the highest Mg# in Terceira (Madureira et al. 2011). These antecryst-bearing rocks have been called ankaramites in São Miguel (Moore 1991) and Pico (França et al. 2006c), or cumulate rocks in Santa Maria (Beier et al. 2012b), Terceira (Madureira et al. 2011) and Faial islands (Machado et al. 2008; Lima et al. 2011; Zanon et al. 2013).

The influence of these antecrysts on the composition of the whole rocks in which they are hosted has been overlooked in most of the islands, but it should be reconsidered, taking into account that these porphyritic rocks can contain up to 70–80% of the total volume fraction of these antecrysts (e.g., Corvo and Pico islands). The antecryst-rich rocks from Corvo have MgO contents of up to 18 wt%; accordingly, all rocks from the archipelago that contain extreme MgO are suspected of being antecryst-bearing rocks and should be re-examined (i.e., São Miguel, Santa Maria, São Jorge, Faial and Pico; Fig. 3a, b). In some cases, more evolved rocks also may be impacted by antecryst accumulation. For example, all rocks from Corvo with MgO >5.5 wt% (Larrea et al. 2013) and rocks from Flores with MgO >5.4 wt% (Larrea et al. 2012) are antecryst-bearing rocks, and in these cases the whole-rock compositions represent mixtures of a quite evolved matrix hosting variable fractions of primitive antecrysts.

In contrast, Graciosa lavas are mostly antecryst-free and present MgO contents <12 wt %. Consequently, all of the distinct crystal populations within a rock (independent of its MgO content) should be evaluated for mineral-melt equilibrium in order to decipher the origin of the different crystal populations. This is particularly critical when using whole-rock compositions to infer and quantify the magmatic processes controlling the evolution of the islands.

However, despite the fact that the magmatic processes and/or plumbing systems controlling the evolution of Graciosa and Corvo islands are clearly different, their lavas show fairly similar trends to other islands in major element compositions (Fig. 3a). Thus, the liquid lines of descent defined by antecryst-bearing rocks and antecryst-free rocks are quite similar. This means that the antecryst accumulation process essentially involves adding back to more evolved melts the crystals that were fractionated out from more primitive magmas in the same magmatic system, in similar proportions. The liquid lines of descent defined by antecryst-bearing and antecryst-free rocks might be similar and equally applicable for estimating the P–T conditions of primary magma evolution.

3.4 Magmatic Processes in the Azores—Fractional Crystallization ± Crystal Accumulation

In all of the islands and volcanic systems for which detailed petrogenetic studies have been performed, fractional crystallization ± crystal accumulation have emerged as the primary processes controlling the evolution of the rocks. Regardless of the inter-island compositional distinctions described above, the fractionating phases that control the liquid lines of descent are consistently similar. In general, olivine, clinopyroxene and plagioclase fractionation controls the evolution of the mafic and intermediate melts at higher pressures and depths, with the most evolved magmas formed by larger degrees of fractional crystallization in shallower magma chambers controlled by fractionation of feldspars, Fe–Ti oxides, amphibole and minor apatite. The alkaline gabbros included in some of these islands represent the crustal-level magma chamber cumulates (Fig. 4; see section above).

METLS modelling can further constrain the fractional crystallization processes operating in the archipelago in terms of pressure, temperature, fO_2 and fractionating mineral assemblages.

The MELTS model previously developed to explain the formation of Graciosa (see Graciosa section above; Larrea et al. 2014a) has been plotted on the bivariate diagrams for the combined archipelago-wide lava major element compositions for comparison (Fig. 6a, b). The Graciosa model has been redone in this study using the new Rhyolite-MELTS algorithm (Gualda et al. 2012), which better characterizes high-silica ryholitic systems, with exactly the same parameters used by Larrea et al. (2014a). Terceira, Faial, Pico, Flores and Corvo lavas show a similar liquid line of descent to the one defined by Graciosa, although there may be slight differences in Terceira Al_2O_3 and Fe_2O_3 contents. In contrast, Santa Maria, São Miguel and São Jorge islands clearly display distinct liquid lines of descent (e.g., lower SiO_2, higher Fe_2O_3 and variable TiO_2 contents).

Four other polybaric and two isobaric fractional crystallization models have been calculated at different initial pressures (as indicated for each model in Fig. 6a, b), with cooling steps of 5 °C, 0.5% water and oxygen fugacity fixed relative to the QFM buffer. The GRZF12 sample was used as the initial composition for all of the Graciosa models (Larrea et al. 2014a). Samples SMA-81 (Anderson 1983) and MA09-019 (Beier et al. 2012b) from Santa Maria, and sample SMB-6 (Widom et al. 1997) from São Miguel, clearly distinct from Graciosa sample, were also considered as starting compositions. These samples have major element compositions that bracket the compositional variability shown by Santa Maria lavas in most major elements, and the high-TiO_2 contents of São Miguel lavas (see discussion below; Fig. 6b). The new Rhyolite-MELTS models starting with the Graciosa GRZF12 sample demonstrate that a change in the initial conditions of the model (e.g., P, T and H_2O content) cannot account for these compositional distinctions among the islands. However, the Santa Maria SMA-81 and MA09-01 polybaric Rhyolite-MELTS fractional crystallization models (Fig. 6b), carried out with the same conditions as the 500–20 MPa polybaric Graciosa model (Larrea et al. 2014a), are able to reproduce most of the compositional variation exhibited by the islands of Santa Maria, São Miguel and São Jorge, except for the high-TiO_2 contents of São Miguel. This distinct high-TiO_2 trend displayed by São Miguel lavas can only be reproduced when a polybaric Rhyolite-MELTS fractional crystallization models (500–30 MPa; Fig. 6b) is carried out with a high-TiO_2 starting compostion (sample SMB-6 from São Miguel; Widom et al. 1997). Prytulak and Elliott (2007) related these high-TiO_2 contents found in São Miguel (and other ocean island basalts) to the addition of less than 10% of recycled mafic crust to the peridotitic mantle.

Accordingly, the evolution of all of these islands is primarily controlled by fractional crystallization ± crystal accumulation, involving compositionally similar yet distinct, non-cogenetic parental magmas for the different islands. As observed in Fig. 6a, b, primitive magmas from the Azores have a wide range in major element compositions (e.g., SiO_2, TiO_2, Na_2O, K_2O) at MgO ~12 wt%; this has to be a function of either lithologically distinct mantle source regions or the conditions of melting (see section below; see Beier et al., Chapter "Melting and Mantle Sources in the Azores"). Future studies focusing comprehensively on the chronostratigraphy of the different volcanoes/volcanic units on the islands (e.g., Santa Maria, Terceira, Corvo and Flores) would be desirable to better establish the petrogenesis of these units, and therefore, the possible intra-island changes in magma generation conditions and evolution through time.

3.5 Crustal Contamination in the Azores

The potential importance of crustal assimilation during the petrogenesis of ocean island magmatic systems has been receiving increasing attention, in part because the use of mafic magmas to infer mantle source compositions and melt generation processes may be compromised by the interaction of ocean-island magmas with the oceanic crust (Davidson and Bohrson 1998). Although the compositional differences between the ocean-island magmas and the basaltic oceanic crust through which they pass en route to the surface may be subtle, it is important to consider the potential effects that shallow level processes may have in modifying primary magma compositions, in order to link their trace element and isotopic signatures to mantle source compositions (see Beier et al., Chapter "Melting and Mantle Sources in the Azores").

Evidence of crustal contamination of mafic magmas has been demonstrated in several detailed and comprehensive studies of individual ocean-island magma systems (e.g., Reisberg et al. 1993; Eiler et al. 1996a, b; Widom and Shirey 1996; Thirlwall et al. 1997; Widom et al. 1999; Harris et al. 2000; Wolff et al. 2000). The Os isotope system frequently reveals a role for crustal assimilation during ocean island basalt petrogenesis, due to its sensitivity to even very minor crustal assimilation, a process that can be elusive in other isotope systems. Only minor amounts (e.g., a few %) of crustal contamination are required to modify the Os isotopic composition of the basaltic magmas, which often show a negative correlation of Os isotopes with indices of differentiation such as MgO content (e.g., Reisberg et al. 1993; Widom 1997). Some highly incompatible trace element ratios (e.g., Nb/U, Ce/Pb and Th/Nb) also can be sensitive to crustal contamination, and a systematic variation of these ratios with indices of differentiation may indicate that sediment or altered oceanic crust played an important role in the ocean island basalt petrogenesis (Weaver 1991). More recently, the role of hydrothermally altered crust contamination of mafic ocean island basalts has been demonstrated through the study of fluid-mobile trace elements and stable isotope ratios (e.g., B and O isotopes), which facilitate the identification of contamination (e.g., Garcia et al. 1998; Gee et al. 1998; Genske et al. 2014, see also Beier et al., Chapter "Melting and Mantle Sources in the Azores").

In the case of the Azores Archipelago, some studies have focused on the study of incompatible trace element ratios and Os, O, Li and B isotope systematics in mafic samples (Widom and Shirey 1996; Yu 2011; Genske et al. 2012a, 2014; Larrea et al. 2014a; see Beier et al., Chapter "Melting and Mantle Sources in the Azores"). Highly incompatible trace element ratios from the Azores do not vary systematically with indices of differentiation such as MgO, arguing against large-scale assimilation of oceanic crust, but nevertheless permitting a minor role for this process in the petrogenesis of these islands (e.g., Madureira et al. 2011; Beier et al. 2012a; Genske et al. 2012b; Larrea et al. 2014a). Os isotope ratios from São Miguel, Graciosa Terceira, São Jorge, Faial and Pico islands correlate with Os concentrations spaning almost the entire range found in ocean island basalts world-wide (Widom and Shirey 1996); at low Os concentrations (<20 ppt), there is a large range in Os isotope signatures that extent to very radiogenic values (up to $^{187}Os/^{188}Os$ $\sim$0.20), suggesting that some samples have experienced at least minor crustal contamination. Oxygen isotope signatures in olivine phenocrysts from some Azores samples, including those from the islands of São Miguel, Graciosa, Terceira, São Jorge, Fail, Pico, Corvo and Flores (Genske et al. 2012a), suggests that the low oxygen isotope ratios may result from combined fractional crystallization and assimilation of hydrothermally altered oceanic crust with a ratio of assimilation to crystallization of $\sim$0.25. The study of Li and B isotope systematics in samples from São Miguel, Terceira, São Jorge, Pico, Corvo and Flores islands reveals that the Li and B concentrations are within the range of what is considered characteristic for ocean island basalts (Genske et al. 2014); however, the variability and heavier isotopic compositions of Li and particularly B in Corvo and Flores islands indicate that assimilation may have been more pronounced underneath the western islands than for the rest of the islands. Therefore, based on the behaviour of incompatible trace element ratios and the highly sensitive Os, O, Li and B isotope systematics in the archipelago, it may be concluded that some Azorean lavas have experienced crustal contamination, especially those from the western group of islands. However, the overall major and trace element compositions of mafic Azorean magmas have not been significantly influenced by crustal contamination, and indeed magma compositions can be used to infer mantle source compositions and partial melting processes beneath the Azores islands.

Assimilation of crustal material has been shown to play a minor but distinct role in the evolution of some silicic rocks from the Azores Archipelago. Detailed petrogenetic studies of the chemically zoned Fogo A trachyte pumice deposit from São Miguel (Snyder et al. 2004) and the Furnas 1630 AD trachyte pumice deposit (Rowland-Smith 2007) both reveal the importance of open-system processes in the formation of silicic magmas on São Miguel. Although the whole-rock chemical variations appear entirely consistent with closed-system processes (Widom et al. 1992; Rowland-Smith 2007), Sr isotopic disequilibrium between single sanidine crystals and host glass separates, as well as variation in the Sr isotopic composition of the trachyte glass within the eruptive deposits, demonstrates that the trachytic magmas assimilated melts of, and incorporated sanidine xenocrysts from, hydrothermally altered syenite. The hydrothermally altered syenite is interpreted as magma chamber wall rock formed during earlier episodes of trachytic magmatism (Widom et al. 1992; Rowland-Smith 2007).

3.6 Melt Generation in the Azores

As demonstrated above, the petrogenetic evolution of all of the Azores islands appears to be controlled by the same magmatic processes, dominated by polybaric fractional crystallization with only minor crustal assimilation. Nevertheless, the existence of distinct parental magma compositions is apparent, and likely controls the compositional variability observed at both the inter- and intra-island scale. Although the Azores islands exhibit significant isotopic heterogeneity, indicative of distinct mantle source regions, the major element compositions and rare earth element patterns of the Azores primitive melts are far less variable (Beier et al. 2010, 2012a and Beier et al., Chapter "Melting and Mantle Sources in the Azores"). Therefore, the major element variation among primitive melts might largely indicate different conditions of melting beneath the archipelago; i.e., spatial variations in the temperature, depth and degree of partial melting of lithologically similar mantle sources.

Recent studies of mafic samples from the Azores (e.g., Beier et al. 2012a; Genske et al. 2012b; Zanon et al. 2013) have demonstrated that variability in the degree of partial melting beneath most of the islands is limited; based on trace element ratios, the degree of partial melting has been estimated to be on the order of 3–5%

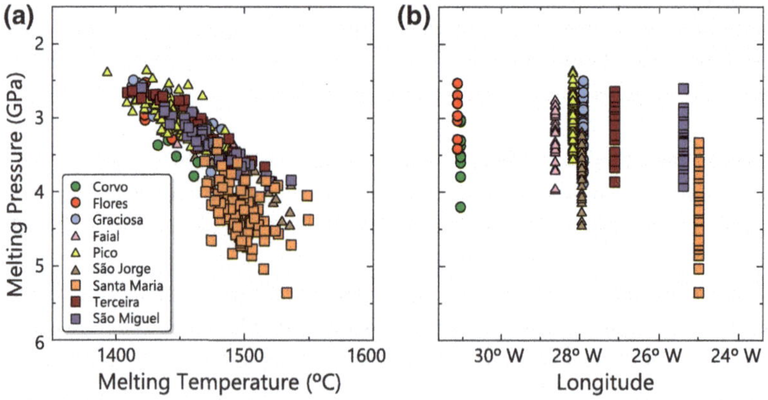

Fig. 7 **a** Melting pressures and temperatures of the Azores lavas with 7–12 wt% MgO calculated using Putirka (2008). **b** Melting pressures (GPa) of the different Azores islands with distance from the Mid-Atlantic ridge. Lava Pressure of melting calculated using the approach described by Putirka (2008)

for all islands (e.g., Beier et al. 2008; Genske et al. 2012b; Zanon et al. 2013), with Santa Maria and São Miguel magmas produced by <4% partial melting (Beier et al. 2008, 2012b).

In order to evaluate whether there are systematic variations in temperature and pressure of melting across the archipelago, the compositions of primary melts for each island have been calculated. In order to minimize effects of multiple-phase fractional crystallization and crystal accumulation, only samples with MgO contents between 8 to 12 wt% have been considered (n = 416; Fig. 3a, b). Primary melt compositions have been calculated using the methodology described by Putirka (2008), which adds olivine to the melt until its composition is in equilibrium with an olivine mantle composition of $Fo_{90.3}$ (O'Hara 1968; Stolper 1980), using olivine-melt K = 0.30 ± 0.03 (Roeder and Emslie 1970). This approach was previously used by Beier et al. (2012a) for most of the islands, except for Corvo and Flores. The olivine compositions added are Fo_{89} for Santa Maria, Graciosa, São Jorge and Faial, Fo_{90} for São Miguel, Terceira and Pico, and Fo_{91} for Corvo and Flores, in agreement with the olivine compositions analysed in each island (Fig. 2). The calculated primary melts for the Azores islands have MgO contents from 15.16 wt% (Pico) to 21.11 wt% (Santa Maria), with mass fractions of added olivine from 0.11 (Graciosa and Terceira) to 0.30 (Santa Maria and São Jorge).

Figure 7a displays the temperatures and pressures of melting calculated using the approach described by Putirka (2008) based on olivine—primary mantle-derived liquid equilibria. Our larger dataset is in agreement with previous calculations (Beier et al. 2012a) but displays a wide range of melting conditions across the archipelago, with the coldest and shallowest melting occurring beneath the Central Group islands (with the exception of São Jorge), and the highest melting temperatures and pressures beneath São Jorge, São Miguel, and Santa Maria islands. Despite the scarcity of available data, the western islands of Corvo and Flores appear to be similar to the Central Group.

The similarity in P and T of melting, and therefore in the primary magma compositions, of the Western and Central Group of islands is in accordance with their similar major element compositional trends (Fig. 6a). In contrast, the distinct liquid lines of descent in some major elements defined by Santa Maria, São Miguel and São Jorge islands can be potentially explained by their different conditions of melting (Fig. 6b). For example, Santa Maria is characterised by anomalously low SiO_2 and high Na_2O contents (in some samples), consistent with its highest pressures of melting and lowest degrees of partial melting, respectively. However, the extremely high TiO_2 and K_2O contents shown by São Miguel lavas might require the existence of a distinct source composition; Beier et al. (2008) suggest that the negative correlation between the $^{143}Nd/^{144}Nd$ isotope ratios and the K_2O contents concentrations are the result of a more enriched mantle source beneath São Miguel, in agreement with the high TiO_2 contents in the island that have been related to the presence of an enriched component from recycled oceanic crust in the peridotitic mantle (Prytulak and Elliott 2007; Beier et al., Chapter "Melting and Mantle Sources in the Azores").

These results show that with increasing distance from the Mid-Atlantic ridge, the islands display higher melting pressures (Fig. 7b), corresponding to an increase in the lithosphere thickness towards the east. Older and thicker lithosphere may limit the extent of mantle decompression, resulting in lower degrees of melting, and magma compositions that are more enriched in incompatible trace elements than those produced beneath thinner and younger lithosphere (Beier et al. 2012a). However, the cause of the anomalous temperatures and pressures displayed by São Jorge, which are much higher than those of the other Central Group islands, is unclear. Beier et al. (2012a) suggested that this may be indicative of the Azores mantle plume rising beneath São Jorge, although He-isotope ratios and U–Th–Pa disequilibria suggest that the Azores plume is centred beneath Terceira (Moreira et al. 1999; Bourdon et al. 2005;

see Beier et al., Chapter "Melting and Mantle Sources in the Azores" and Moreira et al., Chapter "Noble Gas Constraints on the Origin of the Azores Hotspot").

Alternatively, the presence of water in the upper mantle may play an important role in the melting dynamics beneath the Azores, and there has been some debate as to the relative roles of the "Azores Wetspot" versus the thermal hotspot (e.g., Bonatti 1990; Asimow et al. 2004; Métrich et al. 2014; O'Neill and Sigloch, Chapter "Crust and Mantle Structure Beneath the Azores Hotspot—Evidence from Geophysics"). According to the wet melting model developed by Asimow et al. (2004), the Azores magmas form at lower temperatures than the dry peridotite solidus, suggesting volatile-rich partial melting conditions with at least 200 ppm H_2O (Beier et al. 2012a). Recent estimates based on the study of fluid and melt inclusions in olivine crystals from Pico (Métrich et al. 2014) reinforce the hypothesis of anomalously wet mantle beneath the Azores region, suggesting that Pico magmas were generated via decompression melting of a source with 570–680 ppm H_2O. These authors propose a model for Azores magma generation involving the decompression melting of a water-enriched mantle source without the need for significant elevated mantle potential temperatures, and suggest that the link between depth of melting and crustal thickness should be revisited. Future studies focusing on the other islands will be necessary to fully evaluate this and other outstanding archipelago-wide questions (see Beier et al., Chapter "Melting and Mantle Sources in the Azores").

4 Conclusions

This chapter presented a comprehensive review of the existing mineralogical, petrological, and major element geochemical data for the islands of the Azores Archipelago and Formigas Islets. Clearly there is a wide variation in the extent to which the petrology of each island has been documented. The most extensive documentation of the petrology and petrogenesis through time is available for the islands of São Jorge, Graciosa, Faial, Pico, and Corvo. Terceira and São Miguel have been studied extensively, but investigations have focused primarily on individual eruptive sequences or volcanic centres, and the petrology of the older volcanic systems is less well documented. The petrologic development of Santa Maria and Flores, as well as the Formigas Islets, are the least well documented. Future studies focusing comprehensively on the petrology and petrogenesis through time for these islands will be necessary to fully evaluate several outstanding and fundamental, archipelago-wide questions including: (1) What controls the variability in silica saturation and the occurrence of sodic versus potassic series rocks on the different islands; (2) What is the role of mantle heterogeneity versus shallow magmatic evolution processes in controlling the distinct petrogenetic evolution paths documented for some islands; (3) What is the cause of the Daly gap in the volcanic products of some islands; and (4) What is the driving mechanism that results in an exceptionally large fraction of silicic rocks in the Azores as compared to many other ocean islands globally?

References

Abdel-Monem, A., Fernandez, L. A., & Boone, G. M. (1968). Pliocene-Pleistocene minimum K-Ar ages of the older eruptive centers, Eastern Azores. *Transactions America Geophysics Union, 49*(1), 363.

Abdel-Monem, A. A., Fernandez, L. A., & Boone, G. M. (1975). K-Ar ages from the Eastern Azores group (Santa Maria, São Miguel and the Formigas Islands). *Lithos, 8*, 247–254.

Agostinho, J. (1937). Volcanic activity in the Azores, Report for 1933–1936. *Bulletin of Volcanology, 2*(11), 183–192.

Albarède, F. (1992). How deep do common basaltic magmas form and differentiate? *Journal of Geophysical Research: Solid Earth, 97*(B7), 10997–11009.

Allan, J. F., & Dick, H. J. (1996). Cr-rich spinel as a tracer for melt migration and melt-wall rock interaction in the mantle: Hess Deep, leg 147. In C. Mevel, K. M. Gillis, J. F. Allan, & P. S. Meyer (Eds.), *Proceedings of the Ocean Drilling Program, Scientific*

Results, 147 (pp. 157–172). College Station, TX: Ocean Drilling Program.

Allègre, C. J., Treuil, M., Minster, J. F., Minster, J. B., & Albarède, F. (1977). Systematic use of trace element in igneous process. I. Fractional crystallization processes in volcanic suites. *Contributions to Mineralogy and Petrology, 60,* 57–75.

Almeida, M. H. (2001). A fonte mantélica na região dos Açores: constrangimentos impostos pelas características geoquímicas de rochas vulcânicas e de xenolitos ultramáficos. (Ph.D. thesis, Department Geociências, Universidade dos Açores, São Miguel, Azores, Portugal).

Anderson, M. L. (1983). *Santa Maria Azores: Lava petrogenesis and upper mantle geochesmistry* (Ph.D. thesis, Department of Geological Sciences, University of Illinois at Chicago, USA).

Asimow, P. D., Dixon, J. E., & Langmuir C. H. (2004). A hydrous melting and fractionation model for Mid-ocean Ridge Basalts: Application to the Mid-Atlantic Ridge near the Azores. *Geochem. Geophys, 5* https://doi.org/10.1029/2003gc000568.

Asimow, P. D., & Ghiorso, M. S. (1998). Algorithmic Modifications Extending MELTS to Calculate Subsolidus Phase Relations. *American Mineralogist, 83,* 1127–1131.

Azevedo, J. M. M. (1990). Os depósitos vulcanoclásticos do complexo de base sudoeste da ilha das Flores—Estrutura e ambiente de formação. *Memórias e Notícias, Publ. Museu e Laboratório Mineralógico e Geológico, University of Coimbra 109,* 35–63.

Azevedo, J. M. M. (1998). *Geologia e Hidrogeologia da ilha das Flores (Açores—Portugal)* (Ph.D. thesis, Geologia e Hidrogeologia da ilha das Flores (Açores—Portugal). Ciências da Terra, University of Coimbra, Coimbra, Portugal).

Azevedo, J. M. M., Alves, E. I., & Dias, J. L. (2003). Contributo para a interpretação vulcanostrutural da ilha do Corvo, Açores. *Ciências da Terra (UNL), 5,* 2–3.

Azevedo, J. M. M., & Ferreira, M. R. P. (2006). The volcanotectonic evolution of Flores Islands, Azores (Portugal). *Journal of Volcanology and Geothermal Research, 156,* 90–102.

Azevedo, J. M. M., Ferreira, M. R. P., & Martins, J. A. (1986). O complexo de base da ilha das Flores, Açores. *Memórias e Notícias, Publ. Museu e Laboratório Mineralógico e Geológic University Coimbra, 101,* 55–71.

Azevedo, J. M. M., Ferreira, M. R. P., & Martins, J. A. (1991). The emergent volcanism of Flores island, Azores (Portugal). *Arquipélago, 9,* 37–46.

Baker, M. J. (1966). Blocks of plutonic aspect in a basaltic lava from Faial, Azores. *Geological Magazine, 103*(1), 51–60.

Baker, P. E. (1975). Peralkaline acid volcanic rocks of oceanic islands. *Bulletin of Volcanology, 38,* 737–754.

Batanova, V. G., & Savelieva, G. N. (2009). Melt migration in the mantle beneath spreading zones and formation of replacive dunites: A review. *Russian Geology and Geophysics, 50,* 763–778.

Bedemar, V. (1837). Resumo de observações geológicas feitas em viagens às ilhas da Madeira. Porto Santo e Açores nos annos de 1835 e 1836. Lisboa, 14 p.

Beier, C. (2006). *The magmatic evolution of Oceanic Plateaus: a case study from the Azores* (Ph.D. thesis, Faculty of Mathematics and Natural Sciences, University of Kiel, Germany).

Beier, C., Haase, K. M., Abouchami, W., Krienitz, M. S., & Hauff, F. (2008). Magma genesis by rifting of oceanic lithosphere above anomalous mantle: Terceira Rift, Azores. *Geochemistry, Geophysics, Geosystems, 9*(12), Q12013.

Beier, C., Haase, K. M., & Hansteen, T. H. (2006). Magma evolution of the Sete Cidades Volcano, São Miguel, Azores. *Journal of Petrology, 47,* 1375–1411.

Beier, C., Haase, K. M., & Turner, S. P. (2012a). Conditions of melting beneath the Azores. *Lithos, 144–145,* 1–11.

Beier, C., Mata, J., Stöckhert, F., Mattieli, N., Brandl, P. A., Madureira, P., et al. (2012b). Geochemical evidence for melting carbonated peridotite on Santa Maria Island, Azores. *Contributions to Mineralogy and Petrology, 165*(5), 823–841.

Beier, C., Stracke, A., & Haase, K. M. (2007). The peculiar geochemical signatures of São Miguel (Azores) lavas: Metasomatised or recycled mantle sources? *Earth and Planetary Science Letters, 259,* 186–199.

Beier, C., Turner, S., Plank, T., & White, W. (2010). A preliminary assessment of the symmetry of source compositions and melting dynamics across the Azores plume. *Geochemistry Geophysics, Geosystems, 11*(2), Q02004.

Berthois, L. (1953). Contribuition à l'étude lithologique de l'archipel des Açores. *Comunicações dos Serviços Geológicos de Portugal, 34,* 96–116.

Bohrson, W., & Reid, M. R. (1997). Genesis of silicic peralkaline volcanic rocks in an ocean island setting by crustal melting and open-system processes: Socorro island, Mexico. *Journal of Petrology, 38*(9), 1137–1166.

Bonatti, E. (1990). Not so hot "hot spots" in the oceanic mantle. *Science, 250,* 107–111.

Booth, B., Croasdale, R., & Walker, G. P. L. (1978). A quantitative study of five thousand years of volcanism on São Miguel, Azores. *Philosophical Transactions of the Royal Society of London, 288,* 271–319.

Bourdon, B., Turner, S. P., & Ribe, N. M. (2005). Partial melting and upwelling rates beneath the Azores from U-series isotope perspective. *Earth and Planetary Science Letters, 239,* 42–56.

Brousse, R., Bizouard, H., & Metrich, N. (1981). Fayal in the Atlantic and Rapa in the Pacific: Two weakly alkaline series evolving under anhydric conditions. *Bulletin of Volcanology, 44,* 393–410.

Buch, V. L. V. (1825). Physicalische Beschreibungen der Canarischen Inseln, Berlin 1–201.

Calvert, A. T., Moore, R. B., McGeehin, J. P., & Rodrigues da Silva, A. M. (2006). Volcanic history and $^{40}Ar/^{39}Ar$ and ^{14}C geochronology of Terceira Island, Azores, Portugal. *Journal of Volcanology and Geothermal Research, 156,* 103–115.

Camacho, A. G., Montesinos, F. G., Vieira, R., Nunes, J. C., Alves, M. C., & França, Z. (1999). Estudio gravimétrico de la isla de Pico (Azores). In J. M. G. Garcia-Garcia & M. D. R. Romacho (Eds.), *Actas da 1ª Assembleia Luso-Espanhola de Geodesia e Geofísica.* Espanha. Instituto Geográfico Nacional: Almeria.

Canilho, M. H. (1970). Possible origin of olivine nodules in basaltic lavas of Faial, Azores. BOL. Museu Laboratorio Mineral. *Geology Faculdade de Ciências da Universidade de Lisboa, 11,* 193–198.

Cann, J. R. (1967). A second occurrence of dalyite and the petrology of some ajected syenite blocks from São Miguel, Azores. *Mineralogical Magazine, 36*(278), 227–232.

Castro, E. V. P. C. (1888). Recherches micrographiques sur quelques roches de l'Ile de San Miguel, (Açores). *Imprimerie Nationale* 7–91.

Charlier, B. L. A., Wilson, C. J. N., Lowenstern, J. B., Blake, S., Van Calsteren, P. W., & Davidson, J. P. (2005). Magma generation at a large, hyperactive silicic volcano (Taupo, New Zealand) revealed by U-Th and U-Pb systematics in zircons. *Journal of Petrology, 46,* 3–32.

Chovelon, P. (1982). *Évolution volcanotectonique des iles de Faial et de Pico, Archipel des Açores—Atlantique Nord* (Ph.D. thesis, Université de Paris-Sud, Centre D'Orsay, France).

Claude-Ivanaj, C., Joron, J. L., & Allègre, C. J. (2001). 238U–230Th–226Ra fractionation in historical lavas from the Azores: Long-lived source heterogeneity versus metasomatism fingerprints. *Chemical Geology, 176*(1–4), 295–310.

Coutinho R. M. S. (2000) *Elementos para a monitorização sismovulcânica da ilha do Faial (Açores): Caracterização Hideogeológica e Avaliação de anomalias de Rn associadas a fenómenos de desgaseificação* (Ph.D. thesis, Departamento de Geociências, Universidade dos Açores, Portugal).

Cruz, J. V., Nunes, J. C., França, Z., Carvalho, M. R., & Forjaz, V. H. (1995). Estudo vulcanológico das erupções históricas da ilha do Pico—Açores. IV Congresso Nacional de Geologia, Porto. *Mem Mus Lab Min Geol Fac Ciências Porto, 4,* 985–987.

Daly, G. E., Widom, E., & França, Z. (2012). Evolution of silicic magmas and the origin of the Daly Gap at Santa Bárbara volcano, Terceira, Azores. *Mineralogist Magazine, 76*(6), 1613–1674.

Darwin, C. (1844). Geological Observations on Volcanic Islands? Darwin, C. R. 1844. Geological observations on the volcanic islands visited during the voyage of H. M.S. Beagle, together with some brief notices of the geology of Australia and the Cape of Good Hope. Being the second part of the geology of the voyage of the Beagle, under the command of Capt. Fitzroy, R.N. during the years 1832–1836. Smith Elder & Co, London.

Davidson, J. P., & Bohrson, W. A. (1998). Shallow-Level Processes in Ocean-island Magmatism: Editorial. *Journal of Petrology, 39*(5), 799–801.

Davidson, J. P., Morgan, D. J., Charlier, B. L. A., Harlou, R., & Hora, J. M. (2007). Microsampling and isotopic analysis of igneous rocks: Implications for the study of magmatic systems. *Annual Review of Earth and Planetary Sciences, 35,* 273–311.

Davies, G. R., Norry, M. J., Gerlach, D. C., & Cliff, R. A. (1989). A combined chemical and Pb-Sr-Nd isotope study of the Azores and Cape Verde hotspots: The geodynamic implications. *Specical Publication Geological Society, London, 42,* 231–255.

de Assunção, C. F. T. (1961). Estudo petrographico da illha de S. Miguel (Acores). *Communication Services Geologists Portuguesa, 45,* 81–176.

de Assunção, C. F. T., & Canilho, M. H. (1970). Notas sobre petrografia comperada das ilhas Atlanticas. Lisboa Universitatis Faculty Cienc Museum E. Laboratory Mineralogical E. Geological Boletin, 11, 305–342.

Dias, J. L. F. (2001). *Geologia e tectónica da Ilha do Corvo, contributos para o ordenamento do espaço físico* (Ph.D. thesis, University of Coimbra, Portugal).

Dick, H. J. B., & Bullen, T. (1984). Chromian spinel as a petrogenetic indicator in oceanic environments. *Contributions to Mineralogy and Petrology 86,* 54–76.

Eiler, J. M., Kz, Faricy, Valley, J. W., Hofmann, A. W., & Stolper, E. M. (1996a). Oxygen isotope constraints on the sources of Hawaiian volcanism. *Earth and Planetary Science Letters, 144,* 453–468.

Eiler, J. M., Valley, J. W., & Stolper, E. M. (1996b). Oxygen isotope ratios in olivine from the Hawaii scientific drilling project. *Journal Geophysical Research, 101*(11), 807–811.

Elliott, T. R., Blichert-Toft, J., Heumann, A., Koetsier, G. W., & Forjaz, V. H. (2007). The origin of enriched mantle beneath São Miguel, Azores. *Geochimica et Cosmochimica Acta, 71,* 219–240.

Escartín, J., Cannat, M., Pouliquen, G., & Rabain, A. (2001). Crustal thickness of V-shaped ridges south of the Azores: Interaction of the Mid-Atlantic Ridge (36°–39°N) and the Azores hot spot. *Journal Geophysical Research, 106*(10), 21719–21735.

Esenwein, P. (1930). Zur Petrografie der Azores. *Zeitschrift für Vulkanologie, 12*(3), 108–227.

Feraud, G., Gastaud, J., Schmincke, H. U., Pritchard, G., Lietz, J., & Bleil, U. (1981). New K-Ar ages, chemical analyses and magnetic data of rocks from the islands of Santa Maria (Azores), Porto Santo and Madeira (Madeira Archipelago) and Gran Canaria (Canary Islands). *Bulletin of Volcanology, 44,* 359–375.

Feraud, G., Kaneoka, J., & Allègre, C. (1980). K/Ar ages and stress pattern in the Azores: Geodynamic implications. *Earth and planetary science letters, 46*, 275–286.

Fernandez, L. A. (1980). Geology and petrology of the Nordeste volcanic complex, São Miguel, Azores: Summary. *Geological Society of America Bulletin, 91*(12), 675–680.

Fernandez, L. A. (1982). The Petrology and Geochemistry of the Nordeste Volcanic Complex, São Miguel, Azores. *Arquipélago, 3*, 145–158.

Fetter, P. (1981). *Ile de São Miguel (Açores) Structurologie et Petrologie* (Ph.D. thesis, Université de Paris-Sud, Centre d'Orsay, France).

Fidczuk, P. J. (1984). *Chemistry of peralkaline felsic rocks from São Miguel and Terceira, Azores* (Ph.D. thesis, Department of Geology, University of Reading, UK).

Flower, M. F. J., Schmincke, H. U., & Bowman, H. (1976). Rare earth and other trace elements in historic azorean lavas. *Journal of Volcanology and Geothermal Research, 1*(2), 127–147.

Forjaz, V. H. (1966). Carta vulcanológica do sistema vulcânico Faial-Pico-São Jorge à escala 1:20,0000. In F. Machado & V. H. Forjaz (Eds.), *Actividade vulcânica do Faial 1957-67*. Porto: Comissão Regional do Turismo do Distrito da Horta.

Forjaz, V. H. (1976). Carta Vulcanológica da ilha de S. Miguel (1:50,000). Ed. Inst. Geoc. Açores, Ponta Delgada.

Forjaz V. H. (1977a). Carta Vulcanológica da ilha do Faial (1:25,000). Ed. Inst. Geoc. Açores, Ponta Delgada.

Forjaz, V. H. (1977b). Carta Vulcanológica da ilha do Pico (1:50,000). Ed. Inst. Geoc. Açores, Ponta Delgada.

Forjaz, V. H. (1980a). Esboço geológico do sistema Vulcânico Faial-Pico-São Jorge. Rel. Int. 03/80, 1 folha, Lab. Geociências e Tecnologia, S.R.C.I., Ponta Delgada.

Forjaz, V. H. (1980b). Erupções Históricas do Sistema Vulcânico Faial-Pico-São Jorge. Rel. Int. 01/80, 1 folha, Lab. Geociências e Tecnologia, S.R.C.I., Ponta Delgada.

Forjaz, V. H. (1984). Carta geotécnica do Nordeste (S. Miguel) Açores; Ed. Departamento de Geociências da Univ. dos Açores, Ponta Delgada.

Forjaz, V. H. (2008). Vulcão dos Capelinhos—13 histórias. Colecção Piroclástica 14. OVGA Açores—FaiAlentejo. Ponta Delgada.

Forjaz, V. H., & Fernandes, N. S. (1970). Carta Geológica de Portugal—Ilha de S. Jorge (Açores). Folha "B" 1:50,000. Serv. Geology Port., Lisboa.

Forjaz, V. H., Monjardino, J. L., Fernandes, N. S. (1970). Carta Geológica de Portugal—Ilha de S. Jorge (Açores). Folha "A" 1:50,000; Serv. Geology Port., Lisboa.

Forjaz, V. H., & Pereira, V. (1976). *Carta Vulcanológica Preliminar da ilha Graciosa, scale 1:25,000*. São Miguel: Instituto de Geociências dos Açores, Portugal.

Forjaz, V. H., Serralheiro, A., & Nunes, J. C. (1990). Carta vulcanológica dos Açores—Grupo Central (1:200,000). Serviço Regional de Protecção Civil, Universidade dos Açores e Centro de Vulcanologia INIC (Eds). Ponta Delgada.

Fouqué, F. (1873). *Voyages géologiques aux Açores* (p. 28). Paris: Extrait de la Revue des Deux Mondes.

França, Z. (1993). *Contribuição para o estudo dos xenólitos sieníticos do Arquipélago dos Açores*. (MSc thesis, Geosciences Department, Universidade dos Açores, Ponta Delgada, São Miguel, Portugal).

França, Z. (2000). *Origem e evolução petrológica e geoquímica do vulcanismo da ilha do Pico* (Ph.D. thesis, Geosciences Department, Universidade dos Açores, Ponta Delgada, São Miguel, Portugal).

França, Z., Almeida, M. H., & Wallenstein, N. (1995). Ocorrência de xenólitos mantélicos numa lava da ilha do Pico. *Gaia, 11*, 47–52.

França, Z., Cruz, J. V., Nunes, J. C., & Forjaz, V. H. (2005). Geologia dos Açores: Uma perspectiva actual. *Açoreana, 10*(1), 11–140.

França, Z., Forjaz, V. H., Tilling, R., Kuentz, D., Widom, E., & Lago, M. (2009). *Volcanic history of Pico and Faial Islands*. Ponta Delgada, Portugal: Azores. Direcção Regional da Ciência e Tecnologia.

França, Z., Lago, M., Galé, C., Nunes, J. C., Pueyo, O., Gil, A., Arranz, E., Forjaz, V. H., & Pocovi, A. (2006a). Mafic enclaves from the Corvo Island basalts (Azores, Portugal): Preliminary data. IV Pico Island International.

França, Z., Lago, M., Galé, C., Ubide, T., Widom, E., Arranz, E., et al. (2008). Composition of gabbroic xenoliths in Flores island (Azores, Portugal). *Macla, 9*, 103–104.

França, Z., Lago, M., Nunes, J. C., Galé, C., Forjaz, V. H., Pueyo, O., et al. (2006b). Geochemistry of alkaline basalts of Corvo Island (Azores, Portugal): Preliminary data. *Geogaceta, 40*, 87–90.

França, Z., Nunes, J. C., Cruz, J. V., Duarte, H. F., & Forjaz, V. H. (2002). Estudo preliminar do vulcanismo da Ilha do Corvo, Açores. In F. G. Garcia & J. L. Valero (Eds.), *Procceedings da 3ª Assembleia Luso-Espanhola de Geodesia e Geofisica* (Vol. 2, pp. 727–730). Valencia, Spain: University Politécnica Valencia.

França, Z., & Rodrigues, B. (1994) Nota sobre a distribuição espacial de xenólitos sieníticos no arquipélago dos Açores. Geociências. *Reviews University Aveiro, 8*(1–2), 95–115.

França, Z. T. M., Tassinari, C. C. G., Cruz, J. V., Aparicio, A. Y., Araña, V., & Rodrigues, B. N. (2006c). Petrology, geochemistry and Sr-Nd-Pb isotopes of the volcanic rocks from Pico Island- Azores (Portugal). *Journal of Volcanology and Geothermal Research, 156*, 71–89.

Friedlander, I. (1929). *Die Azoren. Zeitschrift für Vulkanologie, 12*, 77–107.

Fructuoso, G. (1583?). Livro Sexto das Saudades da Terra. Ed. Instituto Cultural de Ponta Delgada.

Furnes, H., & Fridleifsson, I. B. (1978). Relationship between the chemistry and axial dimensions of some shallow water pillow lavas of alkaline olivine basalt—

and olivine tholeiite composition. *Bulletin of Volcanology, 41,* 136–146.

Gandino, A., Guidi, M., Merlo, C., Mete, L., Rossi, R., & Zan, L. (1985). Preliminary model of the Ribeira Grande geothermal field (Azores Islands). *Geothermics, 14,* 91–105.

Garcia, M. O., Ito, E., Eiler, J. M., & Pietruszka, A. J. (1998). Crustal contamination of Kilauea Volcano magmas revealed by oxygen isotope analysis of glass and olivine from the Puu Oo eruption lavas. *Journal of Petrology, 39,* 803–817.

Gaspar, J. L. (1996). *Ilha Graciosa (Açores). História vulcanológica e avaliaçao do hazard* (Ph.D. thesis, Department Geociências, Universidade dos Açores, Portugal).

Gee, M. A. M., Thirlwall, M. F., Taylor, R. N., Lowry, D., & Murton, B. J. (1998). Crustal processes: Major controls on Reykjanes Peninsula lava chemistry, SW Iceland. *Journal of Petrology, 39,* 819–839.

Genske, F. S. (2012) *Assessing the heterogeneous source of the Azores mantle plume* (Ph.D. thesis, GEMOC, Department of Earth and Planetary Sciences, Faculty of Science, Macquarie University Sydney, Australia).

Genske, F. S., Beier, C., Haase, K. M., Turner, S. P., Krumm, S., & Brandl, P. A. (2012a). Oxygen isotopes in the Azores islands: Crustal contamination recorded in olivine. *Geology, 41*(4), 491–494.

Genske, F. S., Turner, S. P., Beier, C., Chu, M. F., Tonarini, S., Pearson, N. J., et al. (2014). Lithium and boron isotope systematics in lavas from the Azores islands reveal crustal assimilation. *Chemical Geology, 373,* 27–36.

Genske, F. S., Turner, S. P., Beier, C., & Schaefer, B. F. (2012b). The Petrology and Geochemistry of Lavas from the Western Azores Islands of Flores and Corvo. *Journal of Petrology, 53*(8), 1637–1671.

Ghiorso, M. S., & Sack, R. O. (1995). Chemical mass transfer in magmatic processes. IV. A revised and internally consistent thermodynamic model for the interpolation and extrapolation of liquid-solid equilibria in magmatic systems at elevated temperatures and pressures. *Contributions to Mineralogy and Petrology, 119,* 197–212.

Gill, J., Reagan, M., Tepley, F., & Malavassi, E. (2006). Arenal Volcano, Costa Rica. Magma Genesis and volcanological Processes. *Journal of Volcanology and Geothermal Research, 157,* 1–8.

Girod, M., & Lefèvre, C. (1972). The so-called andesites of the Azores. *Contributions to Mineralogy and Petrology, 35,* 159–167.

Glitsch, L., & Allegre, C. J. (1979). Determination of crystallization temperatures in fractional crystallization series by nickel partitioning equations. *Earth and Planetary Science Letters, 44,* 105–118.

Gualda, G. A. R., Ghiorso, M. S., Lemons, R. V., & Carley, T. L. (2012). Rhyolite-MELTS: A modified calibration of MELTS optimized for silica-rich, fluid-bearing magmatic systems. *Journal of Petrology, 53,* 875–890.

Guest, J. E., Gaspar, J. L., Cole, P. D., Queiroz, G., Duncan, A. M., Wallenstein, N., et al. (1999). Volcanic geology of Furnas Volcano, São Miguel, Azores. *Journal of Volcanology and Geothermal Research, 92*(1–2), 1–29.

Haase, K. M., & Beier, C. (2003). Tectonic control of ocean island basalt sources on São Miguel, Azores? *Geophysical Reseach Letters, 30*(16), 1856.

Harris, C., Smith, H. S., & Le Roex, A. (2000). Oxygen isotope composition of phenocrysts from Tristan da Cunha and Gough Island lavas: Variation with fractional crystallization and evidence for assimilation. *Contrib. Mineral. Petrol., 138,* 164–175.

Hart, S. R., & Ravizza, G. E. (1996). Os partitioning between phases in lherzolite and basalt. In A. Basu & S. R. Hart (Eds.), *Earth processes, reading the isotopic code* (pp. 123–134). Washington DC: AGU.

Hartung, G. (1860). *Die Azoren in ihrer ausseren Erschreinung und nach ihrer geognostichen Natur, Lepzig.*

Hekinian, R., Bideau, D., Francheteau, J., Cheminee, J. L., Armijo, R., Lonsdale, P. & Blum, N. (1993). Petrology of the East Pacific Rise and upper mantle exposed in Hess Deep (Eastern equatorial Pacific). *Journal of Geophysical Research 98,* 8069–8094.

Herrmann, W., & Berry, R. F. (2002). MINSQ—A least squares spreadsheet method for calculating mineral proportions from whole rock major element analyses. *Geochemistry: Exploration, Environment, Analysis, 2,* 361–368.

Hildenbrand, A., Madureira, P., Marques, F. O., Cruz, I., Henry, B., & Silva, P. (2008). Multi-stage evolution of a sub-aerial volcanic ridge over the last 1.3 Myr: S. Jorge Island, Azores Triple Junction. *Earth and Planetary Science Letters, 273,* 289–298.

Hildenbrand, A., Marques, F. O., Costa, A. C. G., Sibrant, A. L. R., Silva, P. F., Henry, B., et al. (2012). Reconstructing the architectural evolution of volcanic islands from combined K/Ar, morphologic, tectonic, and magnetic data: The Faial Island example (Azores). *Journal of Volcanology and Geothermal Research, 241–242,* 39–48.

Hildenbrand, A., Weis, D., Madureira, P., & Marques, F. O. (2014). Recent plate re-organization at the Azores Triple Junction: Evidence from combined geochemical and geochronological data on Faial, S. Jorge and Terceira Volcanic Islands. *Lithos, 210–211,* 27–39.

Jeremine, E. (1957). Etude microscopique des roches de la region de Furmas (S. Miguel, Acores). *Communications Service Geology Port, 38,* 65–90.

Jerram, D. A., Martin, V. M. (2008). Understanding crystal populations and their significance through the magma plumbing system. In C. Annen & G. F. Zellmer (Eds.), *Dynamics of crustal magma transfer, storage and differentiation.* Geological Society, Special Publications. London 304:133–148.

Johansen, T. S. (1996). *Composition and evolution of the upper mantle under São Miguel, the Azores: Evidence from ultramafic xenoliths* (Ph.D. thesis,

Mineralogical-Geological Museum, University of Oslo, Norway).
Johnson, C. L., Wijbrans, J. R., Constable, C. G., Gee, J., Staudigel, H., Tauxe, L., et al. (1998). 40Ar/39Ar ages and paleomagnetism of São Miguel lavas, Azores. *Earth and Planetary Science Letters, 160,* 637–649.
Jónasson, K. (2007). Silicic volcanism in Iceland: Composition and distribution within the active volcanic zones. *Journal of Geodynamics, 43*(1), 101–107.
Lacroix, A. (1893). *Les enclaves des roches volcaniques.* Mâcon: Protat, Impr.
Lago, M., França, Z., Galé, C., Widom, E., Arranz, E., Forjaz, V. H., Pueyo, O., & Ubide, T. (2007). Los enclaves de gabros de la Isla de Corvo (Azores, Portugal). Abstract presented at I Iberian Geochemistry Meeting, Vila—Real (Portugal).
Larrea, P. (2010). *Petrología y geoquímica de los enclaves de la isla Graciosa Azores (Portugal)* (MSc thesis, Department of Earth Sci. University of Zaragoza, Spain).
Larrea, P. (2014). *The magmatic evolution of Graciosa and Corvo oceanic islands azores Archipelago* (Ph.D. thesis, Universidad de Zaragoza). ISBN:978-84-697-0407-3.
Larrea, P., França, Z., Lago, M., Widom, E., Galé, C., & Ubide, T. (2013). Magmatic processes and the Role of Antecrysts in the genesis of Corvo Island (Azores Archipelago Portugal). *Journal of Petrology, 54,* 769–793.
Larrea, P., Galé, C., Ubide, T., Widom, E., Lago, M., & França, Z. (2014a). Magmatic evolution of Graciosa island (Azores Portugal). *Journal of Petrology, 55* (11), 2125–2154.
Larrea, P., Galé, C., Ubide, T., Widom, E., Lago, M., França, Z., & Tierz, P. (2012). Magmatic evolution of the Western Azores Islands (Corvo and Flores). Abstract DI51A-2353 presented at 2012 Fall Meeting AGU San Francisco California (USA).
Larrea, P., Lago, M., França, Z., Widom, E., Galé, C., Ubide, T., et al. (2010). Enclaves in Graciosa Island (Açores Portugal): Preliminary data. *Geogaceta, 48,* 155–158.
Larrea, P., Wijbrans, J. R., Galé, C., Ubide, T., Lago, M., França, Z., et al. (2014b). 40Ar/39Ar constraints on the temporal evolution of Graciosa Island Azores (Portugal). *Bulletin of Volcanology, 76,* 796. https://doi.org/10.1007/s00445-014-0796-8.
Le Bas, M. J., Le Maitre, R. W., Streckeisen, A., & Zanettin, B. (1986). A chemical classification of volcanic rocks based on the total alkali–silica diagram. *Journal of Petrology, 27,* 745–750.
Le Maitre, R. W. (2002). *Igneous rocks: a classification and glossary of terms.* UK: Cambridge Univeristy Press.
Leake, B. E., Wooley, A. R., Arps, C. E. S., Birch, W. D., Gilbert, M. C., Grice, J. D., et al. (1997). Nomenclature of amphiboles: Report of the subcommittee on amphiboles of the international mineralogical association commission on new minerals and mineral names. *Canadian Mineralogist, 35,* 219–246.
Lemarchand, F. (1984). Les séries volcaniques de Fayal (Açores): Étude pétrologique et géochimique. *Canadian Journal of Earth Sciences, 24,* 334–353.
Lemarchand, F. (1987). Les series volcaniques de Fayal (Acores): Étude petrologique et geochimique. *Canadian Journal of Earth Sciences, 24,* 334–353.
Lima, E. F., Machado, A., Nardi, L. V. S., Saldanha, D. L., Azevedo, J. M. M., Sommer, C. A., et al. (2011). Gechemical evidence concerning sources and petrologic evolution of Faial Island Central Azores. *International Geology Review, 53*(14), 1684–1708.
Loucks, R. (1996). A precise olivine–augite Mg–Fe-exchange geothermometer. *Contributions to Mineralogy and Petrology, 125,* 140–150.
Luis, J. F., & Neves, M. C. (2006). The isostatic compensation of the Azores Plateau: A 3D admittance and coherence analysis. *Journal of Volcanology and Geothermal Research, 156*(1–2), 10–22.
Machado, N. (1977). *Géochimie de l'île de São Jorge (Açores)* (MSc thesis, Université de Montréal, Canada).
Machado, A., Azevedo, J. M. M., Almeida, D., Chemale, J., & R. F. (2008). Geochemistry of Volcanic Rocks from Faial Island (Azores). In *e-Terra* 5(1).
Machado, F., & Forjaz, V. H. (1965). A crise sísmica de S. Jorge de fevereiro de 1964. In Bol Soc Geol Port *26* 19–36 Lisboa. In Bol. Mus. e Lab. Min. e Geol. da FCUL 10 n° 2, 89–94, Lisboa.
Machado, F., & Forjaz, V. H. (1968). Actividade vulcânica do Faial 1957-67. Comissão Regional do Turismo do Distrito da Horta Porto.
Madeira, J. (1986). *Geologia estrutural e enquadramento geotectónico da ilha de Santa Maria (Açores).* Provas de Aptidão Pedagógica e Capacidade Cientifica: Universidade de Lisboa, Portugal.
Madeira, J. (1998). *Estudos de neotectónica nas ilhas do Faial Pico e S. Jorge: uma contribuição para o conhecimento geodinâmico da junção tripla dos Açores* (Ph.D. thesis, Lisbon University, Portugal).
Madeira, J. (2005). The volcanoes of Azores Islands: A world-class heritage (examples from Terceira, Pico and Faial Islands). In *IV International Symposium ProGEO on the Conservation of the Geological Heritage*, Field Trip Guide Book.
Madureira, P. (2006). *Geoquímica elementar e isotópica (He e Ne) das lava máficas da ilha Terceira: evidencias para uma origen profunda da pluma dos Açores* (Ph.D. thesis, Evora University, Portugal).
Madureira, P., Mata, J., Mattielli, N., Queiroz, G., & Silva, P. (2011). Mantle source heterogeneity magma generation and magmatic evolution at Terceira Island (Azores Archipelago): Constraints from elemental and isotopic (Sr Nd Hf Pb) data. *Lithos, 126,* 402–418.
Marriner, G. F., Norry, M. J., & Gibson, I. L. (1982). The petrology and geochemistry of the Água de Pau volcano São Miguel Azores. *Arquipélago, 3,* 159–173.
Matias, L., Dias, N. A., Morais, I., Vales, D., Carrilho, F., Madeira, J., et al. (2007). The 9th of July 1998 Faial

Island (Azores North Atlantic) seismic sequence. *Journal of Seismology, 11*(3), 275–298.
Mattioli, M., Upton, B. G. J., & Renzulli, a. (1997). Sub-volcanic crystallization at Sete Cidades volcano, São Miguel, Azores, inferred from mafic and ultramafic plutonic nodules. *Mineralogy and Petrology, 60*, 1–26.
Maund, J. G. (1985). *The volcanic geology petrology and geochemistry of Caldeira volcano Graciosa Azores and its bearing on contemporaneous felsic-mafic oceanic island volcanism.* (Ph.D. thesis, University of Reading United Kingdom, UK).
Metrich, N., Bizouard, H., & Varet, J. (1981). Petrology of the volcanic series of Fayal Azores. *Bulletin of Volcanology, 44*, 71–93.
Metrich, N., Zanon, V., Creon, L., Hildenbrand, A., Moreira, M., & Marques, F. O. (2014). Is the "Azores Hotspot" a wetspot? Insights from the geochemistry of fluid and melt inclusions in olivine of Pico Basalts. *Journal of Petrology, 55*(2), 377–393.
Millet, M. A., Doucelancea, R., Bakerb, J. A., & Schianoa, P. (2009). Reconsidering the origins of isotopic variations in Ocean Island Basalts: Insights from fine-scale study of São Jorge Island Azores archipelago. *Chemical Geology, 265*(3–4), 289–302.
Moore, R. B. (1990). Volcanic geology and eruption frequency São Miguel Azores. *Bulletin of Volcanology, 52*, 602–614.
Moore, R. B. (1991). Geology of three late quaternaty stratovolcanoes on São Miguel Azores. United States Geological Survey Bulletin.
Moreira, M., Doucelance, R., Kurz, M. D., Dupre, B., & Allegre, C. J. (1999). Helium and lead isotope geochemistry of the Azores Archipelago. *Earth and Planetary Science Letters, 169*, 189–205.
Morimoto, N., Fabries, J., Ferguson, A. K., Ginzburg, I. V., Ross, M., Seifert, F. A., et al. (1988). Nomenclature of pyroxenes. *Bulletin Mineralogist, 111*, 535–550.
Morisseau, M. (1987). *Les eruptions hydromagmatiques et les xenoites associes: signification geothermique exemples de Flores e de Faial (Açores)* (Ph.D. thesis, Université de Paris-Sud Paris, France).
Morrisseau, M., & Traineau, H. (1985). Mise en évidence d'une activité hydromagmatique holocène sue l'île de Flores (Açores). *C. R. Acad. Sc. Paris t. 301 Série II, 18*, 1309–1314.
Mügge, O. (1883). Petrographische Untersuchungen na Gesteinen von den Azoren. *Neues Jahrbuch für Mineralogie, Geologie und Paläontologie, 2*, 189–244.
Mungall, J. E., & Martin, R. F. (1995). Petrogenesis of basalt-comendite and basalt-pantellerite suites Terceira Azores and some implications for the origin of ocean-island rhyolites. *Contrib Mineral Petrol., 119*, 43–55.
Nunes, J. C. (2000). Notas sobre a geologia da Ilha Terceira (Açores). *Açoreana, 9*(2), 205–215.
Nunes J. C., Cruz J. V., Carvalho, M. R., França Z., & Forjaz, V. H. (1995). Aspectos vulcanológicos da Ilha do Pico. Poster—Curso de Actualização para Professores de Geociências (XV Edição) e Seminário Geologia Atlântica. Fevereiro. Ponta Delgada, Portugal.
Nunes, J. C., França, Z., Cruz J. V., Carvalho M. R., & Serralheiro, A. (1999) Carta Vulcanológica da Ilha do Pico (Açores)—Versão Preliminar. Escala 1:30,000. Folhas 1 2 e 3.
Nunes, J. C., Lima, E. A., & Fontiela, J. (2006). Caracterização geomorfológica e geológica da fajã lávica das Lajes do Pico (Açores). Condicionantes e vulnerabilidades associadas à arriba fóssil. Relatórios e Comunicações do Departamento de Biologia—XII Expedição Científica do Departamento de Biologia Pico 2005 vol. 34, 15–20.
O'Hara, M. J. (1968). Are ocean floor basalts primary magma? *Nature, 222*, 683–686.
Pacheco, J. M. R. (2001). *Processos associados ao desenvolvimento de erupções vulcânicas hidromagmáticas explosivas na ilha do Faial e sua interpretação numa perspectiva de avaliação do hazard e minimização do risco* (Ph.D. thesis, Departamento de Geociências Universidade dos Açores, Portugal).
Pimentel, A. H. G. (2006). *Domos e Coulées da ilha Terceira (Açores): Contribuição para o estudo de mecanismos de instalação* (MSc thesis, Departamento de Geociências Universidade dos Açores, Portugal).
Prytulak, J., Elliott, T. (2007). TiO_2 enrichment in ocean island basalts. *Earth And Planetary Science Letters, 263*, 388–403. ISSN: 0012-821X.
Prytulak, J., & Elliott, T. (2009). Determining melt productivity of mantle sources from 238U–230Th and 235U–231 Pa disequilibria; an example from Pico Island Azores. *Geochimica et Cosmochimica Acta, 73*, 2103–2122.
Putirka, K. (2008). Thermometers and Barometers for Volcanic Systems. In: K. Putirka & F. Tepley (Eds.), *Minerals, inclusions and volcanic processes. Reviews in mineralogy and geochemistry* (vol. 69, pp. 61–120). Mineralogical Society of America.
Rayleigh, J. W. S. (1896). Theoretical considerations respecting the separation of gases by diffusion and similar processes. *Philosophical Magazine, 42*, 77–107.
Reisberg, L., Zindler, A., Marcantonio, F., White, W., Wyman, D., & Weaver, B. (1993). Os isotope systematics in ocean island basalts. *Earth and Planetary Science Letters, 120*, 149–167.
Renzulli, A., & Santi, P. (2000). Two-stage fractionation history of the alkali basalt-trachyte series of Sete Cidades volcano (São Miguel Island Azores). *European Journal of Mineralogy, 12*, 469–494.
Ribeiro, L. (2011). *Petrologic and Geochemical Characterization of São Jorge Island Volcanism Azores*. (Ph. D. thesis, Department Geociências University of Aveiro, Portugal).
Ridley, W. I., Watkins, N. D., & MacFarlane, D. J. (1974). The oceanic islands: Azores. In A. E. M.

Nairn, F. G. Stehli, (Eds.), *The Ocean Basins and Margins* (Vol. 2, pp. 443–483). New York: Plenum.

Rodrigues, B., Alves, C. M. A., Serralheiro, A., & Forjaz, V. H. (1985). Nota prévia sobre a petrologia e geoquímica da ilha de Santa Maria açores. *MEM. Academy Ciências, 46,* 71–91.

Roeder, P. L., & Emslie, R. F. (1970). Olivine-liquid equilibrium. *Contributions to Mineralogy and Petrology, 29,* 275–289.

Rosenbaum, M. S. (1974). Basaltic volcanism in south-east Terceira Azores. *Geological Magazine, 111*(5), 409–420.

Rowland-Smith, A. (2007). *The 1630 AD eruption of Furnas volcano, São Miguel, Azores (Portugal): chemical variations and magmatic processes* (Ph.D. thesis, Miami University, USA).

Salgueiro, M. A. O. (1991). Estudo paleomagnético e cronología estratigráfica de formações vulcânicas da Ilha de Santa Maria Açores. *Archipélago, 9,* 83–99.

Schmincke, H. U. (1973). Magmatic evolution and tectonic regime in the Canary Madeira and Azores island groups. *Geological Society of America Bulletin, 84,* 633–648.

Schmincke, H. U., & Staudigel, H. (1976). Pillow lavas on central and eastern atlantic islands (La Palma, Gran Canaria, Porto Santo and Santa Maria): Preliminary report. *Bulletin de la Société Géologique de France, 18,* 871–883.

Schmincke, H. U., & Weibel, M. (1972). Chemical study of rocks from Madeira Porto Santo and São Miguel Terceira (Azores). *N. Jb. Miner. Abh., 117*(3), 253–281.

Self, S. (1973). *Recent volcanism on Terceira Azores* (Ph. D. thesis, Geology Department Imperial College London, UK).

Self, S. (1976). The recent volcanology of Terceira Azores. *Journal of the Geological Society, 132,* 645–666.

Self, S., & Gunn, B. M. (1976). Petrology volume and age relations of alkaline and saturated peralkaline volcanics from Terceira Azores. *Contributions to Mineralogy and Petrology, 54,* 293–313.

Serralheiro, A. (2003). A geologia da Ilha de Santa Maria Açores. *Açoreana, 10*(1), 141–192.

Serralheiro, A., Alves, C. A. M., Forjaz, V. H., & Rodrigues, B. (1987). Carta vulcanológica dos Açores: Ilha de Santa Maria. Edited by Serviço Regional de Protecção Civil (Região Autónoma dos Açores) Universidade dos Açores and Centro de Vulcanologia INIC Ponta Delgada 1ª ed. 2 folhas na escala 1:15,000.

Serralheiro, A., Forjaz, V. H., Alves, C. A. M., & Rodrigues, B. (1989). Carta vulcanológica dos Açores Ilha do Faial (1:15,000). Edição do Centro de Vulcanologia do INIC do Departamaneto de Goeciências da Universidade dos Açores e do Serviço Regional de Proteção Civil dos Açores, Portugal.

Serralheiro, A., & Madeira, J. (1990). Stratigraphy and geochronology of the island of Santa Maria (Azores). In Livro jubilar de Carlos Romariz Dep. de Geologia da Faculdade de Ciências de Lisboa pp. 357–376.

Serralheiro, A., & Madeira, J. (1993). Stratigraphy and geochronology of Santa Maria Island (Azores). *Açoreana, 7*(4), 575–592.

Shimizu, N. (1978). Analysis of the zoned plagioclase of different magmatic environments: A preliminary ion-microprobe study. *Earth and Planetary Science Letters, 39,* 398–406.

Shotton, F. W. (1969). Birmingham University radiocarbon dates III: Radiocarbon 11, 266

Shotton, F. W., & Williams, R. E. G. (1971). *Birmingham University radiocarbon dates V: Radiocarbon, 13,* 141–156.

Snyder, D. C. (2005). *Processes and tieme scales of differentiation in silicic magma chambers: chemical and isotopic investigations* (Ph.D. thesis, Miami University, USA).

Snyder, D. C., Widom, E., Pietruszka, A. J., & Carlson, R. W. (2004). The role of open-system processes in the development of silicic magma chambers: A chemical and isotopic investigation of the Fogo a trachyte deposit, São Miguel, Azores. *Journal of Petrology, 45*(4), 723–738.

Stolper, E. (1980). A phase diagram for mid-ocean ridge basalts: Preliminary results and implications for petrogenesis. *Contributions to Mineralogy and Petrology, 74,* 13–27.

Storey, M. (1981). Trachytic pyroclastics from Agua de Pau volcano São Miguel Azores; evolution of a magma body over 4000 years. *Contributions to Mineralogy and Petrology, 78,* 423–432.

Storey, M. (1982). Trachytic pyroclastics from Agua de Pau volcano São Miguel Azores: evolution of a magma body over 4000 years. *Contributions to Mineralogy and Petrology, 78,* 423–432.

Storey, M., Wolff, J. A., Norry, M. J., & Marriner, G. F. (1989). Origin of hybrid lavas from Agua de Pau volcano São Miguel Azores. In A. D. Saunders & M. J. Norry (Eds.), *Magmatism in the Ocean Basins. Geological Society, London, Special Publications, 42,* 161–180.

Thirlwall, M. F., Jenkins, C., Vroon, P. Z., & Mattey, D. P. (1997). Crustal interaction during construction of ocean islands: Pb–Sr–Nd–O isotope geochemistry of the shield basalts of Gran Canaria, Canary Islands. *Chemical Geology, 135,* 233–262.

Turner, S. P., Hawkesworth, C. J., Rogers, N. W., & King, P. (1997). U-Th isotope disequilibria and ocean island basalt generation in the Azores. *Chemical Geology, 139,* 145–164.

Watanabe, S. (2010). *Isotope and trace element investigation of magmatic processes and timescales in the Azores* (Ph.D. thesis, Miami University, USA).

Weaver, B. L. (1991). The origin of ocean island basalt end-member compositions: trace element and isotopic constraints. *Earth and Planetary Science Letters, 104,* 381–397.

Webster, J. W. (1821). *A description of the Island of St.* In R.P. Michael and C. Williams (Ed.), Boston.

White, W. M., Tapia, M. D. M., & Schilling, J. G. (1979). The petrology and geochemistry of the Azores Islands. *Contributions to Mineralogy and Petrology, 69,* 201–213.

Widom, E. (1991). *Petrogenetic processes and timescales of young alkaline volcanic rocks São Miguel* (Ph.D. thesis, UC Santa Cruz, USA).

Widom, E. (1997). Sources of ocean island basalts: A review of the osmium isotope evidence. *Physica A: Statistical Mechanics and its Applications, 244,* 484–496.

Widom, E., Carlson, R. W., Gill, J. B., & Schmincke, H. U. (1997). Th-Sr-Nd-Pb isotope and trace element evidence for the origin of the São Miguel Azores: enriched mantle source. *Chemical Geology, 140,* 49–68.

Widom, E., Gill, J. B., & Schmincke, H. U. (1993). Syenite Nodules as a Long-Term Record of Magmatic Activity in Agua de Pao Volcano São Miguel Azores. *Journal of Petrology, 34*(5), 929–953.

Widom, E., Hoernle, K. A., Shirey, S. B., & Schmincke, H. U. (1999). Os Isotope Systematics in the Canary Islands and Madeira: Lithospheric Contamination and Mantle Plume Signatures. *Journal of Petrology, 40*(2), 279–296.

Widom, E., Schmincke, H. U., & Gill, J. B. (1992). Processes and timescales in the evolution of a chemically zoned trachyte: Fogo A São Miguel Azores. *Contributions to Mineralogy and Petrology, 111,* 311–328.

Widom, E., & Shirey, S. B. (1996). Os isotope systematics in the Azores: Implications for mantle plume sources. *Earth and Planetary Science Letters, 142,* 451–466.

Wolff, J. A., Grandy, J. S., & Larson, P. B. (2000). Interaction of mantle-derived magma with island crust? Trace element and oxygen isotope data from the Diego Hernandez Formation, Las Cañadas, Tenerife. *Journal of Volcanology and Geothermal Research, 103,* 343–366.

Wolff, J. A., & Storey, M. (1983). The volatile component of some pumice-forming alkaline magmas from the Azores and Canary Islands. *Contributions to Mineralogy and Petrology, 82,* 66–74.

Wolff, J. A., & Storey, M. (1984). Zoning in highly alkaline magma bodies. *Geological Magazine, 121,* 563–575.

Wright, T. L., & Doherty, P. C. (1970). A linear programming and least squares computer method for solving petrologic mixing problems. *Geological Society of America Bulletin.* https://doi.org/10.1130/0016-7606(1970)81.

Yu, H. (2011). *Li, Hf and Os isotope systematics of Azores basalts and a new microwave digestion method for os isotopic analysis* (Ph.D. thesis, Miami University, USA).

Zanon, V. (2015). Conditions for mafic magma storage beneath fissure zones at oceanic islands. The case of São Miguel Island (Azores archipelago). In L. Caricchi (Ed.), *Chemical Physical and Temporal Evolution of Magmatic Systems, 22*(1), 85–104. *Journal Geological Society,* London—Special volume. https://doi.org/10.1144/sp422.4.

Zanon V., & Frezzotti, M. L. (2013). Magma storage and ascent conditions beneath Pico and Faial islands (Azores archipelago): A study on fluid inclusions. *Geochemistry Geophysics Geosystems, 14*(9). https://doi.org/10.1002/ggge.20221.

Zanon, V., Kueppers, U., Pacheco, J. M., & Cruz, I. (2013). Volcanism from fissure zones and the Caldeira central volcano of Faial Island Azores archipelago: Geochemical processes in multiple feeding systems. *Geological Magazine, 150*(3), 536–555.

Zanon, V., & Pimentel, A. (2015). Spatio-temporal constraints on magma storage and ascent conditions in a transtensional tectonic setting: The case of the Terceira Island (Azores). *American Mineralogist, 100* (4), 795–805. https://doi.org/10.2138/am-2015-4936.

Zbyszewski, G. (1970). Levantamentos geológicos da ilha Graciosa (Açores) Memórias Academia Ciências Lisboa. *Ciências, 14,* 163–171.

Zbyszewski, G., Assunção, C. T., & Ferreira, V. O. (1961a). La Géologie des îlots de Formigas au NE de l'ile de Santa Maria. *Communication Service Geologique du Portugal, 45,* 513–518.

Zbyszewski, G., & Ferreira, V. O. (1962). Étude géologique de l'île de Santa Maria (Açores). *Communication Service Geologique du Portugal, 46,* 209–246.

Zbyszewski, G., Almeida F. M., Ferreira, V. O., & Assunção, C. T. (1958). Carta Geológica de Portugal na escala de 1:50,000. Notícia explicativa da Folha B S. Miguel (Açores). *Communication Service Geologique du Portugal Lisboa.*

Zbyszewski, G., Almeida F. M., & Ferreira, V. O., Assunção, C. T. (1959a). Carta Geológica de Portugal na escala de 1:25,000. Notícia explicativa da Folha Faial (Açores). *Communication Service Geologique du Portugal Lisboa.*

Zbyszewski, G., Ferreira, V. O., & Assunção, C. T. (1959b). Carta Geológica de Portugal na escala de 1:50,000. Notícia explicativa da Folha A S. Miguel (Açores). Service Geologique du Portugal Lisboa.

Zbyszewski, G., Ferreira, V. O., Assunção, C. T. & (1961b). Carta Geológica de Portugal na escala 1:50,000. Notícia explicativa da folha da ilha de Santa Maria (Açores). Service Geology Portugal Lisboa.

Zbyszewski, G., Medeiros A. C., Ferreira, V. O., & Assunção, C. T. (1968). Carta Geológica de Portugal na escala de 1:25000. Notícia explicativa da Folha Ilha das Flores (Açores). Service Geologique du Portugal Lisboa.

Zbyszewski, G., Ferreira, C. R., & Ferreira, V. O. (1962). Étude géologique de l'ile de Pico (Açores). *Communication Service Geologique du Portugal, 46,* 5–66.

Zbyszewski, G., Medeiros, A. C., & Ferreira, V. O. (1971). Carta Geológica de Portugal na escala de 1:50,000. Notícia explicativa da Folha Ilha Terceira (Açores). *Service Geologique du Portugal Lisboa.*

Zbyszewski, G., Medeiros, A., Ferreira, V. O., & Assunção, C. T. (1967). Carta Geológica de Portugal na escala 1:25,000. Notícia explicative da folha da ilha do Corvo (Açores). Service Geologique du Portugal Lisboa.

Zbyszewski, G., Medeiros, A., Ferreira, V. O. & Assunção, C. T. (1972). Carta Geológica de Portugal na escala 1:25,000. Notícia explicative da folha da ilha da Graciosa (Açores). Service Geologique du Portugal Lisboa.

Melting and Mantle Sources in the Azores

Christoph Beier, Karsten M. Haase and Philipp A. Brandl

Abstract

The Azores archipelago is geochemically distinct amongst the oceanic intraplate volcanoes in that it has trace element and radiogenic Sr–Nd–Pb–Hf isotope signatures that cover much of the global variation observed in Ocean Island Basalts. Thus, it is the prime example of an intraplate melting anomaly preserving the compositional heterogeneity of the Earth's mantle. Here, we review the trace element and radiogenic isotope geochemistry of the Azores islands and few submarine samples analysed and published over the past decades and summarise these findings and conclusions. The volcanoes of all islands erupted lavas of the alkaline series and their compositions broadly range from basalts to trachytes (see also Chapter "Petrology of the Azores Islands" by Larrea et al.). Temperatures and pressures of melting imply that melting in the Azores occurs as a result of both slightly increased temperatures in the mantle (~35 °C) and addition of volatile elements into the mantle source. Basalts from the island of São Miguel show a stronger enrichment in highly incompatible elements like K and the Light Rare Earth Elements than the other islands further to the west. The older and easternmost island Santa Maria has lavas that are more silica-undersaturated than the rocks occurring on the younger islands. Each of the eastern islands shows a different and distinct radiogenic isotope composition and much of this variability can be explained by variably enriched recycled components of different age in their source regions. Amongst the global array, the lavas from eastern São Miguel are uniquely enriched in that they display radiogenic $^{206}Pb/^{204}Pb$, $^{208}Pb/^{204}Pb$ and $^{87}Sr/^{86}Sr$ isotope ratios best explained by a distinct source in the mantle. The implication of the preservation of such unique, enriched sources in the mantle may indicate that stirring processes in the Azores mantle are not efficiently homogenising heterogeneities over the timescales of recycling of 0.1–1 Ga and possibly even up to 2.5 Ga. One possible explanation is the low buoyancy flux of the Azores mantle when compared to other intraplate settings. The preservation of these source signatures in the lavas on the easternmost

C. Beier (✉) · K. M. Haase · P. A. Brandl
GeoZentrum Nordbayern, Friedrich-Alexander Universität Erlangen-Nürnberg, Schloßgarten 5, 91054 Erlangen, Germany
e-mail: christoph.beier@fau.de

P. A. Brandl
Research School of Earth Sciences, The Australian National University, 142 Mills Road, Acton, ACT 2601, Australia

P. A. Brandl
GEOMAR Helmholtz-Zentrum für Ozeanforschung Kiel, Wischhofstraße 1-3, 24148 Kiel, Germany

U. Kueppers and C. Beier (eds.), *Volcanoes of the Azores*, Active Volcanoes of the World, https://doi.org/10.1007/978-3-642-32226-6_11

Azores islands are the result of smaller degrees of partial melting due to a thicker lithosphere. This likely prevents a homogenisation during magma ascent compared to the western islands, preferentially sampling deep, low degree partial melts from the more fertile mantle sources. The geochemical signatures of the two islands west of the Mid-Atlantic Ridge (Corvo and Flores) imply a source enrichment and degrees of partial melting similar to those east of the ridge. Melting underneath the western islands is the result of a source that must be related to the Azores melting anomaly but has been modified by shallow level processes such as assimilation of oceanic crustal material.

1 Introduction

Volcanism along the mid-ocean ridges (MOR) and in the oceanic intraplate environment is the surface expression of melting processes in the mantle. Magmatism associated with ocean islands has often been explained by adiabatic melting of ascending mantle material that has a significant excess temperature compared to the ambient mantle. However, it remains a matter of debate whether oceanic intraplate melting is solely the result of hotter material actively rising into the upper mantle from greater depths ("deep mantle plume"; Morgan 1971); or may also be initiated by compositional heterogeneities (O'Neill and Sigloch, Chapter "Crust and Mantle Structure Beneath the Azores Hotspot—Evidence from Geophysics"). Such geochemical heterogeneities may contain abundant volatiles (e.g., H_2O and CO_2) and would play an important role during mantle melting (e.g., "wetspots"; Bonatti 1990; Schilling et al. 1980; Schilling 1978). A combination of thermal and chemical effects has recently been proposed to be present in the form of thermochemical piles (Farnetani and Samuel 2005; Sobolev et al. 2005), but some intraplate volcanoes in the ocean basins may not be the result of mantle plumes. Instead they could be explained by the interaction of lithospheric structures with the underlying mantle (Farnetani and Samuel 2005; Foulger et al. 2005; Kumagai et al. 2008). Consequently, a critical assessment of the geophysical (O'Neill and Sigloch, Chapter "Crust and Mantle Structure Beneath the Azores Hotspot—Evidence from Geophysics") and geochemical features, particularly the incompatible elements and radiogenic and stable isotopes, is required to explain the excess melt production in these environments.

Basalts erupted at ocean islands (OIB) and those erupted along mid-ocean ridges (MORB) differ significantly in incompatible trace element and radiogenic isotope compositions (e.g., Dupré and Allègre 1983; Hofmann 1997, 2003; McKenzie and O'Nions 1995; Salters and White 1998; Stracke et al. 2003). In contrast to more evolved rocks (e.g., trachytes, rhyolites), basaltic lavas will likely reflect the composition and processes in the mantle (e.g., Hofmann 2003; Sun and McDonough 1989). OIB display signatures enriched in incompatible trace elements like the Rare Earth Elements (REE) that could reflect distinct source compositions, different degrees and depths of partial melting, or influence from shallow level processes (e.g., Assimilation-Fractional Crystallisation, AFC) or, in many cases, a combination of these processes to varying extends. The radiogenic Sr–Nd–Pb–Hf isotopes are less affected by shallow level processes (e.g., AFC) and may thus be used to constrain the mantle source composition. The isotopic ratios of OIB suggest formation from more enriched mantle sources than MORB as a result of element fractionation and changes of the time-integrated parent-daughter ratios during partial melting, recycling of continental and/or oceanic lithospheric material, or metasomatism of the mantle from fluids or melts (Gast 1968; Hofmann 2003).

The mechanism most often invoked to explain the origin of enriched mantle signatures is the subduction and recycling of oceanic crust, with or without sediments and/or continental material (Weaver 1991; Zindler and Hart 1986). Metasomatic processes within the mantle have also been proposed as a potential source of the enrichment in OIB (Donnelly et al. 2004; Halliday et al. 1995; Hart and Staudigel 1989; Hart and Zindler

1989; McKenzie and O'Nions 1995; Niu and O'Hara 2003). The physical and geochemical processes involved in the formation of the enriched sources are complex and manifold. The range of parameters that influences the composition of these sources is large, e.g., the mineral mode of the mantle (peridotite, pyroxenite, eclogite), the source composition and the temperature of the mantle will affect the final composition erupted (Asimow et al. 2001; Hirschmann et al. 2003; Hirschmann and Stolper 1996; Jaques and Green 1979; Pertermann and Hirschmann 2003; Stosch 1982; Stracke et al. 1999; Walter 1998). However, several studies have quantified the potential processes and sources in the mantle using trace element and isotope geochemistry of intraplate and MOR basalts (Brandl et al. 2012; Hart and Staudigel 1989; Kelley et al. 2005; Stracke et al. 2003; Workman et al. 2004).

Here, we review the origin of the Azores intraplate volcanism from a geochemical perspective using published data from submarine and subaerial lavas. We compare these to data from the Mid-Atlantic Ridge (MAR) nearby in order to understand the Azores melting anomaly and the geochemical signatures of rocks erupting on the Azores islands and on the adjacent ridge axis. Lavas from the Azores archipelago and the submarine volcanoes display geochemical compositions that not only cover much of the variation observed globally but are also variable even on the scale of a single volcanic system. Geochemical and petrologic models suggest that the volcanism in the Azores is the result of a combination of melting trace element- and volatile-enriched mantle sources *and* slightly increased mantle temperatures.

2 Geological Setting

The Azores archipelago in the central northern Atlantic Ocean lies on an oceanic plateau with about 350 km diameter between 37° to 40°N and 29° to 31°W. This plateau is situated at the triple junction between the Eurasian, African and American Plates (Fig. 1, Vogt and Jung, Chapter "The "Azores Geosyndrome" and Plate Tectonics: Research History, Synthesis, and Unsolved Puzzles", Miranda et al., Chapter "The Tectonic Evolution of the Azores Based on Magnetic Data"). The Azores archipelago consists of nine islands that emerge from the plateau. Seven of these islands are situated to the east of the Mid-Atlantic Ridge (MAR) that separates the Azores Plateau into a larger eastern and smaller western part. In the eastern part, three islands and a seamount (from west to east Graciosa, Terceira, João de Castro seamount, São Miguel) are aligned along the ultraslow spreading Terceira axis (Georgen 2008; Vogt and Jung 2004) which has a spreading rate of 2–4 mm year^{-1} classifying this ridge as one of the slowest spreading ridges on Earth (Vogt and Jung, Chapter "The "Azores Geosyndrome" and Plate Tectonics: Research History, Synthesis, and Unsolved Puzzles"). The crustal thickness of the Azores Plateau is estimated to be approximately 12–14 km (Dias et al. 2007; Escartín et al. 2001; Matias et al. 2007; O'Neill and Sigloch, Chapter "Crust and Mantle Structure Beneath the Azores Hotspot—Evidence from Geophysics") and thus twice as thick as normal ocean crust of 6–7 km (White et al. 1992). Geophysical and geodynamic modelling indicates that the major portion of the Azores Plateau was emplaced within a relatively short period of time of 4–10 Ma (Beier et al. 2015; Cannat et al. 1999; Gente et al. 2003) which may best be explained by a sudden increase in melt production similar to models for other oceanic plateaus, e.g., the Ontong Java Plateau (e.g., Neal et al. 1997). In addition, an anomaly with slow seismic wave velocities occurs in the upper mantle beneath the Azores (Pilidou et al. 2004; Yang et al. 2006 and O'Neill and Sigloch, Chapter "Crust and Mantle Structure Beneath the Azores Hotspot—Evidence from Geophysics"). The Azores represent an area of increased melt production with geochemical and geophysical features that are consistent with the deep mantle plume model, i.e. magma emplacement within short time-scales (Abdel Monem et al. 1975; Beier et al. 2015; Cannat et al. 1999; Gente et al. 2003) and primitive noble gas isotope ratios (Madureira

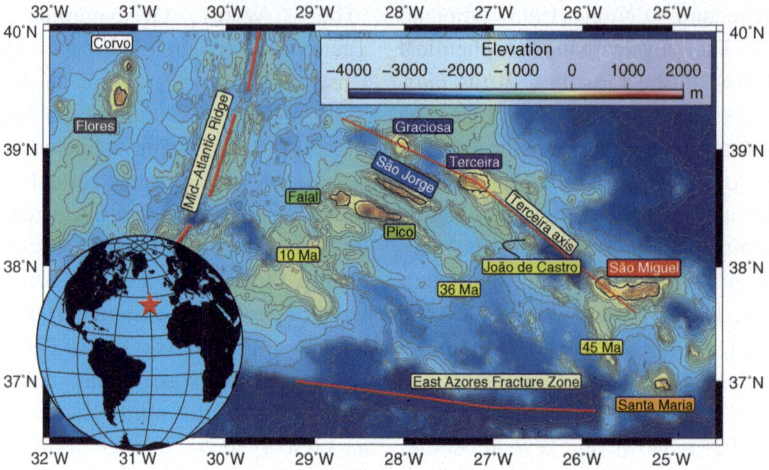

Fig. 1 Bathymetric map of the Azores archipelago. Map generated using Generic Mapping Tools (GMT; Wessel and Kroenke 2009; Wessel and Smith 1991). The ultraslow spreading Terceira axis (Vogt and Jung 2004) separates the Eurasian Plate in the north from the African Plate in the south. The slow spreading Mid-Atlantic Ridge (MAR) marks the plate boundary between the African/Eurasian and the American Plate. Colour coding of the islands is similar to that of the other figures in this chapter

et al. 2005; Moreira et al. 1999) as well as subchondritic and radiogenic Os isotope ratios (Schaefer et al. 2002; Widom and Shirey 1996). On the other hand, several features of Azores volcanism do not agree with the deep mantle plume model consisting of a plume head followed by a tail that is connected to the deeper mantle (Pilidou et al. 2005; Yang et al. 2006, O'Neill and Sigloch, Chapter "Crust and Mantle Structure Beneath the Azores Hotspot—Evidence from Geophysics"). Seismic tomography suggests that the Azores melting anomaly is restricted to the upper mantle (Yang et al. 2006) and the relatively high mantle volatile contents in the Azores magmas (Asimow et al. 2004; Beier et al. 2012, 2013; Metrich et al. 2014) are difficult to reconcile with a deep mantle origin because thermodynamic models indicate that volatiles are unlikely in nominally anhydrous minerals in the deeper mantle (Bolfan-Casanova 2003; Hirschmann 2006; Hirschmann et al. 2005; Litasov et al. 2003; Murakami et al. 2002). Additionally, the occurrence of young volcanism on both flanks of the Mid-Atlantic Ridge across a distance of 200 km to the west and 500 km to the east is difficult to explain with age-progressive volcanic centres above a narrow deep mantle plume. Thus, whereas a melting anomaly exists in the upper mantle beneath the Azores, its origin and the presence of a deep mantle plume beneath the Azores is still a matter of debate and subject of on-going geochemical and geophysical investigations (O'Neill and Sigloch, Chapter "Crust and Mantle Structure Beneath the Azores Hotspot—Evidence from Geophysics").

3 State of the Art

The earliest published studies on the geochemistry and petrology of the Azores lavas have mostly dealt with samples from few islands as a result of the accessibility (Agostinho 1936; Berthois 1953; Girod and Lefevre 1972; Oversby 1971) and at the time of preparation of this chapter few rock samples are available from the submarine Azores Plateau (Beier et al. 2008, 2012, 2015). The petrology of samples from the Azores islands is subject of Chapter "Petrology of the Azores Islands" by Larrea et al. that reviews the petrological and major element variability amongst the Azores islands. Technical improvements since the mid-80s, with the establishment of modern mass spectrometry, have considerably improved the quality of geochemical trace element and isotope data. In order to better compare and present the geochemical data we will only use those data in our figures and in the main part of the discussion that were determined by X-ray fluorescence (XRF) or electron microprobe (EMPA) for major elements of whole rocks and glasses, respectively. For the trace element concentrations, we use data determined by inductively coupled plasma mass spectrometry (ICP-MS) and, for selected elements, by instrumental neutron activation (INAA) and inductively coupled-optical emission spectrometry (ICP-OES). For the Sr–Nd–Pb and Hf isotope ratios we use data analysed by multicollector-ICP-MS (MC-ICP-MS) and thermal ionisation mass spectrometry (TIMS). To ensure consistency between the datasets we have only selected those samples with a complete dataset, i.e. samples that were analysed for radiogenic isotopes were only used if trace elements and major elements are also available. Samples with a loss on ignition greater than 5 wt% have been excluded from our compilation to avoid altered samples. For data sources see Fig. 2.

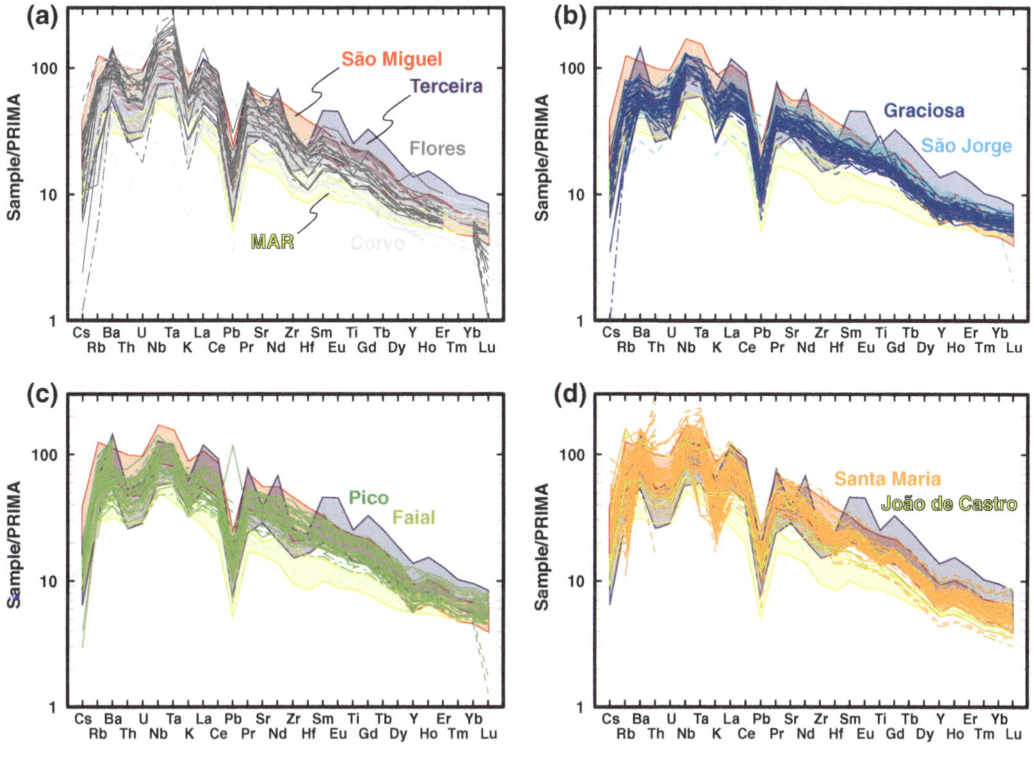

Fig. 2 Primitive mantle normalised trace element abundance pattern of the Azores lavas with >5 wt% MgO. **a** Flores and Corvo, **b** São Jorge and Graciosa, **c** Pico and Faial and **d** Santa Maria and João de Castro. Lavas from São Miguel and Terceira are shown as fields in all panels for reference. Mid-Atlantic Ridge data for the area from 35 to 41°N (Oceanographer to Kurchatov fracture zones) are from Bourdon et al. (1996) and Dosso et al. (1999). Primitive mantle from Lyubetskaya and Korenaga (2007). Data for Santa Maria are from Beier et al. (2013). Data for São Miguel and João de Castro are from Beier et al. (2008), for São Miguel from Elliott et al. (2007), Turner et al. (1997) and Widom et al. (1997), for Terceira from Beier et al. (2008), Madureira et al. (2011) and Turner et al. (1997) and for Graciosa are from Beier et al. (2008) and Larrea et al. (2014). Data for São Jorge from Beier et al. (2012), Millet et al. (2009) and Turner et al. (1997), for Faial from Beier et al. (2012) and Turner et al. (1997), for Pico from Beier et al. (2012), Elliott et al. (2007) and Turner et al. (1997). Data for Flores are from Genske et al. (2012), and data for Corvo are from Genske et al. (2012) and Larrea et al. (2013)

4 Geochemistry of the Rocks from the Azores Islands

The Azores lavas belong to the alkaline volcanic series (see Chapter "Petrology of the Azores Islands" by Larrea et al.) and range from alkali basalts and basanites with few picrobasalts to trachytes and rare rhyolites in a total alkalis ($Na_2O + K_2O$) versus SiO_2 diagram (Le Maitre 1989; LeBas et al. 1986). The variation of the major element compositions implies that fractional crystallisation and mineral accumulation processes are important on each of these islands (Larrea et al., Chapter "Petrology of the Azores Islands"). Here, we aim at describing the geochemical variation and discussing the melting processes and mantle sources in the Azores. In order to minimise the effects of fractional crystallisation and/or assimilation on incompatible element and isotope ratios we focus on those samples with MgO >5 wt% in Figs. 2 and 4, 5, 6, 7, 8 and 9.

For clarity, we have grouped the geochemical description into (from east to west) (a) the island of Santa Maria (b) the island of São Miguel and the seamount João de Castro, (c) the islands of Terceira and Graciosa, (d) the islands of Faial and Pico, (e) São Jorge, (f) Flores and Corvo and (g) samples from the MAR. Submarine samples available from the vicinity of the islands will be discussed along with the respective island nearby.

4.1 Santa Maria

The island of Santa Maria is the easternmost subaerial volcanic island of the Azores (Fig. 1) and Abdel Monem et al. (1975) measured K/Ar-ages of 4.13–8.12 Ma which were then the oldest ages found for lavas of the Azores Plateau and close to the emplacement estimates of the Azores Plateau (Cannat et al. 1999; Gente et al. 2003). In contrast to the other Azores islands that display Pleistocene to recent volcanism (Abdel Monem et al. 1975; Beier et al. 2015; Féraud et al. 1980), Santa Maria is volcanically extinct and no evidence for volcanism younger than 4 Ma is found on the island. The geographical location at the eastern edge of the Azores Plateau allows processes at the outer edge of the Azores melting anomaly to be investigated.

Esenwein (1929) presented the first petrological descriptions of volcanic rocks from Santa Maria. Schmincke and Staudigel (1976) studied the submarine lava series exposed on Santa Maria and suggested that these pillow lavas erupted in shallow water indicating a significant uplift of the island. This has recently been investigated in more detail by Ramalho et al. (2016) who use Ar–Ar age dating to better constrain the uplift history of Santa Maria. White et al. (1979) presented the first major and trace element analyses for two samples from Santa Maria and noted that lavas from Santa Maria are more alkaline than those from the other islands.

A comprehensive major element, trace element and Sr–Nd–Pb isotope dataset on Santa Maria lavas has recently been published by Beier et al. (2013) showing that the major and trace element compositions differ significantly from those observed on the other islands (Figs. 2 and 3 and Larrea et al., Chapter "Petrology of the Azores Islands"). For example, Santa Maria lavas have lower SiO_2, K_2O and TiO_2 (and Rb) concentrations than rocks from other Azores volcanoes (see Larrea et al., Chapter "Petrology of the Azores Islands" and Fig. 2), confirming and extending the observations of White et al. (1979). The trace element ratios of La/Yb and Nb/Zr are more enriched and more variable (20.3 ± 3.3 for La/Yb and $\sim 0.25 \pm 0.03$ for Nb/Zr, respectively) at Santa Maria (Figs. 3 and 4) than compared with any other island of the Azores archipelago (with the exception of La/Yb for São Miguel, Fig. 3). Comparable geochemical variations exist in lavas from the western islands of Corvo and Flores (La/Yb = 21.3 ± 4.1, Nb/Zr = 0.32 ± 0.04, (see below and in Genske et al. 2012). Beier et al. (2013) attributed the more enriched and more variable trace element ratios of Santa Maria lavas to a thickened lithosphere and hence, smaller degrees of partial melting (Fig. 5) in the presence of residual garnet as inferred from increased ratios of the middle REE to heavy REE

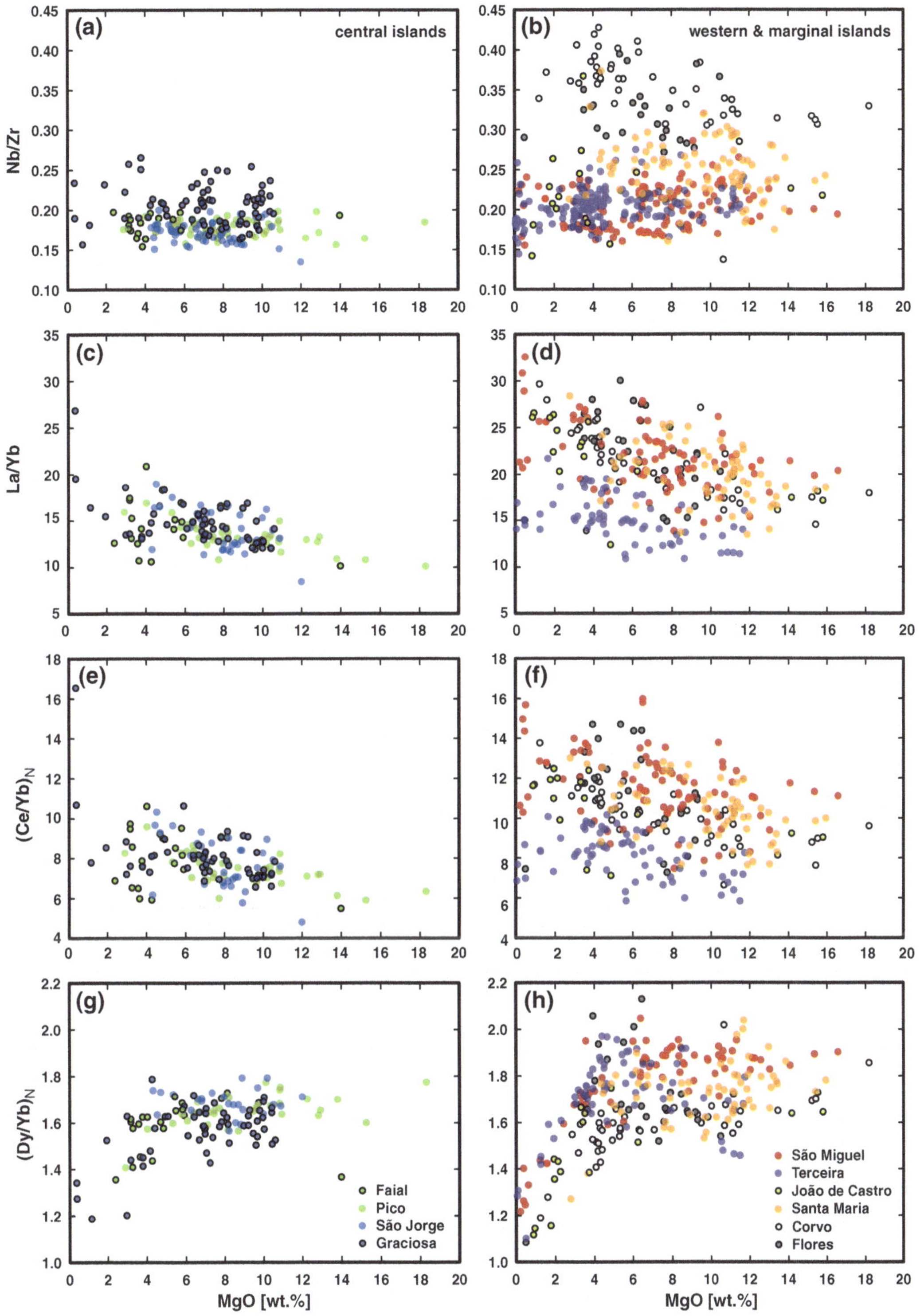

Fig. 3 **a–b** Nb/Zr, **c–d** La/Yb, **e–f** primitive mantle normalised $(Ce/Yb)_N$, and **g–h** primitive mantle normalised $(Dy/Yb)_N$ ratios versus MgO in wt% for the central and western/marginal islands, respectively. Primitive mantle from Lyubetskaya and Korenaga (2007). Data sources as in Fig. 2

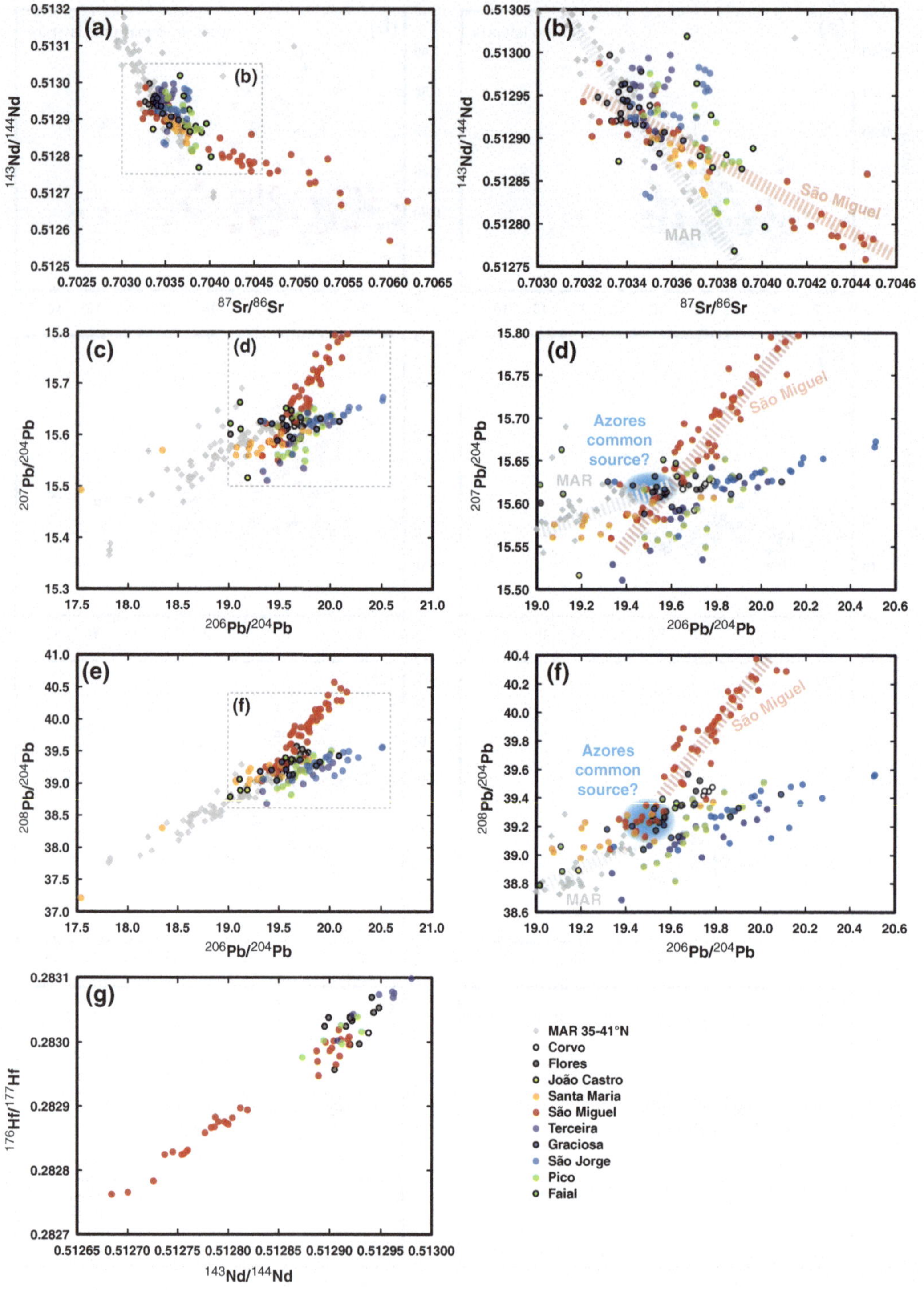

Fig. 4 **a–b** $^{143}Nd/^{144}Nd$ versus $^{87}Sr/^{86}Sr$, **c–d** $^{207}Pb/^{204}Pb$ and **e–f** $^{208}Pb/^{204}Pb$ versus $^{206}Pb/^{204}Pb$ and **g** $^{176}Hf/^{177}Hf$ versus $^{143}Nd/^{144}Nd$ isotope ratios of the Azores lavas >5 wt% MgO. At the time of printing this volume new Hf isotope data were presented by Béguelin et al. (2017) not shown here. Blue shaded area marks approximate location of the Azores common source. Data sources as in Fig. 2

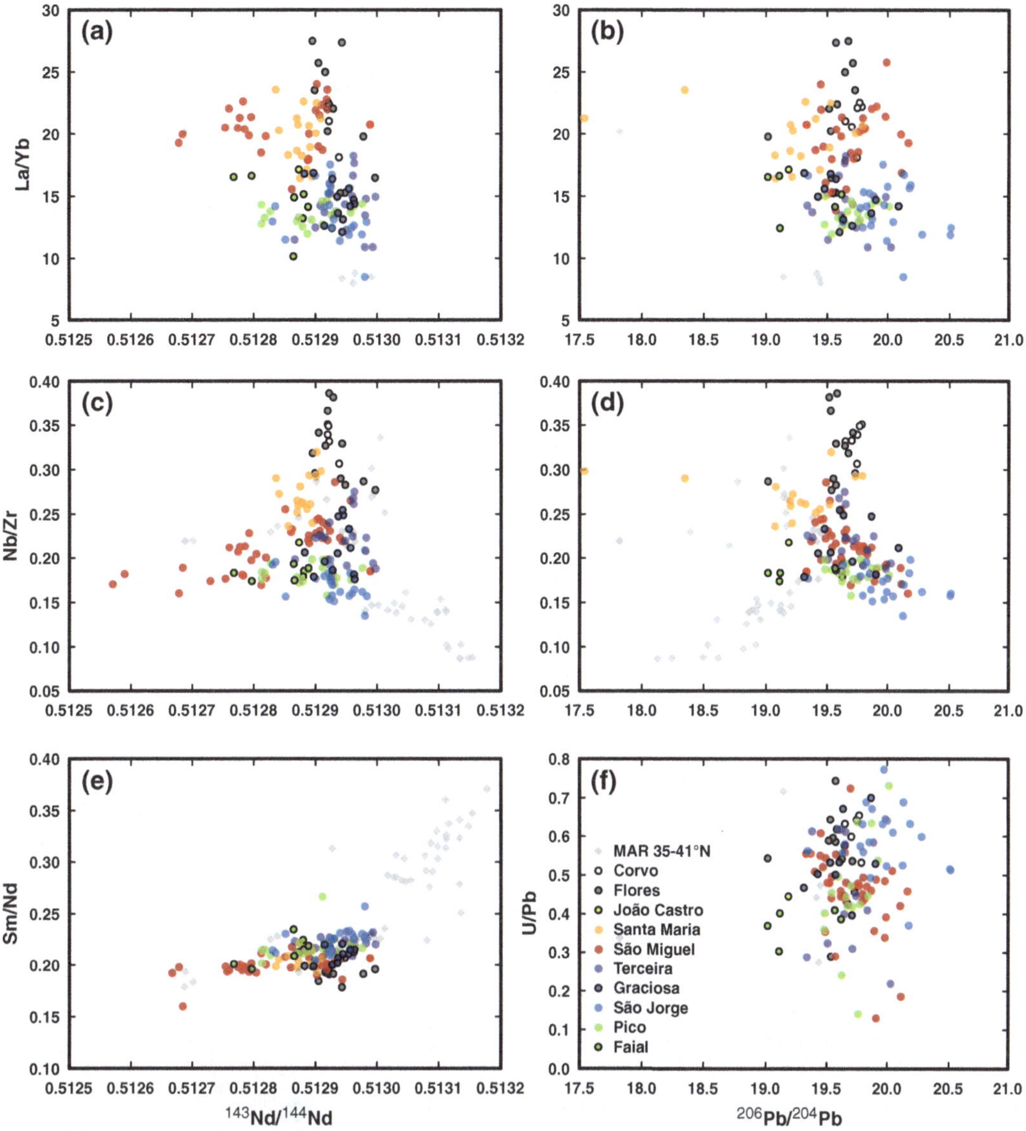

Fig. 5 La/Yb versus **a** $^{143}Nd/^{144}Nd$ and **b** $^{206}Pb/^{204}Pb$. Nb/Zr versus **c** $^{143}Nd/^{144}Nd$ and **d** $^{206}Pb/^{204}Pb$ and **e** Sm/Nd versus $^{143}Nd/^{144}Nd$ and **f** U/Pb versus $^{206}Pb/^{204}Pb$ of the Azores lavas with >5 wt% MgO. Data sources as in Fig. 2. Note the lack of correlation between parent-daughter ratios and isotope ratios in (**e**) and (**f**) with the exception of São Miguel, see discussion in Beier et al. (2007) and Elliott et al. (2007)

(HREE) such as Tb/Yb and Dy/Yb. However, they also noted that this is inconsistent with the lower contents of K_2O, TiO_2 and Rb because these elements behave incompatibly during mantle melting and should be enriched at smaller degrees of partial melting (Larrea et al., Chapter "Petrology of the Azores Islands"). Instead they suggest that the geochemical signatures underneath Santa Maria are due to mixing of melts from a CO_2-bearing peridotite from greater depth with those from a volatile-free garnet peridotite at shallower depth. This is in agreement with the proposed volatile-rich nature of the Azores melting anomaly (Asimow et al. 2004) and pressures and temperatures below the dry peridotite solidus (Beier et al. 2013).

Moreira et al. (1999) showed that the He isotopes in Santa Maria lavas overlap normal MORB ($^4He/^3He = 94{,}376 \pm 3242$) and that the Pb isotopes were relatively unradiogenic compared to other Azores rocks (<19.28 for $^{206}Pb/^{204}Pb$). The $^{206}Pb/^{204}Pb$ isotope ratios reported by Beier et al. (2013) range from 19.07 to 20.22 extending the previous range significantly towards more radiogenic compositions (Fig. 4). The combined Sr–Nd–Pb isotope ratios of Beier et al. (2013), however, indicate a source that is comparable to those observed in lavas from western São Miguel (Sete Cidades volcano; Beier et al. 2007; Elliott et al. 2007) and from the islands of Faial and Pico (Beier et al. 2012; França et al. 2006).

In summary, the island of Santa Maria is the oldest subaerial volcanic edifice in the Azores and its positioning at the easternmost border of the plateau (Fig. 1) appears to reflect melting and source conditions at the edge of the Azores mantle melting anomaly. The volatile-rich source lithology that results in the distinct trace element enrichment underneath Santa Maria may also be present underneath the other Azores islands (that have comparable Sr–Nd–Pb isotope ratios), but may only be visible at Santa Maria as a result of lower degrees of partial melting and a thicker lithosphere preserving the trace element signatures.

4.2 The Terceira Axis

The Terceira axis is an ultraslow spreading axis comprising the volcanoes of São Miguel, Terceira and Graciosa islands and the submarine seamount João de Castro (Vogt and Jung 2004). Due to the similarity in geodynamic setting and the distribution of mantle sources and melting dynamics along the axis, we describe the islands of São Miguel and João de Castro and the islands of Terceira and Graciosa together albeit they display distinct geochemical differences.

4.2.1 São Miguel and João de Castro Seamount

The island of São Miguel and the neighbouring seamount of João de Castro (Fig. 1) are situated at the eastern end of the ultraslow spreading Terceira axis (Vogt and Jung 2004). The active seamount João de Castro situated just west of São Miguel reached subaerial stages in 1638 and 1720 but was later eroded (Nunes et al. 2003). Sampling this young volcanic edifice may provide additional information on the early stages of the formation of the Azores islands.

The seamount was dredged during one research expedition with the German Research Vessel Poseidon (Pos 232) in 1997. While the seamount has been subject of several biological and hydrothermal investigations (Cardigos et al. 2005; Santos et al. 2001), the only geochemical analyses of volcanic rocks are those presented in Beier et al. (2008). Samples from João de Castro range from basalt to trachyte in composition (Larrea et al., Chapter "Petrology of the Azores Islands"). Melting-sensitive trace element ratios (e.g., La/Yb = 22.9 ± 3.2 and $(Ce/Yb)_N = 11.2 \pm 1.2$, Fig. 3) overlap those of Sete Cidades volcano indicating similar degrees of partial melting (1–3% degree of melting; Figs. 3 and 6) at the spinel-garnet transition zone (Beier et al. 2008). The Sr–Nd–Pb isotope analyses of João de Castro imply mixing between a depleted mantle source and a source that is also present beneath the western

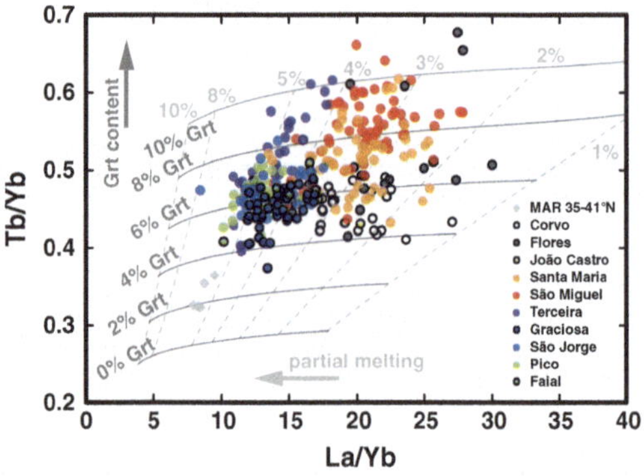

Fig. 6 Tb/Yb versus La/Yb ratios of the Azores lavas with >5 wt% MgO. Data sources as in Fig. 2. Trace element model used from Bourdon et al. (2005) and Beier et al. (2010). Partition coefficients are those compiled in Halliday et al. (1995) and Blundy et al. (1998). Mantle source compositions are those used in Beier et al. (2007) for Sete Cidades volcano. Labelled markers are degrees of partial melting and amount of residual garnet in the mantle lithology

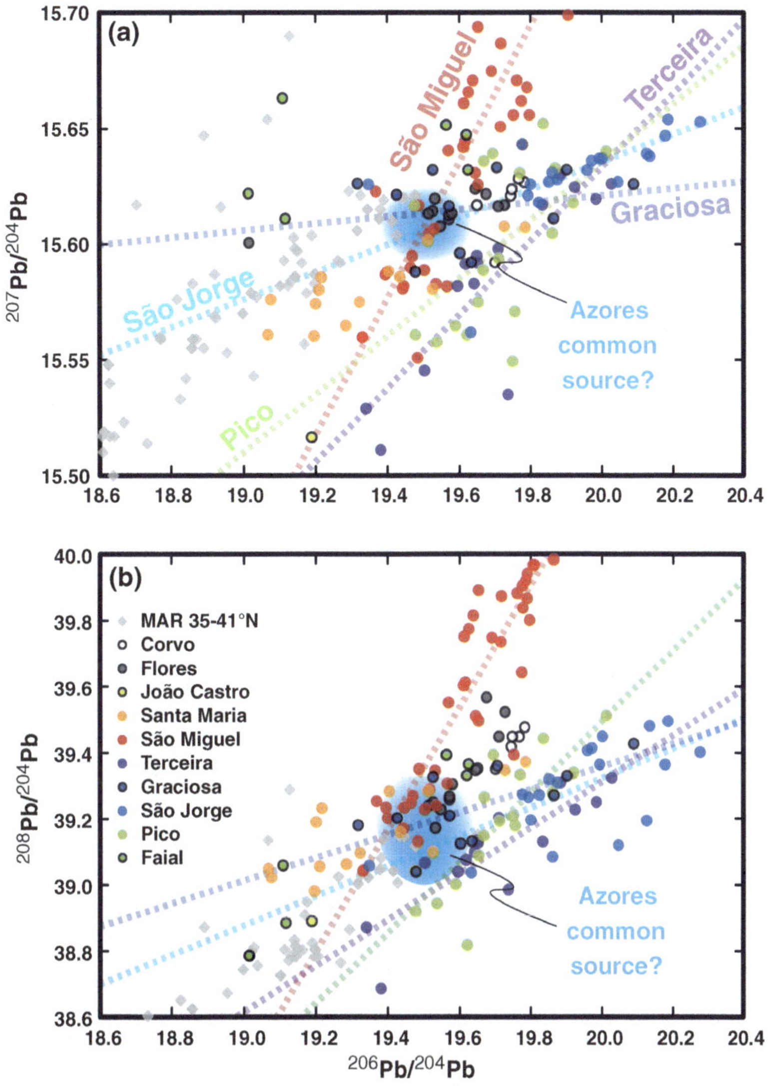

Fig. 7 **a** ^{207}Pb/^{204}Pb and **b** ^{208}Pb/^{204}Pb versus ^{206}Pb/^{204}Pb isotopes of the Azores lavas with >5 wt% MgO. Slopes are mixing arrays of individual islands. Blue shaded area marks approximate location of the Azores common source. Data sources as in Fig. 2

São Miguel volcano Sete Cidades (Fig. 4; Beier et al. 2008). The João de Castro lavas have the lowest Sr and Pb but highest Nd isotope ratios published to date from the Eastern Azores Plateau, i.e. the ^{206}Pb/^{204}Pb ratios reach from 18.8 to 19.7 (Figs. 4 and 7). This suggests the presence of a relatively depleted component underneath the Azores that mixes with a more enriched mantle source that may also be present beneath western São Miguel and underneath Santa Maria. Based on correlated Pb isotopes and U/Nb ratios, Madureira et al. (2011) also proposed a flow of material from João de Castro into the fissural system of Terceira further to the west.

Lavas from São Miguel are unique in that they display a marked correlation between incompatible trace elements, parent-daughter element ratios and Sr–Nd–Hf–Pb isotope ratios implying a strong influence of the mantle source composition not only in the radiogenic isotopes but also in the trace element compositions (Fig. 5). The trace element and isotope signatures of eastern São Miguel are even unique amongst the global ocean island lavas in that they display elevated

$^{87}Sr/^{86}Sr$ and $^{206}Pb/^{204}Pb$, $^{207}Pb/^{204}Pb$ and $^{208}Pb/^{204}Pb$ ratios that are not found anywhere else (Fig. 4). The geochemistry of São Miguel lavas has been a matter of debate since the 1970s particularly as a result of the unusually enriched mantle sources of eastern São Miguel with the extremely enriched Sr–Nd–Pb isotope compositions of Nordeste volcano (Allegre et al. 1987; Beier et al. 2007; Davies et al. 1989; Elliott et al. 2007; Hart 1988; Hawkesworth et al. 1979; Moreira et al. 1999; Turner et al. 1997; White et al. 1976; Widom et al. 1997). Davies et al. (1989) first noted that the radiogenic Sr and Pb isotope ratios are accompanied by increased Ba/Nb and lower U/Nb ratios and concluded that these may represent recycled continental lithosphere possibly involving sediments. Widom et al. (1997) and Turner et al. (1997) published trace element and isotope data on São Miguel lavas but differ slightly in their interpretations of the petrogenesis and mantle sources.

Widom et al. (1997) showed that the increasing enrichment from west to east on São Miguel may represent binary mixing between the less enriched magmas of the Sete Cidades volcano and the more enriched mantle source at the eastern end of the island. They suggested that Nordeste volcano represents an "EM-II-type" source which had generally been attributed to the presence of recycled continental material (Zindler and Hart 1986). However, Nordeste lavas have enriched Sr and Pb isotope compositions different from those of EM-II (Fig. 3). In addition, sedimentary material tends to lower $^{206}Pb/^{204}Pb$ ratios as a result of an increased sediment Pb concentration lowering the parent-daughter U/Pb ratio (Stracke et al. 2003). Widom et al. (1997) also noted that some trace element signatures, particularly Ba/Th and Ti/Zr in the São Miguel lavas are not consistent with a continental lithospheric mantle source involving sediments. These authors concluded that the São Miguel source is the result of subcontinental lithospheric mantle that resided beneath northwestern Africa or Iberia and was delaminated during the initial rifting of the Atlantic Ocean and is now melting as a result of the heat influx from the Azores melting anomaly. Widom and Farquhar (2003) used $\delta^{18}O$ isotopes on olivines from the same sample set to better define the source composition of the São Miguel lavas. They find $\delta^{18}O$ ratios of $<+5.04‰_{(V\text{-}SMOW)}$ and attribute the large variability in the Sr–Nd–Pb isotopes and the small range in $\delta^{18}O$ to either an EM-II rich mantle or to delaminated, metasomatised, lithospheric subcontinental mantle mixing with the (common?) Azores mantle source.

Turner et al. (1997) discussed major and trace element and radiogenic isotope data of samples from the eastern Azores Plateau. They suggested that the Nd model ages of the lavas indicate that the Azores mantle source may have an age of 550–800 Ma and that the radiogenic $^{87}Sr/^{86}Sr$ isotope ratios of São Miguel may only have a shallow mantle origin as a result of its localised occurrence. Moreira et al. (1999) reached a similar conclusion based on Pb and He isotope ratios. In agreement with U–Th disequilibria Turner et al. (1997) concluded that the sources of a deep-seated, regional Azores mantle plume are due to recycled lithospheric mantle. They attributed the localised São Miguel component to a shallow mantle origin and that the radiogenic and isotope signatures indicate addition of 5–10% recycled upper crustal sediments to the mantle source. It should however be noted that Elliott et al. (2007) and Millet et al. (2009) have found significant differences between their own Pb isotope analyses and those of Turner et al. (1997) from similar outcrops (Millet et al. 2009) and within the islands (Elliott et al. 2007) that could imply an analytical bias as also noted by Prytulak et al. (2014) for the Nd isotopes.

Elliott et al. (2007) and Beier et al. (2007) studied the origin of the São Miguel mantle sources based on new Sr–Nd–Pb–Hf isotope and trace element data and both argued that a continental and sedimentary origin of the eastern São Miguel sources cannot explain the trace element and isotope variability in the São Miguel lavas. Particularly, the combined elevated $^{206}Pb/^{204}Pb$ and $^{87}Sr/^{86}Sr$ ratios cannot be explained by the presence of continental crust (Fig. 3). Elliott et al. (2007) proposed that Nd–Hf isotope characteristics can be reproduced by an ancient ($\sim$3 Ga), small-degree melt (~2%) from a garnet peridotite source. They suggested that subduction and recycling of a small amount of this source could explain the isotope characteristics of the São Miguel source. However, they conclude that the incompatible element enrichment is not the result of a subduction and recycling process but may imply that the enriched mantle sources are in fact the result of underplating of basalt that intruded into the oceanic lithosphere. The rarity of such signatures underneath other ocean islands indicates that another process must be homogenising these signatures. Beier et al. (2007) suggested that the source of Nordeste lavas is due to a long-term evolution of a source with high Rb/Sr, U/Pb, Th/Pb, Th/U and low Sm/Nd and Lu/Hf parent–daughter ratios. They observed trace element concentrations similar to those of the HIMU islands (high $\mu = {}^{238}U/{}^{204}Pb$ ratios), with the exception of notably higher alkali element (Cs, Rb, K, Ba) and Th concentrations. The time-integrated parent–daughter source evolution of the western and eastern sources on São Miguel matches the incompatibility sequence commonly observed during mantle melting. Thus, Beier et al. (2007) concluded that the source enrichment must be caused by a basaltic melt as a metasomatic agent or by recycling. Their quantitative model shows, in agreement with the observations of Elliott et al. (2007), that metasomatism involving a small degree basaltic melt is able to explain the isotope compositions of eastern São Miguel but that the trace element signatures are far too enriched to be explained by melt metasomatism. The coherent correlation of parent–daughter and incompatible trace element ratios and Sr–Nd–Hf–Pb isotopes of the eastern and western São Miguel sources implies that both sources share a common origin (Fig. 5). Beier et al. (2007) explain the eastern Nordeste source by recycled oceanic crust that contains small amounts of evolved lavas (1–2%), likely from a subducted seamount, whereas the western Sete Cidades sources can be explained by recycled oceanic crust without such evolved components. Independent of the origin of these sources, the preservation of these compositions underneath Nordeste may be taken as evidence that, in some cases, the homogenisation of mantle sources the length-scales of a single seamount and the timescales of recycling may not be as effective as previously thought.

Widom and Shirey (1996) and Schaefer et al. (2002) both published Os isotope ratios for lavas from São Miguel that generally range from 0.1308 to 0.142 for $^{187}Os/^{188}Os$ and overlap those found in other ocean islands (Schaefer et al. 2002). This radiogenic nature of the São Miguel sources led both to conclude that an enriched component must be involved in the sources of São Miguel in agreement with previous works. Widom and Shirey (1996) suggested that this enriched, radiogenic reservoir may originate from the outer implying a deep mantle origin also proposed by Turner et al. (1997).

More recently, Turner et al. (2007) and Genske et al. (2013, 2014) used $\delta^{18}O$, $\delta^{7}Li$ and $\delta^{11}B$ to evaluate the nature of the São Miguel source. Turner et al. (2007) observed a correlation of Nb/B and $\delta^{11}B$ ratios along with lower $\delta^{18}O$. They attributed the lower Nb/B and $\delta^{11}B$ and higher $^{187}Os/^{188}Os$ and $^{87}Sr/^{86}Sr$ isotope ratios of São Miguel to the presence of ancient (~3 Ga), recycled oceanic crust which is consistent with the trace elements and Hf–Nd isotopes (Beier et al. 2007; Elliott et al. 2007). Genske et al. (2014) however, showed, that with the exception of a single sample from São Miguel, most samples overlap those of other ocean islands in both $\delta^{7}Li$ and $\delta^{11}B$. Genske et al. (2013) found a correlation of $\delta^{18}O$ and forsterite contents of olivines and concluded that some of these signatures may be the result of assimilation of crustal material as also proposed by Prytulak et al. (2014), and in contrast to the interpretations by Turner et al. (2007).

The sources on the island of São Miguel thus represent mixing between a source that is commonly found in the Azores (Sete Cidades volcano, western São Miguel; (for a petrological and geochemical description of Sete Cidades see Beier et al. 2006 and Chapter "Petrology of the Azores Islands" by Larrea et al.) and an unusually enriched, highly radiogenic, ancient source underneath Nordeste (eastern São Miguel). The volcanoes Sete Cidades and Agua de Pau are some 30 km distant but have very different source compositions (Haase and Beier 2003). The two overlapping rift zones between these two volcanoes erupt lava compositions suggesting that the northern rift zone is connected to the eastern Agua de Pau volcano while the southern rift zone is connected to Sete Cidades. The limited mixing between the volcanic systems implies separate ascent paths and magma sources. The co-variation of magma sources and lithosphere tectonics suggests that the tectonic structure of the lithosphere controls the mantle upwelling underneath São Miguel.

On the scale of the Terceira axis the ratios of the Light (LREE) and Middle Rare Earth Elements (MREE) imply that degrees of partial melting underneath São Miguel and João de Castro are on the order of 1–3% with variable amounts of residual garnet as inferred from the MREE and Heavy Rare Earth Element (HREE) ratios (Fig. 6 and Beier et al. 2008). This is in agreement with previous conclusions reached by Schilling (1975) and Schilling et al. (1983) based on the enrichment of the MAR samples and by Turner et al. (1997) inferred from Sr–Nd–Pb isotopes from several of the eastern islands.

In summary, the São Miguel lavas comprise the largest incompatible element and radiogenic isotopic variability amongst the Azores islands. This is largely the result of mixing of the western, less enriched sources (Sete Cidades) with the eastern sources (Nordeste) that are unique amongst global oceanic lavas with radiogenic Sr–Pb–Os and low Nd–Hf isotope ratios. However, the stable isotopes and selected trace element ratios also imply that crustal, shallow level assimilation must have played a significant role for some isotopic systems. The trace element and radiogenic isotopes of the seamount João de Castro west of São Miguel and Sete Cidades imply a common source between the two systems at similar degrees and depths of melting but also a contribution of a João de Castro component to the volcanic system of Terceira.

4.2.2 Terceira and Graciosa

The islands of Terceira and Graciosa are situated along the western continuation of the Terceira axis (Fig. 1) and like João de Castro and São Miguel, are separated by a deep basin. The majority of the samples published are from subaerial outcrops, however a few dredge samples are available from west of Terceira and west of Graciosa (Beier et al. 2008). The two islands share very comparable trace element and isotopic signatures though subtle differences exist.

The island of Terceira in particular has been subject of several geochemical publications. White et al. (1979) presented the first geochemical analyses from Terceira and, additionally, Dupré et al. (1982) complemented a major and trace element dataset by Provost and Allegre (1979) with Sr and Pb isotope ratios indicating mixing of two distinct sources. The trace elements from Terceira are significantly less enriched than those from Santa Maria and São Miguel (Nb/Zr = 0.20 ± 0.02; La/Yb = 14.2 ± 1.8, Figs. 3, 6 and 8) but overlap those of the other, neighbouring islands (Figs. 2 and 3). The incompatible element ratios indicate melting of a garnet peridotite source lithology underneath Terceira with degrees of melting between 2% and 4% (Fig. 6).

Turner et al. (1997) showed that the La/Sm and $^{87}Sr/^{86}Sr$, $^{143}Nd/^{144}Nd$ and $^{207}Pb/^{204}Pb$ ratios overlap those of Faial and Pico and the enriched end of the MAR lavas suggesting that these sources ascend from a deep mantle plume (Figs. 3 and 4). This is in agreement with the He and Pb isotope ratios presented by Moreira et al. (1999) who showed that the Terceira source is likely the result of mixing of recycled oceanic crust (high $^{206}Pb/^{204}Pb$) and entrained lower mantle (high ^{3}He) material. Beier et al. (2008) and Madureira et al. (2011) agreed with Dupré et al. (1982) that lavas from the island of Terceira form binary mixing arrays in radiogenic isotope spaces (Figs. 3 and 6). The enriched end-member of lavas from Terceira comprises a HIMU-like component that has generally been associated with the presence of recycled oceanic crust (Stracke et al. 2005; Zindler and Hart 1986). Madureira et al. (2011) published a comprehensive dataset of major elements, trace elements and Sr–Nd–Pb–Hf isotope data of Terceira lavas and showed that the intra-island heterogeneity observed is the result of mixing of a

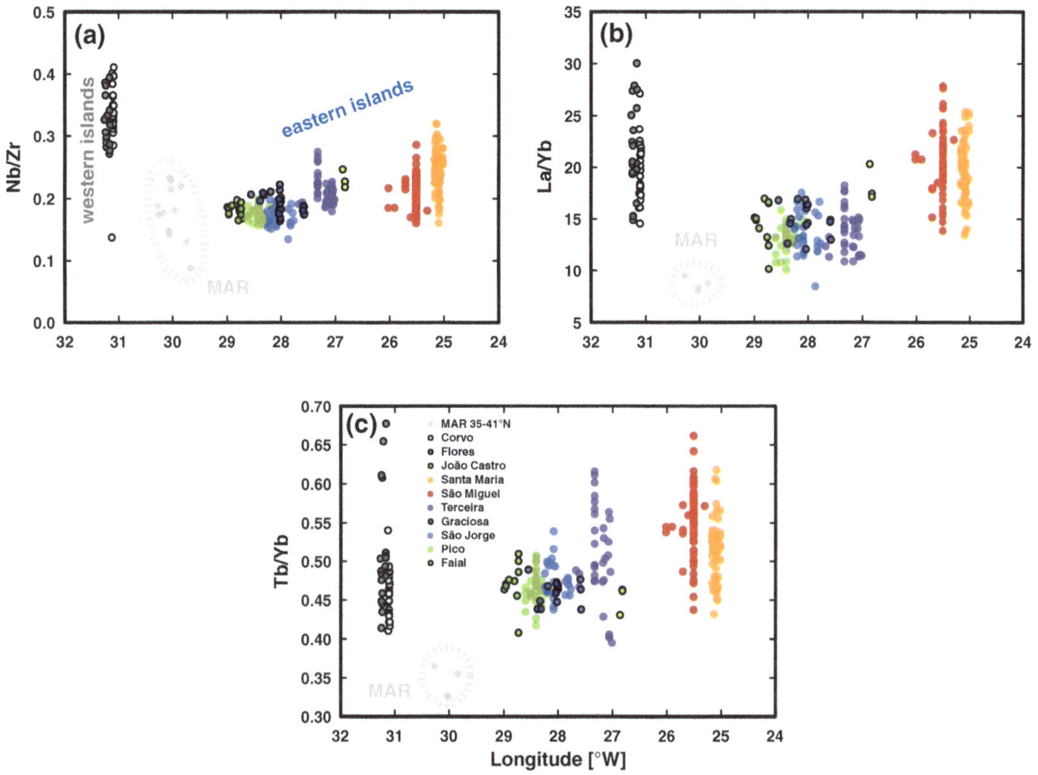

Fig. 8 **a** Nb/Zr, **b** La/Yb and **c** Tb/Yb ratios versus Longitude [°W] of the Azores lavas with >5 wt% MgO. Data sources as in Fig. 2. Note that for samples where the exact longitude is missing in the published data a common longitude was taken for each respective island

plume-derived recycled ancient oceanic crust (HIMU) component with local, depleted sources. In addition, they showed that the melts erupted along the central fissural system contain source material also found at João de Castro seamount, from which they concluded that there also is some westward flow of material from João de Castro towards Terceira.

Widom and Shirey (1996) and Schaefer et al. (2002) reported $^{187}Os/^{188}Os$ isotope ratios from Terceira that vary from 0.125 to 0.195 and cover almost the entire variability of the Azores archipelago, but are mostly radiogenic and overlap those of São Miguel and other ocean islands. Widom and Shirey (1996) calculated the $^{187}Os/^{188}Os$ isotope ratios of the Azores mantle underneath Terceira at 0.128–0.137 overlapping those of most other ocean islands indicating a source enrichment relative to the depleted mantle. δ^7Li, $\delta^{11}B$, and $\delta^{18}O$ isotopes of selected samples from Terceira overlap those of the neighbouring islands of Faial and Pico within the range of normal OIB for δ^7Li and $\delta^{11}B$ (Genske et al. 2014; Turner et al. 2007).

The island of Graciosa is the westernmost and the smallest island along the Terceira axis (Fig. 1). White et al. (1976) presented five samples from Graciosa and Beier et al. (2008) an additional twenty-seven samples from the island and nine submarine samples from volcanic edifices west of Graciosa. Larrea et al. (2014) recently published a complementary dataset of major elements, trace elements and Sr–Nd–Pb–Os isotopes from the island (Figs. 2, 3 and 4 and Larrea et al., Chapter "Petrology of the Azores Islands"). Lavas from Graciosa display trace element ratios less enriched than those at São Miguel and Santa Maria (e.g., Nb/Zr = 0.19 ± 0.02; La/Yb = 15.2 ± 1.6; Fig. 5) but overlap several of the other Azores islands, and, in particular lavas from Terceira (Figs. 3 and 8). Degrees and depth of partial melting of Graciosa are comparable to those from Terceira (Fig. 6) of a lithology within the spinel-garnet transition zone for mantle peridotite (Robinson and Wood 1998). The $^{206}Pb/^{204}Pb$, Sr and Nd isotope ratios in lavas from Graciosa generally overlap those of Terceira but have slightly higher $^{207}Pb/^{204}Pb$ and$^{208}Pb/^{204}Pb$ ratios (Figs. 4 and 7) and form a mixing trajectory extending the lavas from Terceira at radiogenic $^{206}Pb/^{204}Pb$ ratios (at 20.05 in $^{206}Pb/^{204}Pb$) to less radiogenic isotope compositions (19.32 in $^{206}Pb/^{204}Pb$; Fig. 7). The less radiogenic isotope end-member of Graciosa (with the exception of a single sample) is also comparable to that of Sete Cidades volcano at São Miguel with $^{206}Pb/^{204}Pb$ ~ 19.5 (Beier et al. 2007; Elliott et al. 2007) representing a common source in the Azores (Beier et al. 2012; Larrea et al. 2014). Beier et al. (2008) noted that particularly the Nd and Pb isotopes of Graciosa indicate that the geochemical enrichment observed along the MAR (Bougault and Treuil 1980; Schilling 1975; White and Chappell 1977; White and Schilling 1976; White et al. 1975, 1976) may be the result of a mantle flux from Graciosa toward the MAR contrasting with suggestions based on He isotope ratios by Moreira et al. (1999) who suggested that mantle material from Terceira flows towards the MAR. Larrea et al. (2014) showed that much of their Sr–Nd–Pb isotopic signature overlaps that of a common Azores source and that of previous works. However, a single sample from the Serra Branca volcanic complex and an alkaline gabbro display EM-II alike source signatures. In the absence of significant crustal assimilation, they conclude that the preservation of radiogenic Sr–Nd–Pb and Os isotope signatures (up to 0.5043 for $^{187}Os/^{188}Os$) are the result of a three component mixing of a common Azores source with an enriched component that is also present underneath western São Miguel and an EM-II like mantle.

The more radiogenic isotope ratios of $^{207}Pb/^{204}Pb$ and $^{208}Pb/^{204}Pb$ of Graciosa compared to Terceira (Fig. 4) imply that the slope of the mixing array of Terceira in $^{143}Nd/^{144}Nd$ and radiogenic $^{208}Pb^{*}/^{206}Pb^{*}$ and Pb–Pb isotope spaces is slightly steeper than that of Graciosa, but comparable to the slope of João de Castro [Figs. 4 and 7 and Beier et al. (2008)]. The different slopes in Pb–Pb isotope spaces hence may reflect slightly different source ages within each island (see below).

Variable amounts of residual garnet in melting of the magmas of Terceira and Graciosa is inferred from the characteristic fractionation of the HREE ratios, e.g. $(Dy/Yb)_N$ indicating a difference in mean depth of melting beneath the two islands (Figs. 3 and 6). Thus, Beier et al. (2008) suggested an increasing garnet signature from west to east along the Terceira axis and consequently an increasing melting depth due to increasing lithosphere thickness. Madureira et al. (2011) showed that melting is initiated in the garnet stability field but that the variability in amount of garnet observed may be the result of melt segregation through the spinel stability field.

Summarising, the islands of Graciosa and Terceira at the western end of the Terceira axis imply mixing between a common source with radiogenic Pb and Sr–Nd isotope ratios that has commonly been associated with a HIMU source and an enriched source with low $^{206}Pb/^{204}Pb$ ratios. Graciosa in addition also preserves an enriched signature with low $^{206}Pb/^{204}Pb$ isotope ratios but radiogenic Sr–Nd and Os isotopes. Degrees of partial melting are slightly larger than those observed underneath São Miguel and João de Castro and clearly larger than those of Santa Maria but at shallower depth, i.e. melting underneath Terceira and Graciosa occurs in the spinel-garnet transition zone contrasting deeper melting further east.

4.3 São Jorge

Few geochemical data are available from the island of São Jorge, which is situated south of Graciosa and almost west of Terceira (Fig. 1). São Jorge is geologically unique amongst the Azores islands in that it lacks a large stratovolcano but instead comprises three lines of scoria cones with different orientation (Hildenbrand et al. 2008). The first geochemical data were presented by White et al. (1976) and the dataset was expanded by Turner et al. (1997). A single $^{187}Os/^{188}Os$ isotope analysis was presented by Widom and Shirey (1996), within the range of the neighbouring island of Pico. Genske et al. (2013, 2014) measured O, Li and B isotopes for selected samples. Millet et al. (2009) published a first comprehensive dataset including major and trace elements and Sr–Nd–Pb isotope analyses. Millet et al. (2009) extended the known isotopic range of the Azores islands towards more radiogenic $^{206}Pb/^{204}Pb$ isotope ratios, i.e. they showed that the strongest "HIMU-like" signature in the eastern Azores islands is present in the older volcanic units of São Jorge and is mixing with a depleted mantle source (Figs. 4 and 7). In agreement with the observations of Madureira et al. (2011) at Terceira they proposed mixing of a highly enriched source component with depleted upper mantle sources. However, Millet et al. (2009) could also show that samples from the younger geological units of São Jorge imply mixing of the enriched and depleted sources with a third additional enriched MORB-type source. They interpreted this to be the result of shallow level AFC processes which is in agreement with slightly lower $\delta^{18}O$ and forsterite contents in olivines (Genske et al. 2013). Millet et al. (2009) concluded that lavas from São Jorge and São Miguel may represent the only "deep-plume" end-members of the eastern plateau, while the rest of the observed isotopic variability may indicate shallower contamination of the melting anomaly by delaminated, subcontinental mantle material trapped in the asthenosphere after the opening of the Atlantic Ocean. This observation is also consistent with elevated $\delta^{11}B$ and $^{187}Os/^{188}Os$ isotopes indicating the presence of >2.5 Ga fluid and melt depleted oceanic lithospheric mantle (Turner et al. 2007). Beier et al. (2012) presented major element, trace element and Sr–Nd–Pb isotope data from São Jorge but focussed on the melting conditions underneath Faial, Pico and São Jorge based on the major elements. In their calculations, melting underneath São Jorge starts much deeper than underneath the other surrounding islands inferred from the major elements but also evident from the HREE ratios (Fig. 3) and comparable to the melting depth underneath São Miguel. However, the incompatible element ratios of La/Yb and $(Ce/Yb)_N$ overlap those of the other Azores islands (13.6 ± 2.0 and 7.7 ± 1.2, respectively, Fig. 3) indicating comparable degrees of partial melting (Fig. 6). The lithosphere underneath São Jorge has a similar thickness to that beneath Graciosa and Faial/Pico and Beier et al. (2012) concluded, in agreement with Millet et al. (2009) that the main source of upwelling in the eastern Azores may be underneath São Jorge. This contrasts with evidence from U–Th–Pa disequilibria (Bourdon et al. 2005) and He and Ne isotopes (Madureira et al. 2011; Moreira et al. 1999) that the main source of upwelling may be underneath Terceira. The presence of an ancient, deep mantle (plume) source component underneath the neighbouring islands of Faial and Pico (see next section) as inferred from subchondritic $^{187}Os/^{188}Os$ isotope ratios (Schaefer et al. 2002), may be taken as additional evidence that the central conduit of the Azores melting anomaly is situated in the vicinity of the islands of São Jorge, Faial and Pico.

Summarising, lavas from the island of São Jorge represent the most radiogenic end-members of the Azores Plateau with the most pronounced radiogenic Pb isotope signature that may best be explained by the presence of recycled oceanic crust mixing with a depleted mantle source, or, more likely with a source commonly present in the Azores (see Figs. 4 and 7) that is also present underneath the other islands. It should be noted that the mixing trajectory of the São Jorge lavas converges with those of the other Azores islands at a $^{206}Pb/^{204}Pb$ ratio of 19.75 (see above and Figs. 3 and 6) and likely thus has a certain contribution of common Azores mantle source mixing with a radiogenic HIMU-component. The degrees of partial melting are comparable to those of the neighbouring islands but melting starts significantly deeper than underneath Faial, Pico or Graciosa indicating that from a geochemical perspective the main source of upwelling mantle in the Azores could be situated underneath São Jorge.

4.4 Faial and Pico

The islands of Faial and Pico with its submarine extension, the Pico Ridge, form a NW-SE trending lineament south of the island of São Jorge (Fig. 1). However, the volcano stratigraphy and eruption mechanisms are distinct between these two islands with Faial displaying a much greater major element and petrological variability (Larrea et al., Chapter "Petrology of the Azores Islands") than Pico. Hildenbrand et al. (2012) and Zanon et al. (2013) argued that, on a local scale, age constraints on Faial are complex with the large fissure zone of Horta (eastern part of Faial) being younger than the central volcano. The volcanism younger than 1 Ma on Faial and Pico broadly increases in age from west to east, i.e. the youngest volcanic eruptions occur on the western end of Faial (Vulcao dos Capelinhos 1957/1958) and the entire island of Pico is younger than the submarine Pico Ridge indicating that on a larger scale, the islands are moving away from their sources of melting (Beier et al. 2012). Samples from the Pico Ridge, an approximately 200 km long submarine extension east of Pico are rare and to date no comprehensive data have been published (Mitchell et al. 2008). Lavas from Pico are dominantly basaltic while compositions on Faial are more variable and more evolved (Beier et al. 2012; França et al. 2006; Zanon et al. 2013).

The first Pb isotope data for Faial were published by Oversby (1971) and combined data for Pico and Faial were presented by White et al. (1976, 1979) and Turner et al. (1997). The first dataset of major elements, trace elements and Sr–Nd–Pb isotope data for Pico was presented by França et al. (2006) and additional major element, trace element data and Sr–Nd–Pb–Hf isotope data were discussed in Elliott et al. (2007) and, for Ti and Na in Prytulak and Elliott (2007). Widom and Shirey (1996) and Schaefer et al. (2002) presented Os isotope data for selected lavas from both islands. Turner et al. (2007) and Genske et al. (2013, 2014) additionally used B, Li and O isotopes to examine the nature of the mantle source underneath the islands. A larger dataset for the islands of Pico and Faial including submarine data west of Faial were presented by Beier et al. (2012). A recent publication by Zanon and Frezzotti (2013) contributes an additional major element and trace element dataset aiming at understanding the tectonic influence on the distribution of the major element and trace element variability on Faial.

The trace elements of Faial are generally within the range of the neighbouring islands, i.e., La/Yb, Nb/Zr, $(Ce/Yb)_N$ and Tb/Yb ratios overlap those of Terceira, Graciosa, and Pico with 14.3 ± 3.3, 0.18 ± 0.02, 7.8 ± 1.7, and 0.45 ± 0.04, respectively (Fig. 3). The trace element variability of Faial, particularly the La/Yb and Tb/Yb trace element ratios imply melting in the garnet stability field with degrees of partial melting around 4–6% and residual garnet contents of 6–8% (Fig. 6) similar to the range of the surrounding islands. In contrast to the observations from other islands, the Sr–Nd–Pb isotope ratios of Faial range from the Azores common mantle source at $^{206}Pb/^{204}Pb$ at 19.7 to slightly lower ratios (Fig. 4), however, still at more radiogenic $^{87}Sr/^{86}Sr$ and $^{143}Nd/^{144}Nd$ isotope ratios than normal, depleted MORB (Fig. 4; 19.02 for $^{206}Pb/^{204}Pb$). It has to be noted though that these variations are small and possibly represent the variability within the common Azores mantle source, i.e. the $^{206}Pb/^{204}Pb$ variability amongst the common Azores mantle source likely range from 19.7 ± 0.3 (Figs. 4 and 7). The parent–daughter elemental ratios and isotope ratios from Faial are not correlated (Fig. 5) suggesting a decoupling of the trace elements and isotope ratios. However, to date no geochemical difference is observed between the older units on Faial and the youngest Azores eruptions at Capelinhos volcano in western Faial.

The neighbouring island of Pico displays a similar trace elemental and Sr–Nd–Pb isotope signature (Beier et al. 2012; França et al. 2006; Turner et al. 1997; White et al. 1979) which is interesting given the largely basaltic nature of

Pico compared to the more evolved nature of Faial (Beier et al. 2012; França et al. 2006; Zanon and Frezzotti 2013). La/Yb and Nb/Zr ratios are within a range of 13.3 ± 1.5 and 0.18 ± 0.01, respectively thus overlapping those from Faial (Fig. 3). This generally implies similar degrees of partial melting, melting depth and temperatures as also evident from the major elements (Beier et al. 2012). Prytulak and Elliott (2007) used Ti and Na contents of samples from Pico and São Miguel to discuss the influence of recycled mafic components on the melting signatures. They find, in agreement with U–Th–Pa disequilibria (Prytulak and Elliott 2009), that the geochemical signatures underneath Pico are consistent with melting of garnet peridotite that has previously been metasomatised by melts from an enriched eclogitic component.

França et al. (2006) interpret the isotope variability of basalts and selected xenoliths as a binary mixing array between a radiogenic (HIMU) and a depleted mantle source similar to the other islands (Elliott et al. 2007; Millet et al. 2009; Prytulak and Elliott 2007; Turner et al. 1997). The relative homogeneity of trace element and isotope signatures from Pico led several authors to refer to Pico as a reference frame in order to better compare the signatures from the other islands (Beier et al. 2010, 2012; Elliott et al. 2007; Millet et al. 2009; Prytulak and Elliott 2007; Turner et al. 1997). On the other hand, Pico lavas are displaced towards slightly lower $^{207}Pb/^{204}Pb$ and $^{208}Pb/^{204}Pb$ for a given $^{206}Pb/^{204}Pb$, and we will discuss the implications of this below. Metrich et al. (2014) presented trace element, He isotope and H_2O and CO_2 melt inclusion data from five monogenetic cones along the Pico axial fissure system. They showed that melts at Pico may be generated solely by decompression melting of a volatile-rich mantle source that requires little to no elevated temperatures compared to normal ambient mantle. A similar conclusion for melting underneath Faial and Pico was reached by Beier et al. (2012) based on the depths and temperatures of melting. The presence of a volatile-rich source underneath Faial and Pico is consistent with observations made beneath Santa Maria (Beier et al. 2013) and along the MAR (Asimow et al. 2004).

The $^{187}Os/^{188}Os$ isotope ratios for both islands differ significantly between Widom and Shirey (1996) and Schaefer et al. (2002). Widom and Shirey (1996) analysed Os isotopes for Faial and Pico (0.1432–0.1508 and 0.122–0.1629, respectively) that are generally more elevated than those from Schaefer et al. (2002; 0.110–0.130 and 0.1201–0.1388, respectively). The extremely unradiogenic, subchondritic values that are unique amongst the global array of ocean islands led Schaefer et al. (2002) to conclude that these were derived from a depleted harzburgitic mantle. The Re-depletion model ages imply an age of >2.5 Ga for the source of these magmas which may be indicative of a recycled, Archean-Proterozoic oceanic mantle lithosphere. The stable Li, O and B isotopes (Genske et al. 2013, 2014; Turner et al. 2007) indicate a contribution of a low-temperature altered oceanic crust and the subchondritic $^{187}Os/^{188}Os$ isotope ratios (Schaefer et al. 2002; Turner et al. 2007) suggest that it may be Archean in age (>2.5 Ga).

Overall, the islands of Faial and Pico show low degrees of partial melting and melting depths comparable to the islands of Graciosa and Terceira but have much less variable isotope ratios indicative of a more homogenous and volatile-rich source composition. The trace element and isotope homogeneity of the two islands has been taken as reference for comparing and inferring the mantle sources of the Azores islands. The Archean nature of the Faial and Pico sources is evident from their subchondritic Os isotopes and Sr–Nd–Pb and stable isotopes suggest that most of this material originated from recycled oceanic lithosphere in agreement with the common source found elsewhere in the Azores. It is important to note that the petrology and major element geochemistry of the two islands differs significantly (Larrea et al., Chapter "Petrology of the Azores Islands"), but the similarity of the trace element and radiogenic isotopes suggests that the major elements may be influenced by shallower processes, i.e. fractional crystallisation and assimilation in the lithosphere underneath the islands.

4.5 Flores and Corvo

The islands of Flores and Corvo are the only subaerial volcanic edifices west of the MAR and while the eastern islands have been subject of various publications, the availability of geochemical analyses from Flores and Corvo is limited. The first publication on lavas from Corvo and Flores is from White et al. (1979). They presented three samples from Flores and two from Corvo that have also been analysed for U–Th–Pa disequilibria by Bourdon et al. (2005) and the dataset was complemented with trace elements and U–Th–Ra data by Beier et al. (2010) who used these data to infer a potential (a) symmetry of melting and sources across the Azores melting anomaly which will be discussed in more detail below.

More comprehensive major and trace element data from Corvo were presented by França (2006) and by Larrea et al. (2013) who discussed the processes of fractional crystallisation and assimilation (Larrea et al., Chapter "Petrology of the Azores Islands"). Data from Flores are even scarcer. The first comprehensive major element, trace element and Sr–Nd-isotope dataset for both islands was published by Genske et al. (2012). The variability and enrichment in trace elements of both islands matches or even exceeds that of the two easternmost islands of São Miguel and Santa Maria (Figs. 3 and 8). Genske et al. (2012) showed, in agreement with observations made by Beier et al. (2010) that the La/Yb and Nb/Zr ratios of the islands of Flores (22.2 ± 4.5 and 0.32 ± 0.03, respectively) and Corvo (21.4 ± 3.5 and 0.33 ± 0.06), are higher and more variable compared to the eastern islands (Figs. 5 and 8) In addition, Ta/Hf and La/Sm are elevated underneath Corvo (1.03 ± 0.36 and 5.86 ± 1.37) and Flores (1.20 ± 0.27 and 5.59 ± 0.84), compared to any of the eastern islands.

Genske et al. (2012) estimated the degrees of partial melting at 3–5% of a garnet peridotitic source (Fig. 6) with ~6 wt% residual garnet which is consistent with the range observed in the eastern Azores islands (see above). However, they noted that the radiogenic Sr–Nd isotope ratios and trace element systematics do not support the interpretation of França (2006) who, solely based on the trace element data, suggested the presence of HIMU and enriched mantle sources (EM II) underneath Corvo. Instead, Genske et al. (2012) proposed that the mantle source of Corvo and Flores is comparable to those of the least enriched endmembers of the eastern islands Terceira, Graciosa and western São Miguel but displays a more restricted range in source composition ($^{206}Pb/^{204}Pb$ ratios from 19.02 to 19.78; Figs. 4, 5 and 7), and may be generally slightly less enriched than those of the eastern plateau (Figs. 4 and 9). The combined elevated La/Sm and Tb/Yb (5.1 ± 0.2 and 0.64 ± 0.04, respectively) of one particular lava suite on Flores led these authors to conclude that there may be a localised, distinct mantle source present underneath Flores. Combined with the Ta/Hf and higher Nb/Zr ratios (Figs. 4 and 8) this points towards the presence of a unique mantle source component underneath the western islands which is not observed in the Sr–Nd isotope ratios. Two $^{187}Os/^{188}Os$ isotope analyses from Flores (0.132–0.14180) were published by Schaefer et al. (2002) and overlap those from São Miguel and Terceira implying a similarly enriched, radiogenic source usually found in ocean islands which is also consistent with the observations of the Sr–Nd–Pb isotopes.

Genske et al. (2013, 2014) measured $\delta^{18}O$ isotopes from olivine phenocrysts and $\delta^{7}Li$ and $\delta^{11}B$ isotopes from whole rock samples. Using these stable isotopes they suggest that the assimilation of hydrothermally altered material has affected the stable isotope source signature. In addition, Genske et al. (2013) showed that the olivine $\delta^{18}O$ isotope ratios from +4.84 to +5.25‰ are correlated with the forsterite contents of individual phenocrysts which implies assimilation of hydrothermally altered crustal material. The source of the Azores mantle, however, has extremely homogeneous $\delta^{18}O$ of $+5.2 \pm 0.1‰_{(V\text{-}SMOW)}$. Also, the range in $\delta^{7}Li$ and $\delta^{11}B$ ($+3.5‰_{(L\text{-}SVEC)}$ to $+8.2‰_{(L\text{-}SVEC)}$ and $-3.5‰_{(NIST\ SRM\ 951)}$ to $+11.8‰_{(NIST\ SRM\ 951)}$, respectively) underneath the two islands is more extreme than observed underneath the eastern

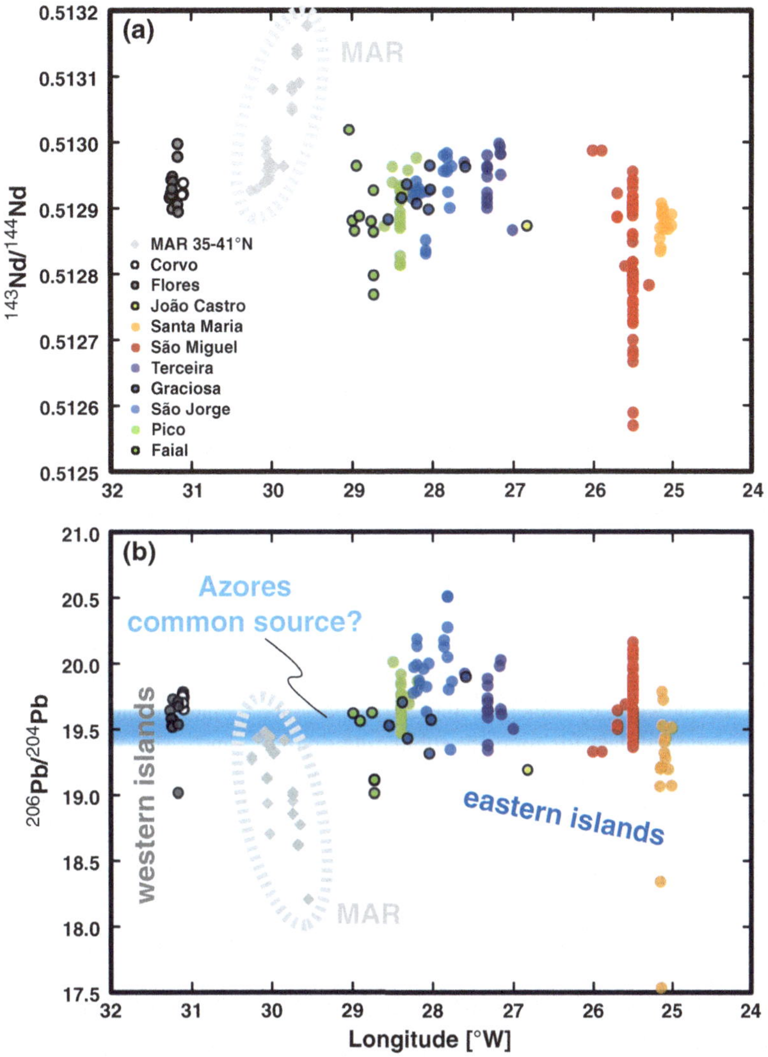

Fig. 9 a ^{143}Nd/^{144}Nd and b ^{206}Pb/^{204}Pb isotope ratios versus Longitude [°W] of the Azores lavas. Data sources as in Fig. 2. There is no clear evidence for a symmetric pattern of the sources underneath the Azores. Blue shaded field marks approximate position of the common Azores source in Pb isotope space

islands indicating assimilation of gabbroic crustal material (Genske et al. 2014). Genske et al. (2013, 2014) finally concluded that the stable isotopes may be used as sensitive tracers for the assimilation of crustal components that could also affect the major and trace elements but that the source signature of both δ^7Li and δ^{11}B for Corvo and Flores is relatively homogenous. The δ^7Li of the western mantle sources are around $+5‰_{(L\text{-}SVEC)}$ commonly found underneath OIB and δ^{11}B ratios range from $-3‰_{(NIST\ SRM\ 951)}$ to $-4‰_{(NIST\ SRM\ 951)}$ and thus are within the range of normal MORB and OIB.

Summarising, the geochemical data from Corvo and Flores suggests that the mantle source enrichment underneath these islands is less pronounced than beneath the eastern islands, however, the composition of the sources may be comparable and thus may be the result of a common Azores mantle source that is present west and east of the Mid-Atlantic Ridge. Trace element ratios show similar dynamics of melting underneath Corvo and Flores as beneath the eastern islands of Faial, Pico, Graciosa and Terceira. Given the ubiquitous global presence of mantle sources with these radiogenic and stable isotopic compositions (Stracke et al. 2005; Zindler and Hart 1986) none of the data available from these two islands can be used to unambiguously confirm or refute a connection to the Azores melting anomaly.

4.6 Submarine Plateau Samples and the Mid-Atlantic Ridge

Submarine samples from the Azores Plateau are rare, but data for some have recently been presented along with subaerial lavas in Beier et al. (2006, 2007, 2008, 2012) and major elements for Pico Ridge are presented in Mitchell et al. (2012, see also Mitchell et al., Chapter "Volcanism in the Azores: A Marine Geophysical Perspective"). These samples, however, are geochemically and geologically clearly related to the subaerial volcanoes nearby and thus do not give additional information on the submarine Azores Plateau. A recent work by Beier et al. (2015) presents geochemical data for fourteen submarine dredge samples across the Azores Plateau. Their data show that samples from Princesa Alice bank south of Faial and Pico display tholeiitic compositions in contrast to the alkaline lavas from the islands. This geochemical signature, along with the Ar–Ar ages, may represent the early plume-related stage of the Azores Plateau volcanism. The tholeiitic compositions are the result of higher degrees of partial melting (up to 15%, La/Yb <15) from an enriched mantle source ($^{206}Pb/^{204}Pb > 19$ for most lavas). The widespread occurrence of volcanism across the Azores Plateau and the higher degrees of partial melting implies that the majority of the Azores Plateau was formed from 10 to 4 Ma (Beier et al. 2015; Cannat et al. 1999; Gente et al. 2003) and possibly represents the initial arrival of a small mantle plume head underneath the lithosphere, but because so few submarine samples are available, a more detailed sampling of the submarine Plateau is required in the future. Recent sampling efforts during several expeditions with the German Research Vessel R/V Meteor (Expeditions M113, M128, M141) will provide numerous sedimentary and igneous submarine samples that will improve the general understanding of the evolution of the submarine Azores Plateau.

The first geochemical analyses from the Mid-Atlantic Ridge (MAR) dealing with the origin of the Azores were published by Schilling (1975) who presented Sr isotopic evidence for the occurrence of enriched material in the Azores. The samples were dredged along the Mid-Atlantic Ridge in the vicinity of the Azores and the elevated $^{87}Sr/^{86}Sr$ isotope ratios indicated the first geochemical evidence for enriched mantle material in the Azores. Schilling (1975) concluded that the Azores are the result of an enriched mantle "blob". Several later publications confirmed these observations using halogens and REE (Schilling et al. 1980, 1983; Schilling 1978). The MAR axis has meanwhile been subject of several publications dealing with the geochemistry and geophysics (Vogt and Jung, Chapter "The "Azores Geosyndrome" and Plate Tectonics: Research History, Synthesis, and Unsolved Puzzles", Fernandes et al., Chapter "The Contribution of Space-Geodetic Techniques to the Understanding of the Present-Day Geodynamics of the Azores Triple Junction ") of the axis and the extent to which the Azores melting anomaly extends south and north along the axis (Asimow et al. 2004; Bougault and Treuil 1980; Bourdon et al. 1996; Cannat et al. 1999; Charlou et al. 2000; Escartín et al. 2001; Gale et al. 2011; Georgen 2008; Goslin et al. 1999; Hamelin et al. 2013; Kingsley and Schilling 1995; Laubier et al. 2012; Luis et al. 1994; Schilling 1975, 1978; White et al. 1975, 1976). Based on new samples Bougault and Treuil (1980) noted that the trace element enrichment along the MAR is not randomly distributed but, similar to Schilling (1975), concluded that this may be the result of a material flux from an Azores mantle plume. Based on new carbon measurements and published trace element and radiogenic isotope data Kingsley and Schilling (1995) concluded that the volatile-rich nature of the Azores mantle also leads to significant amounts of excess melting underneath the adjacent MAR, as recently also observed for Pico (Metrich et al. 2014). Goslin et al. (1999)

restricted the influence of the Azores melting anomaly in the north to 43–44°N, however, the influence to the south is comparatively further and extends south to the FAMOUS segment at 36°N (Cannat et al. 1999). Bourdon et al. (1996) presented a dataset of major elements, selected trace elements and U–Th disequilibria from the MAR and Dosso et al. (1999) presented the first geochemical evidence of a southward extension of the Azores enriched geochemical signature (see their Fig. 6 and data in Fig. 4) based on a comprehensive major element, trace element and Sr–Nd–Pb isotope dataset. They however also noted that some enrichments in the vicinity of the Azores may be associated with shallow level enrichment from continental lithosphere delaminated during the opening of the Atlantic, comparable to the model later proposed by Moreira et al. (1999) for the isotope signature of the Azores islands. More recently, Cooper et al. (2004) presented selected major elements, trace elements and O-isotope data along with Sr–Nd isotope data from Dosso et al. (1999) from the Azores region (Figs. 4 and 5). They inferred that the enriched component in this region is the result of recycled upper oceanic crustal rocks associated with the geodynamic setting prior to the Atlantic opening in abundances of less than a few percent which is also in agreement with the conclusions of Genske et al. (2013) based on O isotopes from Flores and Corvo. The occurrence of recycled upper oceanic crustal rocks is consistent with the presence of comparable isotope signatures underneath the islands as described above. Asimow et al. (2004) showed that melting along the MAR may best be explained by an increase in mantle potential temperatures by about 35 °C, a range in temperatures confirmed by Beier et al. (2012) for Faial and Pico. More recently, Gale et al. (2011), Laubier et al. (2012) and Hamelin et al. (2013) dealt with the major element, trace element and radiogenic isotope anomaly in the MAR segments south of the Azores Plateau. Gale et al. (2011) and Laubier et al. (2012) both showed that the enrichment along the Lucky Strike, Menez Gwen and FAMOUS segments exceeds that of the Azores lavas in incompatible trace elements but that the isotopes are less enriched. They interpreted this to be the result of a low degree partial melt from the Azores mantle metasomatising the MAR mantle prior to melting. The Nd and Pb isotope ratios from Faial, Pico, Graciosa and Terceira (Beier et al. 2008, 2012) indicate that the main material flux from the Azores into the MAR occurs close to Graciosa rather than at Faial and Pico (Fig. 4). Hamelin et al. (2013) presented new trace element and radiogenic isotope data (Sr–Nd–Pb–Hf) for samples from the Lucky Strike segment. They interpreted the enriched geochemical signature and the low Hf isotope ratios to be the result of melting a subcontinental lithospheric mantle component underneath the ridge axis. A similar model was proposed for the island of Terceira by Moreira et al. (1999) and Madureira et al. (2011).

In summary, MORB from the MAR in the vicinity of the Azores are the result of excess melting resulting in a thicker crust and shallower water depth compared to normal MOR. The presence of enriched mantle material leads to the pronounced and distinct enrichment of trace element and radiogenic isotope signatures along the axis. The southward extension of this geochemical signature implies a southward flow of Azores mantle material along the spreading axis, consistent with geophysical observations. However, the processes and physical nature of the enrichment along the MAR are still a matter of active discussion. While an enrichment along the MAR further south than 36° may not directly be linked to the Azores, there is a common agreement that melts from ~44°N to 36°N are clearly influenced by the Azores melting anomaly.

5 Spatial Source Variability Across the Archipelago

The geochemical data presented here provide a unique opportunity to constrain and explore the nature of the Azores melting anomaly and the distribution of the sources across the anomalous Azores mantle. The radiogenic isotopes of the Azores display a large diversity of Sr–Nd–Pb isotope compositions unique amongst most ocean islands but also observed at Samoa (Jackson et al. 2014). The radiogenic Pb isotope ratios (Fig. 7) imply linear mixing arrays on several of the Azores islands indicating mixing of a common Azores mantle source with variably enriched end-members. The enriched compositions range from those commonly associated with a HIMU or FOZO component to slightly lower Pb isotope ratios (with the exception of lavas from Faial, eastern São Miguel and Santa Maria). The origin of FOZO and HIMU sources suggest that these originate from various portions of the recycled oceanic crust (Hanan and Graham 1996; Stracke et al. 2005; Zindler and Hart 1986) but the different slopes in $^{208}Pb/^{204}Pb$ versus $^{206}Pb/^{204}Pb$ space (Fig. 7) may indicate that the sources have slightly different ages. The most enriched sources in the Azores are found underneath the islands of Terceira, São Miguel and São Jorge, which likely is the result of a thicker lithosphere that preserves the ancient, enriched sources underneath Terceira and São Miguel, compared to the islands of Faial and Pico (Figs. 8 and 9) that also have slightly larger degrees of partial melting (Fig. 6). São Jorge is an exception amongst the islands closer to the ridge because it preserves a distinct HIMU-isotope signature. However, the greater melting depths and higher melting temperatures underneath São Jorge, compared to the neighbouring islands could be the result of its extremely enriched source (Beier et al. 2012). The $^{187}Os/^{188}Os$ isotopes across the Azores melting anomaly display a distinct symmetry (Schaefer et al. 2002) with the oldest Re-depletion ages underneath Faial and Pico extending towards younger, more enriched values both east and westward indicative of a deep, ancient, upwelling source underneath these islands.

Generally, the most enriched incompatible trace element signatures in the Azores are found amongst the easternmost islands of São Miguel and Santa Maria due to the thicker lithosphere underneath the eastern plateau which reduces the overall length of the melting column (Figs. 8 and 9). The two western islands of Corvo and Flores display similarly enriched trace element signatures compared to São Miguel and Santa Maria (Fig. 8). However, this may result from generally smaller degrees of partial melting (Fig. 6) but also mantle sources that are isotopically slightly less enriched than those in the East. However, the trace element and isotope signatures imply an enriched source underneath these islands that is comparable to those underneath Faial and Pico. This implies that the Azores mantle source is present both east and west of the Mid-Atlantic Ridge, an observation that is a matter of debate and requires a complex geodynamic history of the Azores mantle and crust.

6 Summary and Perspectives

Here, we review the origin of the Azores mantle melting anomaly from a geochemical perspective. It is obvious from observations on the islands and along the MAR that mantle melting in the Azores is the result of a combination of melting of enriched, volatile-rich mantle sources and potentially also slightly increased temperatures in the mantle ($\sim$35 °C above ambient mantle). Much of the heterogeneity observed in the radiogenic Sr–Nd–Pb–Hf–Os isotopes is the result of mixing ancient, enriched mantle end-members with a ubiquitous Azores mantle source resulting in unique, linear binary mixing arrays in several of the islands (São Miguel, Terceira, Graciosa, João de Castro). Much of the geochemical variability within the Azores dataset results from the isotopic variability of eastern São Miguel with extremely unusual and enriched Sr and Pb isotope ratios. The presence of source compositions with variable radiogenic Pb isotope ratios and subchondritic Os isotopes indicates mixing of ancient, recycled oceanic crust of different ages (2–3 Ga) into a common Azores mantle source. This is particularly important when comparing data from the Azores to those of other ocean islands. The conclusion that may be taken from these observations is that homogenisation processes underneath the Azores are not efficient enough to homogenise the mantle on the length scales of a single volcanic edifice within the timescales of recycling. One potential explanation could be the relatively low buoyancy flux of the Azores mantle compared to other plume-related settings that prevents the mantle from being thoroughly homogenised.

The preferential preservation of these enriched signatures in the eastern islands of Santa Maria, São Miguel and Terceira result from small degrees of partial melting. The dynamics of melting in the eastern Azores Plateau imply that they are influenced by changes in lithosphere thickness, i.e. the increasing lithosphere thickness with increasing distance from the Mid-Atlantic Ridge (decreasing degrees of partial melting). The archipelago displays by far the largest source variability in radiogenic isotope compositions amongst most global ocean islands including the unique sources underneath eastern São Miguel and is the only melting anomaly that rises underneath an ultraslow spreading ridge (Terceira rift axis).

Although all subaerial volcanoes in the Azores have been sampled and studied geochemically, the submarine structures of the Azores Plateau remain largely unknown. Future work should also include more volatile analyses of glasses to better constrain the volatile-rich nature of the Azores mantle. This becomes particularly important for understanding the potential deep (i.e. lower mantle or possibly even core-mantle boundary) origin of the Azores melting anomaly from Ne and He isotopes because the contrasting observation of a volatile-rich nature and a deep origin of the Azores mantle plume requires additional geochemical constraints. The source of the two western islands of Flores and Corvo remains largely unconstrained and needs to be better evaluated and quantified in order to develop a consistent model of the formation of the Azores islands and the relationship of the islands west and east of the Mid-Atlantic Ridge.

Although several geochemical signatures (e.g., Os, Ne and He isotopes) may point towards a deep (plume) origin of the Azores archipelago, other geodynamic features (e.g., lack of age progressive volcanic chain, a volatile-rich source composition) are less well explained by a classic plume model requiring further investigation to solve the long-standing debate on a plume versus non-plume origin in the Azores.

Acknowledgements We acknowledge the extremely constructive and helpful review(s) by Sebastian Wiesmaier and the review by Vittorio Zanon. We acknowledge the help and contributions of various colleagues over the years including linguistic and scientific corrections by Wafa Abouchami, Steve Galer, Dieter Garbe-Schönberg, Felix S. Genske, Thor Hansteen, Folkmar Hauff, Munir Humayun, Michael Joachimski, Niels W. R. Jöns, Stephan Klemme, Mark Krienitz, Stefan Krumm, João Mata, Neil Mitchell, Marcel Regelous, Tracy Rushmer, Bruce Schaefer, Andreas Stracke, Jürgen Titschack, Simon P. Turner, Bill White, and Elisabeth Widom. Christoph Beier and Karsten Haase also acknowledge funding by the Deutsche Forschungsgemeinschaft during various projects in the Azores (BE4459/1-1, BE4459/4-1, HA2568/6–1, HA2568/9–2). Christoph Beier acknowledges the Alexander von Humboldt-Foundation for funding during his stay in Australia and Simon P. Turner for his helpful, entertaining and friendly help. P. A. Brandl and C. Beier acknowledge the inspiring and productive atmosphere at Eurobodalla and Singapore Airlines with sufficient seatspace to be able to revise this manuscript. This manuscript was finalised in the Boilerhouse Café of the RHUL campus.

References

Abdel Monem, A. A., Fernandez, L. A., & Boone, G. M. (1975). K-Ar-Ages from the eastern Azores group (Santa Maria, Sao Miguel and the Formigas islands). *Lithos, 8,* 247–254.

Agostinho, J. (1936). The volcanoes of the Azores Islands. *Bulletin of Volcanology, 8,* 123–138.

Allegre, C. J., Hamelin, B., Provost, A., & Dupre, B. (1987). Topology in isotopic multispace and origin of mantle chemical heterogeneities. *Earth and Planetary Science Letters, 81*(4), 319–337.

Asimow, P. D., Dixon, J. E., & Langmuir, C. H. (2004). A hydrous melting and fractionation model for mid-ocean ridge basalts: Application to the Mid-Atlantic Ridge near the Azores. *Geochemistry, Geophysics, Geosystems, 5*(1), Q01E16.

Asimow, P. D., Hirschmann, M. M., & Stolper, E. M. (2001). Calculation of Peridotite partial melting from thermodynamic models of minerals and melts, IV. Adiabatic decompression and the composition and mean properties of mid-ocean Ridge Basalts. *Journal of Petrology, 42*(5), 963–998.

Béguelin, P., Bizimis, M., Beier, C., & Turner, S. (2017). Rift–plume interaction reveals multiple generations of recycled oceanic crust in Azores lavas. *Geochimica et Cosmochimica Acta, 218*, 132–152. https://doi.org/10.1016/j.gca.2017.09.015.

Beier, C., Haase, K. M., & Abouchami, W. (2015). Geochemical and geochronological constraints on the evolution of the Azores Plateau. Geological Society of America Special Papers.

Beier, C., Haase, K. M., Abouchami, W., Krienitz, M.-S., & Hauff, F. (2008). Magma genesis by rifting of oceanic lithosphere above anomalous mantle: Terceira Rift, Azores. *Geochemistry, Geophysics, Geosystems, 9*(12), Q12013.

Beier, C., Haase, K. M., & Hansteen, T. H. (2006). Magma evolution of the Sete Cidades volcano, São Miguel, Azores. *Journal of Petrology, 47*(7), 1375–1411.

Beier, C., Haase, K. M., & Turner, S. P. (2012). Conditions of melting beneath the Azores. *Lithos, 144–145,* 1–11.

Beier, C., et al. (2013). Geochemical evidence for melting of carbonated peridotite on Santa Maria Island, Azores. *Contributions to Mineralogy and Petrology, 165*(5), 823–841.

Beier, C., Stracke, A., & Haase, K. M. (2007). The peculiar geochemical signatures of São Miguel lavas: Metasomatised or recycled mantle sources? *Earth and Planetary Science Letters, 259*(1–2), 186–199.

Beier, C., Turner, S. P., Plank, T., & White, W. (2010). A preliminary assessment of the symmetry of source composition and melting dynamics across the Azores plume. *Geochemistry, Geophysics, Geosystems, 11*(1), Q02004.

Berthois, L. (1953). Contribution to the lithological study of the Azores archipelago. *Comincacoes Serv Geology Portugal, 24,* 29–198.

Blundy, J. D., Robinson, J. A. C., & Wood, B. J. (1998). Heavy REE are compatible in clinopyroxene on the spinel lherzolite solidus. *Earth and Planetary Science Letters, 160*(3–4), 493–504.

Bolfan-Casanova, N. (2003). Water partitioning at 660 km depth and evidence for very low water solubility in magnesium silicate perovskite. *Geophysical Research Letters, 30*(17), 1905.

Bonatti, E. (1990). Not so hot "hot spots" in the oceanic mantle. *Science, 250*(4977), 107–111.

Bougault, H., & Treuil, M. (1980). Mid-Atlantic Ridge: Zero age geochemical variations between Azores and 22 °N. *Nature, 286*, 209–212

Bourdon, B., Langmuir, C. H., & Zindler, A. (1996). Ridge-hotspot interaction along the Mid-Atlantic Ridge between 37°30' and 40°30'N: the U-Th disequilibrium evidence. *Earth and Planetary Science Letters, 142,* 175–189.

Bourdon, B., Turner, S. P., & Ribe, N. M. (2005). Partial melting and upwelling rates beneath the Azores from U-series isotope perspective. *Earth and Planetary Science Letters, 239*(1–2), 42–56.

Brandl, P. A., et al. (2012). Volcanism on the flanks of the East Pacific Rise: Quantitative constraints on mantle heterogeneity and melting processes. *Chemical Geology, 289–299*(3–4), 41–56.

Cannat, M., et al. (1999). Mid-Atlantic ridge—Azores hotspot interactions: along-axis migration of a hotspot-derived event of enhanced magmatism 10 to 4 Ma ago. *Earth and Planetary Science Letters, 173,* 257–269.

Cardigos, F., et al. (2005). Shallow water hydrothermal vent field fluids and communities of the D. João de Castro Seamount (Azores). *Chemical Geology, 224,* 153–168.

Charlou, J. L., et al. (2000). Compared geochemical signatures and the evolution of Menez Gwen (37° 50'N) and Lucky Strike (37°17'N) hydrothermal fluids, south of the Azores triple junction on the Mid-Atlantic Ridge. *Chemical Geology, 171*(1–2), 49–75.

Cooper, K. M., Eiler, J. M., Asimow, P. D., & Langmuir, C. H. (2004). Oxygen isotope evidence for the origin of enriched mantle beneath the Mid-Atlantic Ridge. *Earth and Planetary Science Letters, 220*(3–4), 297–316.

Davies, G. R., Norry, M. J., Gerlach, D. C., & Cliff, R. A. (1989). A combined chemical and Pb–Sr–Nd isotope study of the Azores and Cap Verde hot-spots: The geodynamic implications. In: A.D. Saunders & M. J. Norry (Eds.), *Magmatism in the oceanic basins* (pp. 231–255). Geological Society Special Publications.

Dias, N. A., et al. (2007). Crustal seismic velocity structure near Faial and Pico Islands (Azores), from local earthquake tomography. *Tectonophysics, 445*(3–4), 301–317.

Donnelly, K. E., Goldstein, S. L., Langmuir, C. H., & Spiegelman, M. (2004). Origin of enriched ocean ridge basalts and implications for mantle dynamics. *Earth and Planetary Science Letters, 226*(3–4), 347–366.

Dosso, L., et al. (1999). The age and distribution of mantle heterogeneity along the Mid-Atlantic Ridge (31-41°N). *Earth and Planetary Science Letters, 170,* 269–286.

Dupré, B., & Allègre, C. J. (1983). Pb-Sr isotope variation in Indian Ocean basalts and mixing phenomena. *Nature, 303*(5913), 142–146.

Dupré, B., Lambret, B., & Allègre, C. J. (1982). Isotopic variations within a single oceanic island: The Terceira case. *Nature, 299,* 620–622.

Elliott, T., Blichert-Toft, J., Heumann, A., Koetsier, G., & Forjaz, V. (2007). The origin of enriched mantle beneath Sao Miguel, Azores. *Geochimica et Cosmochimica Acta, 71*(1), 219–240.

Escartín, J., Cannat, M., Pouliquen, G., & Rabain, A. (2001). Crustal thickness of V-shaped ridges south of the Azores: Interaction of the Mid-Atlantic Ridge (36°–39 °N) and the Azores hot spot. *Journal of Geophysical Research, 106*(10), 21719–21735.

Esenwein, P. (1929). Zur Petrographie der Azoren. *Zeitschrift für Vulkanologie, 3*(12), 128–227.

Farnetani, C. G., & Samuel, H. (2005). Beyond the thermal plume paradigm. *Geophysical Research Letters, 32*(7), L07311.

Féraud, G., Kaneoka, I., & Allègre, C. J. (1980). K/Ar ages and stress pattern in the Azores: Geodynamic implications. *Earth and Planetary Science Letters, 46,* 275–286.

Foulger, G. R., Natland, J. H., Presnall, D. C., & Anderson, D. L. (2005). Plates, plumes and paradigms, 388. Geological Society of America.

França, Z. T. (2006). Geochemistry of alkaline basalts of Corvo Island (Azores, Portugal): Preliminary data. *Geogaceta*(40), 87–90.

França, Z. T. M., et al. (2006). Petrology, geochemistry and Sr–Nd–Pb isotopes of the volcanic rocks from Pico Island-Azores (Portugal). *Journal of Volcanology and Geothermal Research, 156*(1–2), 71–89.

Gale, A., Escrig, S., Gier, E. J., Langmuir, C. H., & Goldstein, S. L. (2011). Enriched basalts at segment centers: The Lucky Strike (37°17′N) and Menez Gwen (37°50′N) segments of the Mid-Atlantic ridge. *Geochemistry, Geophysics, Geosystems, 12*(6), Q06016.

Gast, P. W. (1968). Trace element fractionation and the origin of tholeiitic and alkaline magma types. *Geochimica et Cosmochimica Acta, 32*(10), 1057–1086.

Genske, F. S., et al. (2013). Oxygen isotopes in the Azores islands: Crustal assimilation recorded in olivine. *Geology, 41*(4), 491.

Genske, F. S., et al. (2014). Lithium and boron isotope systematics in lavas from the Azores islands reveal crustal assimilation. *Chemical Geology, 373,* 27–36.

Genske, F. S., Turner, S. P., Beier, C., & Schaefer, B. F. (2012). The petrology and geochemistry of Lavas from the Western Azores Islands of Flores and Corvo. *Journal of Petrology, 53*(8), 1673–1708.

Gente, P., Dyment, J., Maia, M., & Goslin, J. (2003). Interaction between the Mid-Atlantic ridge and the Azores hotspot during the last 85 Myr: Emplacement and rifting of the hot spot-derived plateaus. *Geochemistry, Geophysics, Geosystems, 4*(10), Q8514.

Georgen, J. E. (2008). Mantle flow and melting beneath oceanic ridge-ridge-ridge triple junctions. *Earth and Planetary Science Letters, 270*(3–4), 231–240.

Girod, M., & Lefevre, C. (1972). A propos des "andesites" des Acores. *Contributions to Mineralogy and Petrology, 35*(2), 159–167.

Goslin, J., et al. (1999). Extent of Azores plume influence on the Mid-Atlantic ridge north of the hotspot. *Geology, 27*(11), 991–994.

Haase, K. M., & Beier, C. (2003). Tectonic control of ocean island basalt sources on Sao Miguel, Azores? *Geophysical Research Letters, 30*(16), 1856.

Halliday, A. N., et al. (1995). Incompatible trace elements in OIB and MORB and source enrichment in the sub-oceanic mantle. *Earth and Planetary Science Letters, 133,* 379–395.

Hamelin, C., et al. (2013). Atypically depleted upper mantle component revealed by Hf isotopes at Lucky Strike segment. *Chemical Geology, 341,* 128–139.

Hanan, B. B., & Graham, D. W. (1996). Lead and helium isotope evidence from oceanic basalts for a common deep source of mantle plumes. *Science, 272*(5264), 991–995.

Hart, S. R. (1988). Heterogeneous mantle domains—Signatures, genesis and mixing chronologies. *Earth and Planetary Science Letters, 90*(3), 273–296.

Hart, S. R., & Staudigel, H. (1989). *Isotopic characterization and identification of recycled components, crust/ mantle recycling at convergence zones* (pp. 15–28). Dordrecht-Boston: D Reidel Publishing Company (International).

Hart, S. R., & Zindler, A. (1989). *Constraints on the nature and development of chemical heterogeneities in the mantle, mantle convection; plate tectonics and global dynamics* (pp. 261–387). New York, NY: Gordon & Breach Science Publishers.

Hawkesworth, C. J., Norry, M. J., Roddick, J. C., & Vollmer, R. (1979). $^{143}Nd/^{144}Nd$ and $^{87}Sr/^{86}Sr$ ratios from the Azores and their significance in LIL-element enriched mantle. *Nature, 280,* 28–31.

Hildenbrand, A., et al. (2008). Multi-stage evolution of a sub-aerial volcanic ridge over the last 1.3 Myr: S. Jorge Island, Azores triple junction. *Earth and Planetary Science Letters, 273*(3–4), 289–298.

Hildenbrand, A., et al. (2012). Reconstructing the architectural evolution of volcanic islands from combined K/Ar, morphologic, tectonic, and magnetic data: The Faial Island example (Azores). *Journal of Volcanology and Geothermal Research, 241,* 39–48.

Hirschmann, M. M. (2006). Water, melting, and the deep earth H_2O cycle. *Annual Review of Earth and Planetary Sciences, 34*(1), 629–653.

Hirschmann, M. M., Aubaud, C., & Withers, A. C. (2005). Storage capacity of H_2O in nominally anhydrous minerals in the upper mantle. *Earth and Planetary Science Letters, 236*(1–2), 167–181.

Hirschmann, M. M., Kogiso, T., Baker, M. B., & Stolper, E. M. (2003). Alkalic magmas generated by partial melting of garnet pyroxenite. *Geology, 31*(6), 481–484.

Hirschmann, M. M., & Stolper, E. M. (1996). A possible role for garnet pyroxenite in the origin of the "garnet signature" in MORB. *Contributions to Mineralogy and Petrology, 124*(2), 185–208.

Hofmann, A. W. (1997). Mantle geochemistry: the message from oceanic volcanism. *Nature, 385,* 219–229.

Hofmann, A. W. (2003). Sampling mantle heterogeneity through oceanic basalts: Isotopes and trace elements. In R. W. Carlson (Ed.), *Treatise on geochemistry* (pp. 61–101). Amsterdam: Elsevier.

Jaques, A. L., & Green, D. H. (1979). Determination of liquid compositions in high-pressure melting of peridotite. *American Mineralogist, 64*(11–12), 1312–1321.

Kelley, K. A., Plank, T., Ludden, J., & Staudigel, H. (2005). Subduction cycling of U, Th, and Pb. *Earth and Planetary Science Letters, 234,* 369–383.

Kingsley, R. H., & Schilling, J.-G. (1995). Carbon in Mid-Atlantic Ridge basalt glasses from 28°N to 63°N: Evidence for a carbon-enriched Azores mantle plume. *Earth and Planetary Science Letters, 129,* 31–53.

Kumagai, I., Davaille, A., Kurita, K., & Stutzmann, E. (2008). Mantle plumes: Thin, fat, successful, or failing? Constraints to explain hot spot volcanism through time and space. *Geophysical Research Letters, 35*(16), 5.

Larrea, P., et al. (2013). Magmatic processes and the role of Antecrysts in the genesis of Corvo Island (Azores Archipelago, Portugal). *Journal of Petrology, 54*(4), 769–793.

Larrea, P., et al. (2014). Magmatic evolution of Graciosa (Azores, Portugal). *Journal of Petrology, 55*(11), 2125–2154.

Laubier, M., Gale, A., & Langmuir, C. H. (2012). Melting and crustal processes at the FAMOUS segment (Mid-Atlantic Ridge): New insights from Olivine-hosted Melt Inclusions from Multiple Samples. *Journal of Petrology, 53*(4), 665–698.

Le Maitre, R. (1989). *A classification of igneous rocks and glossary of terms, recommendations of the International union of geological sciences, subcommission on the systematics of igneous rocks*. Oxford, London: Blackwell.

LeBas, M. J., LeMaitre, R. W., Streckeisen, A., & Zanettin, B. (1986). A chemical classification of volcanic-rocks based on the total alkali silica diagram. *Journal of Petrology, 27*(3), 745–750.

Litasov, K., et al. (2003). Water solubility in Mg-perovskites and water storage capacity in the lower mantle. *Earth and Planetary Science Letters, 211*(1–2), 189–203.

Luis, J. F., et al. (1994). The Azores triple junction evolution since 10 Ma from aeromagnetic survey of the Mid-Atlantic Ridge. *Earth and Planetary Science Letters, 125,* 439–459.

Lyubetskaya, T., & Korenaga, J. (2007). Chemical composition of Earth's primitive mantle and its variance: 1. Method and results. *Journal of Geophysical Research, 112,* B03211.

Madureira, P., Mata, J., Mattielli, N., Queiroz, G., & Silva, P. (2011). Mantle source heterogeneity, magma generation and magmatic evolution at Terceira Island (Azores archipelago): Constraints from elemental and isotopic (Sr, Nd, Hf, and Pb) data. *Lithos, 126,* 402–418.

Madureira, P., Moreira, M., Mata, J., & Allègre, C.-J. (2005). Primitive neon isotopes in Terceira Island (Azores archipelago). *Earth and Planetary Science Letters, 233,* 429–440.

Matias, L., et al. (2007). The 9th of July 1998 Faial Island (Azores, North Atlantic) seismic sequence. *Journal of Seismology, 11*(3), 275–298.

McKenzie, D. A. N., & O'Nions, R. K. (1995). The source regions of ocean island basalts. *Journal of Petrology, 36*(1), 133–159.

Metrich, N., et al. (2014). Is the 'Azores Hotspot' a Wetspot? Insights from the geochemistry of fluid and melt inclusions in olivine of pico basalts. *Journal of Petrology, 55*(2), 377–393.

Millet, M.-A., Doucelance, R., Baker, J. A., & Schiano, P. (2009). Reconsidering the origins of isotopic variations in Ocean Island Basalts: Insights from fine-scale study of São Jorge Island, Azores archipelago. *Chemical Geology, 265*(3–4), 289–302.

Mitchell, N. C., Beier, C., Rosin, P., Quartau, R., & Tempera, F. (2008). Lava penetrating water: Submarine lava flows around the coasts of Pico Island, Azores. *Geochemistry, Geophysics, Geosystems, 9.* https://doi.org/10.1029/2007gc001725.

Mitchell, N. C., Stretch, R., Oppenheimer, C., Kay, D., & Beier, C. (2012). Cone morphologies associated with shallow marine eruptions: east Pico Island, Azores. *Bulletin of Volcanology, 74,* 2289–2301.

Moreira, M., Doucelance, R., Kurz, M. D., Dupre, B., & Allegre, C. J. (1999). Helium and lead isotope geochemistry of the Azores Archipelago. *Earth and Planetary Science Letters, 169*(1–2), 189–205.

Morgan, W. J. (1971). Convection plumes in the lower mantle. *Nature, 230,* 42–43.

Murakami, M., Hirose, K., Yurimoto, H., & Nakashima, S. (2002). Water in Earth's lower mantle. *Science, 295,* 1885–1887.

Neal, C. R., Mahoney, J. J., Kroenke, L. W., Duncan, R. A., & Petterson, M. G. (1997). *The Ontong Java Plateau, large igneous provinces; continental, oceanic, and planetary flood volcanism* (pp. 183–216). Washington, DC: American Geophysical Union.

Niu, Y., O'Hara, M. J. (2003). Origin of ocean island basalts; a new perspective from petrology, geochemistry, and mineral physics considerations. *Journal of Geophysical Research, 108*(B4). https://doi.org/10.1029/2002jb002048.

Nunes, J. C., Forjaz, V. H., Alves, J. L., & Bernardes, A. C. (2003). Caracterização vulcanológica do Banco D. João de Castro (Açores): novos dados. Ciências da Terra (UNL), V(CD-ROM): D55–D58.

Oversby, V. M. (1971). Lead in oceanic islands; Faial, Azores and Trindade. *Earth and Planetary Science Letters, 11*(5), 401–406.

Pertermann, M., & Hirschmann, M. M. (2003). Partial melting experiments on a MORB-like pyroxenite between 2 and 3 GPa: Constraints on the presence of a pyroxenite in basalt source regions from solidus location and melting rate. *Journal of Geophysical Research, 108*(B2), 2125.

Pilidou, S., Priestley, K., Debayle, E., & Gudmundsson, Ó. (2005). Rayleigh wave tomography in the North Atlantic: High resolution images of the Iceland, Azores and Eifel mantle plumes. *Lithos, 79,* 453–474.

Pilidou, S., Priestley, K., Gudmundsson, O., & Debayle, E. (2004). Upper mantle S-wave speed heterogeneity and anisotropy beneath the North Atlantic from regional surface wave tomography; the Iceland and Azores plumes. *Geophysical Journal International, 159*(3), 1057–1076.

Provost, A., & Allegre, C. J. (1979). Process identification and search for optimal differentiation parameters from major element data; general presentation with emphasis on the fractional crystallization process. *Geochimica et Cosmochimica Acta, 43*(4), 487–502.

Prytulak, J., & Elliott, T. (2007). TiO_2 enrichment in ocean island basalts. *Earth and Planetary Science Letters, 263*(3–4), 388–403.

Prytulak, J., & Elliott, T. (2009). Determining melt productivity of mantle sources from ^{238}U-^{230}Th and ^{235}U-^{231}Pa disequilibria; an example from Pico Island, Azores. *Geochimica et Cosmochimica Acta, 73*(7), 2103–2122.

Prytulak, J., et al. (2014). Melting versus contamination effects on ^{238}U-^{230}Th-^{226}Ra and ^{235}U-^{231}Pa disequilibria in lavas from Sao Miguel, Azores. *Chemical Geology, 381,* 94–109.

Ramalho, R. S., Helffrich, G., Madeira, J., Cosca, M., Thomas, C., Quartau, R., et al. (2016). Emergence and evolution of Santa Maria Island (Azores)—The conundrum of uplifted islands revisited. *Geological Society of America Bulletin.* https://doi.org/10.1130/b31538.1.

Robinson, J. A. C., & Wood, B. J. (1998). The depth of the spinel to garnet transition at the peridotite solidus. *Earth and Planetary Science Letters, 164*(1–2), 277–284.

Salters, V. J. M., & White, W. M. (1998). Hf isotope constraints on mantle evolution. *Chemical Geology, 145*(3–4), 447–460.

Santos, R. S., et al. (2001). Remote sensing at D. João de Castro bank: Tools for biology and conservation studies. In: *5th Underwater Science Symposium,* Southampton (United Kingdom).

Schaefer, B. F., Turner, S., Parkinson, I., Rogers, N., & Hawkesworth, C. (2002). Evidence for recycled Archaean oceanic mantle lithosphere in the Azores plume. *Nature, 420,* 304–307.

Schilling, J.-G. (1975). Azores mantle blob: Rare-earth evidence. *Earth and Planetary Science Letters, 25,* 103–115.

Schilling, J.-G., Bergeron, M. B., & Evans, R. (1980). Halogens in the mantle beneath the North Atlantic. *Philosophical Transactions of the Royal Society of London, A297,* 147–178.

Schilling, J. G. (1978). F, Cl, Br in MAR glasses and Azores basalts. *Eos Transactions American Geophysical Union, 59*(4), 409.

Schilling, J.-G., et al. (1983). Petrologic and geochemical variations along the Mid-Atlantic ridge from 29°N to 73°N. *American Journal of Science, 283,* 510–586.

Schmincke, H. U., & Staudigel, H. (1976). Pillow lavas on central and eastern Atlantic islands (La Palma, Gran Canrai, Porto Santo, Santa Maria); preliminary report. *Bulletin de la Societe Geologique de France, 18*(4), 871–883.

Sobolev, A. V., Hofmann, A. W., Sobolev, S. V., & Nikogosian, I. K. (2005). An olivine-free mantle source of Hawaiian shield basalts. *Nature, 434,* 590–597.

Stosch, H.-G. (1982). Rare earth element partitioning between minerals from anhydrous spinel peridotite xenoliths. *Geochimica et Cosmochimica Acta, 46*(5), 793–811.

Stracke, A., Bizimis, M., & Salters, V. J. M. (2003). Recycling oceanic crust: Quantitative constraints. *Geochemistry, Geophysics, Geosystems, 4*(3), Q8003.

Stracke, A., Hofmann, A. W., & Hart, S. R. (2005). FOZO, HIMU and the rest of the mantle zoo. *Geochemistry, Geophysics, Geosystems, 6,* Q05007.

Stracke, A., Salters, V. J. M., & Sims, K. W. W. (1999). Assessing the presence of garnet-pyroxenite in the mantle sources of basalts through combined hafnium-neodymium-thorium isotope systematics. *Geochemistry, Geophysics, Geosystems, 1,* 1–15.

Sun, S.-S., & McDonough, W. F. (1989). Chemical and isotopic systematics of oceanic basalts: implications for mantle composition and processes. In A. D. Saunders & M. J. Norry (Eds.), *Magmatism in the ocean basins* (pp. 313–345). London: Geological Society of London Special Publications.

Turner, S., Hawkesworth, C., Rogers, N., & King, P. (1997). U-Th isotope disequilibria and ocean island basalt generation in the Azores. *Chemical Geology, 139*(1–4), 145–164.

Turner, S., Tonarini, S., Bindeman, I., Leeman, W. P., & Schaefer, B. F. (2007). Boron and oxygen isotope evidence for recycling of subducted components over the past 2.5 Gyr. *Nature, 447*(7145), 702–705.

Vogt, P. R., & Jung, W. Y. (2004). The Terceira Rift as hyper-slow, hotspot-dominated oblique spreading axis: A comparison with other slow-spreading plate boundaries. *Earth and Planetary Science Letters, 218,* 77–90.

Walter, M. J. (1998). Melting of garnet peridotite and the origin of komatiite and depleted lithosphere. *Journal of Petrology, 39*(1), 29–60.

Weaver, B. L. (1991). Trace element evidence for the origin of ocean-island basalts. *Geology, 19*(2), 123–126.

Wessel, P., & Kroenke, L. W. (2009). Observations of geometry and ages constrain relative motion of Hawaii and Louisville plumes. *Earth and Planetary Science Letters, 284*(3–4), 467–472.

Wessel, P., & Smith, W. H. F. (1991). Free software helps map and display data. *Eos Transactions American Geophysical Union, 72,* 441.

White, A. J. R., & Chappell, B. W. (1977). Ultrametamorphism and granitoid genesis. *Contributions to Mineralogy and Petrology, 62,* 129–139.

White, R., McKenzie, D., & O'Nions, R. (1992). Oceanic crustal thickness from seismic measurements and rare-earth element inversions. *Journal of Geophysical Research, 97*(B13), 19683–19715.

White, W. M., & Schilling, J. G. (1976). K, Rb, Sr, and Ba, geochemistry of Azores islands and Mid-Atlantic Ridge basalts. *Eos Transactions American Geophysical Union, 57*(4), 343.

White, W. M., Schilling, J. G., & Hart, S. R. (1975). Sr-isotope geochemistry of the Azores and the Mid-Atlantic Ridge; 29°N to 60°N. *Eos Transactions American Geophysical Union, 56*(6), 471.

White, W. M., Schilling, J. G., & Hart, S. R. (1976). Evidence for the Azores mantle plume from strontium isotope geochemistry of the Central North Atlantic. *Nature, 263,* 659–662.

White, W. M., Tapia, M. D. M., & Schilling, J.-G. (1979). The petrology and geochemistry of the Azores Islands. *Contributions to Mineralogy and Petrology, 69,* 201–213.

Widom, E., Carlson, R. W., Gill, J. B., & Schmincke, H. U. (1997). Th–Sr–Nd–Pb isotope and trace element evidence for the origin of the Sao Miguel, Azores, enriched mantle source. *Chemical Geology, 140*(1–2), 49–68.

Widom, E., & Farquhar, J. (2003). Oxygen isotope signatures in olivines from Sao Miguel (Azores) basalts: Implications for crustal and mantle processes. *Chemical Geology, 193,* 237–255.

Widom, E., & Shirey, S. B. (1996). Os isotope systematics in the Azores: Implications for mantle plume sources. *Earth and Planetary Science Letters, 142,* 451–465.

Workman, R. K., et al. (2004). Recycled metasomatized lithosphere as the origin of the Enriched Mantle II (EM2) end-member: Evidence from the Samoan Volcanic Chain. *Geochemistry, Geophysics, Geosystems, 5*(4), Q04008.

Yang, T., Shen, Y., van der Lee, S., Solomon, S. C., & Hung, S.-H. (2006). Upper mantle structure beneath the Azores hotspot from finite-frequency seismic tomography. *Earth and Planetary Science Letters, 250*(1–2), 11–26.

Zanon, V., & Frezzotti, M. L. (2013). Magma storage and ascent conditions beneath Pico and Faial islands (Azores archipelago): A study on fluid inclusions. *Geochemistry, Geophysics, Geosystems, 14*(9), 3494–3514.

Zanon, V., Kueppers, U., Pacheco, J. M., & Cruz, I. (2013). Volcanism from fissure zones and the Caldeira central volcano of Faial Island, Azores archipelago: Geochemical processes in multiple feeding systems. *Geological Magazine*, 1–20.

Zindler, A., & Hart, S. (1986). Chemical Geodynamics. *Annual Review of Earth and Planetary Sciences, 14,* 493–571.

Noble Gas Constraints on the Origin of the Azores Hotspot

Manuel A. Moreira, Pedro Madureira and João Mata

Abstract

Noble gas geochemistry is a valuable tool for assessing the nature of contributors to magma mantle sources. In this chapter, we analyse previously published data regarding helium and neon in the Azores to discuss the origin of the Azores archipelago. After the pioneering works of Kurz et al. (1982a, 1990) examining helium isotopic variations along the Mid-Atlantic Ridge (MAR) between 28°N and 53°N and on São Miguel Island, a systematic study was conducted by Moreira et al. (1999) on several Azores islands. These authors analysed He isotopic ratios from phenocrysts collected in five islands of the archipelago. Helium isotopic data from Terceira (minimum $^4He/^3He$ ratio of ≈63,700; R/Ra ≈11.3) were interpreted as the presence of a relatively primitive component in the mantle source, whereas the radiogenic $^4He/^3He$ ratios found at São Miguel (from 121,600 to 276,800; R/Ra from ≈5.9 to ≈2.6) have been interpreted as resulting from the melting of an enriched and ancient recycled material. Moreira and Allègre (2002) analysed noble gas isotopic data for MORB glasses dredged along a MAR segment (from 21.25°N to 39.9°N), which also includes the Azores triple junction area. The obtained $^4He/^3He$ ratios decrease from 90,000 at 37°N to 75,000 at 38.5°N and later increase to 100,000 at 40.5°N. The low $^4He/^3He$ ratio measured on the ridge at 38.5°N (76,000; R/Ra = 9.5) was interpreted as the result of present-day interaction between the ridge source and the Azores plume (see also Madureira et al. 2014), as sampled by lava erupted at Terceira and São Jorge islands (the latter with $^4He/^3He$ ratios down to 40,000; R/Ra ≈18). Madureira et al. (2005) focused their study on Terceira Island, for which He and Ne isotopic ratios were determined from olivine phenocrysts. Ne isotopic data corroborated the presence of a relatively primitive component in the Terceira mantle source ($^{21}Ne/^{22}Ne_{corr}$ = 0.052), which is, however, dominated by a MORB-type component. Jean-Baptiste et al. (2009) presented helium isotope data for thermal waters

M. A. Moreira (✉)
Institut de Physique du Globe de Paris, Sorbonne Paris Cité, Université Paris Diderot, UMR 7154, CNRS, 75005 Paris, France
e-mail: moreira@ipgp.fr

P. Madureira
Estrutura de Missão para a Extensão da Plataforma Continental, R. Costa Pinto 165, 2770-047 Paço d'Arcos, Portugal

P. Madureira
Instituto de Ciências da Terra/Dep. Geociências da Universidade de Évora, R. Romão Ramalho 59, 7000 Évora, Portugal

J. Mata
Instituto Dom Luiz, Faculdade de Ciências, Universidade de Lisboa, Campo Grande C6, 1749-016 Lisbon, Portugal

U. Kueppers and C. Beier (eds.), *Volcanoes of the Azores*, Active Volcanoes of the World, https://doi.org/10.1007/978-3-642-32226-6_12

and gas emissions sampled at Terceira, Graciosa and São Miguel islands, as well as Faial, Pico and Flores islands. The results were interpreted as the presence of relatively primitive He isotopic ratios at Terceira ($^4He/^3He$ 53,500; R/Ra ≈13.5), which also extend to Graciosa Island ($^4He/^3He$ ≈64,500; R/Ra ≈11.2). These values contrast with those obtained by these authors for São Miguel Island, which are significantly more radiogenic than those typical of MORB ($^4He/^3He$ from 120,400 to ≈138,900; R/Ra from 5.2 to 6.0) and are in agreement with the helium measurements in lavas from the eastern part of São Miguel (Moreira et al. 2012). Noble gas geochemistry suggests, despite the strong dilution of its primitive signature by MORB-like material, that a lower mantle-derived mantle plume is located under the central group of islands, particularly under São Jorge, in agreement with seismic tomographic images. The peculiar helium isotopic ratios observed in São Miguel lavas can be explained by different scenarios invoked for the Sr–Pb isotopic and trace element systematics, which suggest that the São Miguel source contains recycled mafic material ~3 Ga in age (e.g., Beier et al. 2007; Elliott et al. 2007).

1 Introduction

The evolution of the Azores region results from a complex tectono-magmatic interplay between a kinematically unstable RRR triple junction and large scale magmatic processes responsible by the generation of the Azores plateau (400,000 km^2) and islands (Machado 1955; Krause and Watkins 1970; Vogt and Yung 2004, Chapter "The "Azores Geosyndrome" and Plate Tectonics: Research History, Synthesis, and Unsolved Puzzles", Fernandes et al., Chapter "The Contribution of Space-Geodetic Techniques to the Understanding of the Present-Day Geodynamics of the Azores Triple Junction") (Fig. 1). The plateau is part of the Azores platform, which corresponds to a morphological domain bounded by the Hayes Fracture Zone and the Great Meteor seamount chain to the south and the Maxwell Fracture Zone to the north. The excessive magma production processes remained particularly active during the last 20–10 M year (Cannat et al. 2001; Luis and Miranda 2008), has been linked to the existence of the Azores plume (e.g., Schilling 1975; White et al. 1979; Widom et al. 1997; Turner et al. 1997; Moreira et al. 1999; Schaefer et al. 2002; Madureira et al. 2005, 2011; Beier et al. 2013, O'Neill and Sigloch, Chapter "Crust and Mantle Structure Beneath the Azores Hotspot—Evidence from Geophysics") that has been imaged by several seismic tomography models or proposed based on the reduced thickness of the transition zone or on the existence of low S-wave velocities inside the transition zone (Montelli et al. 2004; Yang et al. 2006; Silveira et al. 2010; Saki et al. 2015, O'Neill and Sigloch, Chapter "Crust and Mantle Structure Beneath the Azores Hotspot—Evidence from Geophysics"). Accordingly, mantle temperatures in excess relatively to the MORB source has been calculated for the Azores (Putirka 2008; Herzberg and Gazel 2009). However, several authors have pointed to a possible hydrated mantle source that could restrict the origin of the melting anomaly to the upper mantle (Asimow et al. 2004; Beier et al. 2012; Bonatti 1990; Métrich et al. 2014).

The major and trace element compositions, as well the non-gaseous isotope signatures, are reviewed and discussed in this volume. Most Azores islands are still volcanically active, and subaerial lavas are generally less than 1 Ma old. Older volcanic rocks were found in the eastern part of São Jorge and São Miguel, as shown by radiometric and paleomagnetic data (Hildenbrand et al. 2008, 2012; Johnson et al. 1998; Silva et al. 2012), and at Santa Maria Island, which comprises the oldest volcanic rocks within the archipelago (up to ≈6 Ma (Sibrant et al. 2015; Ramalho et al. 2017).

The major and trace element compositions of the erupted Azorean lavas as well are reviewed and discussed in this volume by Larrea et al. and

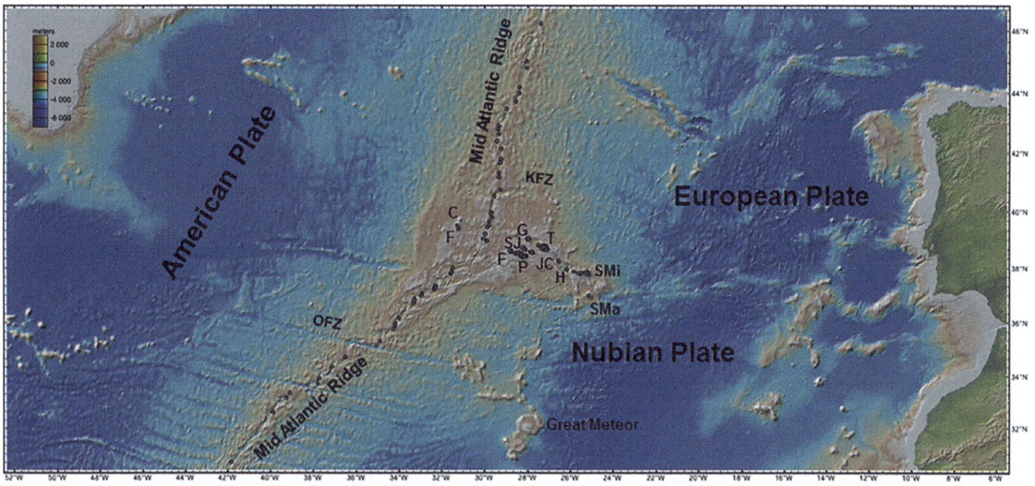

Fig. 1 Location of the samples (black dots) analysed for helium isotopes in the Azores region between 35°N and 40°N. Also shown are the location of the islands (Fr: Flores; C: Corvo; F: Faial; SJ: São Jorge; G: Graciosa; P: Pico; T: Terceira; SMi: São Miguel; SMa: Santa Maria). JC: D. João de Castro bank; H: Hirondelle basin; Fracture zones (OFZ: Oceanographer; KFZ: Kurchatov)

Beier et al. (Chapters "Petrology of the Azores Islands" and Melting and Mantle Sources in the Azores, respectively; see also Zanon 2015). In this chapter, we review the noble gas composition, particularly He and Ne, of the Azores lavas and fluids to discuss the deep origin of the Azores archipelago. Such data also allowed estimation of the present-day location of the Azores mantle plume.

2 Helium and Neon Isotope Geochemistry

We will focus on helium and neon isotopic compositions because these two noble gases are ideally adapted for discussing the depth at which hotspot mantle upwellings are rooted.

2.1 He–Ne Systematics in Oceanic Basalts and the Two-Reservoirs Model

Noble gas elements are characterised by full outer electron shells, which explains their highly unreactive character, being chemically inert for most of situations. Nuclear processes are the major cause of the significant isotopic variations, which can be considered linked to the geochemical processes controlling the distribution of potassium, uranium and thorium, the main heat producing elements in the planet (e.g., Ozima and Podosek 2002; Graham 2002).

Helium and neon isotopic systematics have been used to constrain the deep origin of hotspots. This approach is employed due to the low $^{4}He/^{3}He$ and $^{21}Ne/^{22}Ne$ isotopic ratios observed on oceanic island basalts compared to Mid-Oceanic Ridge Basalts (Kurz et al. 1982b, 2009; Honda et al. 1991, 1993; Moreira and Allègre 1998; Hilton et al. 1999; Trieloff et al. 2000, 2002; Moreira et al. 2001; Stuart et al. 2003). This difference is interpreted as the result of the presence of an undegassed reservoir in the source of OIB (Allègre 1987; Honda and Mc Dougall 1998; Allègre and Moreira 2004b). Indeed, the low $^{4}He/^{3}He$ ratio reflects a low $(U + Th)/^{3}He$ reservoir because ^{4}He is produced in radioactive decay chains of U and Th. As OIB sources are also enriched in incompatible trace elements, such as U and Th, to preserve low $^{4}He/^{3}He$ ratios, the OIB sources must be enriched in ^{3}He and therefore are much less degassed than MORB source. ^{21}Ne is produced in nucleogenic reactions, such as $^{18}O(\alpha,n)$, with a $^{4}He/^{21}Ne$

production ratio of 2×10^7 (Yatsevich and Honda 1997). ^{20}Ne and ^{22}Ne are considered to be stable and non-radiogenic in the mantle. Regarding the nature of the mantle sources a similar interpretation can be applied to neon. Low $^{21}Ne/^{22}Ne$ ratios reflect low $(U + Th)/^{22}Ne$; therefore, the OIB source must also be enriched in ^{22}Ne compared to the MORB reservoir. Therefore, it seems clear that a reservoir rich in noble gases is still present in the deep mantle and is sampled by mantle plumes. The location of these enriched domains within the mantle can also be interpreted from xenon isotopes, particularly ^{129}Xe, which derives from the radioactivity of the extinct ^{129}I (T½ = 17 Ma). It has been suggested that the difference in degassing states between the two reservoirs reflects an early differentiation event linked to the partial degassing of the magma ocean that was present during the first My of Earth's history (Trieloff et al. 2000, 2002; Yokochi and Marty 2006; Coltice et al. 2011; Mukhopadhyay 2012). This partial degassing left a deep reservoir enriched in primordial noble gases, now sampled by mantle plumes. The process of preserving this primordial reservoir for 4.5 Ga in the mantle is still unclear. The whole lower mantle has often been proposed as the undegassed reservoir (e.g., Allègre et al. 1986; Allègre and Moreira 2004a). This two-layer mantle model is often criticized because a whole convection is needed to explain the seismic tomographic images of slabs penetrating the lower mantle (Grand et al. 1997). However, as noted by Allègre (2002), seismic imaging provides a "present-day" picture of the mantle, which contrasts with the time-integrated history since Earth's formation, as interpreted from geochemical data. Geophysical data are dependent on major element composition (fairly uniform in the mantle) and are opaque to the partitioning process of trace elements between different reservoirs.

Recent models propose that the primordial reservoir is at the bottom of the mantle. Deep dense piles formed as the result of deep magma ocean crystallization can be the source of this primordial reservoir, if helium and neon are moderately incompatible elements ($D \sim 10^{-2}$; Coltice et al. 2011). Entrainment of a small percentage of this material can explain the entire spectrum of He and Ne isotopic composition observed in OIB (Coltice et al. 2011). Regardless of the location of this primordial reservoir, it is clear that the low $^4He/^3He$ and $^{21}Ne/^{22}Ne$ ratios measured in OIB are ascribed to the presence of lower mantle derived material, which is primordial for noble gases, pointing to the presence of deep mantle plumes anchored at the lower mantle, eventually at the D″ (e.g. Moreira 2013).

2.2 Type of Samples Available for Helium and Neon Analyses in the Azores

In volcanic areas, magmatic noble gases can be measured in thermal fluids (dissolved and gas), glassy margins of pillow lavas extruded during submarine eruptions, crystals sampled from xenoliths and phenocrysts separated from lava flows. Most of the Azores islands have thermal springs that can be collected for noble gas analyses, and in all of the islands, outcrops of porphyritic basaltic lavas with olivine phenocrysts can be found.

2.2.1 Thermal Springs

A large number of thermal springs are present in the Azores (Cruz and França 2006; Antunes and Carvalho, Chapter "Surface and Groundwater in Volcanic Islands: Water from the Azores Islands") (Fig. 2). All of the islands, except Santa Maria and Corvo, have thermal springs or CO_2-rich water from which helium isotopes can be measured and related to a mantle-derived signal (Greau 2011; Jean-Baptiste et al. 2009). Helium results obtained from thermal springs by these authors are presented in Fig. 3. Figure 3a shows the dissolved helium in water, and Fig. 3b shows the helium isotopic composition in the gas phase. Two distinct isotopic helium signatures have been observed in the Azores based on thermal springs. The springs found at São Miguel clearly have a more radiogenic signature than those from other islands. This finding is in agreement with the helium isotopic ratio obtained from basaltic rocks, which exhibits the same dichotomy (see below).

Fig. 2 Some thermal springs in the Azores and the sampling technique of water samples. Copper tubes are clamped in order to collect few grams of water, which will allow the measurement of dissolved helium and neon. Gas can be sampled by accumulation of gas bubbles under a bucket and then collected either in clamped copper tube or reservoir with valves

A notable difference between helium isotopic compositions in springs and basalts occurs at Graciosa Island. Both rock and water samples at Graciosa display a MORB-like helium signature, which contrasts with the relatively primitive isotopic ratio obtained from gas collected in the Furna do Enxofre fumarole. This discrepancy will be discussed below.

2.2.2 Glassy Margin of the Pillow Lavas

On the Azores plateau, a large number of samples from the MAR have been collected for which volcanic glass has been analysed for noble gases (Kurz et al. 1982a; Moreira and Allègre 2002). The sample locations where the helium isotopic ratios have been measured are reported in Fig. 1.

2.2.3 Xenoliths (Cumulates/Mantle)

Xenoliths are common in the Azores basaltic lavas and can be either mantle xenoliths or cognate cumulates (gabbros, dunites, pyroxenites). Only a few xenoliths have been analysed for helium in (Moreira et al. 1999; Madureira et al. 2014). Interpretation in terms of mantle source signals is not straightforward, and those results are not considered in the present work.

2.2.4 Phenocrysts

In the Azores, the most common samples for measuring helium isotopes are phenocrysts. Helium is incompatible during melting and crystallization and therefore is not present in the crystal matrixes (e.g. Marty 1993). However, during crystallization, melt inclusions containing volatiles are trapped into the mineral lattice. CO_2 shrinkage bubbles contain noble gases because they have approximately the same solubility as CO_2 in magma. Analyses of material ranging from 100 mg to several grams of phenocrysts allow precise measurement of helium content and helium isotopic composition. Most of the islands have porphyritic alkali basalts containing olivine and clinopyroxene phenocrysts suitable for helium analyses. A compilation of existing data

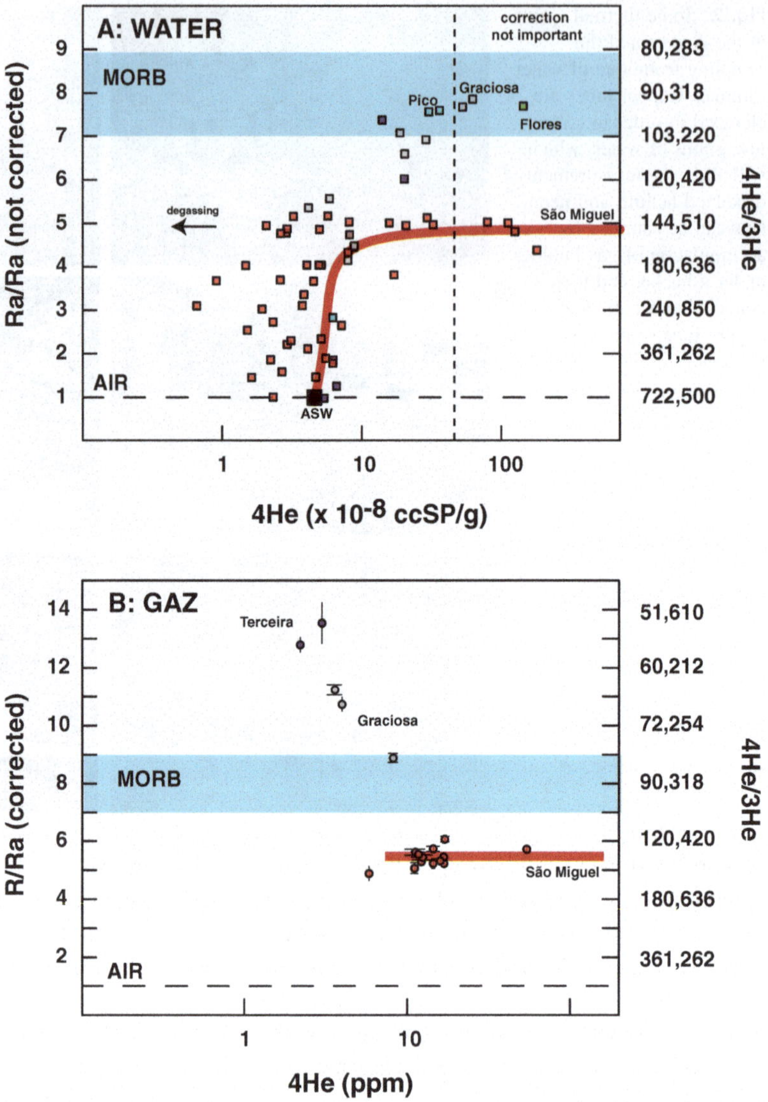

Fig. 3 Helium isotopic ratios of fluid in thermal springs from the Azores. Top figure **a** shows the dissolved helium isotopic composition, not corrected from the atmospheric helium. Bottom figure **b** shows the $^3He/^4He$ ratio in the gas phase, corrected for atmospheric helium. Data are from Jean-Baptiste et al. (2009) and Greau (2011)

on helium isotopic ratios in the Azores islands is shown in Figs. 4 and 5. In these figures, Azores data are compared to the helium isotopic ratios measured on North Atlantic MORB, excluding latitudes above 50°N, where Iceland's plume influence on the MAR becomes detectable (Poreda et al. 1986). Table 1 provides an estimate of the helium isotopic compositions for volcanic systems of the Azores archipelago based on published and unpublished data. Only one study has successfully obtained neon isotope measurements in phenocrysts, which is difficult because atmospheric neon contamination masks the mantle-derived signature (Madureira et al. 2005).

3 Helium and Neon Isotopes in the Azores

3.1 Mid-Atlantic Ridge Between 30°N and 50°N: Is the Azores Plume Detected at the Ridge?

The first evidence for plume-ridge interaction involving the Azores hotspot was proposed by Schilling (1975) on the basis of light rare earth enrichment in MORB near the Azores. A large number of other isotopic systems have provided

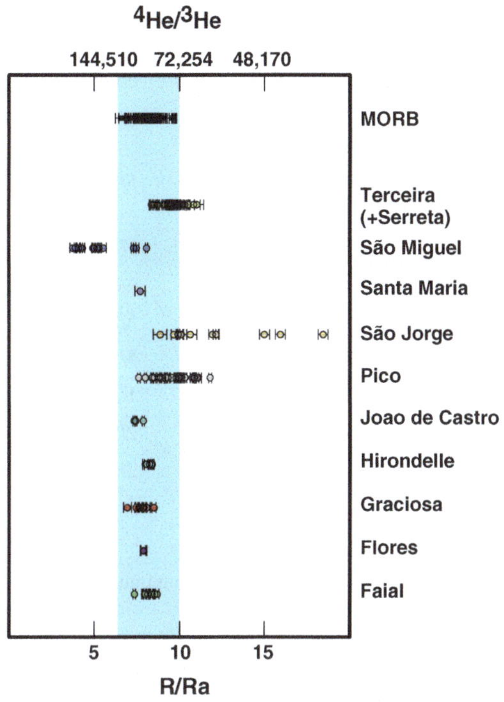

Fig. 4 Helium isotopic ratio for some volcanic structures of the Azores archipelago. The shaded area corresponds to the compositional range of the samples collected along the Mid Atlantic Ridge

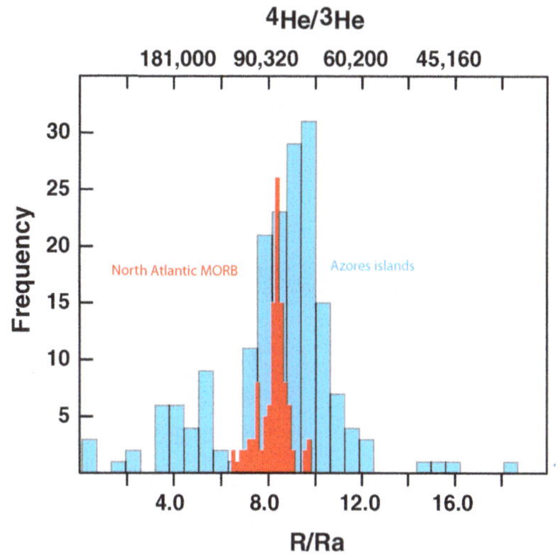

Fig. 5 Histogram of the helium isotopic ratios in lavas from the Azores archipelago, compared to the one of Mid Atlantic MORB. In this figure, Azores islands are not individualized

essential information about plume-ridge interaction in the north Atlantic (White et al. 1976; White and Schilling 1978; Schilling et al. 1980; Dosso et al. 1993, 1999; Bourdon et al. 1996; Kingsley and Schilling 1996; Yu et al. 1997; Gale et al. 2013). With regard to helium, only two articles have focused on the axis crossing the Azores platform. Kurz et al. (1982a) analysed 35 samples from the north Atlantic, but no ^{3}He anomaly was observed near the Azores. Conversely, Moreira and Allègre (2002) observed a decrease in the $^{4}He/^{3}He$ ratio from 90,000 to 75,000 (R/Ra increased from 8 to 9.5) at ~38°N, the latter being compatible with the primitive helium isotopic ratio observed at Terceira-Pico-São Jorge islands (Moreira and Allègre 2002). Accordingly, the data were interpreted as the result of interaction of the Azores plume with the Atlantic ridge (see also Madureira et al. 2014). A recent compilation of north Atlantic MORB helium is provided in Fig. 6 with the $^{206}Pb/^{204}Pb$ ratios of the same samples (Kurz et al. 1982a; Dosso et al. 1999; Gautheron 2002; Moreira and Allègre 2002; Moreira et al. 2011, Dosso unpublished data).

Two extreme cases are proposed in Figs. 6a, b based on whether the plume-ridge interaction is accepted or not as an explanation of the helium signatures along the axis. Figure 6a proposes that the helium anomaly is too small to be statistically suggested. Indeed, several "primitive" helium ratios are observed at 33°N (samples TR 123-5D from Kurz et al. 1982a), far from the Azores, and the anomaly at 38°N could also reflect mantle heterogeneity, rather than interaction with the Azores plume. Figure 6b shows the interpretation of Moreira and Allègre (2002) where the Azores hotspot can be detected at 38.5°N using the helium isotopic composition. This helium peak is also correlated with the variation in lead isotopic ratios (Fig. 6c), supporting the idea that the Azores plume signal is also observed in ridge samples.

Table 1 Helium isotopic ratios estimate for different volcanic centers of the Azores archipelago

Island	$^4He/^3He$	R/Ra	Remark	Level of confidence
São Miguel (excluding Sete Cidades)	131,000	5.5	Mean	5/5
Terceira	Min: 65,700 ± 2400 Mean: 75,265 ± 4700	Max: 11.0 ± 0.4 Mean: 9.6 ± 0.6		5/5
Pico	61,232 ± 520	11.8 ± 0.1		5/5
São Jorge	39,270 ± 640	18.4 ± 0.3		4/5
Graciosa	92,630 ± 4750	7.8 ± 0.4	Mean (9)	5/5
Faial	88,115 ± 4300	8.2 ± 0.4	Mean (9)	5/5
Santa Maria	93,840 ± 3656	7.7 ± 0.3	One value	1/5
Flores	91,460 ± 500	7.9 ± 0.0	Mean (2)	2/5
Joao de Castro	95,070 ± 3750	7.6 ± 0.3	Mean (3)	2/5
Hirondelle	87,050 ± 3150	8.3 ± 0.3	Mean (10)	
Corvo		–	No sample	–

Only samples with uncertainties under 5% (1σ) were used for the estimates of the mean or the maximum

3.2 Helium and Neon Isotopic Ratios in the Azores Islands

We now discuss the helium composition data available either in the literature or from the authors (unpublished data) for each island. Helium compositions of volcanic systems from the Azores are summarised in Table 1 with an indication of the reliability of these ratios. These ratios will be discussed later in light of the origin of the Azores hotspot.

3.2.1 The Western Group: Corvo and Flores

No data regarding lava were published for these two islands. MM has three unpublished helium results obtained on olivines from Flores lavas with $^4He/^3He$ ratios of 91,500, 91,500 and 103,600 (R/Ra = 7.9, 7.9 and 7.0, respectively) suggesting a helium ratio close to the MORB ratio for Flores island (Fig. 4) in agreement with the isotopic measurement of dissolved helium in the Lajedo/Ponta Negra warm springs (R/Ra = 7.7) or in the gas phase (R/Ra = 8.0 ± 0.1) (Jean-Baptiste et al. 2009). Corvo has an unknown noble gas isotopic signature. Due to their close proximity and geochemical affinities (Genske et al. 2012), it is probable that Flores and Corvo share the same helium isotopic composition, although this has yet to be demonstrated.

3.2.2 Central Group

The Central group of the Azores archipelago (Faial, Pico, Terceira, Graciosa, São Jorge) can be divided into two subgroups based on their helium isotopic compositions. Faial and Graciosa islands show MORB-like helium, whereas Pico, Terceira and São Jorge present ratios more primitive than MORB from the north Atlantic (Fig. 4).

The Faial-Graciosa affinity

All lava samples from Faial show MORB-like helium (mean $^4He/^3He$ is 88,115; R/Ra = 8.2 ± 0.4). The same ratio is observed for Graciosa lavas, which shows a mean helium $^4He/^3He$ ratio of 92,630 (R/Ra = 7.8 ± 0.4). However, gas samples from the Furna do Enxofre, located inside the Graciosa caldera, show more primitive helium (up to 11.2Ra; $^4He/^3He$ = 64,340) than MORB, being also different from the isotopic composition of olivine phenocrysts of a lava from nearby subterranean lake water (several meters away) (Jean-Baptiste et al. 2009; Fig. 3). These authors excluded any

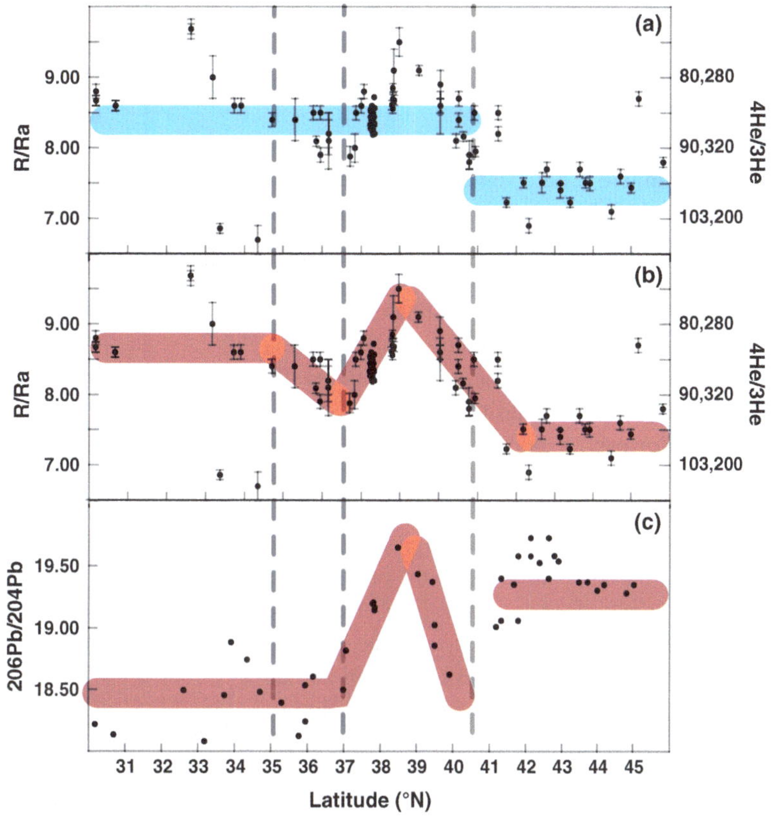

Fig. 6 Variation with latitude of the helium and the $^{206}Pb/^{204}Pb$ isotopic ratios. **a** Case where no plume-ridge interaction can be observed with helium. Only a clear isotopic boundary can be detected at 40.5°N, suggesting two distinct mantle domains in the north Atlantic with different histories. **b** Case in which, as suggested by Moreira and Allègre (2002), a peak of the $^{3}He/^{4}He$ could be observed between 37 and 40.5°N, similarly to the $^{206}Pb/^{204}Pb$ peak **c** (Dosso et al. 1999). Helium data are from Gautheron (2002), Kurz et al. (1982b) and Moreira and Allègre (2002)

isotopic fractionation to explain the discrepancy between lava compositions and the gas from the bubbling mud, suggesting that the measured ratios do reflect the magmatic source. However, it is surprising that such a primitive ratio was never observed in Graciosa lavas. The relationship between the helium content and the helium isotopic ratio observed for Graciosa and Terceira in Fig. 3 could suggest kinetic fractionation of the helium during bubble formation. Indeed, when CO_2 is exsolved forming bubbles due to decompression (either in water or magma), helium diffuses from the water (or magma) to the gas phase because it is non-soluble (see Ruzié and Moreira 2010 for detailed modelling of the vesiculation in disequilibrium in lava). If diffusion is slow enough or vesiculation is fast (rapid decompression), an isotopic fractionation can occur because of the different ^{3}He and ^{4}He diffusivity. The lower the amount of helium in the gas phase, the higher the isotopic fractionation produced. This is observed in Fig. 3 (lower), where the two gas samples from Graciosa with low $^{4}He/^{3}He$ ratios also have low helium abundances, suggesting that such relatively unradiogenic signatures can result from kinetic fractionation. However, this theoretical observation, which offers a possible explanation for the discrepancy between gas from the Furna do Enxofre mud spring and the other Graciosa samples, is difficult to demonstrate in the present case. Therefore, we cannot exclude that Graciosa also has a primitive helium signature. Future studies on this spring should include neon and argon isotopes measurements to test isotopic fractionation during bubble formation.

Terceira-Pico-São Jorge

These three islands show $^{4}He/^{3}He$ ratios lower than MORB (Fig. 4). This property is particularly pronounced at São Jorge, which displays notably primitive helium ratios (down to 39,200;

R/Ra = 18.4). Terceira and Pico show only slightly primitive ratios, although they are significantly higher in R/Ra than MAR basalts (at latitudes below 50°N).

The primitive helium signature was confirmed by the study of neon isotopes in lavas from Terceira (Madureira et al. 2005; Fig. 7). Most samples display neon isotopic ratios similar to those of the present-day atmosphere, which is usually interpreted as a consequence of magma contamination by air and/or hydrothermally altered oceanic crust during its ascent to the surface. However, a portion of olivine samples separated from lavas erupted from the fissural volcanic system still preserve a significant fraction of the mantle source neon signal, providing important constraints to interpret the origin of the Azores magmas. Due to lower $^{3}He/^{22}Ne$ in the primordial reservoir compared to the MORB source, neon is a more sensitive tool than helium to detect the presence of a small proportion of primordial material (Moreira and Allègre 1998; Moreira et al. 2001; Yokochi and Marty 2004; Dixon 2003). Although the helium isotopic composition of Terceira lavas is moderately primitive (e.g., only slightly more primitive than MORB ratio), neon isotope systematics clearly show the presence of a primitive component in the source of this island. Indeed, as illustrated in Fig. 7, the slope of the Terceira line is higher than that of the MORB line defined by Sarda et al. (1988), which clearly suggests the presence of un-nucleogenic material in the mantle source of Terceira ($^{21}Ne/^{22}Ne_{corr}$ = 0.052) in the same proportion as OIB, such as Réunion island (Trieloff et al. 2002). The helium and neon systematics can be interpreted following the hyperbolic mixing model of Moreira et al. (2001) and Madureira et al. (2005), suggesting a significant dilution of the primitive component in the upper mantle beneath Terceira and strengthening the importance of neon data to detect it.

3.2.3 Hirondelle Basin and D. João de Castro Seamount

Phenocryst samples from PM in the Hirondelle basin and D. João de Castro seamount

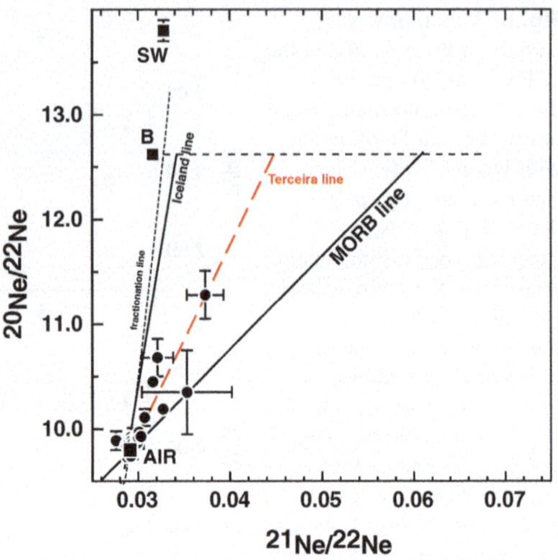

Fig. 7 Three neon isotope diagram showing the primitive signature of Terceira lavas compared to the MORB signature (Madureira et al. 2005). MORB line is from Sarda et al. (1988). The Iceland line is derived from Moreira et al. (2001)

(JdC) display MORB-like helium ($^{4}He/^{3}He$ = 91,460; R/Ra = 7.9 ± 0.8 for JdC and $^{4}He/^{3}He$ = 87,000; R/Ra = 8.3 ± 0.2 for Hirondelle), either not showing the same type of interaction with the primitive Azores plume as interpreted for Terceira island, or evidencing a stronger dilution of MORB component in their mantle sources (Madureira et al. 2014).

3.2.4 São Miguel and Santa Maria Islands

Moreira et al. (1999) published one helium result on olivines from Santa Maria island with a MORB-like signature (R/Ra = 7.7 ± 0.3). This single measurement is not sufficient to determine the mantle source origin of this island. Unfortunately, the noble gases studies from this oldest island of the Azores archipelago have been limited by the high degree of alteration evidenced by most of the olivine phenocrysts.

São Miguel helium compositions show the most interesting feature of the Azores archipelago and also of the OIB systematics. For São Miguel, Kurz (1991) first showed $^{4}He/^{3}He$ variation with longitude. This author observed radiogenic ratios ($^{3}He/^{4}He$ down to ~3.5Ra,

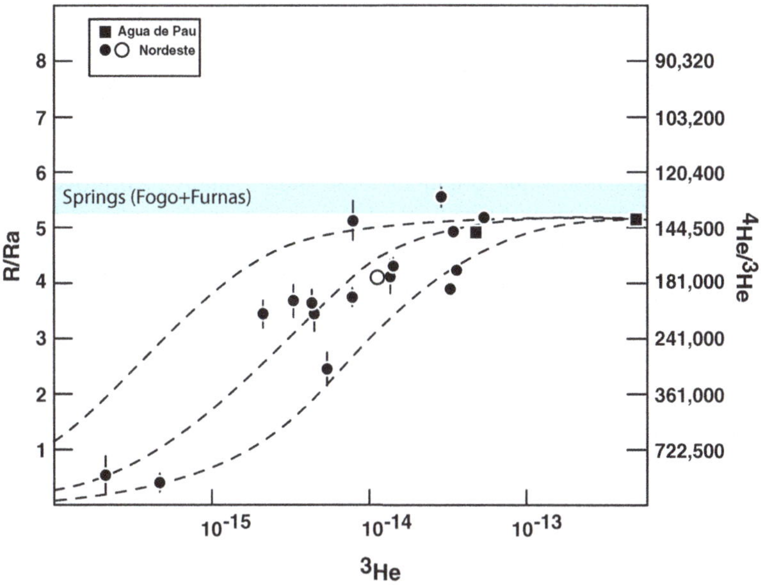

Fig. 8 Helium isotopic signature against the ^{3}He concentrations in samples from the eastern part of São Miguel island, including Agua de Pau and Nordeste volcanoes and compared to the ratio of thermal springs from Furnas and Agua de Pau. Figure derived and modified from Moreira et al. (2012)

$^{4}He/^{3}He$ ~206,000) in the east of São Miguel correlated with radiogenic lead isotopic ratios. Kurz (1991) also showed that the Sete Cidades volcano has MORB-like helium composition. However, no data from this work were published (only a figure). Moreira et al. (1999, 2012) published data on São Miguel lavas, overall confirming the unpublished data of Kurz (1991). Moreira et al. (2012) suggest that the Nordeste, Água de Pau (Fogo) and Furnas volcanoes are derived from the same mantle source, having a $^{4}He/^{3}He$ ratio of 145,000 (R/Ra ~5). This mantle source is different from the fissural zone and Sete Cidades volcano, which have a MORB-like helium composition and seem to be related to the Terceira rift. The most radiogenic helium isotopic ratios ($^{4}He/^{3}He$ ratio of ~1.8 10^{6}) measured by Moreira et al. (2012) reflect post-eruption radiogenic addition in helium-poor and aged phenocrysts (Fig. 8). These authors suggest that alpha particles deriving from the U–Th rich surrounding basalt are implanted at the surface of the phenocrysts and liberated during crushing in vacuum at the moment of the analysis. Thus, these radiogenic ratios do not reflect mantle compositions. The mantle isotopic ratio of the eastern part of São Miguel is proposed to be 145,000 (R/Ra ~5). At Fogo and Furnas, active volcanoes, the helium ratios obtained in lava are fully consistent with thermal springs helium compositions (Figs. 3 and 8; Greau 2011; Jean-Baptiste et al. 2009).

4 Evidence for the Azores Mantle Plume Originating in the Deep Mantle

The Azores helium and neon isotope systematics reflect the presence of three distinct signatures:

(1) A clear primitive signal observed in three islands from the Azores central group: Terceira, Pico and São Jorge. The Graciosa Island can belong to this group but it is still subject to discussion because all lava and dissolved helium in water show a MORB-like signature, whereas the helium in the gas phase from one thermal spring shows a relatively primitive signature (see the discussion above).
(2) Faial, Flores, Santa Maria and Sete Cidades exhibit a MORB-like signature, as well as the submarine João de Castro seamount and Hirondelle Basin.
(3) The eastern part of São Miguel, which includes the active volcanoes of Fogo (Água de Pau) and Furnas and the extinct Nordeste

volcano, shows a third helium isotopic composition, which is clearly more radiogenic (R/Ra ~5) than the previous two, MORB-like signatures included.

We discuss below the origin of these three helium isotopic signatures in light of the presence of a mantle plume deriving from the lower mantle that explains the helium signature of the Central group and passive heterogeneity melting for São Miguel.

4.1 Primordial or Residual Material?

In the classical two-reservoir model, the relatively unradiogenic helium and neon isotope ratios measured in Azores lavas would suggest a lower mantle origin for the Azores hotspot. However, it has been suggested that low $^4He/^3He$ (high R/Ra) could derive from an ancient recycled low $U/^3He$ lithospheric mantle (Anderson 1998; Coltice and Ricard 1999; Meibom et al. 2003; Parman et al. 2005). Indeed, residual mantle after melt extraction could evolve to a low $^4He/^3He$ ratio if uranium is more incompatible than helium, and if the extraction is old enough to "freeze" the helium isotopic composition at the time of depletion, when the mantle was more primitive than the present-day composition of the MORB source. Such a model suggests that the reservoir with unradiogenic helium signatures reflects depletion in U and Th, rather than enrichment in ^3He.

In such a scenario, the Azores lava could sample such a recycled depleted mantle without reflecting a deep mantle plume deriving from a ^3He-rich source. However, more recent experimental data suggest that uranium is not more incompatible than helium (Heber et al. 2007), thus invalidating such hypothesis. Moreover, it is clear that the Azores lavas represents melts from a source enriched in incompatible elements (see Larrea et al., Chapter "Petrology of the Azores Islands", Beier et al., Chapter "Melting and Mantle Sources in the Azores"), which makes improbable that the relatively unradiogenic helium isotopic ratios could result from a mantle domain having a low $(U + Th)/^3He$ stemming from low U and Th contents.

Consequently, it is clear that the relatively primitive helium and neon signatures (Type 1 signatures; see above) observed in Azores lavas imply their provenance from a reservoir enriched in primordial noble gases.

4.2 São Jorge: Mantle Plume Signature

The most primitive helium signatures of the Azores were obtained in São Jorge Island samples from the Serra do Topo volcanic system (R/Ra up to 18.4; see above). These samples also present the highest $^{206}Pb/^{204}Pb$ (up to 20.51; Millet et al. 2009), with those two ratios correlating positively (Fig. 9). These radiogenic lead isotopic signatures, coupled with radiogenic Nd and unradiogenic Sr signatures, were considered by Millet et al. (2009) to result from involvement in the São Jorge mantle source of a local end-member presenting affinities with the HIMU component. This component has been repeatedly assigned to ancient recycling of altered oceanic crust (e.g., Zindler and Hart 1986; Chauvel et al. 1992; Hanyu et al. 2011), which at mantle pressure is present in the form of eclogite-type rocks.

Preferential sub-lithospheric melting of eclogite (sensu lato) can explain the lead isotopic signature of São Jorge and also the deeper melting of S. Jorge source as compared with other islands of the central Azores, as calculated by Beier et al. (2012). Indeed, experimental work has demonstrated that the solidus of pyroxene-rich paragenesis is located deeper than for peridotites (e.g., Dasgupta et al. 2006 and references therein) and that melts ultimately derived from eclogites can reach the surface presenting characteristics typical of alkaline magmas (Mallik and Dasgupta 2012). If we accept the existence of a mantle plume at the origin of the Azores magmatism, it can be proposed that ascending material comprises, dispersed in a peridotitic matrix, stirred pyroxene-rich domains issued from ancient oceanic crust recycling and that São Jorge

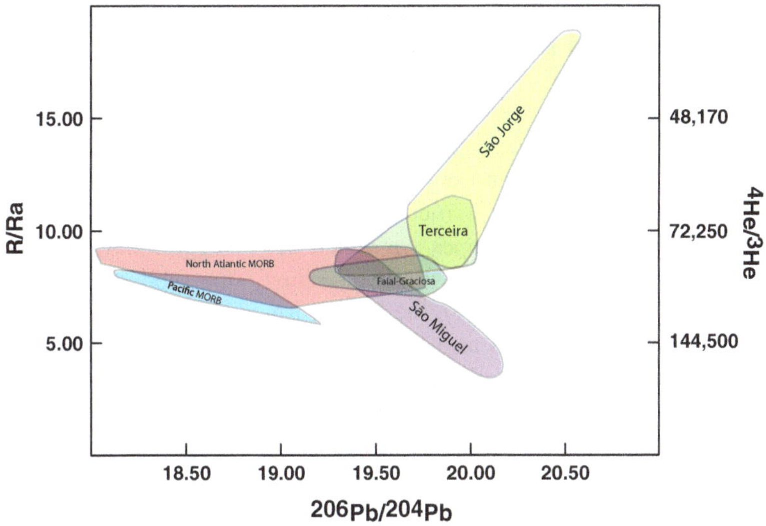

Fig. 9 Helium isotopic ratio reported against $^{206}Pb/^{204}Pb$ for north Atlantic MORB and Azores island (data from Madureira et al. 2005, 2011; Millet et al. 2009; Moreira et al. 1999) and unpublished data from the authors. The data are also compared to Pacific MORB (Hamelin et al. 2011; Moreira et al. 2008)

magmas represent melting of a higher proportion of the eclogite-type lithology than observed for the other Azores islands.

Preferential melting of eclogite can explain the lead isotopic signature of São Jorge but also raises questions regarding the primitive helium signature. Eclogite, as recycled oceanic crust, should have a high $U/^{3}He$ ratio and therefore should present extremely radiogenic helium after 1 Ga. Two explanations are possible. The simplest explanation is that recycled material had been stored in the gas-rich lower mantle and that diffusion of helium homogenised the recycled oceanic crust with the unradiogenic ambient lower mantle, as suggested by Hart et al. (2008). It seems clear from tomographic evidence (e.g., Kárason and Van der Hilst 2001) that recycled slabs deeply penetrate the lower mantle, most likely accumulating at the core-mantle boundary (the slab graveyards of Richards and Engebretson 1992). Residence times of such materials deep into the lower mantle can be significant ($\gg$1 Ga). Given the significant thermal discontinuity, the core-mantle boundary is considered an inception place for mantle plumes and numerical simulations (e.g., Farnetani and Hofmann 2009) have shown that plume heterogeneities inherited from their root zones, can survive during the ascent toward the shallow levels of the mantle.

It is also possible that helium lower mantle signatures are acquired by eclogite-derived melts in the upper mantle if, during melting, helium diffuses rapidly enough to be instantaneously extracted from the primitive helium carrier. The migration of high R/Ra helium, contained in the plume peridotitic matrix and representative of lower mantle material, in direction of the magma, generated from the lower solidus recycled material, would be facilitated by the stretching of lower mantle material at the edges of the ascending conduit (Albarède 2008) and by high He solubility in silicate melts (e.g., Jambon et al. 1986; Broadhurst et al. 1992).

Although the lead isotopic signature of São Jorge lavas suggests an important fraction of recycled eclogite in their source, the primitive helium signature clearly indicates a derivation from the deep mantle, enriched in primordial noble gases. The origin of the ^{3}He enrichment (diffusion to the recycled material in the lower mantle or diffusion to the eclogite during melting) of the HIMU-type magmas under São Jorge requires more detailed research, although it does not change the overall interpretation regarding the deep origin of the Azores plume.

4.3 São Miguel: Piece of ~3 Ga Old Subducted Material

All isotopic systems, including helium, show that the source of Eastern São Miguel has a different composition than the other islands of the archipelago. If this is particularly obvious for lead and strontium isotopes, for which São Miguel is unique among all OIB (e.g., Elliott et al. 2007; Beier et al. 2007), the helium isotopic ratio of its source is also completely different from that of the other Azores islands, and also rare among OIB. Whereas some of the Azores islands (see above) are characterised by a primitive helium signature (R/Ra up to 18.4), the Eastern São Miguel exhibits a significantly more radiogenic helium ratio than MORB, with a primary signature of 145,000 ($^3He/^4He$ ~5 Ra) (Fig. 4). Such a radiogenic signature is not common in OIB, which generally show more primitive helium signatures. However, several hotspots have similar radiogenic helium isotopic ratios, such as Comores, which presents a ratio of 5.2 ± 0.2 Ra (Class et al. 2005), Tristan da Cunha (R/Ra ~5; Class et al. 2005) and also some HIMU-type OIB (Hanyu and Kaneoka 1997; Parai et al. 2009; Hanyu et al. 2011). Moreover, São Miguel is unique because of its Pb–Sr–Nd–He isotopic compositions. To explain the radiogenic lead isotopic ratios of São Miguel, it has been proposed that its source contains enriched material that was subducted in the mantle ~3 Ga ago. Elliott et al. (2007) propose that this enriched material is ancient underplated material subducted with the associated oceanic crust, whereas Beier et al. (2007) suggest that the São Miguel component reflects the subduction of a seamount with the oceanic crust. The radiogenic helium isotopic signature is consistent with an ancient high $(U + Th)/^3He$ material. Moreira et al. (2012) have proposed that the mantle source of this peculiar component could not have been degassed, as it would have been for a seamount; otherwise, the helium isotopic ratio would be purely radiogenic after 2–3 Ga. Therefore, the scenario of underplated material best explains the helium isotopic ratio of São Miguel.

However, we emphasise that the connection between the São Miguel source with the Azores hotpot, as sampled by the Central group, is still uncertain. Does Eastern São Miguel signatures result from an upper mantle passive heterogeneity melted by the plume-induced temperature increase or by adiabatic decompression during extension? Or was the source present within the mantle plume as entrained material from the bottom of the mantle?

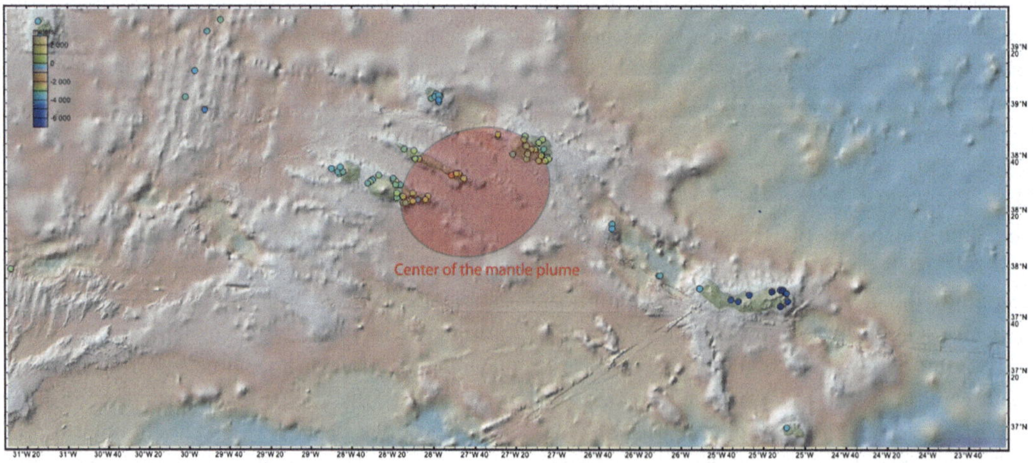

Fig. 10 Location of the centre of the Azores plume deduced from the helium isotopic ratio. The plume conduit is located under São Jorge, at odds with the proposal Moreira et al. (1999) who suggested Terceira as the emerged centre of the Azores plume. Note this is consistent with a higher pressure and higher temperature of melting under São Jorge derived from Beier et al. (2012)

5 Mantle Plume Location

Figure 10 shows the possible location of the present-day Azores plume based on helium isotopic compositions in Azores lava. Moreira et al. (1999) proposed that the plume was centred under Terceira Island, although at that time, no data on São Jorge existed. This is not in opposition with a recent highly resolved tomographic model (Adam et al. 2013). In our present proposal, based on He isotope ratios compilation, we suggest that the plume is located under São Jorge, close to the centre of the Azores plateau. This is in agreement with the proposed conditions of pressures and temperature for melting under Azores islands (Beier et al. 2012). Moreover, Yang et al. (2006) based on seismic tomography locate the plume to the north of Terceira at a depth of 375 km, although at shallower mantle levels, a Southward plume deflection is noticed, placing the plume closer to our proposal at the base of lithosphere. As argued by Yang et al. (2006), such deflection most likely results from the absolute motion of Eurasia and Nubia (236° from north) relative to a hotspot reference frame (Gripp and Gordon 2002). The active spreading along the Terceira rift, inducing larger degrees of partial melting and the consequent homogenization of distinct source signals, could also contribute to a dilution of the plume signal at Terceira (see also Madureira et al. 2011). Moreover, plume deflection would enhance mantle stirring, most likely also contributing to the dilution of the primitive helium component interpreted from most of the Azores islands.

References

Adam, C., Madureira, P., Miranda, J. M., Lourenço, N., Yoshida, M., & Fitzenz, D. (2013). Mantle dynamics and characteristics of the Azores plateau. *Earth and Planetary. Science Letters, 362,* 258–271.

Albarède, F. (2008). Rogue mantle helium and neon. *Science, 319,* 943–945.

Allègre, C. J. (1987). Isotope geodynamics. *Earth and Planetary Science Letters, 86,* 175–203.

Allègre, C. J. (2002). The evolution of mantle mixing, hil. *Transactions of the Royal Society London, 360,* 2411–2431.

Allègre, C. J., & Moreira, M. (2004). Rare gas systematics and the origin of oceanic islands: The key role of entrainment at the 670 km boundary layer. *Earth and Planetary Science Letters, 228,* 85–92.

Allègre, C. J., Staudacher, T., & Sarda, P. (1986). Rare gas systematics, formation of the atmosphere, evolution and structure of the Earth's mantle. *Earth and Planetary Science Letters, 81,* 127–150.

Anderson, D. L. (1998). A model to explain the various paradoxes associated with mantle noble gas geochemistry. *Proceedings of the National Academy of Sciences of the United States of America, 95,* 9087–9092.

Asimow, P. D., Dixon, J. E., & Langmuir, C. (2004). A hydrous melting and fractionation model for mid ocean ridge basalts: Application to the Mid Atlantic Ridge near the Azores. *Geochemistry Geophysics Geosystems, 5*(1).

Beier, C., Haase, K. M., Brandl, P. A. (this volume). Melting and mantle sources in the Azores.

Beier, C., Haase, K. M., & Turner, S. (2012). Conditions of melting beneath the Azores. *Lithos, 144–145,* 1–11.

Beier, C., Mata, J., Stöckhert, F., Mattielli, N., Brandl, P. A., Madureira, P., et al. (2013). Geochemical evidence for melting of carbonated peridotite on Santa Maria Island, Azores. *Contributions to Mineralogy and Petrology, 165,* 823–841.

Beier, C., Stracke, A., & Haase, K. (2007). The peculiar geochemical signatures of São Miguel (Azores) lavas: Metasomatised or recycled mantle sources? *Earth and Planetary Science Letters, 259,* 186–199.

Bonatti, E. (1990). Not so "Hot Spots" in the oceanic mantle. *Science, 250,* 107–110.

Bourdon, B., Langmuir, C., & Zindler, A. (1996). Ridge-hotspot interaction along the Mid-Atlantic Ridge between 37°30′ and 40°30′N: The U-Th disequilibrium evidence. *Earth and Planetary Science Letters, 142,* 175–189.

Broadhurst, C. L., Drake, M. J., Hagee, B. E., & Bernatowicz, T. J. (1992). Solubility and partitioning of Ne, Ar, Kr, and Xe in minerals and synthetic basaltic melts. *Geochimica et Cosmochimica Acta, 56,* 709–723.

Cannat, M., et al. (2001). Mid-Atlantic ridge—Azores hotspot interactions: Along-axis migration of a hotspot-derived magmatic pulse 14 to 4 myrs ago. *Earth and Planetary Science Letters, 173,* 257–269.

Chauvel, C., Hofmann, A. W., & Vidal, P. (1992). HIMU-EM: The French Polynesian connection. *Earth and Planetary Science Letters, 110,* 99–119.

Class, C., Goldstein, S., Stute, M., Kurz, M. D., & Schlosser, P. (2005). Grand Comore island: A well-constrained "low ^{3}He/^{4}He " mantle plume. *Earth and Planetary Science Letters, 233,* 391–409.

Coltice, N., Moreira, M., Labrosse, S., & Hernlund, J. W. (2011). Crystallization of a basal magma ocean recorded by Helium and Neon. *Earth and Planetary Science Letters, 308,* 193–199.

Coltice, N., & Ricard, Y. (1999). Geochemical observations and one layer mantle convection. *Earth and Planetary Science Letters, 174,* 125–137.

Cruz, J. V., & França, Z. (2006). Hydrogeochemistry of thermal and mineral water springs of the Azores archipelago (Portugal). *Journal of Volcanology and Geothermal Research, 151,* 382–398.

Dasgupta, R., Hirschmann, M. M., & Stalker, K. (2006). Immiscible transition from carbonate-rich to silicate-rich melts in the 3 GPa melting interval of eclogite + CO_2 and genesis of silica-undersaturated ocean island lavas. *Journal of Petrology, 47,* 647–671.

Dixon, E. (2003). Interpretation of helium and neon isotopic heterogeneity in Icelandic basalts. *Earth and Planetary Science Letters, 206,* 83–99.

Dosso, L., Bougault, H., & Joron, J.-L. (1993). Geochemical morphology of the North Mid-Atlantic Ridge, 10°–24°N: Trace element-isotope complementarity. *Earth and Planetary Science Letters, 120,* 443–462.

Dosso, L., Bougault, H., Langmuir, C., Bollinge, R. C., Bonnie, R. O., & Etoubleau, J. (1999). The age and distribution of mantle heterogeneity along the Mid-Atlantic Ridge (31–41°N). *Earth and Planetary Science Letters, 170,* 269–286.

Elliott, T., Blichert-Toft, J., Heumann, A., Koetsier, G., & Forjaz, V.-H. (2007). The origin of enriched mantle beneath São Miguel, Azores. *Geochemica Cosmochimica Acta, 71,* 219–240.

Farnetani, C. G., & Hofmann, A. W. (2009). Dynamics and internal structure of a lower mantle plume conduit. *Earth and Planetary Science Letters, 282*(1–4), 314–322.

Fernandes, R. M. S., Catalão, J., & Trota, A. (this volume). The contribution of Space-Geodetic techniques to the understanding of the present-day geodynamics of the Azores Triple Junction.

Gale, A., Laubier, M., Escrig, S., & Langmuir, C. (2013). Constraints on melting processes and plume–ridge interaction from comprehensive study of the FAMOUS and North Famous segments, Mid-Atlantic Ridge. *Earth and Planetary Science Letters, 365,* 209–220.

Gautheron, C. (2002). Le manteau lithospherique sous continental contraint par la systematique des gaz rares: Apport a la geodynamique chimique des reservoirs terrestres, Thesis University Paris VII, 252p.

Genske, F. S., Turner, S. P., Beier, C., & Schaefer, B. F. (2012). The petrology and geochemistry of Lavas from the Western Azores Islands of Flores and Corvo. *Journal of Petrology, 53*(8), 1673–1708.

Graham, D. (2002). Noble gas isotope geochemistry of mid-ocean ridge and ocean island basalts: Characterization of mantle source reservoirs. In: D. Porcelli, C. J. Ballentine, & R. Wieler (Eds.), *Noble gases in geochemistry and cosmochemistry, reviews in mineralogy and geochemistry* (Vol. 47, pp. 247–370). Mineralogical Society of America and Geochemical Society.

Grand, S. P., van der Hilst, R. D., & Widiyantoro, S. (1997). Global seismic tomography; a snapshot of convection in the Earth. *GSA Today, 7*(4), 1–7.

Greau, C. (2011). Apport des gaz rares au suivi hydrogéochimique d'un stockage de CO_2—Application à un analogue naturel de fuites. Thesis University Paris VII.

Gripp, A. E., & Gordon, R. G. (2002). Young tracks of hotspots and current plate velocities. *Geophysical Journal International, 150,* 321–361.

Hamelin, C., Dosso, L., Hanan, B., Moreira, M., Kositsky, A. P., & Thomas, M. Y. (2011). Geochemical portray of the Pacific Ridge: New isotopic data and statistical techniques. *Earth and Planetary Science Letters, 302,* 154–162.

Hanyu, T., & Kaneoka, I. (1997). The uniform and low 3He/4He ratios of HIMU basalts as evidence for their origin as recycled materials. *Nature, 390,* 273–276.

Hanyu, T., Tatsumi, Y., & Kimura, J.-I. (2011). Constraints on the origin of the Himu reservoir from He-Ne-Ar isotope systematics. *Earth and Planetary Science Letters, 307,* 377–386.

Hart, S., Kurz, M. D., & Wang, Z. (2008). Scale length of mantle heterogeneities: Constraints from helium diffusion. *Earth and Planetary Science Letters, 269,* 508–517.

Heber, V. S., Brooker, R. A., Kelley, S. P., & Wood, B. J. (2007). Crystal-melt partitioning of noble gases (helium, neon, argon, krypton, and xenon) for olivine and clinopyroxene. *Geochimica et Cosmochimica Acta, 71,* 1041–1061.

Herzberg, C., & Gazel, E. (2009). Petrological evidence for secular cooling in mantle plumes. *Nature, 458,* 619–622.

Hildenbrand, A., Madureira, P., Marques, F.O., Cruz, I., Henry, B., & Silva, P. (2008). Multi-stage evolution of a sub-aerial volcanic ridge over the last 1.3 Myr: S. Jorge Island, Azores Triple Junction. Earth and Planetary Science Letters, 273, 289–298.

Hildenbrand, A., Marques, F.O., Costa, A., Sibrant, A.L.R., Silva, P., Henry, B., et al. (2012). Reconstructing the architectural evolution of volcanic islands from combined K/Ar, morphologic, tectonic, and magnetic data: The Faial Island example (Azores). *Journal of Volcanology and Geothermal Research, 241–242,* 39–48.

Hilton, D. R., Grönvold, K., Macpherson, C., & Castillo, P. (1999). Extreme $^3He/^4He$ ratios in northwest Iceland: Constraining the common component in mantle plumes. *Earth and Planetary Science Letters, 173,* 53–60.

Honda, M., & Mc Dougall, I. (1998). Primordial helium and neon in the Earth- a speculation on early degassing. *Geophysical Research Letters, 25,* 1951–1954.

Honda, M., McDougall, I., Patterson, D. B., Doulgeris, A., & Clague, D. (1991). Possible solar noble-gas component in Hawaiian basalts. *Nature, 349,* 149–151.

Honda, M., McDougall, I., Patterson, D. B., Doulgeris, A., & Clague, D. (1993). Noble gases in submarine pillow basalt glasses from Loihi and Kilauea, Hawaii: A solar component in the Earth. *Geochimica et Cosmochimica Acta, 57,* 859–874.

Jambon, A., Weber, H., & Braun, O. (1986). Solubility of He, Ne, Ar, Kr and Xe in a basalt melting in the range 1250–1600 °C. Geochemical Implications. *Geochimica et Cosmochimica Acta, 50,* 401–408.

Jean-Baptiste, P., Allard, P., Coutinho, R., Ferreira, T., Fourre, E., Queiroz, G., et al. (2009). Helium isotopes in hydrothermal volcanic fluids of the Azores archipelago. *Earth and Planetary Science Letters, 281,* 70–80.

Johnson, C.L., Wijbrans, J.R., Constable, C.G., Gee, J., Staudigel, H., Tauxe, L., et al. (1998). 40Ar/39Ar ages and paleomagnetism of Sao Miguel lavas,Azores. *Earth and Planetary Science Letters, 160,* 637–649.

Kárason, H., & Van der Hilst, R. D. (2001). Tomographic imaging of the lowermost mantle with differential times of refracted and diffracted core phases (PKP, P (sub diff)). *Journal Geophysical Research, 106,* 6569–6587.

Kingsley, R. H., & Schilling, J. G. (1996). Carbon in Mid-Atlantic Ridge basalt glasses from 28° to 63°N: Evidence for a carbon-enriched Azores mantle plume. *Earth and Planetary Science Letters, 142,* 175–189.

Krause, D. C., & Watkins, N. D. (1970). North Atlantic crustal genesis in the vicinity of the Azores. *Geophysics Journal Royal Astronomical Society, 19,* 261–283.

Kurz, M. D. (1991). Noble gas isotopes in oceanic basalts: Controversial constraints on mantle models. *Mineral Association Canada Short Course, 19,* 259–286.

Kurz, M. D., Curtice, J., Fornari, D., Geist, D., & Moreira, M. (2009). Primitive neon from the center of the Galapagos hotspot. *Earth and Planetary Science Letters, 286,* 23–34.

Kurz, M. D., Jenkins, W. J., & Hart, S. R. (1982a). Helium isotopic systematics of oceanic islands and mantle heterogeneity. *Nature, 297,* 43–47.

Kurz, M. D., Jenkins, W. J., Schilling, J.-G., & Hart, S. R. (1982b). Helium isotopic variation in the mantle beneath the central North Atlantic Ocean. *Earth and Planetary Science Letters, 58,* 1–14.

Kurz, M. D., Kammer, D. P., Gulessarian, A., & Moore, R. B. (1990). Isotopic variations within oceanic islands: He, Sr and Pb isotopes in basalts from São Miguel, Azores. *EOS, 71,* 657.

Larrea, P., França, Z., Widom, E., & Lago, M. (this volume). Petrology of the Azores islands.

Luis, J. F., & Miranda, J. M. (2008). Reevaluation of magnetic chrons in the North Atlantic between 35°N and 47°N: Implications for the formation of the Azores Triple Junction and associated plateau. *Journal of Geophysical Research, 113.*

Machado, F. (1955). The fracture pattern of Azorean volcanoes. *Bulletin of Volcanology, 17,* 119–125.

Madureira, P., Mata, J., Mattielli, N., Queiroz, G., & Silva, P. (2011). Mantle source heterogeneity, magma generation and magmatic evolution at Terceira Island (Azores archipelago): Constraints from elemental and isotopic (Sr, Nd, Hf, and Pb) data, Lithos.

Madureira, P., Moreira, M., Mata, J., & Allègre, C. J. (2005). Primitive helium and neon isotopes in Terceira island (Azores archipelago). *Earth and Planetary Science Letters, 233,* 429–440.

Madureira, P., Moreira, M., Mata, J., Nunes, J.-C., Gautheron, C., Lourenco, N., et al. (2014). Helium isotope systematics in the vicinity of the Azores triple junction: Constraints on the Azores geodynamics. *Chem. Geology, 372,* 62–71.

Mallik, A., & Dasgupta, R. (2012). Reaction between MORB-eclogite derived melts and fertile peridotite and generation of ocean island basalts. *Earth and Planetary Science Letters, 329–330,* 97–108.

Marty. (1993). He, Ar, Sr, Nd and Pb isotopes in the volcanic rocks from Afar: Evidence for a primitive mantle componant end constraints on magmatic sources. *Geochemical Journal, 27,* 219–228.

Meibom, A., Anderson, D. L., Sleep, N. H., Frei, R., Page Chamberlain, C., Hren, M. T., et al. (2003). Are high $^3He/^4He$ ratios in oceanic basalts an indicator of deep-mantle plume components? *Earth and Planetary Science Letters, 208,* 197–204.

Métrich, N., Zanon, V., Créon, L., Hildenbrand, A., Moreira, M., & Marques, F. O. (2014). Is the 'Azores hotspot'a wetspot? Insights from the geochemistry of fluid and melt inclusions in olivine of Pico basalts. *Journal Petrology, 55,* 37–7393.

Millet, M.-A., Doucelance, R., Baker, J. A., & Schiano, P. (2009). Reconsidering the origins of isotopic variations in Ocean Island Basalts: Insights from fine-scale study of São Jorge Island, Azores archipelago. *Chemical Geology, 265,* 289–302.

Miranda, J. M., Luis, J. F., & Lourenço, N. (this volume). The geophysical architecture of the Azores from magnetic data.

Montelli, R., Nolet, F. A., Dahlen, F., Masters, G., Engdahl, E. R., & Hung, S.-H. (2004). Finite-frequency tomography reveals a variety of plumes in the mantle. *Science, 303,* 338–343.

Moreira, M. (2013). Noble gases constraints on the origin and evolution of Earth's volatiles. *Geochemical Perspectives,* 2(2). European Association of Geochemistry.

Moreira, M., & Allègre, C. J. (1998). Helium—Neon systematics and the structure of the mantle. *Chemical Geology, 147,* 53–59.

Moreira, M., & Allègre, C. J. (2002). Rare gas systematics on Mid Atlantic Ridge (37°–40°). *Earth and Planetary Science Letters, 198,* 401–416.

Moreira, M., Breddam, K., Curtice, J., & Kurz, M. (2001). Solar neon in the Icelandic mantle: Evidence for an undegassed lower mantle. *Earth and Planetary Science Letters, 185*(15–23).

Moreira, M., Dosso, L., & Ondréas, H. (2008). Helium isotopes on the Pacific-Antarctic ridge (52.5°–41.5°S). Geophysical Research Letters, *35,* L10306.

Moreira, M., Doucelance, R., Dupré, B., Kurz, M., & Allègre, C. J. (1999). Helium and lead isotope geochemistry in the Azores archipelago. *Earth and Planetary Science Letters, 169,* 189–205.

Moreira, M., Escartín, J., Gayer, E., Hamelin, C., Bezos, A., Guillon, F., et al. (2011). Rare gas systematics on Lucky Strike basalts (37°N, North Atlantic): Evidence for efficient homogenization in a long-lived magma chamber system? *Geophysical Research Letters, 38,* L08304.

Moreira, M., Kanzari, A., & Madureira, P. (2012). Helium and neon isotopes in São Miguel island basalts, Azores Archipelago: New constraints on the "low 3He" hotspot origin. *Chemical Geology, 322–323,* 91–98.

Mukhopadhyay, S. (2012). Early differentiation and volatile accretion recorded in deep mantle Neon and Xenon. *Nature*. https://doi.org/10.1038/nature11141.

Ozima, M., & Podosek, F. A. (2002). *Noble gas geochemistry* (2nd ed., 286pp). Cambridge: Cambridge University Press.

Parai, R., Mukhopadhyay, S., & Lassiter, J. C. (2009). New constraints on the HIMU mantle from neon and helium isotopic compositions of basalts from the Cook-Austral Islands. *Earth and Planetary Science Letters, 277,* 253–261.

Parman, S. W., Kurz, M. D., Hart, S. R., & Grove, T. L. (2005). Helium solubility in olivine and implications for high $^{3}He/^{4}He$ in ocean island basalts. *Nature, 437,* 1140–1143.

Poreda, R., Schilling, J. G., & Craig, H. (1986). Helium and hydrogen isotopes in ocean-ridge basalts north and south of Iceland. *Earth and Planetary Science Letters, 78,* 1–17.

Putirka, K. (2008). Excess temperatures at ocean islands: Implications for mantle layering and convection. *Geology, 36,* 283–286.

Ramalho, R. S., Helffrich, G., Madeira, J., Cosca, M., Thomas, C., Quartau, R., et al. (2017). Emergence and evolution of Santa Maria Island (Azores)—The conundrum of uplifted islands revisited. *Geological Society of America Bulletin, 129*, 372–391.

Richards, M. A., & Engebretson, D. C. (1992). Large-scale mantle convection and the history of subduction. *Nature, 355,* 437–440.

Ruzié, L., & Moreira, M. (2010). Magma degassing process during Plinian Eruptions. *Journal of volcanology and geothermal researches, 192,* 142–150.

Saki, M., Thomasa, C., Nippress, S. E. J., & Lessinga, S. (2015). Topography of upper mantle seismic discontinuities beneath the North Atlantic: The Azores, Canary and Cape Verde plumes. *Earth and Planetary Science Letters, 409,* 193–202.

Sarda, P., Staudacher, T., & Allègre, C. J. (1988). Neon isotopes in submarine basalts. *Earth and Planetary Science Letters, 91,* 73–88.

Schaefer, B. F., Turner, S., Parkinson, I. J., Rogers, N. W., & Hawkesworth, C. J. (2002). Recycled oceanic mantle lithosphere in the Azores plume. *Nature, 420,* 324–326.

Schilling, J. G. (1975). Azores mantle blob: Rare earth evidence. *Earth and Planetary Science Letters, 25,* 103–115.

Schilling, J.-G., Bergero, M. B., & Evans, R. (1980). Halogens in the mantle beneath the north atlantic. *Philosophical Transactions of the Royal Society of London, 297,* 147–178.

Sibrant, A.L.R., Hildenbrand, A., Marques, A., & Costa, A., (2015). Volcano-tectonic evolution of the Santa Maria Island (Azores): Implications for paleostress evolution at the western Eurasia–Nubia plate boundary. *Journal of Volcanology and Geothermal Research, 291*, 49–62.

Silva, P., Henry, B., Marques, F.O., Hildenbrand, A., Madureira, P., Mériaux, C. et al. (2012). Palaeomagnetic study of a subaerial volcanic ridge (São Jorge Island, Azores) for the past 1.3 Myr: Evidence for the CobbMountain Subchron, volcano flank instability and tectonomagmatic Implications. *Geophysical Journal International.*

Silveira, G., Vinnik, L., Stutzman, E., Farra, V., Kiselev, S., & Morais, L. (2010). Stratification of the Earth beneath the Azores from P and S receiver function. *Earth and Planetary Science Letters, 229,* 91–103.

Stuart, F. M., Lass-Evans, S., Fitton, J. G., & Ellam, R. M. (2003). High $^{3}He/^{4}He$ ratios in picritic basalts from Baffin Island and the role of a mixed reservoir in mantle plumes. *Nature, 424,* 57–59.

Trieloff, M., Kunz, J., & Allègre, C. J. (2002). Noble gas systematics of the Reunion mantle plume source and the origin of primordial noble gases in Earth's mantle. *Earth and Planetary Science Letters, 200,* 297–313.

Trieloff, M., Kunz, J., Clague, D. A., Harrison, D., & Allègre, C. J. (2000). The nature of pristine noble gases in mantle plumes. *Science, 288,* 1036–1038.

Turner, S., Hawkesworth, C., Rogers, N., & King, P. (1997). U-Th isotope disequilibria and ocean island basalt generation in the Azores. *Chemical Geology, 139,* 145–164.

Vogt, P. R., & Jung, W. (this volume). The "Azores Geosyndrome" and plate tectonics: Research history, present understanding, and unsolved puzzles.

Vogt, P., & Yung, W. Y. (2004). The Terceira Rift as a hyper-slow, hotspot-dominated oblique spreading axis: A comparison with other slow-spreading plate boundaries. *Earth and Planetary Science Letters, 218,* 77–90.

White, W. M., & Schilling, J.-G. (1978). The nature and origin of geochemical variation in Mid-Atlantic Ridge basalts from the Central North Atlantic. *Geochimica et Cosmochimica Acta, 42,* 1501–1516.

White, W. M., Schilling, J. G., & Hart, S. R. (1976). Evidence for the Azores mantle plume from strontium isotopes geochemistry of the Central North Atlantic. *Nature, 263,* 659–663.

White, W. M., Tapia, M. D. M., & Schilling, J.-G. (1979). The petrology and geochemistry of the Azores Islands. *Contribution Mineralogy Petrology, 69,* 201–213.

Widom, E., Carlson, R. W., Gill, J. B., & Schmincke, H. U. (1997). Th-Sr-Nd-Pb isotope and trace element evidence for the origin of the São Miguel, Azores, enriched mantle source. *Chemical Geology, 140,* 49–68.

Yang, T., Shen, Y., van der Lee, S., Solomon, S. C., & Shu-Hue, H. (2006). Upper mantle structure beneath the Azores hotspot from finite-frequency seismic tomography. *Earth and Planetary Science Letters, 250*(1–2), 11–26.

Yatsevich, I., & Honda, M. (1997). Production of nucleogenic neon in the Earth from natural radioactive decay. *Journal of Geophysical Research, 102,* 10291–10298.

Yokochi, R., & Marty, B. (2004). A determination of the neon isotopic composition of the deep mantle. *Earth and Planetary Science Letters, 225,* 77–88.

Yokochi, R., & Marty, B. (2006). Geochemical constraints on mantle dynamics in the Hadeen. *Earth and Planetary Science Letters, 238,* 17–30.

Yu, D., Fontignie, D., & Schilling, J.-G. (1997). Mantle plume-ridge interaction in the Central North Atlantic: A Nd isotope study of Mid-Atlantic Ridge basalts from 30° to 50°N. *Earth and Planetary Science Letters, 146,* 256–272.

Zanon, V. (2015). The magmatism of the Azores Islands. In: J. L. Gaspar, J. E. Guest, A. M. Duncan, F. J. A. S. Barriga, & D. K. Chester (Eds.), *Volcanic geology of São Miguel Island (Azores Archipelago)* (Vol. 44, pp. 51–64). Geological Society, London, Memoirs. http://doi.org/10.1144/M44.5.

Zindler, A., & Hart, S. (1986). Helium: Problematic primordial signals. *Earth and Planetary Science Letters, 79,* 1–8.

Surface and Groundwater in Volcanic Islands: Water from Azores Islands

Paulo Antunes and M. Rosário Carvalho

Abstract

Small volcanic islands face unique challenges in the assessment and management of their hydrological systems and water resources, which distinguish them from other regions around the world. This chapter summarizes the latest knowledge of the water resources in the Azores Archipelago. Although precipitation is abundant, the Azores islands have poorly developed watersheds and few perennial streams. Most water is contained in volcanic lakes whose geomorphological characteristics are being controlled by volcano-tectonic activity. Some lakes interact with volcanic fluids. Several lakes are subject to anthropogenic pollution, mainly related to agricultural production and/or open air grazing. Lakes on the Azores are under the influence of various chemical processes: (1) sea-salt spraying effect, (2) water-rock interaction due to hydrolysis, and (3) contamination with volcanic fluids. Groundwater is a vital resource in Azores islands for human consumption, agriculture and industry. The occurrence, movement, and storage of groundwater show that heterogeneity and anisotropy directly depend on volcanic eruptions and secondary processes, such as fracturing and weathering. The total volume of groundwater resources in the Azores equals 1.6×10^9 m^3/year. Groundwater occurs in two main aquifer systems: perched and basal aquifer systems. The spatial distribution of springs is very heterogeneous, with densities varying between 0.01 to 0.72 springs/km^2 and median discharge values varying between 1.73 and 36.29 m^3/day. 160 wells were known, located at distances between 300 and 5825 m from the coastline and had capture depths from 25 to 284 m. The specific flow rates (q) range from 1.4×10^{-2} to 266.7 L/sm, with a median value equal to 22 L/sm, and transmissivity varies between 1.43 and 34809 m^2/day, with a median of 3162 m^2/day. Groundwater chemical composition facies varies from sodium chloride to sodium bicarbonate waters and show low mineralization. Volcanic degasification and hydrothermal steam rising from deeper geothermal systems contribute to the contamination of shallower aquifers resulting in the occurrence of a large number of mineral and

P. Antunes (✉)
Department of Geosciences, University of Massachusetts Amherst, Amherst, MA, USA
e-mail: paullusunes@gmail.com

P. Antunes
Centro de Investigação e Tecnologia Agrária dos Açores (CITA-a), Azores University, Ponta Delgada, Portugal

M. Rosário Carvalho
Departamento de Geologia, Faculdade de Ciências da Universidade de Lisboa and Instituto D. Luiz, 1749-016 Lisbon, Portugal
e-mail: mdrcarvalho@ciencias.ulisboa.pt

U. Kueppers and C. Beier (eds.), *Volcanoes of the Azores*, Active Volcanoes of the World, https://doi.org/10.1007/978-3-642-32226-6_13

thermal waters. With the exception of those sites, associated with active faults and volcanoes, the groundwater quality in some of the Azores islands is mainly deteriorated by salt water intrusion in the coastal aquifers, inland agricultural activities, and the lack of sanitation system.

Keywords
Surface water · Volcanic lakes
Groundwater · Perched and basal aquifers
Water resources

1 Introduction

Freshwater supply in small volcanic islands around the world is driven through the hydrological cycle by heat from the sun. Generally, small islands have limited freshwater resources that occur as precipitation water, surface water and groundwater.

Surface water occurs on small and high elevation islands in the form of ephemeral and perennial streams, springs, and as freshwater swamps, lagoons and lakes. Perennial streams and springs occur mainly on old rock formations on high volcanic islands where different past eruption styles has given rise to less permeable bedrock. In many cases, watersheds in volcanic islands are small due to the islands' recent formation and small dimensions. This makes the assessment of surface water resources difficult and laborious (Peterson 1972; Falkland 1991; LNEC 1998, 1999; Falkland 2002). Freshwater lagoons and lakes are not common but are found on some small islands at different latitudes. These aquatic systems are generally thought of as strategic sources of freshwater (Wetzel 2001; Saunders et al. 2002; Antunes and Rodrigues 2011; ILEC 2011). Lakes can occur in the craters of extinct volcanoes or depressions in the topography (Falkland 1991; Cerezal 2013). Lakes emplaced in active volcanoes are not unusual around the world (Africano and Bernard 2000; Christenson 2000; Delmelle and Bernar 2000; Manville et al. 2007; Antunes 2009; Antunes and Rodrigues 2011; Rouwet et al. 2014; Christenson et al. 2015), and water quality varies significantly due to contamination by natural processes (magmatic fluids interaction; Delmelle and Bernar 2000; Varekamp et al. 2000; Van Hinsberg et al. 2010; Christenson and Tassi 2015; Varekamp 2015) or through anthropogenic nutrient sources.

Groundwater on small islands can be the main source of freshwater, since surface water supplies on small islands generally are unreliable on a yearly basis, and groundwater is less affected by droughts and seasonal changes. Groundwater is a vital source for terrestrial ecosystems and for domestic, agriculture and industrial consumption (Peterson 1972; Macdonald et al. 1983; Bijlsma et al. 1995; White et al. 2007).

On small volcanic islands perched aquifers commonly occur over horizontally less permeable layers than those above and below them. The "impermeable" layer may be a stratum of dense lava, alluvium or volcanic ash, or a paleosol. When erosional valleys cut across perched aquifers, groundwater occurs as springs. The perched water bodies usually are limited; hence storage is limited and the flow of these springs tend to be unstable and could dry in the summer season (Macdonald et al. 1983; Menezes 1993; LNEC 1998; Rodrigues 2002). Dyke-confined aquifers are a less common form of a perched aquifer. They are formed by vertical volcanic dykes that are less permeable than the layers between them and trap water in the intervening compartments. The volume of water stored in dyke-confined aquifers is large compared with other perched aquifers. It is a relative stable source of water supply, but the amount of water is small compared to the total volume of basal aquifers (Peterson 1972; Macdonald et al. 1983).

Basal aquifers consist of unconfined, partially confined or confined freshwater aquifers which form at or below sea level, except where permeability is very low (Falkland 2002). Generally small volcanic islands are composed of extremely permeable rocks, primarily basalts, which are one of most permeable rocks on earth. Rainfall moves downward through the soil and rocks and the freshwater forms a lens-shaped body that floats above the saltwater. The transition zone, a zone of mixing, separates the saltwater from the freshwater lens due to differences in viscosity between fluids and a change in density caused mainly by salinity differences, and change due to the hydrodynamic characteristics of the aquifer (Peterson 1972; Custodio et al. 1988; Falkland 1991, 2002; Join et al. 2005). Anisotropy, climate, and size of the island are among a number of factors that impart unique characteristics to each aquifer (Izuka and Gingerich 2002). Springs from basal aquifers discharge above or below sea level. The groundwater discharge into the coastal waters can be diffuse and difficult to detect. However, when groundwater flows in structures such as lava tubes and cracks and discharges in coastal water, basal springs can be easily detected (Macdonald et al. 1983; Rodrigues 2002).

Rainfall is the most common form of precipitation on Azores islands. Evaporation from the surrounding warm ocean is the main source of surface water and groundwater recharge. Rain water that reaches the ground surface has high infiltration rates into young basaltic formations (Peterson 1972; Macdonald, et al. 1983; Falkland 1991; LNEC 1998; Rodrigues 2002). Infiltration recharges aquifers or feeds streams. Drainage capacity is higher at high elevations where steep slopes prevail and water flows overland (runoff), runs into streams, and can rich the sea. Rain water can also feed endorheic basins, which generally correspond to volcanic craters. The occurrence, movement, and storage of groundwater shows heterogeneity and anisotropy as direct effect of volcanic eruptions and of secondary processes such as fracturing and weathering (Fig. 1).

This chapter shall give an overview of the water resources in the Azores Archipelago. We outline the surface and subsurface aspects of the hydrologic cycle within the volcanic environment of small islands. We address contamination, water supply, and hydrogeological properties of aquifer materials. The Fresh Water

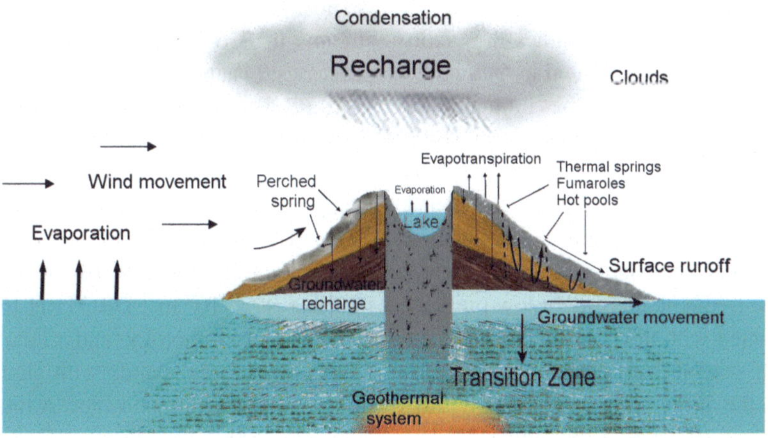

Fig. 1 Ilustration of hydrologic cycle on a volcanic island. Water evaporated from the ocean condenses and falls as rain on land or forms fog that is an important source of water at high altitudes. Rainfall goes to atmosphere by evapotranspiration, returns to the sea, and some infiltrates the ground. Meteoric water in geothermal aquifers can return to the surface through thermal springs, fumaroles, hot springs, and geysers (the latter does not exist on the Azores)

chapter is based on previous studies in the Azores Archipelago.

1.1 Geological Setting

The Azores archipelago is located in the North Atlantic Ocean between the latitudes of 37°–40° N and the longitudes of 25°–31°W. It can be divided in three groups according to the geographical distribution of the nine inhabited islands. The Azores are located at the junction between the North American, Eurasian and African lithospheric plates forming a complex geodynamic setting reflected by the high level of seismicity and volcanic activity. The Mid-Atlantic Ridge (MAR) crosses the archipelago between the islands of Flores (West Group) and Faial (Central Group) (Krause and Watkins 1970; Laughton and Whitmarsh 1974; Steinmetz et al. 1976; Searle 1980; Forjaz 1983; Lourenço et al. 1998). Flores Island lies west of the Mid-Atlantic Ridge (MAR) on the American Plate (Fig. 1 in Vogt and Jung, Chapter "The "Azores Geosyndrome" and Plate Tectonics: Research History, Synthesis, and Unsolved Puzzles").

1.2 Climate

The archipelago's latitude determines the Azorean climate. The Azores are located in the middle North Atlantic Ocean under the influence of subtropical high pressure (Azores anticyclone) resulting in a humid subtropical climate. It is possible to distinguish two seasons: (1) a cold and relative humid season with high precipitation between September and March, and with steady wind due to the frequent crossing of low pressure systems associated with the polar front, and (2) a dry and warm season during the other months controlled by the influence of Azorean anticyclone (Bettencourt 1979; Azevedo 1996a, b). The climate is largely influenced by the Gulf warm current and the ocean that has an important role in air temperatures (Agostinho 1938; Santos and Miranda 2006). The temperature variation with 100 m altitude is approximately 0.6 °C and a 2.4% increase of humidity saturation (Agostinho 1938; Bettencourt 1979). The average annual rainfall in the Azores is 1585 L/m^2 and evapotranspiration is 597 L/m^2. 75% of rainfall occur during the winter season, the amount of precipitation increases from east to west and in altitude. Humidity is related precipitation and higher humidity values occur when the amount of the monthly rainfall is higher than the average temperature of the correspondent month (Bettencourt 1979; Azevedo 1996a, b). The summer season begging when the anticyclone moves to north of the Azores Archipelago and rainfall decreases (Azevedo 1996a, b). Temperature shows a regular variation during the year with high average values during August (23.2 °C) and lower average values during February (9.7 °C; Bettencourt 1979).

2 Surface Water

In the Azores islands, watersheds are poorly developed due to the islands' recent formation and limited dimensions. Watersheds have areas less than 30 km^2 and the largest stream is less than 10 km long (Table 1). Generally, streams have a flashy regime with more discharge in winter. Permanent streams only exist on the Santa Maria, São Miguel, São Jorge, Faial and Flores islands. The chemical compositions of permanent streams in São Miguel Island result from three processes: (1) atmospheric inputs by oceanic type rains, (2) weathering of volcanic rocks, and, (3) in some cases, inflows of solute-rich thermal spring waters (Louvat and Allègre 1998). The Azores islands have an average runoff rate of 322×10^6 m^3 with São Miguel island having the highest runoff (1.731×10^6 m^3) and Graciosa the least (8×10^6 m^3; DROTRH-INAG 2001).

The geomorphology and climate have provided the Azores islands with significant quantities of interior surface water. A total of 88 lakes (Porteiro 2000) are distributed throughout the islands of São Miguel (Eastern Group), Terceira, Pico (Central Group), Flores and Corvo (Western Group; Fig. 2). Small ponds are present on all of the islands. The total volume of water stored in the crater lakes is about 90×10^6 m^3, 93% of

Table 1 Simplified classification of ten major watersheds and streams (DROTRH-INAG 2001; *A*—watershed area, *L*—length, D_d—drain density, T_c—time of concentration, K_c—coefficient of compactness)

Stream	Island	A	L	D_d	T_c	K_c
Povoagao	São Miguel	29.1	8.9	5.0	2.3	1.5
Quente	São Miguel	26.1	14.4	4.6	5.6	1.5
Areia	Terceira	25.7	17.0	1.1	5.7	1.9
P. Santo	Terceira	19.0	10.7	1.8	3.8	1.7
Grande	São Miguel	18.3	13.3	5.3	3.8	1.9
Grande	Flores	16.0	8.0	5.2	2.3	1.9
Flamengos	Faial	16.0	10.9	3.8	2.8	1.7
F. Terra	São Miguel	15.4	8.3	5.5	2.3	1.4
Seca	São Miguel	15.3	8.6	3.5	2.3	1.5

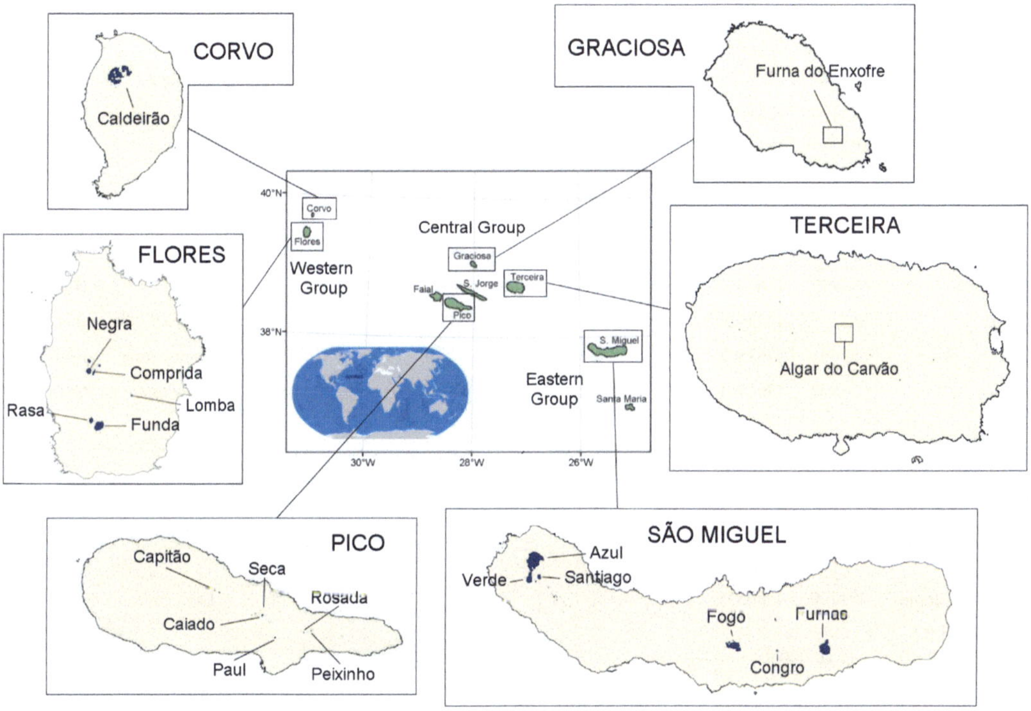

Fig. 2 Location of the Azores lakes contains in Table 2

which is on São Miguel. Flores lakes contribute 5% of the total water and the remaining 2% corresponds to lakes located on the islands of Terceira, Pico and Corvo.

Physical characteristics of the lakes are constrained by the geological setting associated with the volcano-tectonic activity that formed the Azores islands. However, these lakes are predominantly located inside craters. Generally, lakes located inside calderas have the largest areas and water volume. However, the greater depths are related with lakes placed inside maars (Table 2). Lakes represent strategic sources of fresh water and the water quality of these systems varies largely, mainly due to anthropogenic pollution. High nutrient input into some lakes due to fertilizers from agricultural development results in the eutrophication of the aquatic systems (Table 2; Azevedo et al. 2005; Gonçalves et al. 2006a, b; Antunes and Rodrigues 2011).

Table 2 Physical characteristic of volcanic lakes

Island/Lake	Location		Altitude	Area	Lenght	Width	Depth	Volume	Geological classification	Water quality[b]
	M	P	(m)	(km^2)	(m)	(m)	(m)	(10^3 m^3)[a]		
São Miguel										
Azul	607723	4192669	260	3.6	2590	2093	29	47361	Caldera	Moderate
Verde	606686	4189300	260	0.9	1540	777	26	10679	Caldera	Poor
Santiago	607844	4189891	355	0.2	705	445	36		*Maar (S.L)*[c]	
Fogo	634260	4180919	587	1.5	2203	1010	32	23443	Caldera	High
Congro	640255	4179964	418	0.05	276	236	21	143	*Maar (S.L)*[c]	Bad
Furnas	647045	4179743	280	1.9	2045	1485	13	14334	Caldera	Bad
Pico										
Capitao	384991	4260744	790	0.03	300	122	4	58	Tect. Depression[d]	Bad
Caiado	390813	4257201	810	0.05	302	237	6	110	Und. Dep.[d]	Moderate
Peixinho	397544	4254688	867	0.02	214	135	7		Und. Dep.[d]	Bad
Rosada	396490	4254474	922	0.01	187	81	7		Cinder cone[d]	Moderate
Paul	392412	4254082	785	0.4	342	255	1.7		Cinder cone[d]	Moderate
Seca	391340	4257186	776	0.004	99	55				
Flores										
Negra	652681	4367372		0.1	451	389	122		*Maar (S.L)*[d]	
Comprida	652973	4367020	515	0.05	496	155	17	378	*Maar (S.L)*[d]	High
Funda	653325	4363026	371	0.4	873	635	34		*Maar (S.L)*[d]	Bad
Rasa	652795	4363508	527	0.1	423	323	17	754	Cinder cone[d]	High
Lomba	655802	4365461	650	0.02	177	152	16	143	*Maar (S.L)*[d]	Good
Corvo										
Caldeirao			398	0.032	914	560			Caldera	Moderate
Graciosa										
Furna do Enxofre	415874	4320017	92				11		Lava cave	
Terceira										
Algar do Carvao	481299	4286535	500				15		Lava cave	

[a]Data from PRA
[b]Data from Gonçalves, et al. 2006b
[c]Data from Nunes 1999
[d]Data from Morrisseau 1987; Und. Dep.—undifferentiated depression

Lake water temperature is fundamental for the equilibrium of aquatic systems, directly affecting chemical reactions such as the solubility of gases and minerals. Samples from Azorean lakes have water temperatures ranging between 5.2 °C (Winter) and 23.5 °C (Summer). Lakes exposed to periods of fog and intense wind can occasionally have surface water temperatures lower than those at the bottom of the lake (Antunes and Cruz 2005; Antunes 2009). The gradual increase of water temperature, commonly associated with the spring season, contributes to the onset of the

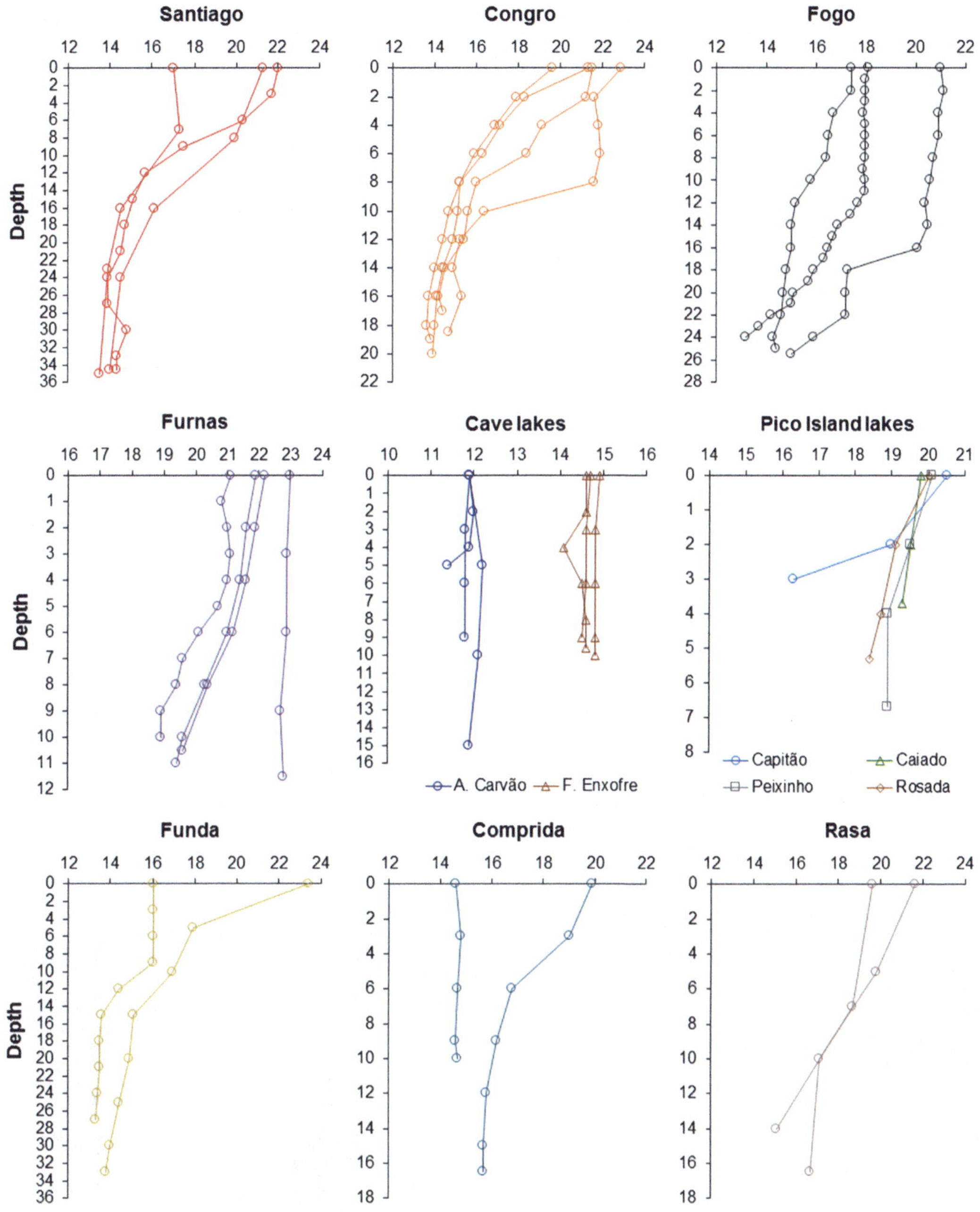

Fig. 3 Summer profiles from different lakes showing the evolution of water column temperature (°C) gradient in depth (m)

water stratification, in the deepest lakes (Fig. 3). Water becomes less dense with increasing temperature. Also, the decrease in wind intensity, a major agent in the induction of convection currents in lakes, contributes to water stratification.

Thermal stratification inhibits water circulation at depth dividing the water column into three distinct layers. The epilimniom, a warm and less dense surficial water layer. Beneath the epilimnion, a stratum is located that is characterised by a strong temperature decrease in depth. This so-called thermocline or metalimnion isolates the epilimnion from the hypolimnion, a layer of cooler water with relatively stable temperature (Fig. 3). In summer, eutrophication reduces water quality (INOVA 1999; Gonçalves 1997, 2008; Gonçalves et al. 2005; Aguiar et al. 2008). High biological activity contributes to increased turbidity of the water, which contributes to greater heat retention in the lake's superficial layer. During late summer, air temperature deceases and winds increase, both of which cause decreases in surface water temperature. Thermal stratification weakens, resulting in a water turnover. In the Azores, lakes shallower than 12 m show small temperature changes with depth throughout the year and do not stratify.

Water from the Azores volcanic lakes showed a Na–Cl, Na–HCO_3 or Mg–HCO_3 composition (Antunes and Cruz 2004, 2009). Considering the calculated total dissolved solids, their water can be classified as freshwater. Chemical composition of lake water is controlled by different processes, but marine contributions to lake water from sea spray is the main process that controls lake water with a Na–Cl water typology. In general, these lakes have small water volume and the residence time is low (Antunes 2009; Antunes and Rodrigues 2014). Lakes located in calderas and maars are influenced by another process affecting chemical composition. Na–HCO_3 water typology, results from sea salt contribution and other processes, such as water-rock interaction and hydrothermal seepage into the lakes. Furna do Enxofre Lake, due to volcanic fluid input, reflects a Mg–HCO_3 water trend (Fig. 4).

Carbon dioxide concentration has a wide range of expression in the volcanic lakes. Volcanic degassing in Furna do Enxofre Lake (Graciosa Island), a small lake located inside a dome shaped lava cave, explains the large CO_2 concentration (474 mg L^{-1}). The fumarolic field inside the cave is formed by a boiling pool and steaming ground and degasification in the lake creates the highest CO_2 concentration (Fig. 5) and the lowest pH (Fig. 6) of all Azores lakes. Lakes with small areas and depths (Pico Island lakes) have the lowest concentrations of CO_2 due to substantial water circulation throughout the year (Fig. 5). Volcanic fluid input supports high CO_2 concentrations in the Congro and Furnas lakes and contributes to the Verde and Santiago Lakes (Antunes 2009).

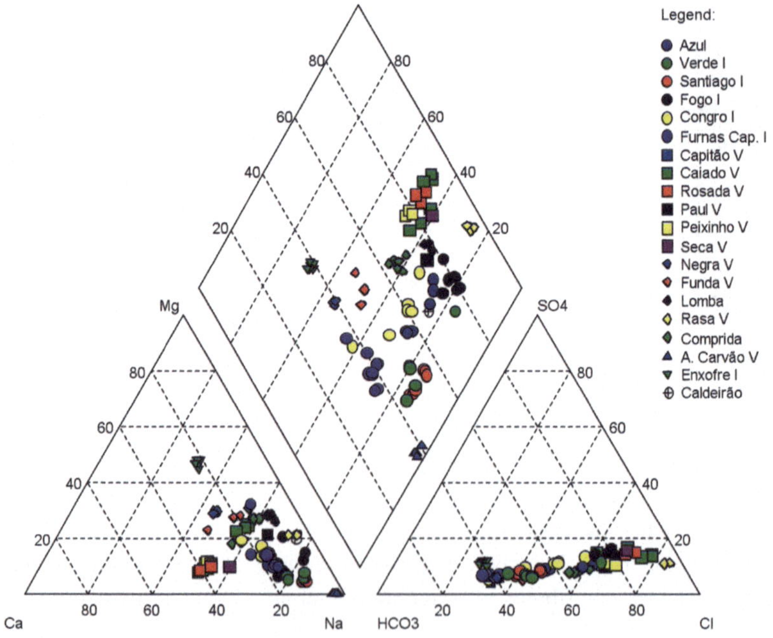

Fig. 4 Piper diagram of water chemistries for Azores volcanic lakes. Azul, Verde, Congro and Furnas lakes show an enrichment in both HCO_3 and Mg with respect to water rock interaction and volcanic fluids discharged

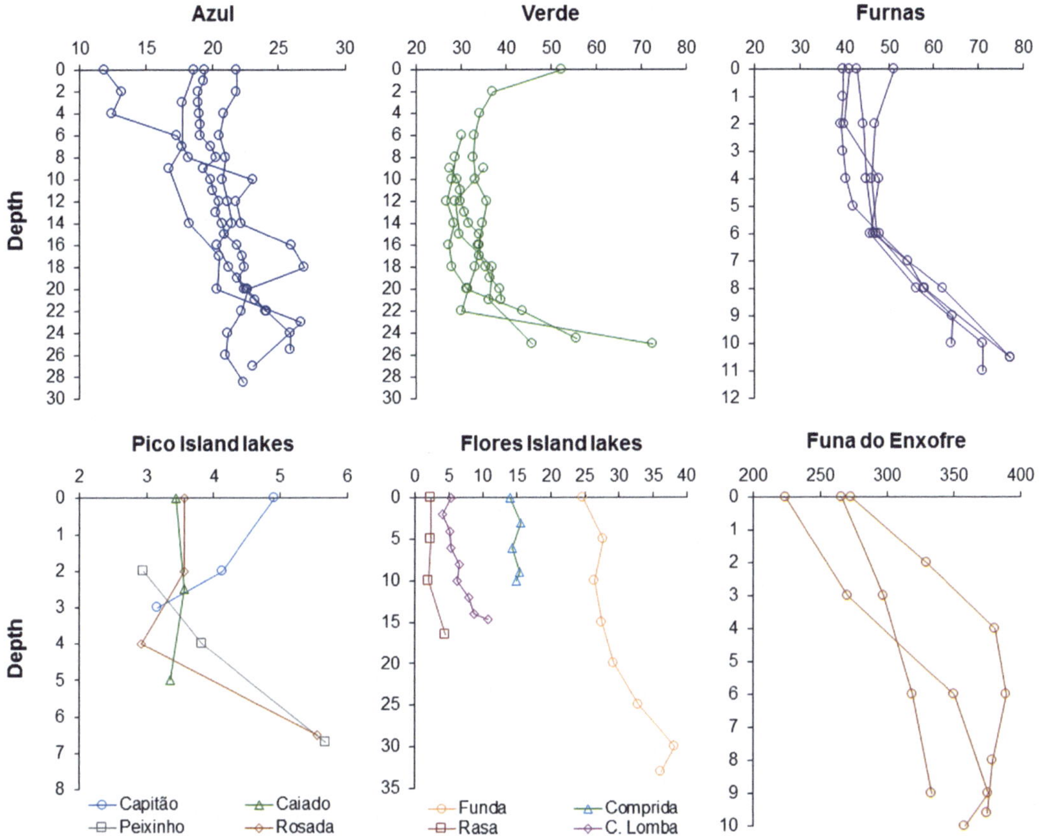

Fig. 5 CO_2 concentrations (mg L^{-1}) observed in summer profiles. Verde lake shows a CO_2 higher concentration compared to Azul lake (Sete cidades lake) in depth (m). Verde and Azul lakes summarised the larger CO_2 concentration compare with Flores lakes, mainly with Funda lake which presents a higher eutrophication state. Furna do Enxofre lake reveal the highest CO_2 concentrations (474 mg L^{-1})

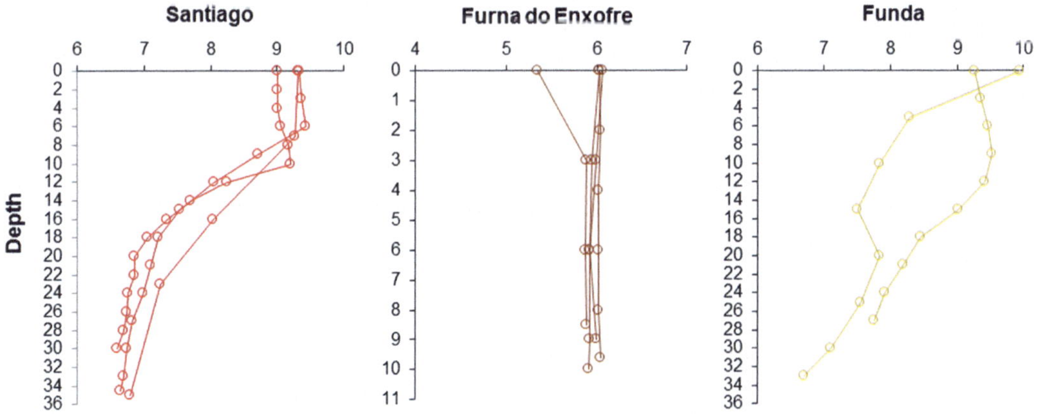

Fig. 6 Summer pH profiles show a good agreement between water column temperature and CO_2 concentration

Deep lakes with water circulation throughout the water column show a CO_2 concentration distribution with depth. When water stratification occurs, CO_2 concentration increases in the hypolimnion of different lakes. The lakes on São Miguel are located in active volcanos and show higher CO_2 concentrations than lakes on Flores (Fig. 5). Eutrophication in highly productive lakes can also cause higher CO_2 concentration, mainly in the hypolimnion. Organic matter decomposition contributes to the increase of CO_2 concentration in the lakes' hypolimnion, as stratification prevents mixing throughout the water column. In summer, anaerobic conditions prevail due to temperature stratification and organic matter decomposition is not as efficient as when it occurs under aerobic conditions. Lagoa Funda (Funda Lake) has a significantly higher concentration of CO_2 in the hypolimnion, compared with other lakes on Flores Island, consistent with the higher productivity of the lake as indicated by the higher concentration of chlorophyll ***a*** and phaeopigments at the epilimnion as shown in Table 3. In general, CO_2 in Azores lakes is controlled by pH. Bicarbonate is the main species which contributes to total CO_2 as demonstrated from the correlation between total-CO_2 and bicarbonate (Fig. 7). Enrichment of SiO_2 at depth is compatible with HCO_3 enrichment through water-rock interaction, which contributes to the overall mineralization at this depth (Fig. 8). Assimilation of large amounts of silica by some groups of phytoplankton for the synthesis of internal structures causes SiO_2 decreases in the photic zone (Wetzel 2001; Fig. 8).

Azores aquatic systems are under the influence of sea-salt spraying effect, the hydrolysis of water-rock interaction, and a restricted group of lakes show a contamination by volcanic fluids. The interaction between magmatic fluids and lakes gives us insight into volcanic processes (Takano and e Watanuki 1990; Pasternack and Varekamp 1994; Christenson 2000; Delmelle and Bernar 2000; Martínez et al. 2000; Ohba et al. 2000; Rice 2000; Varekamp 2002, 2015; Tassi et al. 2009; Christenson and Tassi 2015; Christenson et al. 2015). In the Azores there is evidence and documentation for the occurrence of extremely explosive volcanic eruptions in and near Fogo Lake (1563) and Furnas Lake (1630; Moore 1991; Jones et al. 1999; Wallenstein 1999; Wallenstein et al. 2005).

3 Groundwater

Groundwater is a vital resource in the Azores islands, playing an important role as a public water supply, estimated at about 98% (Cruz and Coutinho 1998), and as an ecosystem support.

3.1 Aquifer Systems and Conceptual Models

In the Azores islands groundwater occurs in two main aquifer systems (LNEC 1998; Cruz 2003,

Table 3 Chlorophyll ***a*** and phaeopigments determined for different depth in three Flores island lakes (Aguiar et al. 2008)

	Depth (m)	Chlorophyll (ug/L)	Phaeopigments
Lagoa Funda	6	29.24	7.23
	11	6.8	2.72
	23	2.27	0.62
Lagoa Comprida	0	2.68	0.64
	7.5	3.09	1.09
	10	2.88	1.3
Lagoa Rasa	0	0.82	0.04
	7	1.03	0.41
	14	1.85	1.17

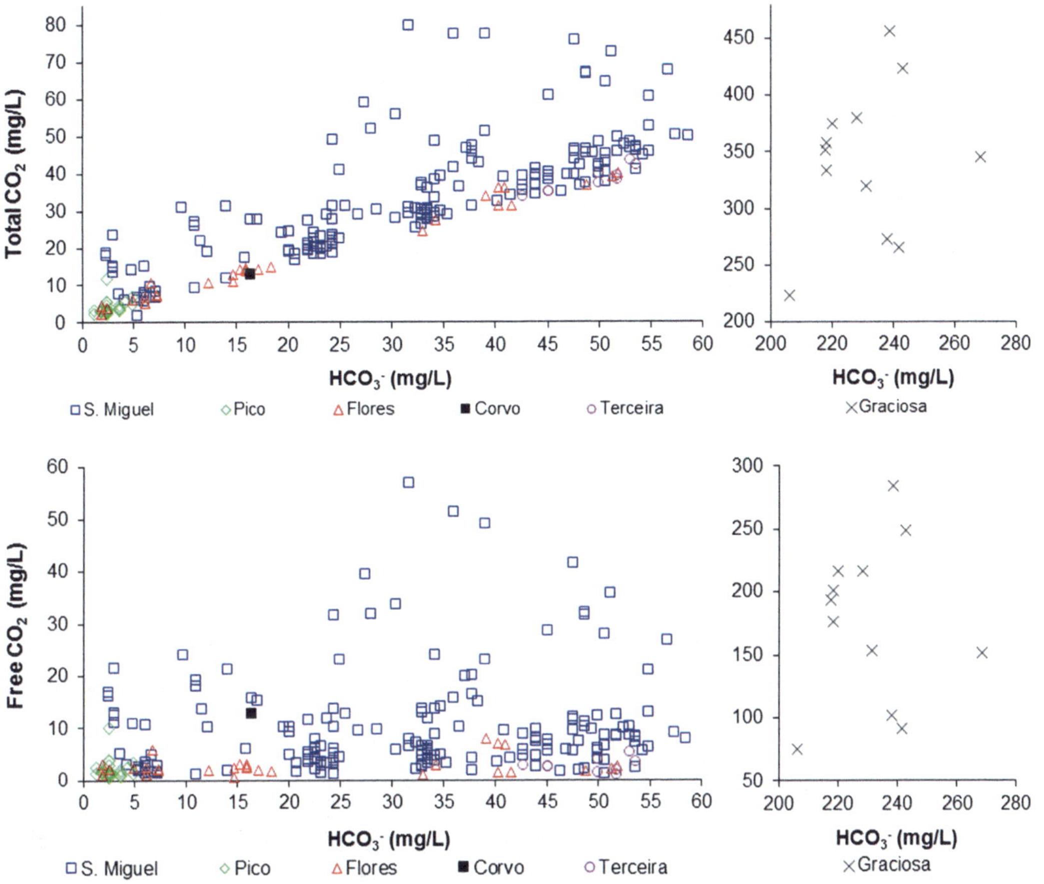

Fig. 7 Total-CO_2 and free-CO_2 vs HCO_3. Samples was select from surface, middle and bottom of water column. Total-CO_2 show a good correlation with HCO_3

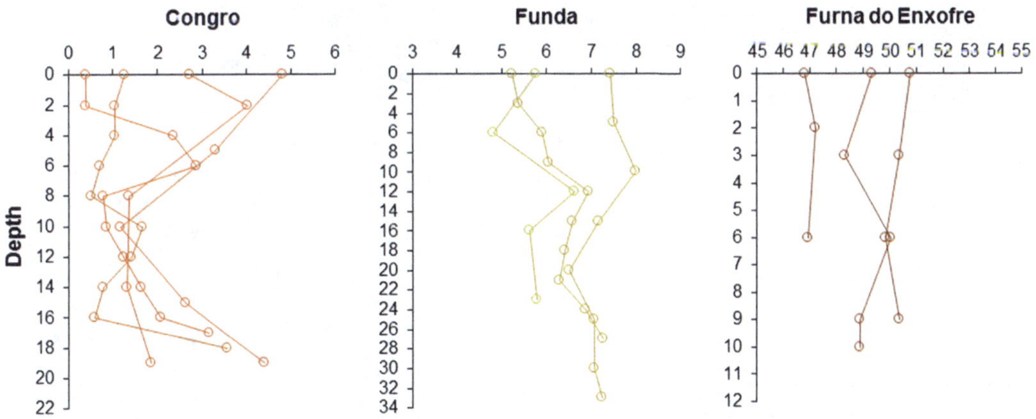

Fig. 8 Water column SiO_2 concentration (mg/L) in summer sampling. Silicon, which is used in the cell structure of photosynthesising organisms, shows a profile variation along the photic zone (epilimnion) due to the assimilation of large silicon. Silicon concentration at Cave lakes, as Furna do Enxofre Lake, show a constant concentration along the water column

2004), (Fig. 1): (1) perched-water bodies, which correspond to unconfined or confined aquifers with leakage, at elevation; (2) basal aquifer systems, corresponding to aquifers developed close to sea level where fresh-water lenses float on underlying salt water.

The perched-water bodies correspond to aquifers developed at altitude, with impermeable to very low-permeability layers at the bottom formed by paleosols, pyroclastic layers or compact lava flows. Some aquifers lose water through the bottom level and are important for the recharge of deeper aquifers. When the topographic conditions are favourable, these aquifers are drained by a large number of springs, and waters generally have a small residence time, enhanced by low mineralized water.

The basal aquifer systems present, generally, a very low hydraulic gradient and at coastal regions the freshwater lenses float on underlying saltwater. According to Cruz et al. (2010a) there are no evidences of a continuous basal water lens under each island, but that can exist in certain aquifers. Recharge is a direct result of effective rainfall or water transfer from the perched aquifers. Basal aquifer systems are exploited for water consumption through wells, which replace old hand dug wells along the coast. Due to overexploitation basal aquifer salinization is common mainly related to the proximity of salt water interface or in wells near the sea (Cruz and Silva 2001).

Based on geological criteria, such as stratigraphy, lithology and tectonic constraints, 54 aquifers systems were defined under the Regional Water Management Plan (PRA) in all the Azores islands (DROTRH-INAG 2001). Figures 9, 10, 11, 12, 13, 14, 15, 16 and 17 show the referred aquifer systems. These systems have been characterised through their recharge areas and resources, limits, dominant lithologies, hydraulic parameters, productivity and water composition.

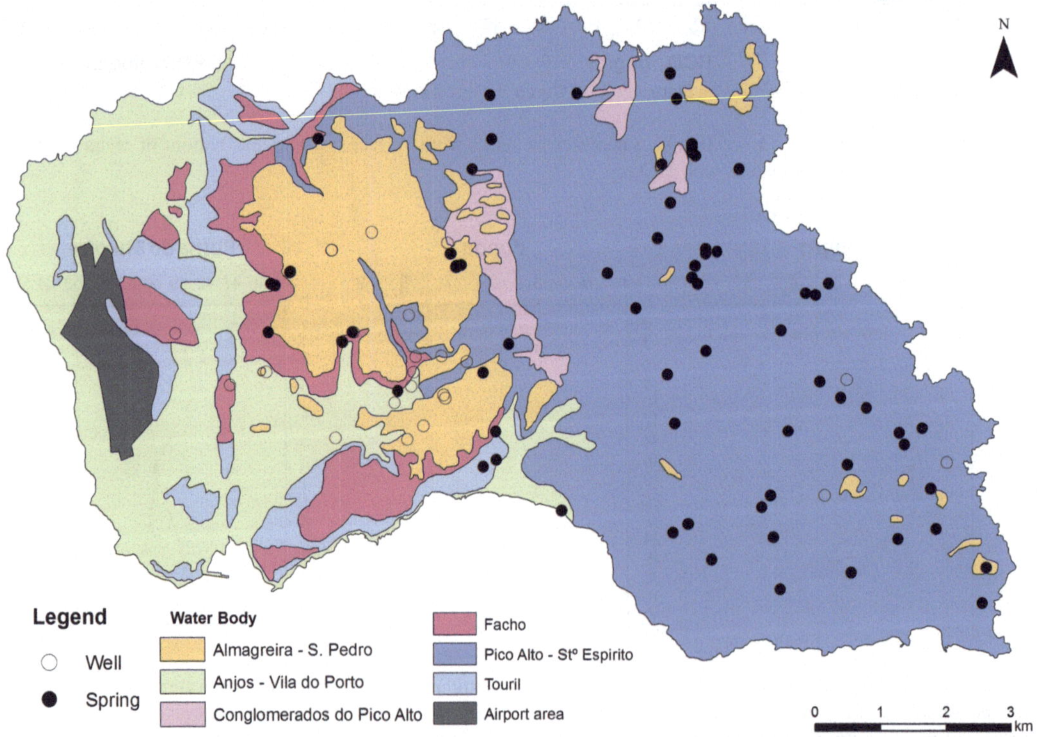

Fig. 9 Groundwater bodies (aquifers systems, DROTRH-INAG 2001) in the Santa Maria Island (adapted from SRAM 2012)

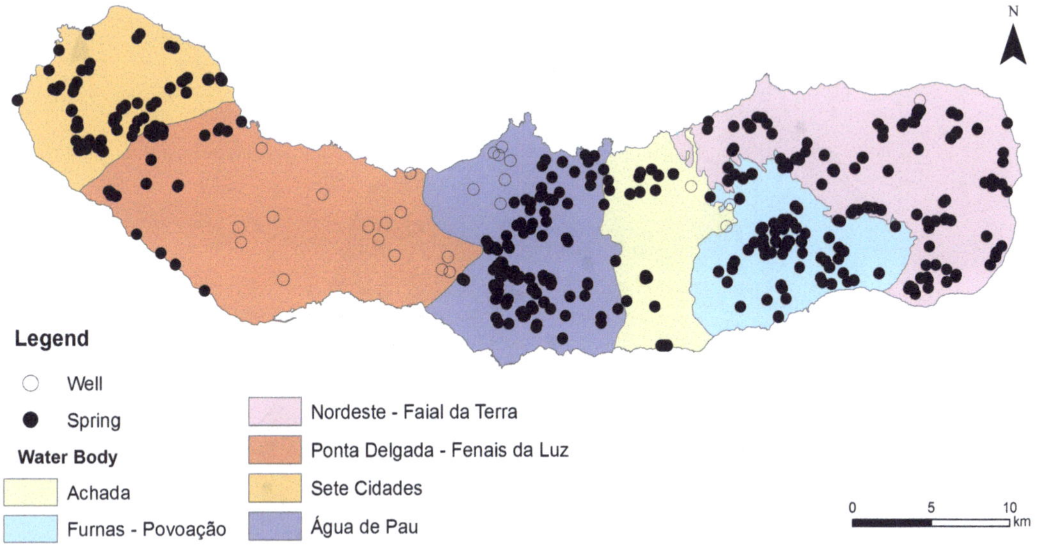

Fig. 10 Groundwater bodies (aquifers systems, DROTRH-INAG 2001) in São Miguel Island (adapted from SRAM 2012)

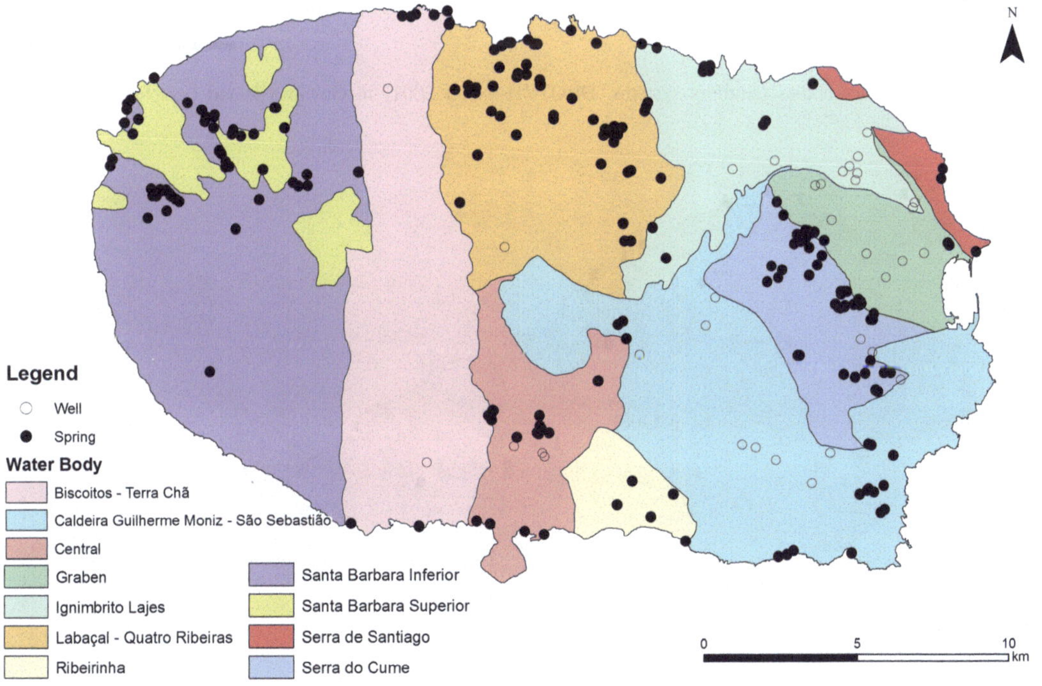

Fig. 11 Groundwater bodies (aquifers systems, DROTRH-INAG 2001) in Terceira Island (adapted from SRAM 2012)

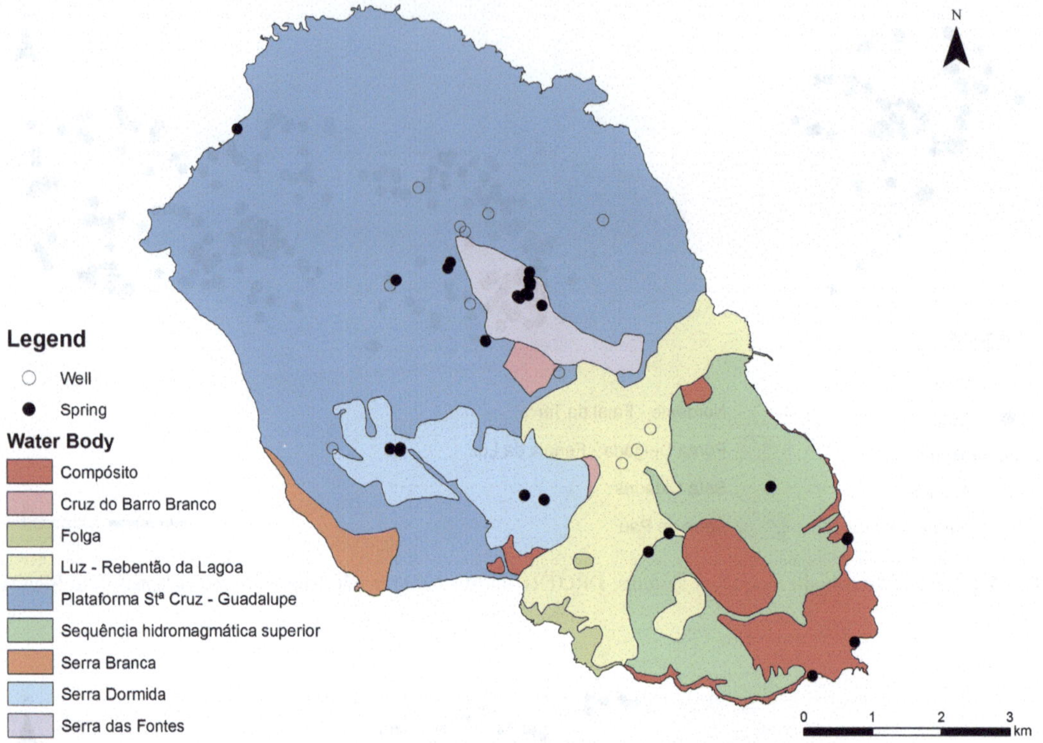

Fig. 12 Groundwater bodies (aquifers systems, DROTRH-INAG 2001) in Graciosa Island (adapted from SRAM 2012)

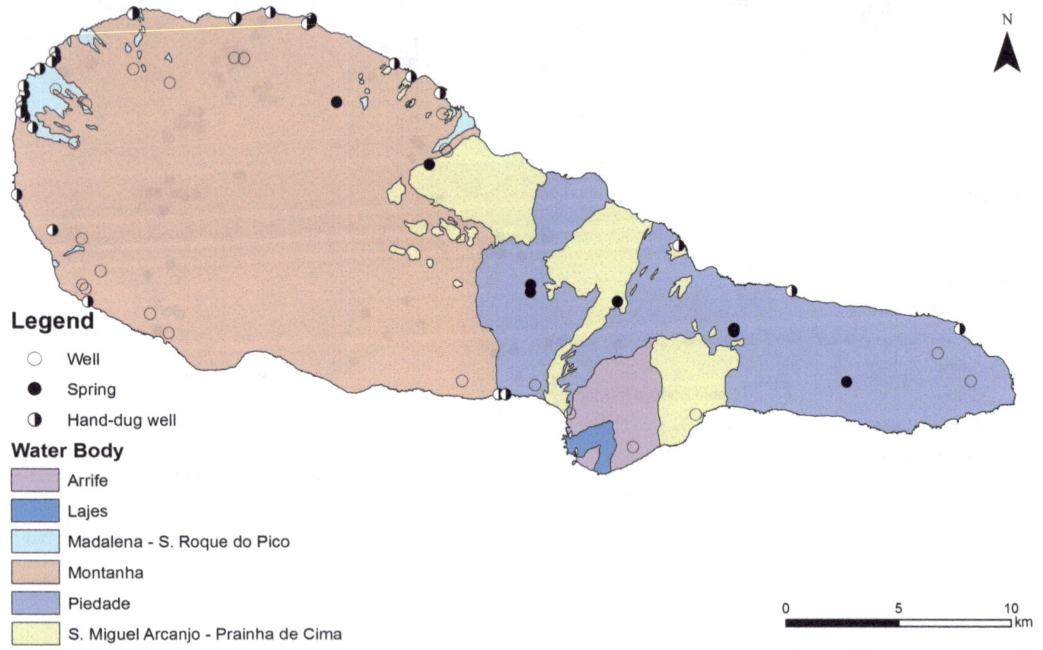

Fig. 13 Groundwater bodies (aquifers systems, DROTRH-INAG 2001) in Pico Island (adapted from SRAM 2012)

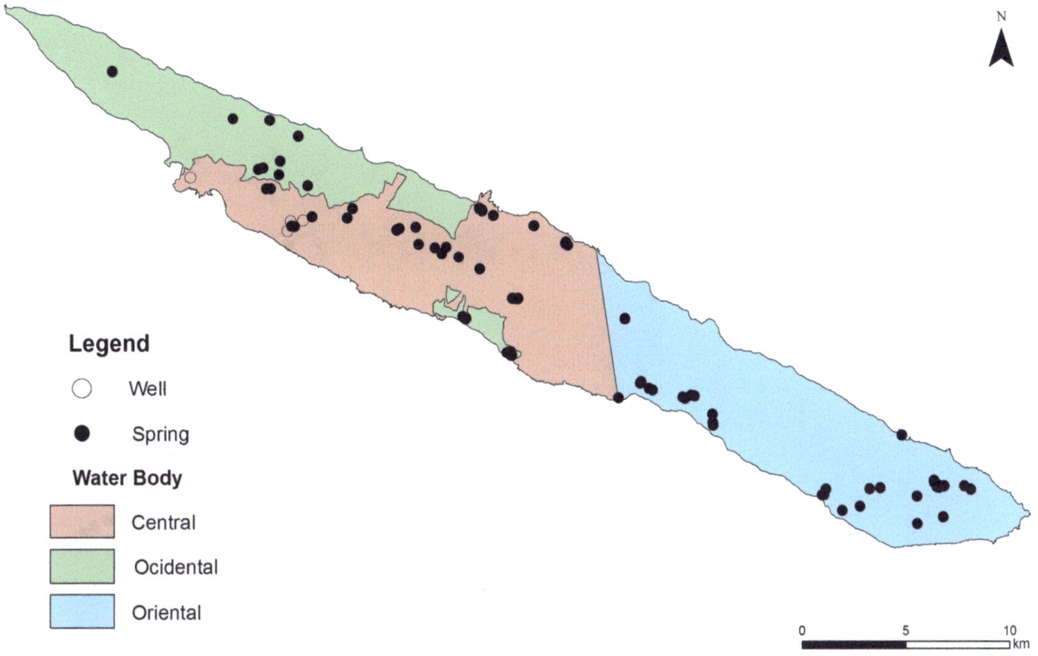

Fig. 14 Groundwater bodies (aquifers systems, DROTRH-INAG 2001) in São Jorge Island (adapted from SRAM 2012)

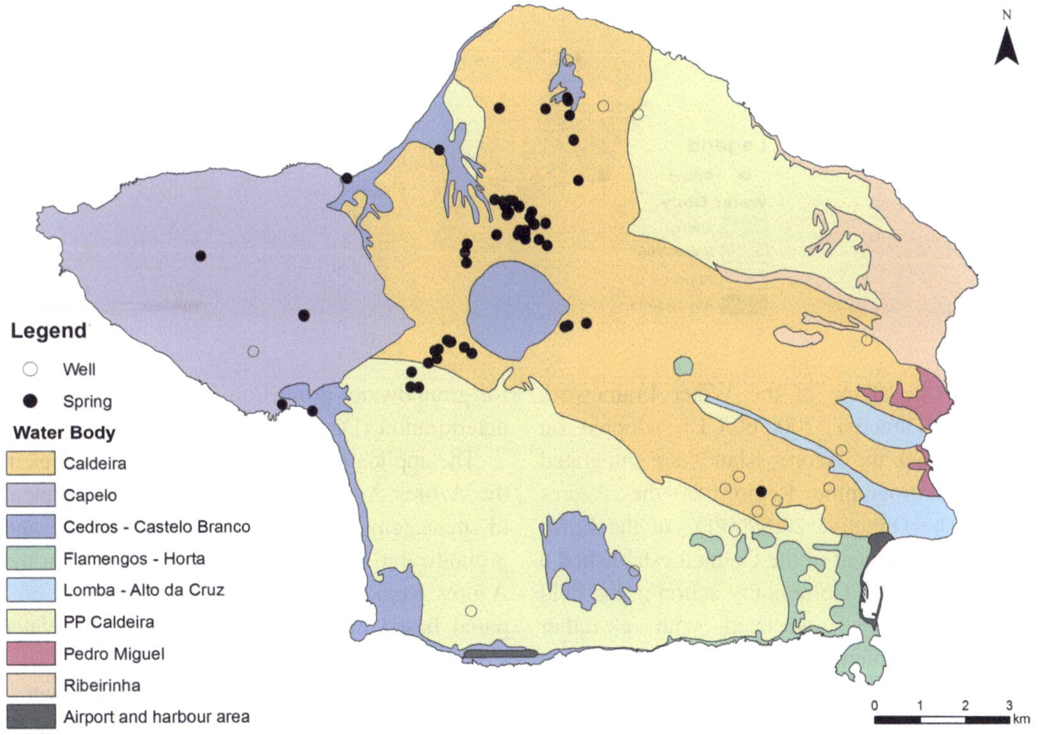

Fig. 15 Groundwater bodies (aquifers systems, DROTRH-INAG 2001) in Faial Island (adapted from SRAM 2012)

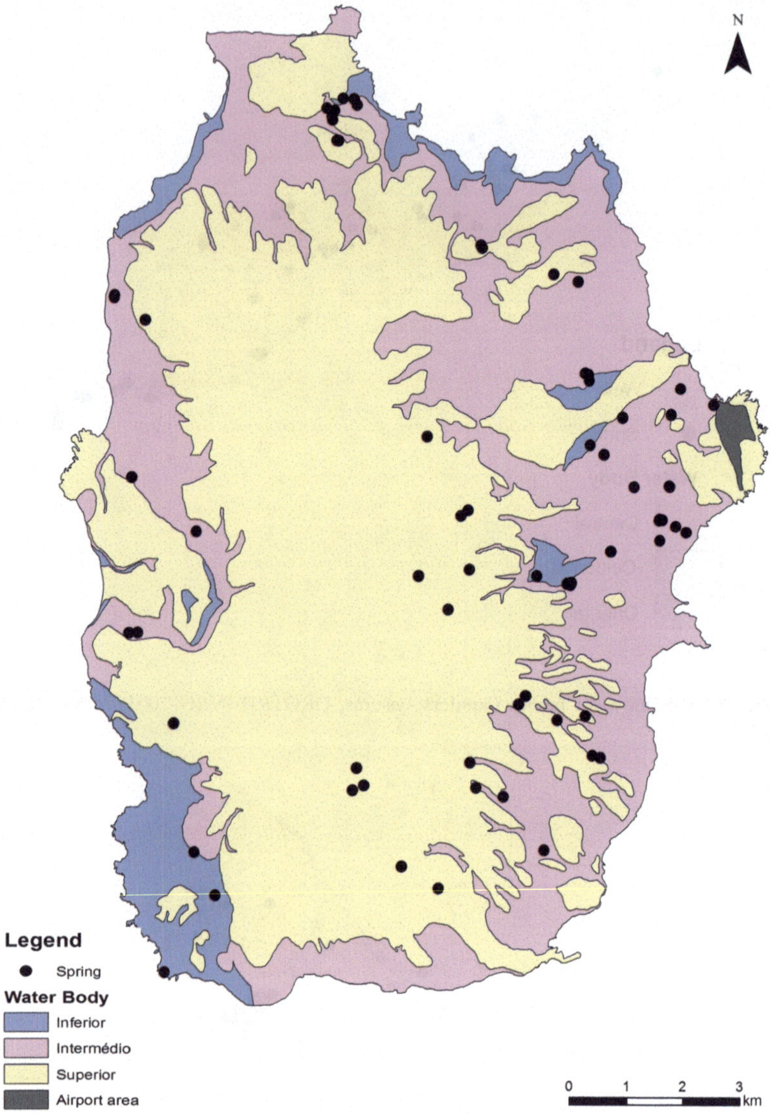

Fig. 16 Groundwater bodies (aquifers systems, DROTRH-INAG 2001) in Flores Island (adapted from SRAM 2012)

Under the scope of the Water Framework Directive (Directive 2000/60/CE), adopted on October 2000, the Azores islands are integrated in the Hydrographic Region of the Azores (RH9). The Directive 2000/60/EC of the European Parliament and of the Council established a framework for the Community action in the field of water policy, associated with a rather demanding implementation strategy regarding quantitative and chemical aspects. The water Directive was complemented in December 12, 2000 by the adoption of technical specifications for groundwater protection against pollution and deterioration (Directive 2006/118/EC).

The application of the EU Water Directives in the Azores Archipelago led to the development of management plans for surface water and groundwater (SRAM 2012; SRAA 2015). In the Azores Region the management plan was prepared based on studies of the Regional Water Resources Plan (PRA, DROTRH-INAG 2001). In the characterization report of the Hydrographic Region of the Azores (RH9), (SRAM 2012; SRAA 2015) groundwater bodies took the

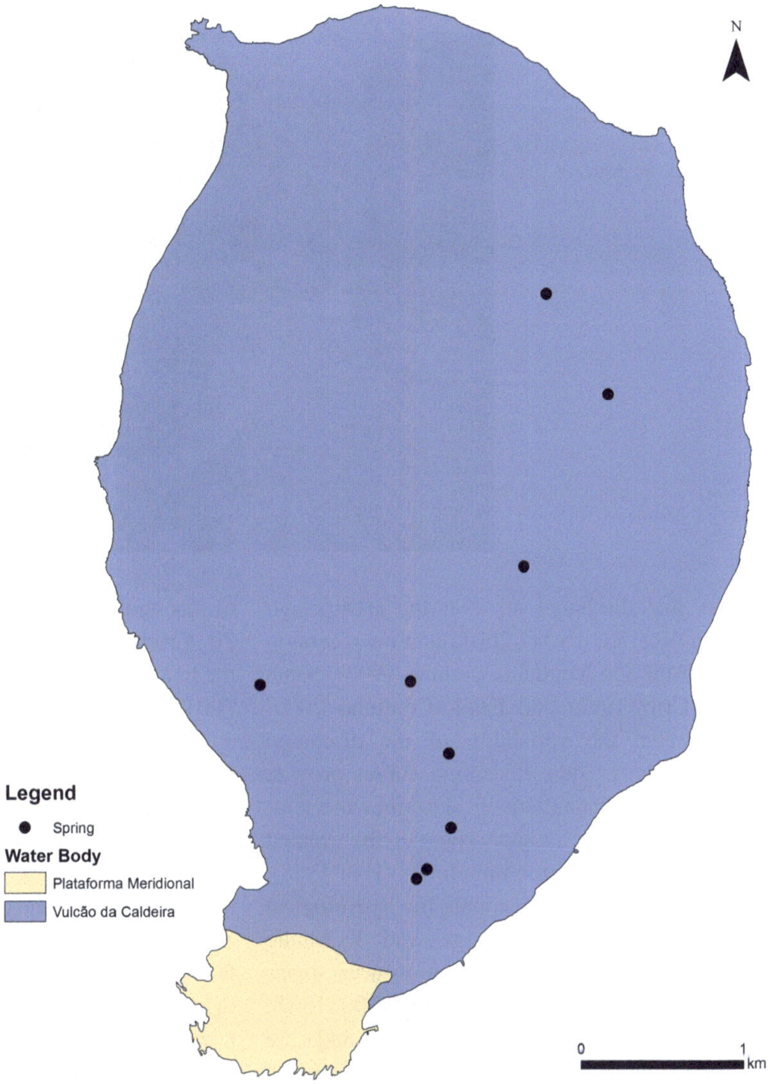

Fig. 17 Groundwater bodies (aquifers systems, DROTRH-INAG 2001) in Corvo Island (adapted from SRAM 2012)

same geographical boundaries as those aquifers systems defined by the PRA.

3.2 Aquifers Hydraulic Characterization

The anisotropy and heterogeneities of the volcanic formations and secondary factors such as faulting and weathering determine groundwater movement and a large variation in hydraulic parameters of the aquifers. Under the scope of the regional water Plans (PRA, DROTRH-INAG 2001; SRAM 2012; and SRAA 2015), 1,692 springs and 160 wells were identified in the Azores islands (Figs. 9, 10, 11, 12, 13, 14, 15, 16, 17 and 18).

The areal distribution of springs is very heterogeneous, with densities varying between 0.01 springs/km^2 at Pico and 0.72 springs/km^2 observed at Santa Maria island (DROTH-INAG 2001). This pattern highlights the importance of the volcanism type and age of formations controlling groundwater movement. Pico Island consists mainly of recent basaltic lava flows in contrast to Santa Maria aquifers that are associated to the oldest volcanic formation of the archipelago. Investigations carried out on the

Fig. 18 Groundwater is a vital resource in the Azores islands, playing important roles as public water supply; on the left, some springs captured in a thalweg; on the right, a well for water exploitation

springs that discharge all over the archipelago (PRA, DROTRH-INAG 2001) and on recession curves from São Miguel (Coutinho 1990), Santa Maria (Cruz 1992) and Faial (Coutinho 2000) demonstrated the variability of the discharge rates between islands. Recession curves provide a function that quantitatively describes the temporal discharge decay and expresses the drained volume between specific time limits (Hall 1968). This analysis allows to estimate the hydrological significance of the discharge and hydraulic aquifer parameters. The recession curve exponential coefficients (α) vary between 0.00116 (S. Miguel) and 0.065 (Santa Maria) and are approximately constant for each spring (DROTRH-INAG 2001), reflecting the hydraulic conductivity of different media through which groundwater flows to the spring. The main factors that affect α are the aquifer lithology and the geometry of the water conduits. The highest coefficients correspond to fast flow, whereas the lower coefficients reflect the slow flow through the porous medium or baseflow; the α values are more dispersed in Santa Maria. The natural groundwater discharge (Q) in the Azores islands has median values (summer data) from 1.73 m^3/day (Santa Maria) to 36.29 m^3/day (S. Jorge). Spring discharges are scattered in Santa Maria, S. Miguel and Terceira. Basic statistics of the spring discharges and exponential coefficients of recession curves are presented in Figs. 19 and 20, through box diagrams (minimum, maximum, median, 1° and 3° quartile, DROTRH-INAG 2001).

The well characteristics and hydraulic parameters of the considered aquifer systems, estimated by pumping tests, are expressed in some articles and reports (Cruz 2004; DROTRH-INAG 2001; SRAM 2012). The wells known in the islands are located between 300 and 5825 m from the coast and capture depths from 25 to 284 m. The specific flow rates (q) range from 1.4×10^{-2} to 266.67 L/sm, with a median value equal to 22 L/sm (Fig. 21). The variability of q is greater in Graciosa and Terceira islands; the highest values were registered in Graciosa (median value of 185 L/sm) and Pico (median of 45.8 L/sm), and are associated with aquifers developed in recent or very fractured basaltic lava flows (sensu lato), presenting frequently interbedded slag layers.

The transmissivity of the Azores aquifers varies between 0.23 and 34,819 m^2/day, with a median of 1,340 m^2/day (SRAA 2015). According to Krásný classification (1993, in DROTRH-INAG 2001) for the aquifer's transmissivity classification, 63.5% of Azores aquifers have very high transmissivity (values higher than 1,000 m^2/day); 20.6% of the aquifers have high transmissivity (100–1,000 m^2/day); 14.3% have

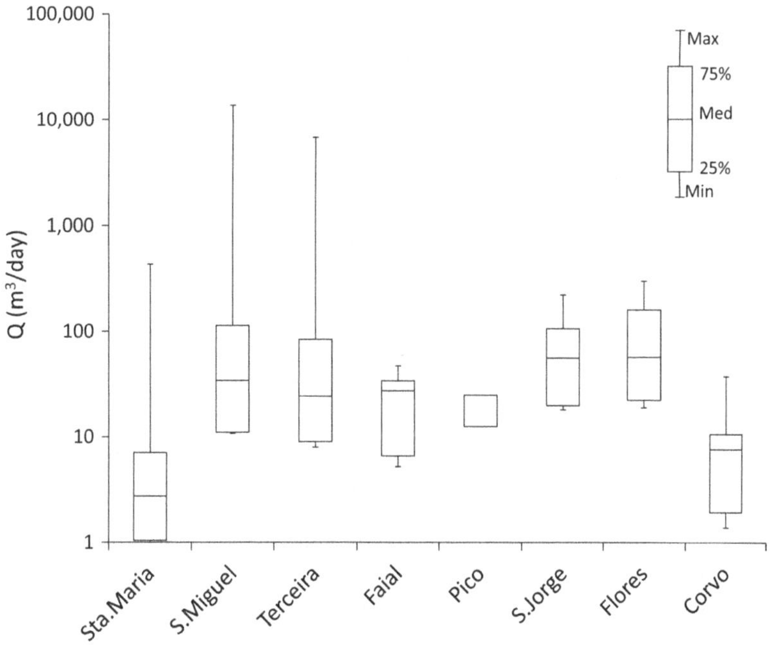

Fig. 19 Box diagram showing the spread of the spring's discharge (Q) in the Azores islands (in DROTRH-INAG 2001)

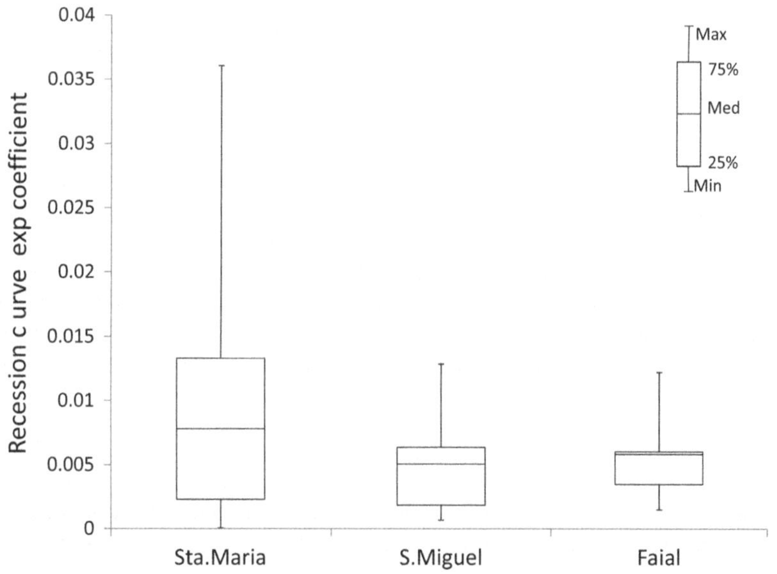

Fig. 20 Box diagram showing the spread of the spring's recession curve coefficient (α) in the Azores islands (in DROTRH-INAG 2001)

intermediate transmissivity (10–100 m^2/day); 1.6% has low values (1–10 m^2/day). The highest median values of transmissivity were observed in Graciosa (14,086 m^2/day), S. Jorge (6,957 m^2/day, with only 3 values) and Pico (9,592 m^2/day) islands. The referred islands and Faial are the islands with major aquifer heterogeneities (Fig. 22). In contrast, Santa Maria island presents the lowest transmissivity values (median of 59 m^2/day); 50% of the estimated values in Santa Maria and 45.4% in São Miguel island can be classified in the intermediate transmissivity category (DROTRH-INAG 2001).

Graciosa, São Jorge, Pico and Faial islands have the aquifers with the highest hydraulic

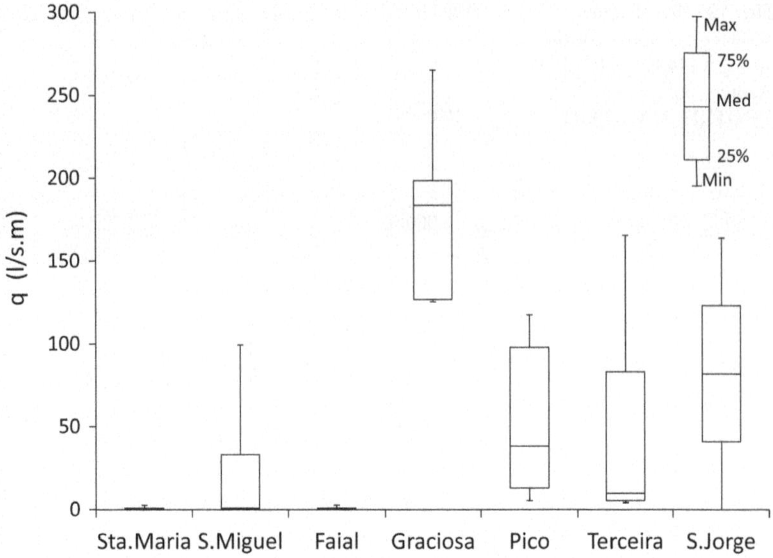

Fig. 21 Box diagram showing the spread of the specific flow rates (q) of wells in the Azores islands (in DROTRH-INAG 2001)

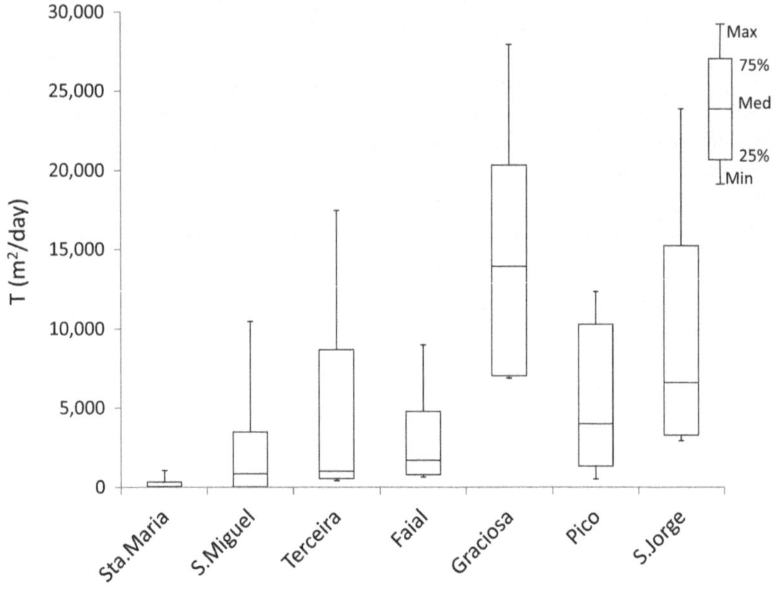

Fig. 22 Box diagram showing the spread of aquifers transmissivity (T) of the aquifers in the Azores islands (DROTRH-INAG 2001)

conductivity. This parameter was obtained from analyses of well tests (Cruz 1992, 1997; Coutinho 2000; DROTRH-INAG 2001) and it has similar behaviour as the transmissivity (Fig. 23), pointing out the slight influence of the aquifers thickness. The hydraulic conductivity can be considered generally high ranging from a minimum of 0.016 m/day up to 21,081.6 m/day (DRAOTH-INAG 2001).

The study of sinusoidal variations of the piezometric levels, due to the influence of the ocean tides on Pico (Cruz 1997) and Faial (Coutinho 2000) islands, also enabled the hydrodynamic characterization of basal aquifers through the estimation of hydraulic diffusivity. Cruz (1997) calculated the hydraulic diffusivity for Pico Island considering two types of aquifer models, confined aquifers and confined aquifers with leakage. The results obtained by this author show great variability and an average value of 114.58 m^2/s (median of 52.11 m^2/s) for the confined aquifer, and an average of 38.08 m^2/s

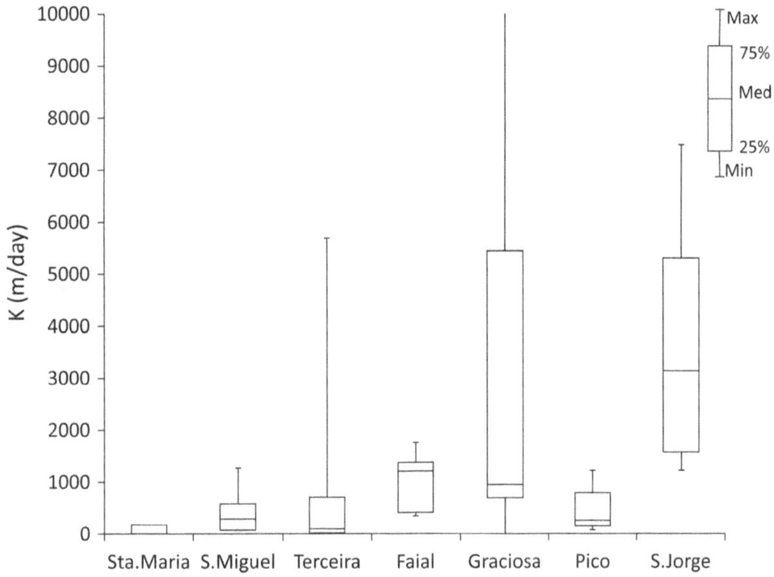

Fig. 23 Box diagram showing the spread of the hydraulic conductivity (K) of the aquifers in the Azores islands (DROTRH-INAG 2001)

(median of 6.09 m^2/s) for the confined aquifer with leakage model. In Faial Island Coutinho (2000) obtained 31.17 and 61.85 m^2/s, in both cases considering the confined aquifers solution.

There are very few estimations for the storage coefficient parameter in the Azores aquifers. Cruz (1997) obtained values from 1.03×10^{-1} to 7.54×10^{-5} for Pico Island basal aquifers.

3.3 Groundwater Chemistry

Groundwater chemical composition on the Azores islands, as on other active volcanic islands, is strongly influenced by rock-water interaction and the environment acidity degree (CO_2(g) input). Other common factors that contribute to the groundwater composition variability include: rain composition, volcanic rock type, rock composition, rock alteration, residence time and moreover the temperature, pressure and thermodynamic and kinetic equilibrium of the dissolution/dissociation chemical reactions. Anthropogenic influences also contribute to the natural groundwater composition, namely with the input of contaminants.

The predominant hydrogeochemical facies in the Azores waters corresponds to sodium chloride or sodium bicarbonate waters. The chloride type is present in low mineralized waters flowing from perched aquifers at high altitude or in mineralized water influenced by sea water intrusion (Cruz et al. 1999; Cruz 2004; DROTRH-INAG 2001); the bicarbonate type is associated with $CO_{2(g)}$ gas dissolution, from the atmosphere, soil and volcanic degassing (Cruz et al. 1999; Carvalho 1999; DROTRH-INAG 2001; Cruz 2004; Cruz and França 2006). According to those authors, groundwater of perched-water bodies is usually cold, slightly acidic to slightly alkaline, has low mineralization, with electric conductivity up to 725 μS/cm (median value of 158 μS/cm), (DRAOTH-INAG 2001; Cruz 2004; SRAM 2012; SRAA 2015); the main water types are Na–Cl to Na–HCO_3 (Fig. 24), depending on the proximity to the sea and the total inorganic carbon content. Despite the predominance of the Na ion, the lithology of the aquifers can be determined based on water chemistry. Ca and Mg ions are enriched when the groundwater flows through basaltic rocks while the sodium waters are very well correlated with trachytic aquifers.

The major-ion composition projected by Piper diagrams shows spring waters from perched aquifers in São Miguel, São Jorge, Pico and Faial islands corresponding mainly to Na-HCO_3 type waters (Fig. 24). In Santa Maria, Flores and

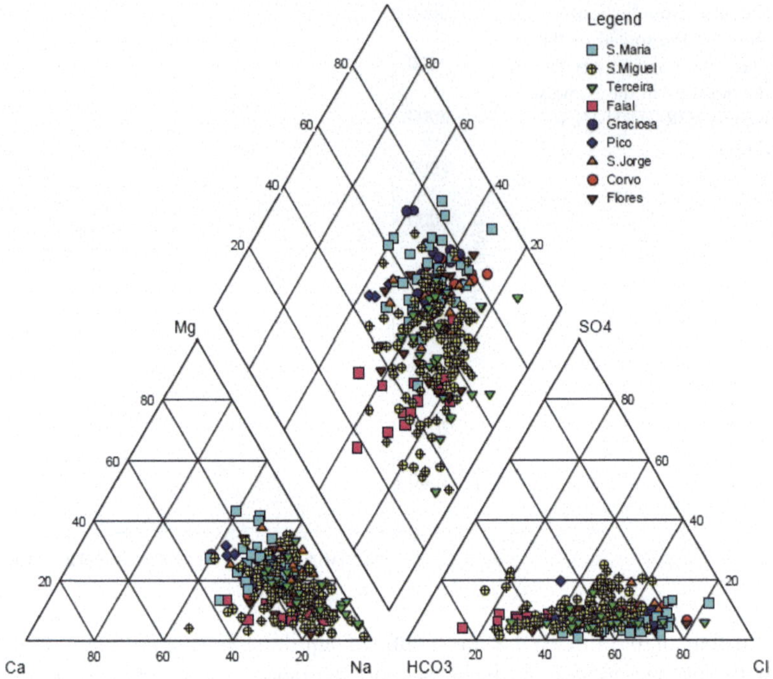

Fig. 24 Piper diagram representing the major chemical composition of spring water from the perched aquifers in the Azores islands (in Cruz 2004)

Corvo islands spring waters are mainly from the Cl–Na type, while at Graciosa Island the intermediate Cl–HCO_3 and HCO_3–Cl facies are dominant (DROTRH-INAG 2001; Cruz 2004).

Groundwater chemistry of the basal aquifer system is predominantly of sodium-chloride type (Fig. 25) and has a variable degree of mineralization, mainly due to seawater intrusion or sea salts from rain and soils. The electric conductivity varies between 123 and 9,670 μS/cm (with median value of 1,044 μS/cm); higher value are generally observed in waters from Santa Maria, São Miguel, Pico, Graciosa and S. Jorge islands (DROTRH-INAG 2001; Cruz et al. 2014; SRAM 2012; SRAA 2015); the waters are usually cold and slightly acidic to alkaline. The Piper diagram related to major-ions composition of water from the basal aquifers (in Cruz 2004) shows that water mineralization is mainly controlled by Cl and Na content, which accounts for 10.8 to 45.1% and 15.7 to 53.2%, respectively, of the relative ions content in the groundwater (Cruz et al. 2010a). The magnitude of the seawater intrusion in the basal aquifers reaches a maximum of 22.5%, but frequent values are lower than 3.3% (Cruz et al. 2010a).

Some of the perched and basal aquifers are affected by extensive water-rock interaction promoted by the input of volcanic gases and/or hydrothermal steam arising from geothermal systems at depth (Fig. 1; Cruz et al. 1999; Carvalho 1999; Cruz and França 2006; Freire 2006; Carvalho et al. 2011, 2014; Freire et al. 2014; Mateus et al. 2014). The contamination of the aquifers by the steam and gases results in the occurrence of a large number of mineral and thermal springs with temperatures between 27 and 97 °C, particularly in São Miguel Island, and to a lesser extent, in Terceira, Faial, Pico, São Jorge, Graciosa and Flores islands (Cruz et al. 1999; Carvalho 1999; Cruz and França 2006; Freire 2006; Carvalho et al. 2011, 2014; Freire et al. 2014; Mateus et al. 2014).

The mineral waters are cold when the volcanic/hydrothermal input is only gas (mainly $CO_{2(g)}$), resulting in acidic (pH range of 4.7–6.3) $CO_{2(g)}$-rich waters, that can be bubbling. The composition is usually of sodium bicarbonate type and the mineralization degree is very variable depending on the water acidity and the amount of rock leaching. The aquifers contaminated by volcanic/hydrothermal steam (rich in

Fig. 25 Piper diagram representing the major chemical composition of water from the base aquifers in the Azores islands (in Cruz 2004)

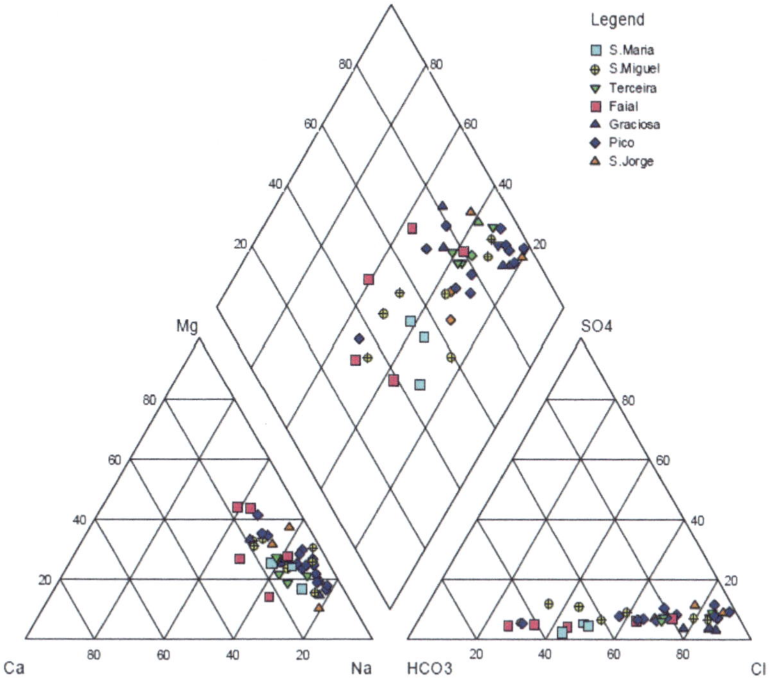

Fig. 26 Caldeira Velha spring, a steam heated water from a perched aquifer

$CO_{2(g)}$ and minor amounts of $H_2S_{(g)}$), with steam heated (spring temperatures between 27 and 75 °C, Cruz and França 2006) and aggressive waters (acidic waters with pH between 4.7 and 7.2, Cruz and França 2006), promote variable degree of rock leaching. These waters usually have very high mineralization, Na–HCO_3 to Na–SO_4 type composition (due to magmatic $CO_{2(g)}$ and $H_2S_{(g)}$ dissolution) and are very enriched in metals, such as iron and aluminium (Fig. 26; Cruz et al. 1999, 2010a; Carvalho 1999; Cruz and França 2006) as consequence of acidic rock leaching.

3.4 Groundwater Quality

Beside groundwater being contaminated by natural volcanic/hydrothermal activity, groundwater quality in the Azores islands is sometimes negatively impacted by salt-water intrusion into coastal aquifers, agricultural activities or the lack of sewage plants.

According to Cruz et al. (2010b) and SRAM (2012), 20% and 22% of the chemical analyses from basal aquifer waters from Santa Maria and Faial islands, respectively, have chloride contents above 250 mg/L. This is the threshold value defined by Portuguese water quality standards for groundwater used as a water source, similar to values proposed by EU Directive 98/83/CE and World Health Organization. Water from wells in the basal aquifer of the Pico Island exceeds the referred maximum value in 42% of the chemical analyses.

Nitrogen contents in groundwater exceeding the 50 mg/L of NO_3 (imposed by the Portuguese water quality standards), were recorded on Santa Maria (1.4% of the chemical analysis) and São Miguel islands (4.7% of the chemical analysis) (Cruz et al. 2010b). The high nitrogen content is derived from agricultural practices and from animal waste leaching.

3.5 Groundwater Resources

The total volume of groundwater resources in the Azores is equal to 1,588 × 10^6 m^3/year. More specifically, Corvo has 8.3 × 10^6 m^3/year, Pico 582 × 10^6 m^3/year (Fig. 27; Table 4, DROTRH-INAG 2001; Cruz 2004).

The groundwater resources were estimated from the aquifers recharge rates calculated for the entire archipelago through the mathematical model Cielo (Azevedo 1996a, b in DROTRH-INAG 2001; SRAM 2012; SRAA 2015).

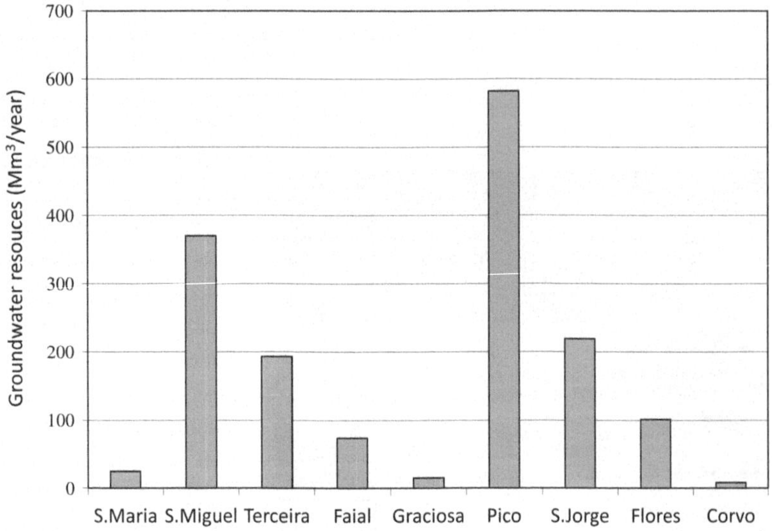

Fig. 27 The total volume of groundwater resources in the Azores islands (adapted from DROTRH-INAG 2001)

Table 4 The total volume of groundwater resources in the Azores islands; maximum and minimum of the aquifers recharge rate in % the rainfall (in DROTRH-INAG 2001)

Island	Groundwater resources	Recharge rate	
	(Mm^3/year)	Max. (%)	Min. (%)
Santa Maria	25.2	33.2	14.6
São Miguel	369.7	45.0	16.0
Terceira	193.1	48.6	16.2
Faial	74.1	47.5	12.4
Graciosa	15.0	36.2	8.5
Pico	582	62.1	18.5
São Jorge	219	54.0	19
Flores	101.3	14.0	32.0
Corvo	8.3	25.9	15.9

The aquifer recharge rates vary between 8.5 and 62.1% of the rainfall (Table 4 in DROTRH-INAG 2001). The highest values are reached in Pico, Terceira, Faial, São Miguel and Graciosa islands, particularly in aquifer systems developed in recent basaltic lava flows and sparse soil (DROTRH-INAG 2001).

4 Summary

The geomorphology of the Azores islands and the humid subtropical climate have given rise to significant amounts of freshwater. Large rivers are non-existent as a result of poorly developed watersheds due to the recent formation of the islands. Just a few perennial streams exist on different islands with lengths less than 10 km.

All islands have a significant number of lakes, compared with other small oceanic islands as a result of the geomorphology and the climate. São Miguel Island stores 93% of the total volume of freshwater in the crater lakes due to numerous calderas and *maars*. In general, lake waters correspond to freshwater with low mineralization. The smaller lakes are Na–Cl dominated, whereas the larger lakes show a Na–HCO_3 chemical trend. Three major processes control the hydrogeochemical properties of the lake waters: (1) marine contribution due to atmospheric transportation and deposition that influences the water chemistry of all lakes, but is most dominant in small volume lakes; (2) water-rock interaction processes contributing to water composition in larger and deeper lakes (Na–HCO_3 type); (3) and lakes that show interaction with volcanic fluids, such as Furnas do Enxofre which shows a clear Mg–HCO_3 trend.

In lakes with depth >13 m, the water column stratifies in the summer season, although lakes do not a show marked chemical stratification. Commonly, CO_2 concentration increase with depth in larger lakes. This enrichment is particularly noticeable in the profiles made in the Furnas and Congro lakes, where concentrations at the lake bottom are approximately two times higher than at the surface. Due to the volcanic origin of the Azores archipelago, diffusive or concentrated gaseous emanations occur at several locations, where CO_2 is the dominant component.

The concentrations of major chemical species in all of the studied lakes, except in Furnas do Enxofre Lake, are well below the European Union standards and Portuguese law for drinking-water quality. Lake water quality degradation is related to anthropogenic influences, with the input of artificial nutrients causing the increase in biological productivity due to eutrophication. In the Azores islands, groundwater occurs in perched-water bodies and basal aquifers. Both types of aquifers are classified in 54 aquifer systems and play an important role, contributing to 98% of the public water supply.

The anisotropy and heterogeneity of the volcanic formations and secondary factors such as faulting and weathering determine the groundwater movement and a large variation in the hydraulic parameters of the aquifers. According to Krásný classification (1993, in DROTRH-INAG 2001), 63.5% of Azores aquifers have very high transmissivity (values higher than 1,000 m^2/day). 20.6% of the aquifers have high transmissivity (100–1,000 m^2/day), 14.3% have intermediate transmissivity (10–100 m^2/day), and 1.6% have low values (1–10 m^2/day).

Perched-water bodies are usually cold, slightly acidic to slightly alkaline, with low mineralization, electric conductivity up to 725 μS/cm, and mainly of Na–Cl to Na–HCO_3 types, depending on the proximity to the sea and the total CO_2 content. Groundwater chemistry from the basal aquifer system is predominantly of Na–Cl type with great variability in total mineralization, mainly due to seawater intrusion or sea salts from rain and soils. In these waters the electric conductivity varies between 123 and 9,670 μS/cm;

In general, the perched water bodies have good quality, excluding the water affected by volcanic/hydrothermal processes or contaminations mainly related to agricultural and urban activities. The coastal aquifers are very vulnerable due to salt water intrusion caused by overexploitation.

The total volume of groundwater resources in the Azores is about to $1,588 \times 10^6$ m^3/year, but varies strongly from island to island dependent on the volcanism type and formation age. Values range from 8.3×10^6 m^3/year on Corvo Island to 582×10^6 m^3/year on Pico Island.

References

Africano, F., & Bernard, A. (2000). Acid alteration in the fumarolic environment of Usu volcano, Hokkaido, Japan. *Journal of Volcanology and Geothermal Research, 97,* 475–495.

Agostinho, J. (1938). Clima dos Açores. *Açoreana, 1*(2), 35–65.

Aguiar, P., Antunes, P., Mestre, R., Raposeiro, P.M., & Costa, A. (2008). *Dinâmica biogeoquímica de sistemas aquáticos da ilha das Flores* (pp. 15–27). XIII expedição científica do departamento de biologia – Flores e Corvo 2007, Rel. e Comum. Dep. Biol., nº 35

Antunes, P. & Cruz, J. V. (2004) Lagos Vulcânicos de São Miguel e Terceira: caracterização hidrogeoquímica e comparação com outros lagos do Arquipélago dos Açores. In Atlântida, Revista do Instituto Açoriano de Cultura, Angra do Heroísmo (Vol. XLIX, pp. 255–278).

Antunes, P., & Cruz, J. V. (2005). Estudo hidrogeológico em lagos Vulcânicos dos Açores: perfis realizados nos lagos da ilha de São Miguel. *Comum. Geol.* (t. 92, pp. 79–106), INETI

Antunes, P. (2009). *Estudo hidrogeoquímico e vulcanológico de lagos no arquipélago dos Açores: aplicações para a mitigação de riscos naturais* (Ph.D. thesis, University of the Azores, Ponta Delgada), 306 p.

Antunes, P. & Rodrigues, F. C. (2011). Azores Volcanic Lakes: Factors affecting water quality. In N. E. Peters, V. Krysanova, A. Lepisto, R. Prasad, M. Thoms, R. Wilby, & Zandaryaa, (Eds.), *Water quality: Currents trends and expected climate change impacts* (pp. 106–114). Oxfordshire, UK: IAHS Publication 348. ISSN 0144-78-15

Antunes, P., & Rodrigues, F. C. (2014). Hydrogeochemistry assessment of volcanic lakes in the Flores Island protected areas (Azores, Portugal). *Journal of Integrated Coastal Zone Management, 14*(2), 321–334. https://doi.org/10.5894/rgci494.

Azevedo, E. M. V. B. (1996). *Modelação do clima Insular à escala local* (Ph.D. thesis, University of Azores, Angra do Heroísmo, Portugal)

Azevedo, J. M. N., Gonçalves, V., Raposeiro, P., Couto, A. I., & Costa, A. C. (2005). Contribuição para o conhecimento biológico das águas interiores da Graciosa. *Relatórios e Comunicações do Departamento de Biologia 32,* 143–149.

Azevedo, E. B. (1996). *Modelação do Clima Insular aÁ Escala Local. Modelo CIELO aplicado à ilha Terceira* (Ph.D. thesis, University of Azores), 247 p.

Bettencourt, M. L. (1979). *O clima dos Açores como recurso natural, especialmente em agricultura e indústria do turismo.* O clima de Portugal, Lisboa: Instituto Nacional de Meteorologia e Geofísica.

Bijlsma, L., Ehler, C. N., Klein, R. J. T., Kulshrestha, S. M., McLean, R. F., Mimura, N., et al. (1995). Coastal zones and small islands. In R. T. Watson, M. C. Zinyowera, & R. H. Moss (Eds.), *Climate Change 1995: Impacts, adaptations and mitigation of climate change: Scientific—technical analyses.* Contribution of Working Group II to the Second Assessment Report of the Intergovernmental Panel on Climate Change (pp. 289–324). Cambridge, UK: Cambridge University Press.

Carvalho M. R. (1999). *Hidrogeologia do Maciço Vulcânico de Água de Pau/Fogo (S. Miguel – Açores).* Dissertação para obtenção do grau de Doutor em Geologia, na Especialidade de Hidrogeologia. Universidade de Lisboa, Lisboa, 445 p.

Carvalho, M. R., Mateus, A., Nunes, J. C., & Carvalho, J. M. (2011). Chemistry of the Ferraria thermal water, S. Miguel Island, Azores: Mixing and precipitation processes. *Environmental Earth Sciences, 64*(2), 539–547. https://doi.org/10.1007/s12665-010-0877-8. WOS: 000294347100023; ISSN: 1866-6280.

Carvalho, M. R., Mateus, A. Nunes, J. C., Carvalho, E. J. M. (2014). Origin and chemical nature of the thermal fluids at Caldeiras da Ribeira Grande (Fogo Volcano, S. Miguel Island, Azores). *Environmental Earth Sciences,* Agosto, 16 p. https://doi.org/10.1007/s12665-014-3585-y. ISSN 1866-6280.

Cerezal, J. C. S. (2013). *Avances en la investigación de los recursos hídricos en islas y terrenos volcánicos* (269 p). Madrid, Spain: Colegio de Ingenieros de Montes.

Christenson, B. W. (2000). Geochemistry of fluids associated with the 1995-1996 eruption of Mt. Ruapehu, New Zealand: Signatures and processes in the magmatic-hydrothermalsystem. *Journal of Volcanology and Geothermal Research, 97,* 1–30.

Christenson, B. W., & Tassi, F. (2015). Gases in volcanic lake environments. In D. Rouwet, B. W. Christenson, F. Tassi, & J. Vandemeulrbrouck (Eds.), *Volcanic lakes* (533 p). Heidelberg: Springer.

Christenson, B. W., Rouwet, D., Németh, K., Tassi, F., Vandemeulrbrouck, J., & Varekamp, J. (2015). Volcanic lakes. In D. Rouwet, B. W. Christenson, F. Tassi, & J. Vandemeulrbrouck (Eds.), *Volcanic lakes* (533 p). Heidelberg: Springer.

Coutinho, R. M. (1990). Estudo hidrogeológico do maciço das Sete Cidades (MSc thesis, Universidade de Lisboa).

Coutinho, R. M. (2000). Elementos para a monitorização sismovulcânica da ilha do Faial (Açores): caracterização hidrogeológica e avaliação de anomalias de Rn associadas a fenómenos de desgaseificação (Ph.D. thesis, University of Azores, Ponta Delgada, Portugal).

Cruz, J. V. (1992). *Hidrogeologia da ilha de Santa Maria.* Dissertação para a obtenção do Grau de Mestre

em Geologia Económica e Aplicada pela FCUL (2 Vols., 486 p). FCUL, Lisboa.
Cruz, J. V. (1997). *Hidrogeologia da Ilha do Pico.* Dissertação para a obtenção do grau de Doutor em Geologia (2 Vols., 433 p). University of Açores, Ponta Delgada.
Cruz, J. V. (2003). Groundwater and volcanoes: Examples from the Azores archipelago. *Environmental Geology, 44,* 343–355.
Cruz, J. V. (2004). *Ensaio sobre a água subterrânea nos Açores. História, ocorrência e qualidade* (288 p). Ponta Delgada: SRA.
Cruz, J. V., & Coutinho, R. M. (1998). Breve nota sobre a importância dos recursos hídricos subterrâneos no arquipélago dos Açores. *Açoreana, 8,* 591–594.
Cruz, J. V., Coutinho, R. M., Carvalho, M. R., Oskarsson, N., & Gislason, S. R. (1999). Chemistry of waters from Furnas volcano, São Miguel, Azores: Fluxes of volcanic carbon dioxide and leached material. *Journal of Volcanology and Geothermal Research, 92,* 151–167.
Cruz, J. V., & Silva, M. O. (2001). Hydrogeologic framework of the Pico island (Azores, Portugal). *Hydrogeology Journal, 9,* 177–189.
Cruz, J. V., & França, Z. (2006). Hydrogeochemistry of thermal and mineral water springs of the Azores archipelago (Portugal). *Journal of Volcanology and Geothermal Research, 151*(2006), 382–398.
Cruz, J. V., Coutinho, R., Pacheco, D., Cymbron, R., Antunes, P., Freire, et al. (2010a) *SWIM21—21st Salt Water Intrusion Meeting, 21–26 June,* Azores, Portugal (pp. 109–112).
Cruz, J. V., Freire, P., & Costa, A. (2010b). Mineral waters characterization in the Azores archipelago (Portugal). *Journal of Volcanology and Geothermal Research, 190*(2010), 353–364.
Cruz, J. V., Andrade, C., Coutinho, R. (2014). Estado quÍmico das massas de água subterrânea nas ilhas do Grupo Central do arquipélago dos Açores (Portugal). *Comunicações Geológicas, 101,* 665–670.
Custodio, E., Lopez-Garcia, L., & Amigo, E. (1988). Simulation par modele mathèmatique de l'île volcanique de Ténérife (Canaries). *Hydrogéologie, 2,* 153–167.
Delmelle, P., & Bernar, A. (2000). Volcanic lakes. In H. Sigurdsson, S. R. McNutt, H. Rymer, & D. Six (Eds.), *Encyclopedia of volcanoes* (pp. 877–895). San Diego: Academic Press.
DROTRH/INAG. (2001). *Plano Regional da Água. Relatório técnico* (575 p). Ponta Delgada: DROTRH-INAG.
Falkland, A. (1991). *Hydrology and water resources of small islands: A practical guide* (435 p). Paris: UNESCO.
Falkland, A. (2002). Tropical island hydrology and water resources current knowledge and future needs. Hydrology and water management in the humid tropics. In IHP (Ed.), *Proceedings of the Second International Colloquium in Panama, Technical Documents in Hydrology* (N° 52, 487 p). Paris: UNESCO.
Forjaz, V. H. (1983). *Azores tectonic sketch.* Centro de Vulcanologia: University of Azores.
Freire, P. (2006). *Águas minerais da Ilha de S. Miguel (Açores): caracterização hidrogeológica e implicações para a monitorização vulcanológica* (Dissertação de Mestrado, Universidade dos Açores, Ponta Delgada), 272 p.
Freire, P., Andrade, C., Viveiros, F., Silva, C., Coutinho, C., & Cruz, J. V. (2014). Mineral water occurrence and geochemistry in the Azores volcanic archipelago (Portugal): Insight from an extended database on water chemistry. *Environmental Earth Sciences.* https://doi.org/10.1007/s12665-014-3157-1.
Gonçalves, V. (1997). *Estrutura da Comunidade Fitoplanctónica da Lagoa das Furnas. PAPCC* (229 p). Ponta Delgada: Departamento de Biologia, Universidade dos Açores.
Gonçalves, V., Costa, A., Raposeiro, P., & Marques, H. (2005). Caracterização biológica das massas de água superficiais das ilhas de São Miguel, e Santa Maria. Cent. Cons. Prot. Amb., Dep. Biologia, Universidade dos Açores, 76 p.
Gonçalves, V., Raposeiro, P., Couto, A. I., Costa, R. M., Rocha, F., Wattiez, X., et al. (2006a). Contribuição para a caracterização das águas interiores de superfície da ilha do Pico. *Rel. Com. Dep. Biol., 34,* 93–106.
Gonçalves, V., Raposeiro, P., Porteiro, J., Alves, M. H., & Medeiros, M. C. (2006b). Proposta de definição de ecótipos e classificação preliminar do estado ecológico das lagoas dos Açores. VIII Congresso da Água Rel. *Com. Dep. Biol., 34,* 1–16.
Gonçalves, V. (2008) *Contribuição para o estudo da qualidade ecológica das lagoas dos Açores: Fitoplâncton e diatomáceas bentónicas* (Ph.D. thesis, Universidade dos Açores, Ponta Delgada), 331 p.
Hall, F. R. (1968). Base flow recessions. A review. *Water Resources Research, 4*(5), 973–983.
ILEC. (2011). Methodology for the GEF transboundary waters assessment programme. Methodology for the Assessment of Transboundary Lake Basins, (Vol. 3, p. viii, 69 p). UNEP.
INOVA. (1999). *Análise das Águas das Lagoas da Região Autónoma dos Açores. Relatório final* (p. 291). Ponta Delgada: Instituto de Inovação Tecnológica dos Açores.
Izuka, S. K., & Gingerich, S. B. (2002). A thick lens of fresh groundwater in the southern Lihue Basin, Kauai, Hawaii, USA. *Hydrogeology Journal, 11,* 240–248.
Join, J. L., Folio, J. L., & Robineau, B. (2005). Aquifers and groundwater within active shield volcanoes. Evolution of conceptual models in the Piton de la Fournaise volcano. *Journal of Volcanology and Geothermal Research, 147,* 187–201.
Jones, G., Chester, D. K., & Shooshtarian, F. (1999). Statistical analysis of the frequency of eruptions at FurnasVolcano, São Miguel, Azores. *Journal of Volcanology and Geothermal Research, 92,* 31–38.
Krause, D., & Watkins, N. (1970). North Atlantic crustal genesis in the vicinity of the Azores. *Geophysical Journal of Royal Astronomical Society, 19,* 261–283.

Laughton, A., & Whitmarsh, R. (1974). The Azores-Gibraltar plate boundary. In L. Kristjansson (Ed.), *Geodinamics of Iceland and North Atlantic Area* (pp. 63–81). Dordrecht: Reidel Publishing Company.

LNEC. (1998). Desenvolvimento de um inventário das águas subterrâneas dos Açores. Relatório de progresso – fase1, Rel. 105/98-GIAS.

LNEC. (1999). Desenvolvimento de um inventário das águas subterrâneas dos Açores. Relatório de progresso – fase 2, rel. 124/99-GIAS, 301 p.

Lourenço, N., Miranda, J. M., Luis, J. F., Ribeiro, A., Victor, L. A. M., Madeira, J., et al. (1998). Morpho-tectonic analysis of the Azores Volcanic Plateau from a new bathymetric compilation of the area. *Marine Geophysical Researches, 20*(3), 141–156. https://doi.org/10.1023/A:1004505401547.

Louvat, P., & Allègre, C. J. (1998). Riverine erosion rates on São Miguel volcanic island, Azores archipelago. *Chemical Geology, 148*(3–4), 177–200. https://doi.org/10.1016/S0009-2541(98)00028-X.

Macdonald, G. A., Abbott, A. T., Peterson, F. L. (1983). *Volcanoes in the sea: the geology of Hawaii* (2^a ed.). Honolulu, Hawaii: University of Hawaii Press, 517 p. ISBN-10: 0824808320.

Manville, V., Hodgson, K. A., & Nairn, I. A. (2007). A review of break-out floods from volcanogenic lakes in New Zealand. *New Zealand Journal of Geology and Geophysics, 50*(2), 131–150. https://doi.org/10.1080/00288300709509826.

Mateus, A., Carvalho, M. R., Nunes, J. C., Carvalho, E. J. M. (2014). Influence of wall-rock alteration and fluid mixing mechanisms in the chemistry of thermal fluids and mud-pool sediments at Caldeiras da Ribeira Grande (S. Miguel Island, Azores). *Environmental Earth Sciences,* Setembro; 23 p. https://doi.org/10.1007/s12665-014-3439-7. ISSN 1866-6280.

Martínez, M., Fernández, E., Valdés, J., Barboza, V., Van Der Laat, R., Duarte, E., et al. (2000). Chemical evolution and volcanic activity of the active crater lake of Poás volcano, Costa Rica, pp. 1993–1997. *Journal of Volcanology and Geothermal Research, 97,* 127–141.

Menezes, J. G. (1993). Hidrologia e Hidrogeologia da ilha Terceira: grandes condicionantes. In *Angra do Heroísmo, Conference "Recursos Hídricos e o Ambiente na Região Autónoma dos Açores"*, 56 p.

Moore, R. B. (1991). Geology of three late quaternary stratovolcanoes on São Miguel. *U.S. Geological Survey* bulletin, *1900*, 46 p.

Morrisseau, M. (1987) *Les eruptions hydromagmatiques et les xenolites associes: signification geothermique (exemples de Flores et de Faial, Açores)* (Ph.D. thesis – Geotermia, Universidade de Paris – Sul Centro D'orsay, França), 487 p.

Nunes, J. C. (1999). *A actividade vulcânica da ilha do Pico do Plistócénico Superior ao Holocénico: mecanismo eruptivo e hazard vulcânico*. Dissertação para a obtenção do grau de Doutor em Geologia. Universidade dos Açores, Ponta Delgada, 357 p.

Ohba, T., Hirabayashi, J., & Nogami, K. (2000). D/H and $^{18}O/^{16}O$ ratios of water in the crater lake at Kusatsu-Shirane volcano, Japan. *Journal of Volcanology and Geothermal Research, 97,* 329–346.

Pasternack, G. B., & Varekamp, J. C. (1994). The geochemistry of the Keli Mutu crater lakes, Flores, Indonesia. *Geochemical Journal., 28,* 243–262.

Peterson, F. L. (1972). Water development on tropic volcanic islands—Type example: Hawaii. *Groundwater, 10*(5), 18–23.

Porteiro, J. (2000). *Lagoas dos Açores: elementos de suporte ao planeamento integrado* (Ph.D. thesis, Ponta Delgada, University of Azores), 344 p.

Rice, A. (2000). Rollover in volcanic crater lakes: a possible cause for Lake Nyos type disasters. *Journal of Volcanology and Geothermal Research, 97,* 233–239.

Rodrigues, F. C. (2002) *Hidrogeologia da Ilha Terceira (Açores - Portugal).* Dissertação para a obtenção do grau de Doutor em Engenharia do Ambiente. Universidade dos Açores, Ponta Delgada, 404 p.

Rouwet, D., Tassi, T., Mora-Amador, R., Sandri, L., & Chiarini, V. (2014). Past, present and future of volcanic lake monitoring. *Journal of Volcanology and Geothermal Research, 272,* 78–97. https://doi.org/10.1016/j.jvolgeores.2013.12.009.

Santos, F. D., & Miranda, P. (2006). *Alterações climáticas em Portugal cenárions, imactos e medidas de adaptação, projecto SIAM II* (89 p). Lisboa: Gradiva.

Saunders, D. L., Meeuwig, J. J., & Vincent, A. C. J. (2002). Freshwater protected areas: Strategies for conservation. *Conservation Biology, 16*(1), 30–41. https://doi.org/10.1046/j.1523-1739.2002.99562.x.

Searle, R. (1980). Tectonic pattern of the Azores spreading centre and triple junction. *Earth and Planetary Science Letters, 51,* 415–434.

SRAM. (2012). *Plano de Gestão de Região Hidrográfica dos Açores - RH9. Relatório Técnico. Caracterização da Situação de Referência e Diagnóstico* (381 p). Açores: SRAM.

SRAA. (2015). A água nos Açores. Direção de Serviços de Recursos Hídricos e Ordenamento do Território, Nova Gráfica, Ponta Delgada, 43 p. ISBN 978-989-98731-1-7.

Steinmetz, L., Hirn, A., Sapin, M., Whitmarsh, R., & Moreira, V. (1976). Zone of P wave attenuation beneath the crest of Mid-Atlantic Ridge. *Bulletin de la Société Géologique de France, 4*

Takano, B., & e Watanuki, K. (1990). Monitoring of volcanic eruptions at Yugama crater lake by aqueous sulfur oxyanions. *Journal of Volcanology and Geothermal Research, 40,* 71–87.

Tassi, F., Vaselli, O., Fernandez, E., Duarte, E., Martinez, M., Huertas, A. D., et al. (2009). Morphological and geochemical features of crater lakes in Costa Rica: An overview. *Journal of Limnology, 68*(2), 193–205. https://doi.org/10.3274/JL09-68-2-04.

Van Hinsberg, V., Berlo, K., Sumarti, S., Van Bergen, M., & Williams-Jones, A. (2010). Extreme alteration by hyperacidic brines at Kawah Ijen volcano, East

Java, Indonesia: II: Metasomatic imprint and element fluxes. *Journal of Volcanology and Geothermal Research, 196*(3–4), 169–184. https://doi.org/10.1016/j.jvolgeores.2010.07.004.

Varekamp, J. C., Pasternack, G. B., & Rowe, G. L., Jr. (2000). Volcanic lake systematics II. Chemical constraints. *Journal of Volcanology and Geothermal Research, 97,* 161–179.

Varekamp, J. C. (2002). *Volcanic lakes: Coloured windows into magma chamber* (pp. 71–72). Pico, Portugal: II Jornadas Internacionais de Vulcanologia da Ilha do Pico.

Varekamp, J. (2015). The chemical composition and evolution of volcanic lakes. In D. Rouwet, B. W. Christenson, F. Tassi, & J. Vandemeulrbrouck (Eds.), *volcanic lakes* (533 p). Heidelberg: Springer.

Wallenstein, N. (1999). Estudo da história recente e do comportamento eruptivo do Vulcão do Fogo (S. Miguel, açores). Avaliação preliminar do hazard, Tese de doutoramento no ramo de Geologia, especialidade de Vulcanologia. Universidade dos Açores, Departamento de Geociências, 266 p.

Wallenstein, N., Chester, D. K., & Duncan, A. M. (2005). Methodological implications of volcanic hazard evaluation and risk assessment: Fogo Volcano, São Miguel, Azores. *Z. Geomorph. N.F., 140,* 129–149

Wetzel, R. G. (2001). *Limnology: Lake and river ecosystems* (1006 p). San Diego: Academic Press.

White, I. M., Falkland, T., Metutera, T., Metai, E., Overmars, M., Perez, P., et al. (2007). Climatic and Human Influences on Groundwater in Low Atolls. *Vadose zone Journal, 6*(3), 581–590.

Where to Go? A Selection and Short Description of Geological Highlights in the Azores

Ulrich Kueppers, Christoph Beier, Felix S. Genske and Diogo Caetano

1 Introduction

This chapter is a selection of geological outcrops from each of the Azores islands that we have selected to be worthwhile visiting to get an overview of prominent geological features of each island. This chapter will significantly differ from the must-see/must-do lists of many guidebooks. We do not intend to cover all geological features and the entire diversity of the complex geology of the Azores but rather aim at presenting representative geological features on each island. We have selected comprehensive outcrops on each island, some of which have been subject to detailed geological descriptions and are regularly visited by researchers and students. We will use data from the peer-reviewed literature wherever available, however, most of the outcrops described here have been incorporated into a larger geological context and thus more detailed information will be widespread amongst the peer-reviewed and grey literature. Note that here, we focus on citations of papers in English, however, a large number of Portuguese publications are available. Particularly for the islands of Faial and Pico, a detailed and handy field guide is available by França et al. (2009, Volcanic history of Pico and Faial islands, Azores) to which the reader is referred to for more detailed descriptions of the geology and outcrops. Each island has well-marked and -described hiking paths that allow for a close contact with the beauty of these islands (e.g. *Amigos dos Açores* or *Os Montanheiros*) and most outcrops described here are easily accessible usually by car or via short hikes. When we refer to *historic eruptions*, please note that this term refers to eruptions since settlement (Beier and Kramer, Chapter "A Portrait of the Azores: From Natural Forces to Cultural Identity") that were observed either directly or that indirect evidence exists for. Geochemical and petrological analyses from each of the islands described here are subject of Chapter "Petrology of the Azores Islands" (Larrea et al.) and Chapter "Melting and Mantle Sources in the Azores" (Beier et al.) to which the reader is referred to for more detail.

Order of islands

1. Corvo
2. Faial
3. Flores

U. Kueppers (✉)
Department für Geo- und Umweltwissenschaften, Ludwig-Maximilians-Universität (LMU) München, Munich, Germany
e-mail: u.kueppers@lmu.de

C. Beier
GeoZentrum Nordbayern, Friedrich-Alexander Universität Erlangen-Nürnberg, Erlangen, Germany

F. S. Genske
Institut für Mineralogie, Westfälische Wilhelms-Universität Münster, Münster, Germany

D. Caetano
LabGeo—Engenharia E Geotecnologia, Ponta Delgada, Azores, Portugal

U. Kueppers and C. Beier (eds.), *Volcanoes of the Azores*, Active Volcanoes of the World, https://doi.org/10.1007/978-3-642-32226-6_14

4. Graciosa
5. Pico
6. Santa Maria
7. São Jorge
8. São Miguel
9. Terceira

1. Corvo

Corvo is the smallest inhabited island of the Azores. It is the northernmost island and one (of two, together with Flores) of the islands situated to the West of the Mid-Atlantic Ridge (see Fig. 1 in Vogt and Jung, Chapter "The "Azores Geosyndrome" and Plate Tectonics: Research History, Synthesis, and Unsolved Puzzles"). In terms of plate tectonics, both Corvo and Flores are part of the American plate. There is no evidence for historical eruption on Corvo or the surrounding sea, however, U-series dating reveals that volcanic activity in the caldera may be younger than 370 ka (Beier et al. 2010). Samples from Corvo have been described in more detail in Genske et al. (2012) and Larrea et al. (2013).

Coastline at Corvo

Name: Eastern and western coastline at Corvo
Coordinates: Eastern coastline from 39° 40′ 25.27″N, 31° 6′ 29.34″W to 39° 43′ 14.46″N and western coastline from 39° 43′ 35.49″N, 31° 7′ 8.52″W to 39° 40′ 56.34″, 31° 7′ 6.75″W
Access: Apart from the southern part, the coastline of Corvo is only accessible by boat (boats and drivers may be rented on either Corvo or the neighbouring island of Flores). Access to the individual outcrops may be hampered by waves and currents. Most of the beach sections underneath the cliffs contain large pebbles and boulders even underneath the waterline. Beware of overhanging cliffs (some of which are several hundred's meters high).
Geology: The eastern coastline exhibits numerous lava flows of basaltic to intermediate composition, intercalated by thin layers of clastic material (minor explosive eruptions or lava flow autobrecciation) and soil deposits implying periods of smaller magmatic activity. The rise and formation of ocean islands can explicitly be studied in these outcrops. (Sub-)vertical feeder dykes can be followed to their respective lava flows (e.g., at 39° 40′ 30.93″N, 31° 6′ 13.30″W).

Fig. 1 Vertical coastal cliff of Corvo island, showing the layered structure and the countless cross-cutting feeder dykes and sills. Width of the image approx. 200 m. (Picture credit: Christoph Beier)

Several dozens of lava flows are exposed in the eastern cliffs, belonging to the shield-building stage of the island. The western coastline offers spectacular views of a cliff reaching up to the caldera rim at an elevation of almost 700 m. These cliffs are almost vertical and provide impressive insights into the pre-caldera shield building stage of volcanic islands, exposing numerous dykes and sills, many of which range from centimetres to meters in size forming complex sill-dyke networks within the shield stage (Fig. 1). The eastern and western coastlines contain large areas of landslides and numerous fault systems. Lava flows are generally basaltic to intermediate in composition; however, several dykes and lava flows are also ankaramitic and may contain crustal and ultramafic xenoliths. The ages of the shield building stage ranges from 1.0 to 1.5 Ma (França et al. 2006).

Author's hot tip: A trip around the island of Corvo offers stunning views of the island itself, however, the wave and weather conditions must be taken into account. Binoculars are essential on this trip and may also provide stunning views of large bird's colonies and goats living on these cliffs.

Caldera of Corvo

Name: Caldeirão do Corvo

Coordinates: Parking and view at 39°42′39.45″ N, 31° 5′56.55″W, young cinder cones along the crater lake at and around 39°42′36.40″N, 31° 6′ 40.51″W

Access: Taxi from Vila do Corvo and driving up the caldera rim, walk into the caldera from the parking lot (unpaved, unmarked path but easy to hike).

Geology: The *Caldeirão do Corvo* (Fig. 2) is one of the Azores prime examples of caldera formation (besides the calderas on São Miguel and at Faial). Early lavas during the shield-building stage are mostly basaltic, the relatively young cones along the ring fault are of intermediate composition. The compositional range of Corvo lavas is generally small and less variable than those from the other islands (Genske et al. 2012), however, it has to be noted that this could also result from a sampling bias due to limited accessibility of the coastline. The caldera forming mechanism at Corvo still remains a matter of active debate as no evidence is yet found for deposits related to a major explosive eruption.

Fig. 2 Panoramic view of the summit caldera of Corvo looking West. The caldera is filled by a shallow lake surrounding some small heaps, generated during post-caldera activity. The caldera diameter is 2 km. (Picture credit: Christoph Beier)

Harbourside at Corvo

Name: Vila do Corvo flank eruption and cumulates
Coordinates: 39° 40′ 24.30″N, 31° 7′ 18.97″W
Access: Follow the road along the airport runway westward, at the end of the runway walk down to the beach.
Geology: The rocks exposed here belong to the Vila do Corvo flank eruption, the youngest and most recent volcanic activity on the island of Corvo (França et al. 2006). The lavas exposed here contains numerous gabbroic, crustal enclaves. The host lava is intensely altered and aphyric, while the gabbroic xenoliths contain xenomorphic shapes. The host lavas are the most evolved rocks found to date on Corvo (trachyandesites, Genske et al. 2012). Gabbroic xenoliths range in size up to 30 cm and may provide additional information on the crustal evolution of the western islands.

2. Faial

The island of Faial hosts one of the most famous and best-studied eruptions of the 20th century. The eruption of Capelinhos (1957/58) at the western end of the island certainly is one of the youngest and most spectacular historic volcanic edifices accessible. Contrasting the neighbouring island of Pico, Faial displays the entire range from explosive and effusive basaltic to trachytic lavas and pyroclastic deposits. Two eruptions took place on Faial in historic times, in 1672/73 (*Cabeço do Canto*, today called *Cabeço da Gordo*) and in 1957/58 (Capelinhos).

Trachytic dome at Castelo Branco

Name: Castelo Branco
Coordinates: 38° 31′ 27.11″N, 28° 44′ 53.33″W
Access: Leave Horta along the south coast on EN1-1A westward. Pass the airport and turn left at 38° 32′ 6.91″N and 28° 44′ 29.62″W into a narrow road to Castelo Branco. Have a look from the cliff at the Eastern end of the bay that starts from Castelo Branco to the West (Fig. 3, direction of view: SW).
Geology: The peculiar half-sphere shaped, white-grey rock mound is connected to the main island by a small isthmus. The coastline to the north of Castelo Branco exposes primarily basaltic lavas. Castelo Branco is a trachytic dome, i.e. a cooled mass of fairly viscous, dense lava. Based on its overall shape, it is likely that the Castelo Branco dome intruded into shallow crustal levels. A distinct cleavage pattern can be observed; in the lower part, the cleavage planes are vertical whereas the cleavage planes at the top are curved and sub-parallel to the surface morphology. Trachyte is formed by crystal fractionation from a more primitive, basaltic lava and can be observed in several outcrops in the Azores. The higher viscosity of trachytic melts favours the formation of domes or low-aspect ratio lava flows. In the case of Faial, weathering has led to the exposed positioning of Castelo Branco along the coastline because most of the surrounding more basaltic lavas have been eroded away. The rocks that build the isthmus (central left part of Fig. 3) show a sharp contact (dipping westward) between the dense trachytic lava and an underlying pyroclastic sequence. These are possibly the remains of a tuff cone from the submarine start of the eruption.

Capelinhos (1957–1958)

The eruption of Capelinhos surely is one of the most spectacular, but also most fragile, volcanic edifices in the entire Azores. Erosion has been significantly affecting the area and formed several steep and dangerous cliffs, affected by landslides. Note that the area is protected and access is limited, thus permission must be sought in the museum at the lighthouse.

Name: Capelinhos—The 1957–58 eruption
Coordinates: 38° 35′ 48.16″N, 28° 49′ 24.10″W
Access: Continue along the South coast to the village of Capelo and follow the signs to the lighthouse and museum (*Farol dos Capelinhos*). A visit to the museum provides a great overview over the last eruption that took place in the Azores on land. From the top of the old lighthouse, you have a stunning overview of the old Capelo and younger Capelinhos eruptions.

Fig. 3 The trachytic dome of Castelo Branco, exposed by coastal erosion. The grey rock is in stark contrast to the dominating darker basaltic rocks of the northern and southern coastlines. The two distinct cleavage patterns in the lower (central) and the upper (former outside) part show the relative cooling history. The sea-side cliff face is approx. 25–30 m high. (Picture credit: Ulrich Kueppers)

Geology: The famous Capelinhos eruption lasted for 13 months and started in a shallow submarine setting that subsequently built a small island of poorly consolidated pyroclastic material (Machado et al. 1962). This Surtseyan eruption, predating the eponymous Surtsey eruption South of Iceland some 4 years, showed several phases of variable eruption intensity, based on the degree of interaction of the rising magma with the seawater. At phases of significant interaction with sea water, the eruption was characterised by pulsating explosions that eventually built a volcanic edifice above sea level (Cole et al. 2001). The dominant deposits are golden-coloured and strongly stratified ash and lapilli tuffs (Fig. 4). Abundant cross-stratification indicates that the growing volcanic edifice was repeatedly strongly modified by partial gravitational collapse events and erosion due to the passage of energetic density currents (Zanon et al. 2009; Douillet et al. 2013). The growing cone was capable of significantly reducing the degree of magma-water interaction, leading to a shift in eruption style towards strombolian type eruptions. The related deposits are black or reddishly oxidised and partially welded. The prevailing winds led to significant deposition of ash deposits in the area. The entire island was affected by ash fall and earthquake activity and caused an important wave of emigration that reduced the population of Faial to a third.

After the museum visit, climb the remnants of the golden coloured tuff ring (magma-water contact) and further onto the red strombolian cone (dry phase). Numerous black volcanic bombs from the strombolian phase may be found

Fig. 4 The remnants of the 1957/58 volcanic edifice of the Capelinhos eruption. **a** View from the old lighthouse with the strombolian cone (red) inside the golden-coloured tuff ring of palagonitized ash and lapilli tuffs. **b** View of Capelinhos from the West, showing the strombolian deposits inside the tuff ring. (Picture credit: Ulrich Kueppers)

on the golden tuffs and give a nice colour contrast. The lavas mostly consist of fine-grained to aphyric lavas and scoria with plagioclase, olivines and clinopyroxenes. Looking eastwards one can observe the Capelo peninsula formed from several fissural eruptions e.g., during the last eruption prior to Capelinhos (1672/73). Directly adjacent to Capelinhos is *Costado da Nau*, the pre-Capelinhos coastal cliff, formed by a similarly complex Surtseyan eruption. The outcrop displays a cross section through a tuff ring that was covered by strombolian spatter agglomerates at least twice. These facies alternations show that the *Costado da Nau* eruption showed a repeated change in eruption style with alternating presence and absence of water similar to Capelinhos. Be aware that as a result of the unconsolidated nature of the rocks landslides are common especially on steeper sections of the cliffs.

Praia do Norte—Type locality of Faialite

Name: Fajã da Praia do Norte
Coordinates: 38° 36′ 42.69″N, 28° 45′ 43.95″W
Access: Following the main ring road (EN1-1A) clockwise around the island, passing Capelinhos and continuing towards Praia do Norte, follow the steep road through the village and once you see the black beach on your right-hand side, follow the gravel road to the left, where parking is available at the end of the road. The outcrop starts just directly behind the parking lot towards the ocean.
Geology: This outcrop is part of the historic lava flow (1672/73) from the volcanic system of *Mistério da Praia do Norte* just on the top of the hill towards the south. The lava outcrop on the coastline consists of an olivine-pyroxene-plagioclase phyric lava. The iddingsitized brownish olivines are the first faialite crystals that have been collected and thus this locality is the type locality of the Fe-rich endmember of the olivines. The mafic lava flow contains numerous crustal-gabbroic and ultramafic xenoliths variable in size. Both ultramafic and gabbroic xenoliths are mostly crustal in nature, there is little evidence for the presence of mantle xenoliths in this area.
Author's hot tip: Visit this outcrop during the afternoon hours. Have a break at the black sandy beach of Praia do Norte and enjoy the stunning view, however, be careful when aiming for a swim, the currents are strong and water depth increases very quickly. Often the waves are rough.

3. Flores

The island of Flores is well known for its lavish vegetation. Flores, being the larger of the two western islands on the American plate, is characterized by a landscape formed by craters, lakes and streams. The seven major volcanic craters have, over time, changed into individual lakes, with the *Caldeira Negra* being more than 100 metres deep. No historic eruptions have taken place on Flores, however, similar to Corvo, U-series disequilibria imply that eruptions have taken place less than 8000 years ago (Beier et al. 2010). Samples from Flores have been described in Genske et al. (2012).

Fajã Grande—westernmost lava flow of (political) Europe

Name: Fajã Grande lava flow
Coordinates: 39° 27′ 12.1″N, 31° 16′ 1.8″W
Access: The shore at Fajã Grande is easily accessible by car. At a long stretch of road there is plenty of parking available, so the outcrops at the shoreline can be visited very conveniently. However, the surface of this massive lava flow is very rough, proper footwear is strongly recommended.
Geology: This massive lava flow has made its way into the ocean and forms the whole coastline of Fajã Grande, Europe's westernmost settlement. Indeed, it appears that this eruption serves as the basement of this village, as an entire platform was created by this single lava flow. The rocks are trachybasalts with commonly observed xenocrysts of olivine. Individual olivines are up to centimetre in size. Other minerals that can easily be identified in this flow are plagioclase and clinopyroxene, both of which are also a few millimetres in size. The samples were intensely described and discussed in Genske et al. (2012).
Author's hot tip: A trip from Fajã Grande to Fajazina roughly 5 km to the south offers stunning views of this beautiful coastline, with numerous waterfalls intersecting the green wall. The largest draining the Ribeira Grande falls for almost 300 metres.

Columnar jointing at Rochas dos Bordões

Name: Rochas dos Bordões
Coordinates: Parking and view at 39° 24′ 25.96″ N, 31° 14′ 34.14″W.
Access: By car along the major road ER1-2 towards the southwest. There are several good spots from where this outcrop can be seen. The authors advise here, that these columns must not be destroyed by sampling, since the *Rocha dos Bordões* are a designated regional monument.
Geology: This impressive and symbolic outcrop displays a set of vertical columns of basalt that are up to 30 metres high (Fig. 5). They represent one of the islands younger volcanic features; as such they belong to the upper volcanic complex (see Azevedo and Ferreira et al. 1999). The characteristic jointing of the basalt columns occurred during the cooling of this massive lava, where the direction of heat loss is perpendicular to the direction of the columns. In essence, during cooling the lava shrinks, resulting in polygonal columns. The reader is also referred to the columnar joints at Santo Espirito on the island of Santa Maria.

Northern shore between Farol de Albarnaz and Ponta Delgada

Name: Northern coast of Flores
Coordinates: Between 39° 31′ 11.4″N, 31° 14′ 9.5″W and 39° 31′ 15.0″N, 31° 12′ 43.8″W.
Access: Start at the eastern point (above), where a steep set of stairs leads down to the coastline. Once returned from this part of the cliff, take the car towards the western tip, where the lighthouse forms a prominent landmark.
Geology: The rocks exposed here belong to the basic volcanic complex with the transition into the upper volcanic zone (Azevedo and Ferreira 1999). Starting at sea level, one can climb through the exposed youngest volcanic history of the island. The first flows of this outcrop are some of the oldest rocks of the islands subaerial history and have been dated at around 2 Ma (Azevedo and Ferreira 1999). The lowest flow in the cliff section shows features of Pāhoehoe lava and is covered by another five flows. Once back on top of the 25–30 m high cliff, one can then continue in the upper volcanic complex. The Moinho creek, approximately two thirds along the way to the lighthouse at Farol de Abarnaz, cuts through the stratigraphy effectively allowing further insights into the volcanic evolution of this

Fig. 5 Columnar jointing in the Rocha das Bordões. (Picture credit: Christoph Beier)

beautiful island, up to an elevation of around 600 metres above sea level.

4. Graciosa

Despite its small size, Graciosa comprises a large range of volcanic rocks from basaltic to trachytic composition. No historic eruptions have taken place here but there is still notable fumarolic activity in the *Furna do Enxofre*.

Fumarole field at Furna do Enxofre

Name: Furna do Enxofre
Coordinates: 39° 1′ 30.00″N, 27° 58′ 17.05″W
Access: Take the road to the Caldeira Volcano (note that you will have to go through a small tunnel in order to enter the Caldeira). Once inside the Caldeira drop by at the visitor centre and you will receive additional instructions for your visit of the *Furna do Enxofre*.
Geology: *Furna do Enxofre* is a volcanic pit located in the southeastern part of the caldera of Caldeira Volcano (Fig. 6)—the smallest one of the archipelago. With a maximum length of 194 meters and a maximum height of 40 m it is the uppermost part of a volcanic conduit (see *Algar do Carvão* in the Terceira section, this chapter). It is possible to see the convex ceiling with prismatic jointing and a lake with about 130 m diameter in the deepest part of the cave. The fumarole field consists of gas emanations on the cave's floor and a muddy fumarole. Inside the *Furna do Enxofre* pit, there is a gas monitoring and warning system installed for permanent monitoring. At times, access is limited to the pit due to the presence of high levels of CO_2.

Volcanic sequence at Porto Afonso

Name: Porto Afonso
Coordinates: 39° 3′ 58.31″N, 28° 4′ 1.85″W
Access: From the centre of Santa Cruz village take the regional road along the coastline towards the airport and continue along the same road until reaching Porto Afonso.

Fig. 6 The entrance area of the Furna do Enxofre, a drained volcanic edifice with active degassing inside. The cave is constantly monitored. (Picture credit: Diogo Caetano)

Geology: Outstanding coastal outcrop, cutting through a polygenetic basaltic scoria cone and exposing several feeding dykes (Fig. 7). The layered and stratified proximal basaltic fall deposits were generated by a mildly explosive, pulsating eruptions. Several discordant contact planes prove the erosion separating eruptive episodes. During one eruption, a lava flow poured from the vent. The thermal imprint of the flowing lava on the underlying soil and volcaniclastic deposits has caused a layer of fritted deposits directly below the lava flow.

Volcanic sequence at Ponta da Restinga

Name: Ponta da Restinga
Coordinates: 39° 0′ 50.36″N, 27° 57′ 17.83″W
Access: Proceed to Carapacho, a place known for its natural pools and thermal spa. From there you will have a few sightseeing points of the Ponta da

Fig. 7 Coastal cliff at the little port Porto Afonso, exposing a cross section through fall deposits of explosive basaltic volcanism. The outcrop is dissected by several feeding dykes as well as a 2 m-thick lava flow. (Picture credit: Diogo Caetano)

Restinga coastal cliff. From land, the best spot to observe the nearby islets is the viewpoint overlooking Carapacho. Nonetheless the best way to observe these geological structures is from the sea.
Geology: In the coastal cliff at Ponta da Restinga outcrops a volcanic succession that includes submarine pyroclasts of basaltic nature, undifferentiated pyroclastic deposits and lava flows, and hydromagmatic deposits along with a series of dykes. The islets located southeast of the Ponta da Restinga are part of this sequence, preserving the remnants of a tuff cone, seen in the stratified tuff deposit (Fig. 8). Lavas from these sections have been subject to a recent publication by Larrea et al. (2014). Most of Graciosa's lavas consist of a series of basaltic to intermediate rocks that likely form a fractional crystallization series.

5. Pico

Pico is the largest island of the central group and hosts the active (currently fumarolic stadium) Pico volcano, the tallest mountain of Portugal and fourth highest mountain in the Atlantic (after the volcanoes on Tenerife, Fogo and La Palma islands).

Gruta das Torres lava cave

Name: Gruta das Torres
Coordinates: 38°29′39.51″N, 8°30′08.63″W
Access: Leave Madalena southward direction on the EN1. In the small village of Criação Velha, turn left onto *Largo 20 de Novembro* and follow the indications to *Gruta das Torres*.
Geology: *Gruta das Torres* is the longest and largest lava tube that is known in the Azores (more than 5 km). It is situated in a complex lava flow field that was generated during a basaltic eruption between 500 and 1500 years BP that formed the Cabeço Bravo lava cone. A small but beautiful portion of the tube has been made accessible to visitors. The tour consists of a short

Fig. 8 Remnants of an ancient tuff ring off the Ponta da Restinga. (Picture credit: Diogo Caetano)

movie about lava tube formation in general and continues through the entrance of the tube via stairs that descend approximately 15 m to the bottom of the lava tube through a hole in its roof. These openings can be original (the roof was never entirely closed) or secondary (partial roof collapse). Lava tubes are peculiar features in cooled lava flows that are generated when several conditions are full-filled: (1) lava is flowing at a fairly constant rate in a geographically stable location, (2) cooling is most efficient in contact with the atmosphere and a crust is formed (as rock is a bad heat conductor, the crust prevents efficient cooling of the flowing lava below), (3) lava supply is reduced or has come to a stop, (4) lava viscosity is low enough to allow for a draining of the lava flow and (5) the crust is sufficiently stable so the roof will not sink in but remain stable and an empty elongate cavity has been formed. Inside the *Gruta das Torres*, a plethora of features proof the complex history of the cave with varying lava levels, meandering, lateral plastering and overflowing. In case the Gruta das Torres is closed or all visits are overbooked, the road towards Pico mountain provides a smaller exposure of such lava tunnel (Fig. 9).

Author's hot tip: Reservations are recommended particularly for larger groups (http://parquesnaturais.azores.gov.pt/en/pico-eng/what-visit/interpretation-centers/visitors-center-of-gruta-das-torres).

Pico Mountain

Name: Pico Mountain
Coordinates: 38° 28′ 14.43″N, 28° 25′ 35.19″W
Access: Leave Madalena's central square in a southward direction and turn left after a short while onto EN3 and follow the road on a steady climb. Above the forest region, you will see various volcanic landforms along the road, including hornitos and lava tunnels. After approximately 13 km turn right at the indication to *Montanha do Pico* (38° 29′ 51.51″N, 28° 24′ 56.16″W) and follow the road to the parking lot at the end. In the information centre you have to

Fig. 9 Inside the Gruta das Torres. Tube width and height are approx. 4 and 2.5 m, respectively. Where daylight can enter, plants are colonizing the inside of the tube. (Picture credit: Christoph Beier)

register and start your climb after having watched a video explaining the specific risks of the Pico climb. Follow the signposts uphill.

Geology: Pico (2351 m) is the fourth highest mountain in the Atlantic after Teide volcano and the *Roque de los Muchachos* peak (Tenerife and La Palma islands, Canary Islands) and Fogo volcano (Fogo island, Cape Verde Islands). Pico is the only central volcano in the Azores that has exclusively erupted basaltic lavas and not undergone a significant collapse event forming a caldera. It is more that 1000 m higher than any other Azorean peak above sea level and has a total height of >4500 m from the ocean floor. Along the path uphill you will recognise numerous parasitic cones from flank eruptions. Many of them display fairly steep flanks owing to the welding of the spatter clasts upon deposition. Higher up on the slopes you will be passing many small lava flows that reflect three major rock types: (1) aphyric, (2) rich in feldspar and (3) rich in olivine and pyroxene. These different lithologies can be explained by chemically distinct magma batches or magma ascent conditions (the lower the crystal content or the smaller the crystals' density, the smaller the average ascent speed). Sometimes, rocks with a crystal framework of olivine and pyroxene of up to 1 cm in size and only a 1 mm thin lava coating can be found. This proves that the lava viscosity at eruption was very high indicative for high eruptive temperatures, allowing for the lava to "drain" from the crystal framework at a very late stage of the eruption. The summit of Pico is built by the smaller cone Picinho (Fig. 10), emerging by approximately 80 m from inside the summit crater (diameter 550 m). Picinho grew by effusion of small-volume lava flows from a single vent. At the summit, fumaroles are still active, evidenced by approx. 40 °C warm steam emission. To the west of the summit an eruptive fissure has cut the crater rim and the Picinho lava flows that cover the crater floor (the exact eruption date is unknown but must be younger than

Fig. 10 Picinho, the summit of Pico island, a little cone emerging from Pico's summit crater. (Picture credit: Ulrich Kueppers)

Picinho, Nunes 1999). This fissure eruption was mildly explosive and left a partially welded spatter deposit in the vicinity of the fissure.

Author's hot tip: Pico is the fourth highest mountain in the Atlantic and weather conditions may change quickly and unexpectedly. You will have to climb approx. 1200 m to the top on a small, rough footpath and there are no facilities on the way. Bring plenty of water, some food and sun/rain/wind protection. Mountain guides are recommended but you can also walk at your own risk. Everybody is signing a declaration that rescue is at your own expenses.

More info here: http://parquesnaturais.azores.gov.pt/en/pico-eng/what-visit/interpretation-centers/mountain-house

Nariz de Ferro

Name: Nariz de Ferro ankaramite quarry

Coordinates: 38° 32′ 8.74″N, 28° 19′ 47.43″W

Access: Leave Madalena's central square in a northward direction and follow road EN1 towards Santa Luzia, continue towards Santo Antonio, just on the entrance to the village of Almas (38° 32′ 0.24″N, 28° 19′ 53.10″W) and follow the road towards the harbour (turn right at 38° 32′ 3.44″N, 28° 19′ 57.68″W) until reaching the old quarry. Parking is available alongside the road at the quarry (38° 32′ 4.59″N, 28° 19′ 50.37″W). Follow the footpath down into the quarry and climb over the concrete wave breakers. On the way back, walk on top of the lava flow and just back to the road.

Geology: The outcrop extends approximately 125 m in an east-west direction and consists of a large porphyritic lava flow that ranges from almost aphyric on the top to a spectacular ankaramitic crystal-rich lava flow on the bottom. Olivines and pyroxenes have sizes up to 2–3 cm, cleavages of the pyroxenes may be seen in hand specimen. The increasing viscosity of the crystal-rich lava flow and the accumulation of mafic minerals are clearly evidenced in this outcrop and it is one of the most spectacular of its kind. The origin of these ankaramitic lavas is still a matter of active debate. The olivine and clinoyproxene crystals show strong evidence that they have formed by fractional crystallisation, however, the size and volume of these idiomorphic crystals dispersed in a fine-grained plagioclase-bearing matrix is difficult to reconcile.

Northern coastline along Arcos to Cabrito

Name: Arcos to Cabrito
Coordinates: from 38° 33′ 36.83″N, 28° 24′ 32.76″W to 38° 33′ 26.11″N, 28° 23′ 19.68″W
Access: Leave Madalena's central square in a northward direction and follow to Laje, turn left towards the airport and on the coast (just underneath the airport's runway) turn right to the village of Cachorro (you will recognise several UNESCO protected vineyards to your right) until arcos. Park your car at the roadside and walk along the coastline from Arcos to Cabrito. The walk is approximately 1 km one way.
Geology: This walk along the coastline covers almost any geological feature that *Paho'eho'e* lava flows exhibit. Pico's northern coastline is one of the most spectacular *Paho'eho'e* lava exposures to be found in the northern hemisphere (with the notable exception of the Hawaiian type examples. Tree trunks, tree molds, as well as charcoal are found close to Arcos; toes, tumuli, pressure ridges and ropy lava flows are visible on the way towards Cabrito. Further west, the young Santa Luzia lava flow (eruption in 1718) is evident. Along the walk *Paho'eho'e* lava frequently is overlain by *A'a'* lava.
Author's hot tip: Enjoy the beautiful and stunning views along the coastline and towards Pico mountain. The lava flows have quite sharp edges, so proper footwear is recommended. On sunny days, this outcrop might become quite hot, however, this walk certainly is one of the best places to study the structures in *Paho'eho'e* lava and the associated *A'a'* lava flows.

6. Santa Maria

Santa Maria, the easternmost and southernmost island, is the oldest volcanic edifice known. Santa Maria is the only Azores island with a significant paleontological record (see also Ávila et al. 2017) interbedded into a series of pillow lavas and lava flows that imply eruption close to the sea surface. Avila et al. (Chapter "The Marine Fossil Record at Santa Maria Island, Azores" this book) note that Santa Maria is one of the ideal places to study the Neogene marine fossil record in the Northern Atlantic. We have selected three localities at Santa Maria covering the sedimentary, fossil record, the subsurface pillow lava sections and a series of ankaramite dykes intruding into lava flows.

Marine fossils

Name: Marine fossils south of Almeigra
Coordinates: 36° 56′ 46.20″N, 25° 7′ 9.67″W
Access: Leave Vila do Port in easterly direction. After approximately 2 km turn south on a road that then turns west and park close to the quarry and continue walking to the trail along the coast.
Geology: 3–4 m thick succession of marine fossiliferous sandstones (see Ávila et al. 2017 for more detail) at a coastal cliff (Fig. 11). The fossiliferous sediments contain numerous large bivalves but also echinoids, barnacles, brachiopods, bryozoans, calcareous algae and corals.
Short description: This outcrop is one of a series of fossil rich outcrops on Santa Maria and represents the fossil diversity of the Azores marine habitat. These deposits have been subject of numerous scientific projects and are unique in their appearance in the Azores. Note that most outcrops are vigorously protected and must not be sampled.

Pillow lavas at Farol de Gonçalo Velho

Name: Gonçalo Velho
Coordinates: 36° 55′ 44″N 25° 1′ 5″W
Access: Walk down the pathway from the lighthouse at *Farol de Gonçalo Velho—Maia* and turn westward along the coastline. The next small bay will exhibit pillow lavas covering a small layer of carbonate fossiliferous sediments.
Geology: This outcrop contains a series of volcaniclastic beds and pillow lavas from Santa Maria's early, yet still submarine formation

Fig. 11 One of several outcrops showing (volcani-)clastic sandstones rich in marine fossils. Person for scale. (Picture credit: Ulrich Kueppers)

phase which cover sedimentary carbonates (see Ramalho et al. 2016). The lavas are intercalated by ash layers and are cut by several dykes. The pillow lavas appear fresh and still contain glass rims. Interspersed in between the pillows are primarily hyaloclastites (glassy lava fragments because of cooling in contact with water) and subordinate marine sediments that intercalated. Some pillows are broken and their interior is exposed, showing peculiar bubble textures because of degassing. Stratigraphically higher, on a steep cliff of several tens of metres high, shallow submarine and subaerial lava flows are exposed and are interpreted as the period of formation of the island close to sea surface. The positioning of the outcrop several tens of metres above sea level is being interpreted as reflecting uplift of the island.
Author's hot tip: On the way back to the car, have a swim in the old harbour underneath the lighthouse. Alternatively, if you prefer sand beaches, stop at Praia Formosa.

Columnar jointing at Santo Espirito

Name: Calçada do Gigante
Coordinates: 36° 56′ 5.7″N, 25° 3′ 40.4″W
Access: (to where you start your hike): leave Vila do Porto in an eastward direction on EN1-2A towards Santo Espirito. Before reaching the village, turn right twice (first at 36° 57′ 28.68″N, 25° 3′ 42.31″W then at 36° 57′ 16.52″N, 25° 3′ 19.07″W). Follow the road for approximately 2 km in a southward direction. Some 500 m after the road starts turning towards the West, park your car at the "intersection" with an unpaved road and walk, first in an east-south-eastward direction (approximately 150 m) then in a south-westward direction. After reaching a grass field gently sloping to the west (~500 m),

follow the gradient towards the commonly dried out riverbed accessing the cliff of basaltic columnar joints.

Geology: Columnar jointing (Fig. 12) is a cooling feature of lavas and welded ignimbrites, most commonly observed in basaltic lava flows. The cooling lava body shrinks and the volumetric stress is accommodated by the formation of cracks. The long axis of these columns is perpendicular to the strongest cooling gradient, i.e. for this particular outcrop the cooling directions were from the top and the bottom of the flow unit. In cross section, the columns are commonly hexagonal. The basalts here are mostly aphyric with few feldspars and contain numerous filled and unfilled degassing vesicles giving evidence for a volatile rich composition of this lava flow.

7. São Jorge

São Jorge Island is a 54 km long volcanic fissure system of exclusively basaltic volcanism and is characterized by its elongated shape with a WNW-ESE trend. The island can be largely divided in two complexes, an older eastern part and a younger western part with >100 volcanic cones. By area, the younger part comprises approximately 2/3 of the island. Historic eruptions on land took place in 1580 and 1808 (Urzelina). In February 1964, during a phase of strong seismic activity, sulphur smell was noted in Velas, likely related to a submarine eruption.

Volcanic sequence at Ponta dos Rosais

Name: Ponta dos Rosais

Coordinates: 38° 45′ 15.77″N, 28° 18′ 55.41″W

Fig. 12 Columnar jointing of basaltic lava flow on Santa Maria. The columns show a fairly even cross-sectional area from top to bottom. Height of the cliff is approx. 8–10 m. (Picture credit: Ulrich Kueppers)

Access: This outcrop can only be seen from the sea. It is the sea cliff below the lighthouse at the NW tip of São Jorge.
Geology: This steep sea cliff shows a thick sequence of deposits associated with the basaltic Rosais volcanic system (Fig. 13). The cliffs expose a very heterogeneous lithofacies. The tip of the island itself is made up of many layers of pyroclastic deposits, some of which have been altered to palaeosol. These layers that vary in thickness, grainsize and colour, are cut by several feeding dykes. Some 200 m further to the east, approximately below the lighthouse, there are several layers of 1–15 m thick lava flows. In between the lava flows are only fairly thin layers of fritted soil or clastic units, most likely generated from autobrecciation during lava flow activity. The lava flows seem to be younger than the volcaniclastic layers at the island's tip.

Volcanic axis at Cordilheira Central

Name: Cordilheira Central
Coordinates: 38° 38′ 58.21″N, 28° 4′ 24.58″W
Access: From the transversal road which connects the North and the South coast of the island follow along the road that reaches the central part of the island (Planalto Central).
Geology: Corresponds to a WNW-ESE-trending alignment of cinder cones—scoria and spatter cones—expressing the fissure system along which magma erupted, evidence of the fissural volcanism (Fig. 14). This volcanic axis includes the eruptive centres of the historical eruptions.

Tuff cones and lava delta at Velas

Name: Velas
Coordinates: 38° 41′ 15.42″N, 28° 13′ 8.21″W

Fig. 13 View of the westernmost tip of São Jorge from a boat. Lighthouse for scale. Pyroclastic units and proximal fall deposits, cut by dykes, are building most of the cliff. From the lighthouse on to the East, the top of the island and also a significant portion of the cliff's vertical height is made up by lava flows. (Picture credit: Diogo Caetano)

Fig. 14 View of the dorsal ridge of São Jorge with several volcanic cones lined up. Houses in the foreground for scale. (Picture credit: Diogo Caetano)

Access: This outcrop is accessible from the historical centre of Velas village by a short walk.
Geology: Two tuff cones—*Morro Grande* and *Morro de Lemos*—are situated near Velas village, evidence for submarine eruptions (Fig. 15). The *Morro Grande* tuff cone is the younger and is much better preserved. The formerly unconsolidated ash and lapilli layers have been turned into a coherent unit by palagonitization, a process of alteration of the basaltic glass fragments to palagonite, a yellowish clay mineral. This process is starting very quickly after deposition under the presence of fluids and latent heat (some tens of degrees). It is only because of this efficient "cementing" that primarily loose volcanic ash and lapilli tuff cones have been preserved and give evidence for the frequent eruptions taking place offshore.
The village of Velas is emplaced on a lava delta that is covered by the tephra from *Morro Grande*. The lava flows that compose this lava delta erupted from *Pico dos Loiros* cone. Lava deltas (new land built by lava flows advancing into the sea) are commonly flat ground that is frequently used for urbanisation (e.g., Mosteiros on São Miguel) or agriculture (e.g., on La Palma, Canary Islands).

Historic (1808 AD) Lava Flow "Mistério" at Urzelina

Name: "Mistério" da Urzelina
Coordinates: 38° 38′ 46.19″N, 28° 7′ 40.85″W
Access: From the centre of Velas village take the regional road towards the airport. Once in the Urzelina parish you will easily find its church. It is near the coastline that you have the best view of this very recent lava flow.
Geology: The *Mistério da Urzelina* is made up by the lava flows from the eruption at "Caldeirinhas" along the central volcanic axis of São Jorge. Lava was very low viscous and quickly flowed down the slopes and inundated the village of Urzelina, now called São Mateus. It destroyed

Fig. 15 View of the city of Velas with Pico (left) and Faial (center) islands in the background. Velas was founded in a little bay, formed by a lava delta and tuff rings. (Picture credit: Diogo Caetano)

a large part of the village but, more importantly, covered rural land, contributing to the famine in the years thereafter. Of the old village centre, only the bell tower of the church has survived, emerging from the lava flow field that surrounded and invaded the church. Another left-over bell tower can be found in Mexico, in the lava flow field of the 1943 Parícutin eruption.

8. São Miguel

São Miguel is one of the geologically most diverse islands of the Azores and comprises several volcanic systems. The Eastern part is older and volcanically inactive (Nordeste and Povoação, Abdel Monem et al. 1975; Johnson et al. 1998). There are three active central volcanoes that have developed large caldera systems. These trachytic centers are separated by tectonically controlled fissure systems that are also responsible for basaltic eruptions. Rock types range from basaltic to trachytic with the post-caldera eruptions being dominated by intermediate to trachytic compositions. The western Sete Cidades volcano and the surrounding coastline contain several spectacular outcrops that are geologically diverse (see Beier et al. 2006). Amongst those localities that are not described here due to the abundance of geological outcrops on São Miguel but that the interested reader may want to visit are Ponta da Ferraria at the western Sete Cidades coastline, a spectacular outcrop made up primarily by deposits of pyroclastic density currents and pumice fall as well as a fairly young lava delta. Three historic eruption took place, in 1563/1564, 1630 and 1652. The oldest historical eruption had taken place from two vents, one located inside today's caldera of Fogo volcano and the other one at *Pico Queimado*. The eruption at *Pico Queimado* was particular as it was a basaltic eruption cutting through a trachytic dome. Two features prove the lava had a very low viscosity: the lava flow extends to the North in a very narrow creek. In *Ribeira Seca*, one can see a spring catchment that was engulfed and covered by this lava flow. The *Pico Queimado* itself hosts an algar (− > see Terceira), a vertical empty cavity when the uppermost plumbing system of the eruption was drained towards the end when the magmatic pressure from below was reduced. The 1630 eruption in the Furnas caldera was a classical explosive, trachytic eruption, starting with a highly energetic phase with possibly some interaction with external water and eventually, once dried out, building a pumice cone and a thick sheet-like pumice cover. The last eruption in 1652 took place north of the village of Lagoa. Although this dome-forming eruption lasted for less than 10 days, the impact was catastrophic for the island at that time. Additionally, submarine eruptions were observed off the W and SW coast at the beginning of the 19th and 20th centuries.

Sete Cidades—Vista do Rei

Name: Vista do Rei
Coordinates: 37° 50′ 21.82″N 25° 47′ 42.45″W

Access: From Ponta Delgada on São Miguel on road EN1-1A in a westward direction along the South Coast. After approximately 15 km, just before arriving in the village of Feteiras, right onto EN9-1A, following the road to a parking next to the ruins of the abandoned hotel Monte Palace.

Geology: Sete Cidades is one of three central volcanoes on São Miguel that has experienced a major caldera forming collapse in the past. The Sete Cidades caldera has a multiple-stage history and the last major caldera forming eruption took place approx. 16 ka ago (Queiroz et al. 2015; Porreca et al. 2018). Caldera collapse does not necessarily terminate the end of a volcano's life but rather often mark a significant change in eruption style and composition (and sometimes also volume). The view from *Vista do Rei* (Fig. 16) into the caldera shows two large lakes [*Lagoa Verde* (the green lake) and *Lagoa Azul* (the blue lake)]. To the West of these two lakes a series of volcanic post-caldera cones (*Caldeira Seca*, *Caldeira do Alfares* and an eroded,

Fig. 16 View of the Sete Cidades caldera with several post-caldera-collapse cones. The diameter of the caldera is 5 km. (Picture credit: Christoph Beier)

nameless cone) are aligned parallel to the cliff of the caldera wall. These three craters show characteristics of explosive magmatic eruptions that have not been in contact with significant amounts of groundwater whereas *Lagoa Rasa* and *Lagoa do Santiago* have presumably been generated by eruptions resulting from the interaction of an aquifer or a lake (wet, phreatomagmatic eruptions) with the rising magma. These five eruptive centres together with a presumed centre at the eastern end of the caldera are situated on the "caldera ring fault" and proof that magmatic activity had not come to a halt after the caldera collapse. Further eruptive centres in *Lagoa Azul* and *Lagoa Verde* are likely. North of the bridge crossing the isthmus between the two lakes (*Ponta dos Rêgos*), right in the centre of the caldera, strongly layered deposits indicate that a small-volume eruption took place in shallow water. As the individual craters show different degrees of erosion-induced degradation, the have been formed by different events in the past 5.000 years (Queiroz et al. 2008). At least nine eruptive centres have been active since the last major collapse.
Author's hot tip: Have a walk around the caldera rim, either clock- or counterclock wise. The views are stunning into the caldera but also onto the caldera flanks. Note that you will have to take a sufficient supply of water with you.
or:
Leave the road 8-2 at *Lagoa do Canario* and follow an unpaved road that initially leads North. Follow the road for approx. 1 km and then walk on a path leading to the West to "Sombreiros" and you will have yet another spectacular view over the caldera.

Furnas—Fumaroles

Name: The fumaroles in Furnas
Coordinates: 37° 46′ 21.72″N, 25° 18′ 14.85″W
Access: Drive into Furnas village, car parking is available and walk to the fumarole field.
Geology: There are several fumarole fields in Furnas village (Fig. 17) and at the shoreline of Furnas lake. Based on the area affected by warm/hot fumaroles, Furnas looks like the most active volcanic system. Fumaroles generally release hot gas and steam and are often accompanied by a foul smell (sulphur gas H_2S) and different sublimates. The sublimates are mineral deposits that form when the gas is cooled in contact with the atmosphere and becomes oversaturated. In areas with significant precipitation (like in the Azores) fumaroles are often accompanied by boiling pools of water and variable degrees of mud. The liberated steam here is mostly heated meteoric and ground water that had been heated up by the heat of a magma reservoir at depth. In the vicinity of the fumaroles, several natural springs of sparkling, CO_2-rich water can be found and it is certainly worth to taste them. Spring water varies between hot and cold and some even have a very salty taste, possibly from contamination with sea water (Viveiros et al. 2008).
Author's hot tip: Eat maize boiled in the hot springs.

Rochas de Cascalho

Name: Rochas de Cascalho
Coordinates: 37° 45′ 50.66″N, 25° 44′ 6.21″W
Access: Follow road EN1-1A at the western end of *Relva Velha* and continue on an unpaved road westward until reaching the parking at *Rochas de Cascalho*.
Geology: Descend from the parking and within less than 200 m, leave the path continuing straight ahead to *Rochas de Relva* and turn left and continue downwards crossing fall deposits of basaltic lapilli. After a canyon (commonly dry unless severe rainfalls appear), several subhorizontal basaltic lava flow units appear on the left. Continuing down to the coast passing the small vineyards where the grapes are growing on big lava boulders. At the coastline, turn right and walk on the pebbles edge of the bay. On the cliff, a large volcanic dyke appears that has cut through thick layers of pyroclastic density currents and fall units (Fig. 18). The dyke mainly consists of fine-grained basalt with some olivine phenocrysts. The intruded pyroclastic rocks are discoloured and welded with an increasing heat signature as they approach the dyke. These dykes

Fig. 17 Pool of boiling water in the fumarole field in Furnas village. Diameter approx. 4 meters. In areas of high precipitation, fumaroles (emitting steam and gas) and boiling water/mud pools are co-existing in close vicinity. (Picture credit: Fabian Goldstein)

are a common feature of magma transport to the surface and in several outcrops across all Azores islands a large variety of sills and dykes may be found (e.g., see coastline of Corvo and Flores). The intrusion of sills into the volcanic edifice of an ocean island has also been taken as one important feature in the rise of volcanic islands (e.g., at Santa Maria).

9. Terceira

Contrasting lavas from São Miguel and also from the neighbouring islands of Pico and Graciosa, the alkaline rocks at Terceira are dominated by evolved lava flows and pyroclastic deposits. The name of the island derives from it being the third island that has been discovered. Similar to Pico, most settlements are situated along the coast while the central parts are used for dairy farming and bull raising. One historic eruption has taken place in 1761. In a fairly shallow (<400 m) submarine setting, two submarine eruptions have taken place in 1867 and 1998–2001. The latter of the two could be directly observed as large eruptive products (up to 3 m in length and cigar-shaped) were observed floating at the sea surface (Kueppers et al. 2012).

Gruta do Algar do Carvão

Name: Algar do Carvão
Coordinates: 38° 43′ 41.97″N, 27° 12′ 58.34″W
Access: Leave Angra do Heroismo towards the NE on the Via Rapida, leave and take the EN 5-2 towards west. Keep after approx. 2.5 km and continue for approx. 5 km. Then turn right and follow the indications to the *Algar do Carvão*.
Geology: The *Algar do Carvão* is a vertical cave in a volcanic edifice (Fig. 19), some 50 m deep

Fig. 18 View of the basaltic Rochas do Cascalho dyke, obliquely cutting through a several tens of meters thick sequence of trachytic fall and PDC layers of Sete Cidades volcano. The dyke is approx. 6 m wide and has thermally impacted on the surrounding older and colder units. (Picture credit: Ulrich Kueppers)

and with a small lake at the bottom. It represents the drained (and now empty) uppermost part of a volcanic plumbing system, generated when the magma level lowered towards the end of the eruption. During the eruption, magma was ponding here through which gas bubbles were rising. As a result of an increasing (over)pressure, magma was disrupted at variable depth and pyroclasts were ejected from the volcanic cone. After touching ground, they were still hot enough to allow for welding and forming this steep cone. Similar features exist in Graciosa (*Furnas do Enxofre*), São Miguel (*Pico Queimado*) or Iceland (*Thrihnukagigur*).

Farol da Serreta

Name: Lighthouse of Serreta
Coordinates: 38° 45′ 57.28″N, 27° 22′ 25.48″W

Access: The lighthouse is sitting at the Western tip of Terceira island. You can reach it via the coastal roads or the EN 5-2 crossing the island. Shortly north from Serreta, take the road going down to the Coast.

Geology: The lighthouse is sitting on an approx. 400 m into the sea extending lava tongue. It was formed by an ancient eruption of viscous lava that was slowly flowing into the sea. At the sea level, flow, squeeze and shear features can be observed. Just a few kilometres west from here, submarine eruptions took place in 1867 and from 1998 to 2001 (Gaspar et al. 2003; Kueppers et al. 2012), proving the still ongoing but very localised volcanism in the Azores. The submarine deposits are subject to recent research and were mapped, imaged and sampled during research cruises M113 and M128 with the German R/V Meteor.

Fig. 19 View from inside the Algar do Carvão towards up. Persons for scale. An algar is formed when the uppermost plumbing system is drained at the end of a volcanic eruption. The low viscosity of the erupted lava is further proven by the high degree of welding of proximal spatter and bombs, making the cones very steep. (Picture credit: João Fontiela)

Authors hot tip: Enjoy the sunset while observing Graciosa, São Jorge and possibly the peak of Pico volcano.

Welded pyroclastic units

Name: Ponta da Forcada
Coordinates: 38° 46′ 59.3″N, 27° 8′ 22.89″W
Access: Drive to Vila Nova at the North coast. Approx. 700 m east (N 38° 46′ 53.16″, W 27° 8′ 27.81″) turn into a dead-end road towards the coast. After approx. 200 m, park your car and take the road to the left leading down to the coast towards a boat ramp.
Geology: Along this road, different lithofacies are cropping out, including pumice fall deposits, lava flows (in places fritting the pumice layers) and deposits of pyroclastic density currents (PDCs) that are partially welded. These PDCs derive from the large central volcano in the East of the island that has collapsed after a climactic eruption to form the *Caldeira de Guilherme Moniz*. Welded PDC deposits indicate explosive eruptions of high mass eruption and deposition rate. Although PDCs have been generated during basically each caldera-forming event (e.g. Pimentel et al. 2015; Porreca et al. 2018), there are few welded PDC deposits in the Azores. Another well-known deposit can be found in Povoação on São Miguel.

Acknowledgements ChB acknowledges a nice field trip with Andreas Stracke and Stephan Klemme to Pico and Faial and a memorable trip with Bruce Schaefer and Felix Genske to Flores and Corvo along with numerous field trips with Karsten M. Haase. UK lived and worked for nearly two years on São Miguel and acknowledges financial support through several projects funded by the *Fundaçao Gaspar Frutuoso* foundation of the Azorean government, primarily the projects M1.1.2/I/009B/2005 and M2.1.2/I/005/2007. He is grateful to colleagues from the *Centro de Vulcanologia e Avaliação de Risgos Geologicos (CVARG,* now *IVAR)* and friends for sharing spectacular sights and sites and much more of this archipelago. Special thanks to Pedro Cerqueira, João Fontiela, Catarina Goulart, Robert Hoge, Adriano Pimentel, Norbert Schnitzer, Fatima Viveiros, Frank Wirth and Vittorio Zanon and many others for making me understand the meaning of *saudades*.

References

Abdel Monem, A. A., Fernandez, L. A., & Boone, G. M. (1975). K-Ar-Ages from the eastern Azores group (Santa Maria, Sao Miguel and the Formigas islands). *Lithos, 8,* 247–254.

Ávila, S.P., Ramalho, R., Habermann, J., & Titschack, J., (2017). The marine fossil record at Santa Maria Island (Azores), this volume.

Azevedo, J. M. M., & Ferreira, M. R. P. (1999). Volcanic gaps and subaerial records of palaeo-sea-levels on Flores Island (Azores): tectonic and morphological implications. *Journal of Geodynamics, 28,* 117–129.

Beier, C., Haase, K. M., & Hansteen, T. H. (2006). Magma evolution of the Sete Cidades volcano, São Miguel, Azores. *Journal of Petrology, 47,* 1375–1411. https://doi.org/10.1093/petrology/egl014.

Beier, C., Turner, S. P., Plank, T., & White, W. (2010). A preliminary assessment of the symmetry of source composition and melting dynamics across the Azores plume. *Geochemistry, Geophysics, Geosystems, 11,* Q02004. https://doi.org/10.1029/2009GC002833.

Cole, P. D., Guest, J. E., Duncan, A. M., & Pacheco, J.-M. (2001). Capelinhos 1957–1958, Faial, Azores: Deposits formed by an emergent surtseyan eruption. *Bulletin of volcanology, 63,* 204–220.

Douillet, G. A., Pacheco, D. A., Kueppers, U., Letort, J., Tsang-Hin-Sun, E., Bustillos, J., et al. (2013). Dune bedforms produced by dilute pyroclastic density currents from the August 2006 eruption of Tungurahua volcano. *Ecuador. Bulletin of Volcanology, 75,* 762.

França, Z. T. M., Lago, M., Nunes, J. C., Gale, C., Forjaz, V. H., Pueyo, O., & Arranz, E. (2006). Geochemistry of alkaline basalts of Corvo Island (Azores, Portugal): preliminary data. Geogaceta. 40.

Gaspar, J. L., Queiroz, G., Pacheco, J. M., Ferreira, T., & Wallenstein, N. (2003). Basaltic lava balloons produced during the 1998–2001 Serreta Submarine Ridge eruption (Azores). In: J. White, D. Clague, J. Smellie, (Eds.), *Subaqueous explosive volcanism* (Vol. 140, pp. 205–212). AGU Monograph.

Genske, F. S., Turner, S. P., Beier, C., & Schaefer, B. F. (2012). The petrology and geochemistry of lavas from the western Azores Islands of Flores and Corvo. *Journal of Petrology, 53*(8), 1673–1708.

Johnson, C. L., Wijbrans, J. R., Constable, C. G., Gee, J., Staudigel, H., Tauxe, L., et al. (1998). $^{40}Ar/^{39}Ar$ ages and paleomagnetism of Sao Miguel lavas, Azores. *Earth and Planetary Science Letters, 116,* 637–649.

Kueppers, U., Nichols, A. R. L., Zanon, V., Potuzak, M., & Pacheco, J. M. R. (2012). Lava balloons—peculiar products of basaltic submarine eruptions. *Bulletin of volcanology, 74,* 1379–1393.

Larrea, P., França, Z., Lago, M., Widom, E., Galé, C., & Ubide, T. (2013). Magmatic processes and the role of antecrysts in the genesis of Corvo Island (Azores Archipelago, Portugal). *Journal of Petrology, 54,* 769–793. https://doi.org/10.1093/petrology/egs084.

Machado, F., Parsons, W. H., Richards, A. F., & Mulford, J. W. (1962). Capelinhos eruption of Fayal Volcano, Azores, 1957–1958. *Journal of Geophysical Research, 67*(9), 3519–3529.

Nunes, J. C. (1999). *A actividade vulcânica na ilha do Pico do Plistocénico Superior ao Holocénico: Mecanismo eruptivo e hazard vulcânico.* Ph.D. Thesis, Departamento de Geociências, Universidade dos Açores, Ponta Delgada. 357 pp.

Pimentel, A., Pacheco, J., & Self, S. (2015). The ~1000-years BP explosive eruption of Caldeira Volcano (Faial, Azores): The first stage of incremental caldera formation. *Bulletin of Volcanology, 77,* 42.

Porreca, M., Pimentel, A., Kueppers, U., Pacheco, J., Queiroz, G., & Izquierdo, T. (2018). Large eruptions on small volcanic islands: Lessons learned from incomplete eruption deposits of Sete Cidades volcano (São Miguel Azores). *Bulletin of Volcanology.*

Queiroz, G., Gaspar, J. L., Guest, J. E., Gomes, A., & Almeida, M. H. (2015). Eruptive history and evolution of Sete Cidades Volcano, São Miguel Island, Azores. In: J. L. Gaspar, J. E. Guest, A. M. Duncan, F. J. A. S. Barriga, & D. K. Chester (Eds.), *2015. Volcanic Geology of São Miguel Island (Azores Archipelago). Geological Society, London, Memoirs,* vol. 44, pp. 87–104.

Queiroz, G., Pacheco, J. M., Gaspar, J. L., Aspinall, W. P., Guest, J. E., & Ferreira, T. (2008). The last 5000 years of activity at Sete Cidades volcano (São Miguel Island, Azores): Implications for hazard assessment. *Journal of Volcanology and Geothermal Research, 178,* 562–573.

Ramalho, R. S., Helffrich, G., Madeira, J., Cosca, M., Thomas, C., Quartau, R., et al. (2016). Emergence and evolution of Santa Maria Island (Azores)—The conundrum of uplifted islands revisited. *Geological Society of America Bulletin.* https://doi.org/10.1130/b31538.1.

Viveiros, F., Ferreira, T., Cabral Vieira, J., Silva, C., & Gaspar, J. L. (2008). Environmental influences on soil CO_2 degassing at Furnas and Fogo volcanoes (São Miguel Island, Azores archipelago). *Journal of Volcanology and Geothermal Research, 177,* 883–893.

Zanon, V., Pacheco, J., & Pimentel, A. (2009). Growth and evolution of an emergent tuff cone: Considerations from structural geology, geomorphology and facies analysis of São Roque volcano, São Miguel (Azores). *Journal of Volcanology and Geothermal Research, 180,* 277–291.

The manufacturer's authorised representative in the EU is Springer Nature Customer Service Centre GmbH, Europaplatz 3, 69115 Heidelberg, Germany. If you have any concerns regarding our products, please contact ProductSafety@springernature.com

Printed and bound by CPI Group (UK) Ltd, Croydon, CR0 4YY
30/09/2026
02225591-0020